QUADRATIC FORMULA

The solutions of $ax^2 + bx + c = 0$ $(a \neq 0)$ are

$$x = \frac{-b \pm \sqrt{b^2 - 4ac}}{2a}.$$

FUNCTIONS

Composition: $(g \circ f)(x) = g(f(x))$

Linear function: $f(x) = ax + b \quad (a \neq 0)$

Quadratic function: $f(x) = ax^2 + bx + c \quad (a \neq 0)$

Polynomial function of degree n:

$$f(x) = a_n x^n + a_{n-1}x^{n-1} + \cdots + a_1 x + a_0 \quad (a_n \neq 0)$$

Rational function: $f(x) = \dfrac{p(x)}{q(x)}$,

where $p(x)$ and $q(x)$ are polynomial functions.

Exponential function: $f(x) = b^x \quad (b > 0,\ b \neq 1)$

Logarithmic function: $f(x) = \log_b x \quad (b > 0,\ b \neq 1)$

$\log_b a = c$ if and only if $a = b^c$.

LAWS OF LOGARITHMS

$\log_b b = 1 \qquad \log_b 1 = 0$

$\log_b xy = \log_b x + \log_b y$

$\log_b \left(\dfrac{x}{y}\right) = \log_b x - \log_b y$

$\log_b x^r = r \log_b x$

BINOMIAL THEOREM

$$(a + b)^n = \binom{n}{0}a^n + \binom{n}{1}a^{n-1}b + \binom{n}{2}a^{n-2}b^2 + \cdots + \binom{n}{k}a^{n-k}b^k + \cdots + \binom{n}{n}b^n,$$

where $\dbinom{n}{k} = \dfrac{n!}{k!(n-k)!}$.

COLLEGE ALGEBRA

COLLEGE ALGEBRA

John R. Durbin
The University of Texas at Austin

John Wiley & Sons
NEW YORK CHICHESTER
BRISBANE TORONTO SINGAPORE

Library of Congress Cataloging in Publication Data:

Durbin, John R.
College algebra.

Includes index.
1. Algebra. I. Title.
QA154.2.D87 512.9 81-11379
ISBN 0-471-03368-5 AACR2

Printed in the United States of America

10 9 8 7 6 5 4 3 2

PREFACE

This book presents the standard topics of college algebra, preceded by a thorough review of the essential prerequisites.

In writing the book I was guided especially by the following convictions.

- Algebra is learned primarily by studying examples and working exercises.
- Applications, which provide one of the best motivations for learning algebra, should be presented at every good opportunity.
- Word problems deserve special attention from the start.
- In most college algebra classes the preparation levels of the students vary widely. Therefore, a textbook should begin with a careful and moderately paced review, even if time limitations dictate that the students must cover part of this material on their own.

Here are some suggestions on how to use the book.

Exercises. For a representative sample of the exercises that conclude each section, simply choose every third one beginning with either 1, 2, or 3. Each chapter concludes with a set of review exercises, and either the even- or the odd-numbered exercises will give a representative sample of these.

Answers. An appendix gives answers to most of the odd-numbered exercises from each section, and to nearly all of the review exercises.

Applications. Chapter II offers tips on how to approach problems given in words. It is designed to build confidence through exercises for which the algebraic demands are slight, and also to review such essential ideas as percentage, ratio, proportion, and variation. The applications in the book range from the diversionary (like running and baseball) to the more serious (like demography and finance).

Review material. The pace of the first four chapters is intentionally slower than the rest of the book. This is for the benefit of students who have inadequate mathematics backgrounds or who have forgotten what they once knew. The amount of material considered as review will vary, of course, depending on the prerequisites and goals of the course. One topic in Chapter I that deserves special attention is the solution of inequalities, especially absolute value inequalities.

Flexibility. Nearly all sections of the book can be covered in one lecture each. This implies that a one-semester course could start with Chapter I and continue through Chapter IX. Or, with well-prepared students, a one-semester course could cover everything from Chapter V through the end of the book. The book is flexible enough to allow for many other possibilities, as well. For example, by spending only one week on Chapter I, and by omitting conic sections (in Chapter V), partial fractions (in Chapter VII), and computation with logarithms (in Chapter IX), an instructor could begin a one-semester course with Chapter I and continue through Chapter X.

Calculators. Although algebra can be learned without a calculator, more interesting examples and exercises can be considered if a calculator is used for some of the computations. The symbol [C] marks the steps where a calculator has been used in examples; it also marks the exercises for which a calculator is suggested. In examples, the results of computations marked with [C] can be taken on faith. If a calculator isn't available, then the exercises marked [C] can be either completed without a calculator or solved with the final computations only indicated and not carried out. Where the text gives advice on how to solve a problem with a calculator, the instructions will apply to many calculators but not to all of them; consult an owner's manual if you have questions.

It is a pleasure to acknowledge the advice I have received from the following persons: Chris Avery (De Anza College), Ronald Bates (Hartnell College), Rita Bordano (San Antonio College), Garrett Etgen (University of Houston), G. S. Gill (Brigham Young University), Jerome Kaufman (Western Illinois University), Charles Lanski (University of Southern California), Merry McDonald (Northwest Missouri State University), Ann Megaw (The University of Texas at Austin), David Rosen (Swarthmore College), and Dale Walston (The University of Texas at Austin). It is also a pleasure to thank Elaine Schneider, who helped with the answer appendix, and Gary Ostedt of John Wiley, who understands carrots and sticks.

John R. Durbin

CONTENTS

COLLEGE ALGEBRA

CHAPTER I
BASIC ALGEBRA

The first two sections of this chapter give a quick but thorough review of the laws that govern the fundamental operations with numbers. Section 1 includes decimal numbers and the general properties concerning signs and parentheses. Section 2 reviews fractions. Sections 3 through 6 introduce basic ideas that include exponents and how to solve the simplest kinds of equations and inequalities. The amount of time to be spent on this material will depend on your background; you should study the chapter till you can handle its exercises with ease.

SECTION

1 Real Numbers

A. Some Important Sets of Numbers

We begin with descriptions of the following sets of numbers, which are used throughout mathematics:

natural numbers
integers
rational numbers
irrational numbers
real numbers

Two other sets of numbers, imaginary and complex, will be described in Chapter VIII.

The **natural numbers** are the numbers we count with:

$$1, 2, 3, 4, 5, \ldots$$

The **integers** are the natural numbers together with their negatives and zero:

$$\ldots, -5, -4, -3, -2, -1, 0, 1, 2, 3, 4, 5, \ldots$$

The **positive integers** are the same as the natural numbers.

A number is **rational** if it is equal to a fraction in which both the numerator and denominator are integers with the denominator not zero. Examples are

$$\frac{2}{3}, \quad \frac{-43}{11}, \quad 0.5 = \frac{1}{2}, \quad 5\tfrac{1}{3} = \frac{16}{3}, \quad \text{and} \quad \frac{2\pi}{5\pi} = \frac{2}{5}.$$

Thus the rational numbers are the numbers that can be written in the form a/b, where a and b are integers and $b \neq 0$. Every integer is rational. For example,

$$-17 = \frac{-17}{1} \quad \text{and} \quad 0 = \frac{0}{1}.$$

(The word *rational* is used because of the connection between fractions and ratios. If, for instance, one object weighs $\frac{2}{3}$ as much as another, then their weights are in the *ratio* 2 to 3.)

Before defining real numbers and irrational numbers, it will be useful to look more closely at rational numbers. By division, we can represent any rational number as a decimal number. Here is the calculation for $\frac{2}{27}$.

$$
\begin{array}{r|l}
 & 0.074074\cdots \\
27 & 2.000000\cdots \\
 & 1\,89 \\
 & \;\;110 \\
 & \;\;108 \\
 & \;\;\;\;200 \\
 & \;\;\;\;189 \\
 & \;\;\;\;\;\;110 \\
 & \;\;\;\;\;\;108 \\
 & \;\;\;\;\;\;\;\;\;2
\end{array}
\qquad \frac{2}{27} = 0.074074\cdots = 0.\overline{074}
$$

To indicate that a block of digits is to be repeated, such as 074 above, we write a bar over the block, as indicated on the right. Here are some other examples.

$$
\begin{array}{ll}
3 = 3.0 & \frac{1}{3} = 0.333\cdots = 0.\overline{3} \\
\frac{-7}{2} = -3.5 & -\frac{103}{330} = -0.31212\cdots = -0.3\overline{12} \\
\frac{15}{8} = 1.875 & \frac{16}{7} = 2.\overline{285714}
\end{array}
$$

In the three examples on the left, the decimal representation *terminates:* there is a place after which only 0's would appear. In the three examples on the right, the decimal representation *repeats:* there is a place after which only the repetitions of some block that is not all 0's would appear. These two cases—terminating and repeating—are the only ones that can arise from rational numbers:

> A decimal number represents a rational number
> iff*
> it either terminates or repeats.

Appendix A explains why that statement is true. The statement gains interest because many decimal numbers neither terminate nor repeat. A simple example is $0.1010010001\cdots$, in which the number of 0's between the 1's increases by one each time.

Any number that has a decimal representation is called a **real number.** The real numbers that are not rational are called **irrational numbers.** Thus every real number is either rational or irrational. From what has already been said we can conclude this:

> A decimal number represents an irrational number
> iff
> it neither terminates nor repeats.

---

*We follow the practice, now widely accepted in mathematics, of using "iff" to denote "if and only if." This expression can be used only to connect things that are entirely equivalent. Here are some examples.

TRUE	A number is a natural number *iff* it is a positive integer.
TRUE	A number is a natural number *if* it is a positive integer.
TRUE	A number is rational *if* it is an integer.
FALSE	A number is rational *iff* it is an integer.

The example 0.1010010001 ⋯ is irrational. Also, a real number is irrational iff it *cannot* be written as a quotient of two integers. Figure 1.1 summarizes the relation between the sets of numbers discussed thus far; each set contains the sets that appear beneath it.

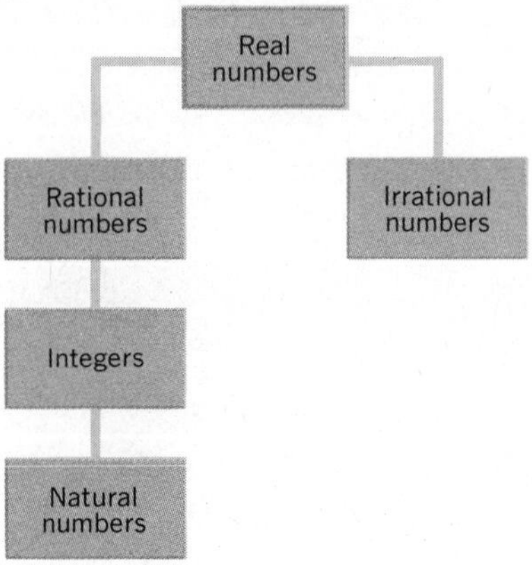

Figure 1.1

It is reasonable to ask if we really need all of the real numbers. Why not use only rational numbers, for instance? The answer was discovered by the Pythagoreans, nearly 2500 years ago, when they came to realize that even very simple problems can force us to go beyond the rational numbers. Specifically, they showed that $\sqrt{2}$, which gives the length of the hypotenuse of a right triangle having each leg of length 1, is irrational (Figure 1.2).* This means that if we restricted our attention to rational numbers, we would have no number to measure the length of this hypotenuse. A proof that $\sqrt{2}$ is irrational is given in Appendix A. It can also be proved that $\sqrt{3}$, $\sqrt{5}$, $\sqrt{6}$, $\sqrt{7}$, and $\sqrt{8}$ are irrational; in fact, *if n is a positive integer, then $\sqrt{n}$ is irrational unless n is a perfect square* (1, 4, 9, 16, 25, . . .). The number π, which represents the ratio of the circumference of any circle to its diameter, is also irrational (but the proof is not easy). These examples show that irrational numbers arise naturally, and not just from somewhat artificial constructions like 0.1010010001 ⋯.

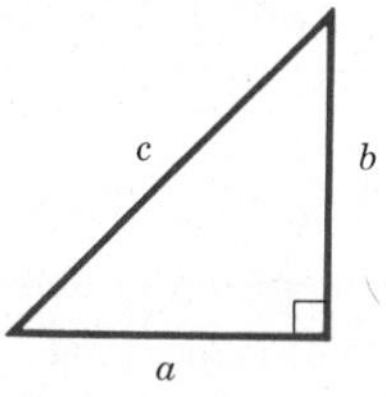

Figure 1.2 By the Pythagorean Theorem, $a^2 + b^2 = c^2$. If $a = 1$ and $b = 1$, then $c^2 = 2$ so $c = \sqrt{2}$.

EXAMPLE 1.1 Which of the following are natural numbers? integers? rational numbers? irrational numbers?

---

*Square roots will be reviewed in Section 3. All that's needed here is that $\sqrt{2}\sqrt{2} = 2$, $\sqrt{3}\sqrt{3} = 3$, and so on.

$$8, \quad 1.414, \quad \frac{\pi}{2}, \quad \frac{\sqrt{25}}{3}, \quad 22.\overline{34}, \quad \sqrt{20}, \quad -6\tfrac{2}{3}, \quad \frac{-85}{17}$$

Solution

Natural numbers:	8	
Integers:	8	(Every natural number is an integer.)
	$\frac{-85}{17}$	$(\frac{-85}{17} = -5)$
Rational numbers:	8, $\frac{-85}{17}$	(Every integer is a rational number.)
	1.414	(Terminating decimal.)
	$\sqrt{25}/3$	$(\sqrt{25}/3 = \frac{5}{3})$
	$22.\overline{34}$	(Repeating decimal.)
	$-6\frac{2}{3}$	$(-6\frac{2}{3} = \frac{-20}{3})$
Irrational numbers:	$\pi/2$	(If $\pi/2$ were a fraction of integers, say $\pi/2 = a/b$, then π would also be a fraction of integers, $\pi = 2a/b$.)
	$\sqrt{20}$	(20 is not a perfect square.)

Because $\sqrt{2}$ and π are irrational, no calculator can represent them exactly. A calculator with eight-digit accuracy will use the rounded-off values 1.4142136 for $\sqrt{2}$ and 3.1415927 for π. Such approximations are perfectly adequate for most practical work, but even more accurate values like 1.41421356237309504880 and 3.14159265358979323846 are not exact, since a terminating decimal cannot be exactly equal to an irrational number. Whenever the word *number* is used without qualification in this book, it will mean *real number.*

B. Real Lines

Much of the importance of the real numbers derives from their use for measurement. This, in turn, is based on the assumption of a one-to-one correspondence between the set of real numbers and the set of points on a line. This correspondence can be described as follows: Choose two points on a straight line and label them 0 and 1 (Figure 1.3). Make the line into a *directed line* by taking the direction from 0 toward

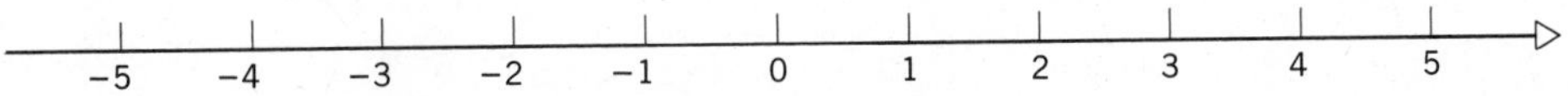

Figure 1.3

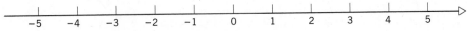

1 to be positive. The integers correspond to the points on this line in the obvious way shown in the figure. The other rational numbers arise when the segments between integer points are divided into equal parts (halves, thirds, fourths, and so on). For example, $\frac{1}{2}$ corresponds to the point midway between 0 and 1. In this way the rational numbers will account for a vast number of the points on the line. Our remarks show, however, that there will still be points—such as those corresponding to $\sqrt{2}$ and π—that will not be accounted for by rational numbers. Again, the basic assumption about measurement and numbers is that there is precisely one real number for each point and precisely one point for each real number. A line together with such a correspondence is called a **real line** (or **number line** or **coordinate line**). The

number associated with a point on a real line is called the **coordinate** of the point. Figure 1.4 shows examples of points with rational coordinates.

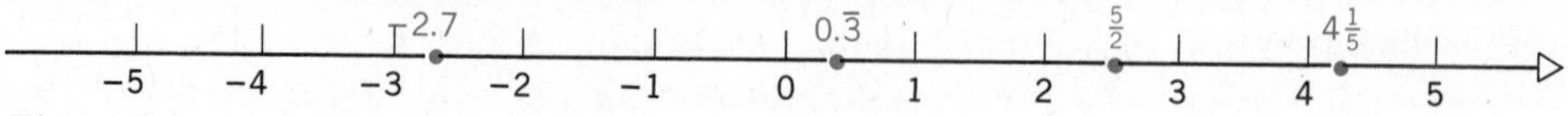

Figure 1.4

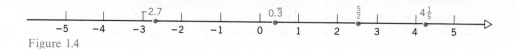

Points with irrational coordinates are located using decimal or fractional approximations. Figure 1.5 shows several examples. Figure 1.6 shows a magnified piece of Figure 1.5 with several additional points labeled to indicate the position of π; notice that $\frac{22}{7}$, which is often used as an approximation for π, is slightly larger than π. Usually, 3.14 will suffice as an approximate decimal coordinate for π.

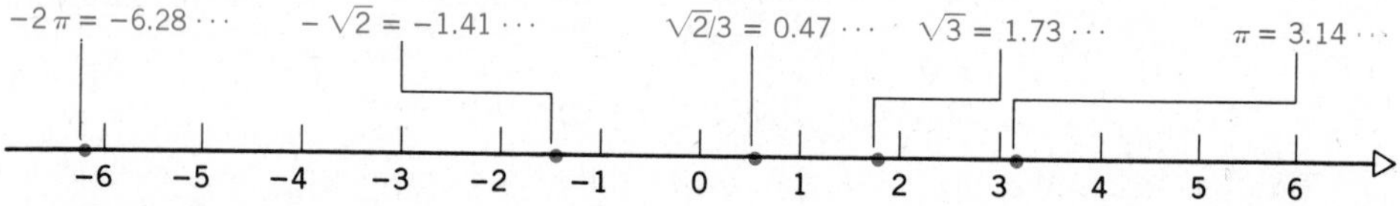

Figure 1.5

Figure 1.6

Instead of saying "the point with coordinate c" when referring to a point on a real line, we generally just say "the point c" if there is no chance of confusion.

Consider a horizontal real line directed to the right. If a and b are real numbers, and a is to the *left* of b on the line, then we say that a is **less than** b, and we write $a < b$. Thus

$$3 < 6, \quad -17 < 2, \quad -20 < -17, \quad \text{and} \quad 0 < \pi.$$

The notation $b > a$ means the same as $a < b$. Thus

$$5 > 2, \quad 5 > -2, \quad 2 > -5, \quad \text{and} \quad -2 > -5.$$

We say that b is **greater than** a if $b > a$.

The notation $a \le b$ means that either $a = b$ or $a < b$. Similarly for $b \ge a$. Thus

$$4 \le 9, \quad 9 \le 9, \quad -1 \ge -3, \quad \text{and} \quad -1 \ge -1.$$

The notation $a \le b$ is read "a is less than or equal to b." Similarly for $b \ge a$.

A number c is **positive** if $c > 0$, and **negative** if $c < 0$. Statements involving $<$, $>$, $\le$, or $\ge$ are called **inequalities.** More will be said about inequalities in Section 6.

C. Properties of the Real Numbers

This subsection summarizes most of the basic properties of the real numbers. The first few of these properties are *axioms* (statements that are assumed to be true); the other properties are *theorems* (statements that can be proved from the axioms). Although the theorems will be distinguished from the axioms, the proofs of the theorems will be omitted. The important point for our purposes is to be able to use the properties correctly. In all of the properties the letters a, b, and c denote real numbers.

Commutative laws

$a + b = b + a$ for all a and b.

$ab = ba$ for all a and b.

Example for addition

$$3 + 5 = 8 \quad \text{and} \quad 5 + 3 = 8$$

Associative laws

$a + (b + c) = (a + b) + c$ for all a, b, and c.

$a(bc) = (ab)c$ for all a, b, and c.

Example for multiplication

$$3 \cdot (8 \cdot 4) = 3 \cdot 32 = 96 \quad \text{and} \quad (3 \cdot 8) \cdot 4 = 24 \cdot 4 = 96$$

Distributive laws

$a(b + c) = ab + ac$ for all a, b, and c.

$(a + b)c = ac + bc$ for all a, b, and c.

Example of the first distributive law

$$3(2 + 4) = 3 \cdot 6 = 18 \quad \text{and} \quad 3 \cdot 2 + 3 \cdot 4 = 6 + 12 = 18$$

Zero and one

$a + 0 = 0 + a = a$ for every a.

$a \cdot 1 = 1 \cdot a = a$ for every a.

Negatives. For each number a there is a unique number denoted $-a$, and called the **negative** of a, such that

$$a + (-a) = (-a) + a = 0.$$

Reciprocals. For each number a except 0 there is a unique number denoted $1/a$, and called the **reciprocal** of a, such that

$$a \cdot \frac{1}{a} = \frac{1}{a} \cdot a = 1.$$

Equality

$a = a$ for every a. *Reflexive law*

If $a = b$, then $b = a$. *Symmetric law*

If $a = b$ and $b = c$, then $a = c$. *Transitive law*

The preceding properties in this subsection are axioms, called the **field axioms** (because any system that satisfies them, like the real numbers, is called a **field**). The remaining properties in this subsection are theorems that can be proved from the field axioms.

Properties of zero

$a \cdot 0 = 0 \cdot a = 0$ for every a.

$ab = 0$ iff either $a = 0$ or $b = 0$.

Examples

$$0 + 8 = 8, \quad \text{but} \quad 0 \cdot 8 = 0.$$

$$\text{If} \quad 3x = 0, \quad \text{then} \quad x = 0.$$

Subtraction. This property is defined in terms of addition and negatives, as follows:

$$a - b = a + (-b) \qquad \text{for all } a \text{ and } b.$$

Example

$$4 - 9 = 4 + (-9) = -5$$

Other properties of negatives. For all real numbers a, b, and c,

$$-(-a) = a$$

$$-(a + b) = (-a) + (-b)$$

$$(-a)b = a(-b) = -(ab)$$

$$(-a)(-b) = ab$$

$$a(b - c) = ab - ac.$$

Examples

$$-(-7) = 7$$

$$-(3 + 8) = (-3) + (-8) = -11$$

$$(-2)3 = 2(-3) = -(2 \cdot 3) = -6$$

$$(-3)(-2) = 3 \cdot 2 = 6$$

$$5(8 - 2) = 5 \cdot 6 = 30 \quad \text{and} \quad 5 \cdot 8 - 5 \cdot 2 = 40 - 10 = 30$$

Division. This property is defined in terms of multiplication and reciprocals, as follows: For all real numbers a and b with $b \neq 0$,

$$\frac{a}{b} = a \cdot \frac{1}{b} = \frac{1}{b} \cdot a.$$

The number a/b is called a **fraction.**

Example

$$\tfrac{5}{2} = 5 \cdot \tfrac{1}{2} = \tfrac{1}{2} \cdot 5$$

Other properties of fractions. For all real numbers a and b with $b \neq 0$,

$$\frac{-a}{b} = \frac{a}{-b} = -\frac{a}{b}$$

$$\frac{-a}{-b} = \frac{a}{b}.$$

Examples

$$\tfrac{-12}{3} = \tfrac{12}{-3} = -\tfrac{12}{3} = -4$$

$$\tfrac{-12}{-3} = \tfrac{12}{3} = 4$$

Division by 0 is undefined. Remember that $a/b = c$ iff $a = bc$. (For example, $\frac{10}{2} = 5$ because $10 = 2 \cdot 5$). If $a \neq 0$, then $a/0$ is undefined because there is *no* c such that $a = 0c$. If $a = 0$, then $a/0$ is undefined because *every* c satisfies $0 = 0c$.

Other properties of equality

If $a = b$, then $a + c = b + c$ and $ac = bc$ for every c.

If $a = b$ and $c = d$, then $a + c = b + d$ and $ac = bd$.

If $a + c = b + c$, then $a = b$. *Cancellation for addition*

If $ac = bc$ and $c \neq 0$, then $a = b$. *Cancellation for multiplication*

D. More Examples

The examples that follow contain several more reminders about handling numbers, including a review of the fundamental operations with decimals.

EXAMPLE 1.2 In an expression of the form $ab + c$ the product must be computed first. Thus

$$7 \cdot 2 + 3 = 14 + 3 = 17.$$

This contrasts with

$$7(2 + 3) = 7 \cdot 5 = 35.$$

EXAMPLE 1.3 The distributive law $a(b + c) = ab + ac$ can be extended to such cases as

$$a(b + c + d) = ab + ac + ad.$$

The law $(a + b)c = ac + bc$ can be extended in the same way.

EXAMPLE 1.4 Rewrite $-3(2a - b + 5c) + 4(a + 2b)$ without parentheses and simplify.

Solution

$$\begin{aligned}&-3(2a - b + 5c) + 4(a + 2b)\\&= (-3)(2a) + (-3)(-b) + (-3)(5c) + 4a + 4(2b)\\&= -6a + 3b - 15c + 4a + 8b = -2a + 11b - 15c.\end{aligned}$$

Notice that the negative sign in front of the first parentheses affects the sign of *every* term in the parentheses ($2a$, $-b$, and $5c$).

EXAMPLE 1.5 Operations inside of parentheses should be performed first. When there are parentheses inside of parentheses (or brackets), compute inside the innermost parentheses first.

(a) $5(8 - 2) - 4(3 - 1) = 5(6) - 4(2) = 30 - 8 = 22$

(b) $4(3 - 4.7) + 0.1(2.1 + 3.07) = 4(-1.7) + 0.1(5.17)$
$= -6.8 + 0.517 = -6.283$

→ subtract $\begin{array}{r}6.800\\ \underline{0.517}\\ 6.283\end{array}$

(c) $3 - [2.1(-5 + 1) + 6] = 3 - [2.1(-4) + 6]$
$= 3 - [-8.4 + 6] = 3 - (-2.4) = 5.4$

Even though computations like those in the next two examples can be carried out most easily with a calculator, it is important to understand how to do them without one. These examples will refresh your memory in case you've been away from arithmetic for awhile.

EXAMPLE 1.6 To multiply 3.14 by 5.6:

```
   3.14     2 digits to the right of the decimal point
 ×  5.6     1 digit to the right of the decimal point
 ------
   1884
  1570
 ------
  17.584    2 + 1 = 3 digits to the right of the decimal point
```

EXAMPLE 1.7 To divide 0.6867 by 3.27:

```
           .2 1
 3.27^ ) 0.68^67
           65 4        0.6867 ÷ 3.27 = 0.21
           ----
            3 27
            3 27
            ----
```

First, move the decimal point in the divisor (here 3.27) to the right so that the divisor becomes an integer; now move the decimal point in the dividend (here 0.6867) to the right the same number of places (the new positions of the decimal points are indicated by $_\wedge$). Now divide as with integers. Finally, place the decimal point in the quotient above the new position of the decimal point in the dividend.

EXERCISES FOR SECTION 1

In each of Exercises 1–3, make four lists, one of the *natural numbers,* one of the *integers,* one of the *rational numbers,* and one of the *irrational numbers.*

1. $17,\ -3.14,\ \pi,\ -16\frac{2}{3},\ 4.\overline{07},\ -6/\sqrt{4},\ 6,\ \sqrt{10}$

2. $\frac{0}{2},\ 16.\overline{6},\ -\pi,\ -\sqrt{5},\ 16.24,\ \sqrt{81}/\sqrt{16},\ \frac{-10}{2},\ \sqrt{7}\,\sqrt{7}$

3. $0\,\sqrt{2},\ \sqrt{3},\ 4\frac{1}{7},\ -13.\overline{5},\ 2\pi,\ \sqrt{64},\ -2.81,\ -15/\sqrt{9}$

Rewrite each of the following fractions as a decimal number.

4. $\frac{11}{8}$ **5.** $\frac{-31}{16}$ **6.** $\frac{3}{7}$ **7.** $\frac{-5}{13}$ **8.** $\frac{35}{6}$ **9.** $\frac{-7}{200}$

In each of Exercises 10–12, draw a real line and label the points with the given numbers as coordinates.

10. $4\frac{1}{2},\ -2.\overline{3},\ \pi,\ -2\sqrt{2},\ \frac{5}{7},\ -2.3$ **11.** $-\frac{3}{2},\ 2\pi,\ \sqrt{2}/2,\ 1.6,\ -3.\overline{6},\ -3.7$

12. $\pi/2,\ 1.3,\ -\sqrt{2}/3,\ \frac{-7}{4},\ 0.45,\ 1.\overline{3}$

Replace each □ with the appropriate sign, either $<$ or $>$.

13. 4 □ 9 **14.** −2 □ 5 **15.** 0.7 □ $\frac{3}{4}$ **16.** −2 □ −5

17. −9 □ −4 **18.** 0.3 □ 0.03 **19.** 0.2 □ −20 **20.** −0.2 □ $-0.\overline{2}$

21. $-\frac{1}{4}$ □ $-\frac{1}{5}$

Simplify.

22. $-2(3-5)$ **23.** $-(-7)(5)$ **24.** $(1-6)(-3)$

25. $(-45) \div 5$ **26.** $0(0-12)$ **27.** $-[-(-5)]$

28. $4 \cdot 5 + 7$ **29.** $3 + 2(-9)$ **30.** $-6 + 2 \cdot 5$

31. $(-6)(-8) \div (-12)$ **32.** $121 \div (-11)$ **33.** $(12 + 3) \div (-5)$
34. $(-4) \div (-4 - 1)$ **35.** $(-4 + 4) \div (4 + 4)$ **36.** $0 \div 6$
37. $-0(14 \div 2)$ **38.** $(0.08)(-12.5)$ **39.** $(-0.39) \div 1.3$
40. $(-1.21) \div (1.1)$ **41.** $0.5[(0.27) \div 0.3]$ **42.** $1 - 3[5 - 7(11 - 13)]$
43. $2.4[2 - 3(4 - 5)] \div (-0.5)$ **44.** $3.5[1 - 3(5 - 2)] \div (-0.4)$
45. $\{8[5 - 3(2 - 1)] \div [4(3 - 4)]\} \div 2$ **46.** $\{-4[-5 - 2(-3)] \div 2\} \div 2$
47. $(-15.41) \div (-0.67)$ **48.** $2(78.3 - 19.51) \div 0.2$
49. $(1.234 - 3.21) \div 0.04$ **50.** $[(6.594) \div (-0.21)] \div 3.14$
51. $(4.192 - 5.82 - 1.7074) \div 0.9$

Rewrite each expression without using parentheses and simplify.

52. $-(2a - b)$ **53.** $-2(-a + b)$ **54.** $-(-a - b)$
55. $3[-(2b + c) + (c - b) - b]$ **56.** $-5[a - 2(a - b) - (2a + b)]$
57. $-[-2(a + c) - (b - c) + 3(-a + c)]$
58. $-\{-2[-(a - b) + 2(-a + b)] + a\}$
59. $4\{-[(a - b) + 2(b - a)] - (a + b)\}$
60. $-3\{4(-a + b) - 3[-(2a - b) - 5(-a + b)]\}$

In each exercise that follows, one of the statements is true and the other is false. Decide which statement is true and explain (with an example, if appropriate) why the other statement is false.

61. (a) $a - (b - c) = (a - b) - c$ for all a, b, and c.
(b) If a, b, and c denote real numbers such that $abc = 0$, then either $a = 0$ or $b = 0$ or $c = 0$.

62. (a) If a number is not irrational, then it must be an integer.
(b) The reciprocal of every nonzero rational number is rational.

63. (a) A real number r is rational iff $2r$ is rational.
(b) If a and b denote irrational numbers (perhaps equal), then ab must also be irrational.

64. (a) A real number is irrational iff it is not rational.
(b) If $ab = 0$, then $a = 0$ and $b = 0$.

65. (a) If r is a nonzero real number, then $r/2 < r$.
(b) A decimal number represents a rational number if it terminates.

66. (a) A number is a natural number if it is a positive integer.
(b) $\sqrt{2} > \frac{71}{50}$.

SECTION 2

More about Real Numbers

A. Fractions

Section 1 gave several laws (properties) concerning fractions:

$$\frac{a}{b} = a \cdot \frac{1}{b} = \frac{1}{b} \cdot a$$

$$\frac{-a}{b} = \frac{a}{-b} = -\frac{a}{b}$$

$$\frac{-a}{-b} = \frac{a}{b}.$$

The set of basic laws of fractions is completed by the list that follows. Each law can be proved from the definition $a/b = a(1/b)$ and other properties of the real numbers. *As always, each denominator must be nonzero.*

Laws of fractions

$$\frac{a}{b} = \frac{c}{d} \text{ iff } ad = bc$$

$$\frac{a}{b} \cdot \frac{c}{d} = \frac{ac}{bd}$$

$$\frac{ac}{bc} = \frac{a}{b} \quad (c \neq 0)$$

$$\frac{\frac{a}{b}}{\frac{c}{d}} = \frac{a}{b} \cdot \frac{d}{c} = \frac{ad}{bc}$$

$$\frac{a}{d} + \frac{b}{d} = \frac{a+b}{d}$$

$$\frac{a}{d} - \frac{b}{d} = \frac{a-b}{d}$$

EXAMPLE 2.1

(a) $\frac{12}{20} = \frac{3}{5}$ because $12 \cdot 5 = 20 \cdot 3$

(Multiply as indicated: $\frac{12}{20} \rightleftarrows \frac{3}{5}$.)

(b) $\dfrac{2}{3} \cdot \dfrac{5}{7} = \dfrac{2 \cdot 5}{3 \cdot 7} = \dfrac{10}{21}$

(To multiply fractions, multiply their numerators and multiply their denominators.)

(c) $\dfrac{2 \cdot \cancel{5}}{3 \cdot \cancel{5}} = \dfrac{2}{3}$

(A nonzero common factor of the numerator and denominator can be canceled from a fraction.)

(d) $\dfrac{\frac{2}{5}}{\frac{3}{7}} = \dfrac{2}{5} \cdot \dfrac{7}{3} = \dfrac{2 \cdot 7}{5 \cdot 3} = \dfrac{14}{15}$

(To divide fractions, invert the denominator and multiply.)

(e) $\dfrac{3}{5} + \dfrac{11}{5} = \dfrac{3 + 11}{5} = \dfrac{14}{5}$

(To add fractions with a common denominator, add their numerators.)

(f) $\dfrac{2}{3} - \dfrac{7}{3} = \dfrac{2 - 7}{3} = \dfrac{-5}{3}$

(To subtract fractions with a common denominator, subtract their numerators.)

Each part of the following example uses more than one law. As with any other example, you should convince yourself of the reason behind each step.

EXAMPLE 2.2 The object is to simplify each expression.

(a) $-\dfrac{1}{6} \cdot \dfrac{-3}{4} \cdot 10 = \dfrac{\overset{1}{\cancel{3}} \cdot \overset{5}{\bcancel{10}}}{\underset{2}{\cancel{6}} \cdot \underset{2}{\bcancel{4}}} = \dfrac{5}{4}$ (The pair of /'s corresponds to one cancellation and the pair of \'s corresponds to another. Hereafter, cancellations will not be shown in detail like this.)

(b) $(\frac{1}{8} + \frac{5}{8}) \div \frac{3}{10} = \frac{6}{8} \div \frac{3}{10} = \frac{6}{8} \cdot \frac{10}{3} = \frac{5}{2}$

(c) $\dfrac{1}{a/b} = 1 \cdot \dfrac{b}{a} = \dfrac{b}{a} \quad (a \neq 0,\ b \neq 0)$

(d) $\dfrac{3/-4}{-2} = \dfrac{3}{-4} \div \dfrac{-2}{1} = \dfrac{3}{-4} \cdot \dfrac{1}{-2} = \dfrac{3}{8}$

A number made up of an integer and a fraction, such as $6\frac{2}{3}$, is called a **mixed number.** Before trying to combine numbers given in this form they should nearly always be converted to the form a/b. For example,

$$6\tfrac{2}{3} = 6 + \tfrac{2}{3} = \tfrac{18}{3} + \tfrac{2}{3} = \tfrac{20}{3}.$$

EXAMPLE 2.3 Simplify.

(a) $1\frac{2}{3} \div 5\frac{1}{4} = \frac{5}{3} \div \frac{21}{4} = \frac{5}{3} \cdot \frac{4}{21} = \frac{20}{63}$

(b) $(3\frac{1}{3} - 5\frac{2}{3})(-\frac{9}{2}) = (\frac{10}{3} - \frac{17}{3})(-\frac{9}{2}) = (\frac{-7}{3})(\frac{-9}{2}) = \frac{21}{2}$

B. Factors and Least Common Multiples

If n is a positive integer and a is a real number, then

$$a^n = \overbrace{a \cdot a \cdot \cdots \cdot a}^{n \text{ factors}}.$$

In particular, $a^1 = a$. We call a the **base,** n the **exponent,** and a^n the ***n*th power** of a.

EXAMPLE 2.4

(a) $2^3 = 2 \cdot 2 \cdot 2 = 8$
(b) $(-5)^4 = (-5)(-5)(-5)(-5) = 625$

An integer b is said to be a **multiple** of an integer a if $b = aq$ for some integer q. Thus 6 is a multiple of 2 because $6 = 2 \cdot 3$. But 6 is not a multiple of 4. If b is a multiple of a, we also say that a is a **factor** of b.

EXAMPLE 2.5 The positive factors of -45 are 1, 3, 5, 9, 15, and 45.

An integer p is a **prime** if $p > 1$ and p has no positive integral factor other than 1 and p itself. Thus the first ten primes are

2, 3, 5, 7, 11, 13, 17, 19, 23, and 29.

The ancient Greeks proved that the list of primes is infinite; the largest known prime has over 6000 digits. The primes are the building blocks of the integers in the sense that each positive integer is equal to a product of primes. Furthermore, if we use only positive exponents, then each integer greater than 1 can be written as a product of prime powers in only one way (provided we do not distinguish between different arrangements of the factors).

EXAMPLE 2.6 Each of the following integers has been expressed as a product of powers of distinct (that is, unequal) primes.

$$12 = 2^2 \cdot 3$$
$$504 = 2^3 \cdot 3^2 \cdot 7$$
$$19 = 19$$

(Think of 19 as a "product" with one factor.) To write an integer in this way we can reduce the integer one step at a time as we divide out successive prime factors. It helps to remember that

2 is a factor iff the last digit is even;

5 is a factor iff the last digit is 0 or 5;

3 is a factor iff 3 is a factor of the sum of the digits (for example, 3 is a factor of 627 because 3 is a factor of $6 + 2 + 7$).

Divide out factors in whatever order you discover them.

EXAMPLE 2.7 Write 1530 as a product of powers of distinct primes.

Solution The work can be arranged as follows, where the quotient is written below the dividend at each step. The prime factors of 1530 are the divisors, and each new divisor is obtained by inspecting the quotient from the preceding step.

$$\begin{array}{r|l} 5 & 1530 \\ \hline 2 & 306 \\ \hline 3 & 153 \\ \hline 3 & 51 \\ \hline & 17 \end{array} \qquad 1530 = 2 \cdot 3^2 \cdot 5 \cdot 17$$

The **least common multiple (l.c.m.)** of a set of integers is the least positive integer that is a multiple of each integer in the set.

EXAMPLE 2.8

(a) The l.c.m. of 3 and 2 is 6.
(b) The l.c.m. of 6, 15, and 12 is 60.

The l.c.m.'s of some sets can be computed by inspection, as in the preceding example. Here is a systematic scheme (followed by an example) for more difficult cases.

Step 1. Factor each integer in the set into powers of distinct primes.

Step 2. List all of the primes that appear in the factorizations in Step 1, and on each prime place the greatest exponent that appears on that prime in Step 1.

Step 3. The product of the prime powers chosen in Step 2 is the l.c.m. of the original set.

EXAMPLE 2.9 To compute the l.c.m. of 18, 75, and 168, first write

$$18 = 2 \cdot 3^2$$
$$75 = 3 \cdot 5^2$$
$$168 = 2^3 \cdot 3 \cdot 7.$$

Next,

2 appears in 2 and 2^3: choose 2^3
3 appears in 3^2, 3, and 3: choose 3^2
5 appears in 5^2: choose 5^2
7 appears in 7: choose 7.

Finally, the l.c.m. is

$$2^3 \cdot 3^2 \cdot 5^2 \cdot 7 = 12{,}600.$$

C. More about Fractions

To add fractions with common denominators we simply add their numerators:

$$\frac{a}{d} + \frac{b}{d} = \frac{a+b}{d}.$$

Similarly for subtraction. If fractions have unequal denominators, however, then we must first convert to fractions with common denominators. This is done by multiplying by 1 in a judiciously chosen form. For instance, we can write

$$\begin{aligned} \tfrac{5}{6} + \tfrac{1}{10} &= \tfrac{5}{6}\cdot\tfrac{10}{10} + \tfrac{1}{10}\cdot\tfrac{6}{6} \quad (\text{multiply by } 1 = \tfrac{10}{10} \text{ and } 1 = \tfrac{6}{6}) \\ &= \tfrac{50}{60} + \tfrac{6}{60} \\ &= \tfrac{56}{60} \\ &= \tfrac{14}{15}. \end{aligned}$$

Although the common denominator *can* be the product of the separate denominators, as $6 \cdot 10$ in the illustration just given, computations are often easier if the least common denominator is used: The **least common denominator (l.c.d.)** of a set of fractions is the least common multiple (l.c.m.) of the set of denominators of the fractions.

EXAMPLE 2.10 The l.c.d. of $\frac{1}{6}$ and $\frac{7}{10}$ is the l.c.m. of 6 and 10, which is 30. Therefore, we write

$$\frac{1}{6} - \frac{7}{10} = \frac{1}{6}\cdot\frac{5}{5} - \frac{7}{10}\cdot\frac{3}{3} = \frac{5-21}{30} = \frac{-16}{30} = -\frac{8}{15}.$$

In fractions that use letters to denote numbers, the l.c.d. is determined by modifying the idea for l.c.m.'s from Example 2.9. Treat each letter like a prime; in this way the l.c.d. will be the product of powers of the various letters—the power on each letter will be the greatest exponent that appears on that letter in the various denominators.

EXAMPLE 2.11 Write $\dfrac{1}{x} + \dfrac{1}{y}$ as a single fraction.

Solution The l.c.d. is xy, which includes x as a factor and y as a factor. Thus we write

$$\frac{1}{x} + \frac{1}{y} = \frac{1}{x}\cdot\frac{y}{y} + \frac{1}{y}\cdot\frac{x}{x} = \frac{y}{xy} + \frac{x}{xy} = \frac{y+x}{xy}.$$

EXAMPLE 2.12 Write

$$\frac{3}{ab^2} - \frac{1}{a^2c} + \frac{4}{bc^2}$$

as a single fraction.

Solution The l.c.d. must include a^2, b^2, and c^2 as factors. Therefore, we adjust each fraction so that its denominator becomes $a^2b^2c^2$. This gives

$$\frac{3}{ab^2}\cdot\frac{ac^2}{ac^2} - \frac{1}{a^2c}\cdot\frac{b^2c}{b^2c} + \frac{4}{bc^2}\cdot\frac{a^2b}{a^2b} = \frac{3ac^2 - b^2c + 4a^2b}{a^2b^2c^2}.$$

EXAMPLE 2.13 Write each expression as a single fraction and simplify.

(a) $\frac{1}{2a} - \frac{1}{6b}$ **(b)** $1 + \frac{1}{a} - \frac{1}{a^2}$

Solution

(a) The l.c.d. is $6ab$. Thus

$$\frac{1}{2a} - \frac{1}{6b} = \frac{1}{2a}\cdot\frac{3b}{3b} - \frac{1}{6b}\cdot\frac{a}{a} = \frac{3b - a}{6ab}.$$

(b) The l.c.d. is a^2. Thus

$$1 + \frac{1}{a} - \frac{1}{a^2} = 1\cdot\frac{a^2}{a^2} + \frac{1}{a}\cdot\frac{a}{a} - \frac{1}{a^2} = \frac{a^2 + a - 1}{a^2}.$$

Following are some illustrations with **complex fractions**—that is, fractions in which the numerator or denominator (or both) are also fractions. In each part of Example 2.14 we first write both the denominator and numerator as single fractions, next invert the denominator and multiply, and then simplify.

EXAMPLE 2.14 Simplify.

(a) $$\frac{4 + \frac{1}{2}}{\frac{1}{3} - 1} = \frac{\frac{8+1}{2}}{\frac{1-3}{3}} = \frac{\frac{9}{2}}{\frac{-2}{3}} = \frac{9}{2}\cdot\frac{3}{-2} = -\frac{27}{4}$$

(b) $$1 - \frac{1}{1 - \frac{1}{2}} = 1 - \frac{1}{\frac{2-1}{2}} = 1 - \frac{1}{\frac{1}{2}} = 1 - 2 = -1$$

(c) $$\left(1 + \frac{1}{t}\right) \div \frac{1}{t} = \frac{t+1}{t}\cdot\frac{t}{1} = t + 1$$

(d) $$\frac{\frac{1}{x} - \frac{1}{x^2}}{x} = \frac{\frac{x-1}{x^2}}{x} = \frac{x-1}{x^2}\cdot\frac{1}{x} = \frac{x-1}{x^3}$$

Another method for simplifying complex fractions is to multiply both the numerator and denominator by the l.c.d. of all the fractions in both the numerator and denominator, as in the following examples.

EXAMPLE 2.15 The l.c.d. of $\frac{5}{2}$, $-\frac{2}{3}$, $\frac{3}{4}$, and $\frac{1}{6}$ is 12, so we multiply by $\frac{12}{12}$ in the following case.

$$\frac{\frac{5}{2}-\frac{2}{3}}{\frac{3}{4}+\frac{1}{6}} = \frac{\frac{5}{2}-\frac{2}{3}}{\frac{3}{4}+\frac{1}{6}} \cdot \frac{12}{12} = \frac{(\frac{5}{2})12-(\frac{2}{3})12}{(\frac{3}{4})12+(\frac{1}{6})12}$$

$$= \frac{30-8}{9+2}$$

$$= \frac{22}{11}$$

$$= 2.$$

EXAMPLE 2.16 Simplify.

$$\frac{\frac{1}{a}-\frac{2}{b}}{\frac{3}{b}-\frac{4}{a}} = \frac{\left(\frac{1}{a}-\frac{2}{b}\right)}{\left(\frac{3}{b}-\frac{4}{a}\right)} \cdot \frac{ab}{ab} = \frac{\left(\frac{1}{a}\right)ab-\left(\frac{2}{b}\right)ab}{\left(\frac{3}{b}\right)ab-\left(\frac{4}{a}\right)ab} = \frac{b-2a}{3a-4b}.$$

EXERCISES FOR SECTION 2

Write each of these integers as a product of powers of distinct primes.

1. 18 **2.** 45 **3.** 30 **4.** 264 **5.** 17 **6.** 1001 **7.** 31
8. 225 **9.** 83

Determine the least common multiple of the integers in each set.

10. $\{6, 9\}$ **11.** $\{10, 15\}$ **12.** $\{12, 10\}$ **13.** $\{14, 21, 10\}$
14. $\{12, 66, 99\}$ **15.** $\{25, 40, 70\}$

Simplify each of the following fractions, writing the final answer in the form a/b, where a and b are integers.

16. $(\frac{1}{2})(\frac{4}{3})$ **17.** $(\frac{5}{2})(-2\frac{1}{3})$ **18.** $(1\frac{1}{5})(1\frac{2}{3})$
19. $(-6\frac{3}{7})(\frac{14}{-5})$ **20.** $(-\frac{7}{10})(\frac{5}{6})(\frac{9}{-14})$ **21.** $(\frac{-3}{28})(1\frac{2}{5})(\frac{10}{-9})$
22. $\frac{1}{3}+\frac{1}{4}$ **23.** $\frac{2}{5}-\frac{1}{3}$ **24.** $\frac{5}{6}-\frac{1}{4}$
25. $\frac{1}{3}-\frac{1}{6}+\frac{1}{9}$ **26.** $1\frac{1}{2}+2\frac{1}{3}-3\frac{1}{4}$ **27.** $1\frac{3}{8}-2\frac{5}{6}-1$
28. $\frac{-8}{5}/\frac{15}{2}$ **29.** $2\frac{1}{2} \div (\frac{1}{3})$ **30.** $1 \div (-4\frac{1}{7})$
31. $(-4\frac{2}{3})/(-3)$ **32.** $(-1\frac{2}{7})/3$ **33.** $(6\frac{2}{3})/(-5)$
34. $(\frac{3}{2}) \div (\frac{7}{4}-3)$ **35.** $(-\frac{1}{4}) \div (2-\frac{4}{3})$ **36.** $(\frac{-12}{5}) \div (\frac{1}{3}-\frac{7}{3})$
37. $(4\frac{1}{3}+1\frac{2}{5}) \div (2\frac{1}{2})$ **38.** $(\frac{7}{50}-\frac{7}{30}) \div (\frac{1}{6}+\frac{1}{5})$ **39.** $(3+\frac{1}{2}) \div (2-\frac{1}{3})$
40. $1+\dfrac{1}{1+\frac{1}{3}}$ **41.** $1-\dfrac{2-\frac{3}{4}}{5}$ **42.** $[(1\frac{1}{2}) \div (2\frac{1}{3})] \div (2\frac{1}{4})$

Write each expression as a single fraction and simplify.

43. $\dfrac{1}{a} - \dfrac{1}{b}$ 44. $\dfrac{1}{r} + \dfrac{1}{s} + \dfrac{1}{t}$ 45. $\dfrac{a}{2} + \dfrac{b}{3}$ 46. $\left(\dfrac{1}{2a} - \dfrac{1}{3a}\right) \div a$

47. $\left(\dfrac{2}{y} - \dfrac{3}{x}\right) \div xy$ 48. $\left(1 - \dfrac{m}{n}\right) \div (m - n)$

49. $\left[\left(r + \dfrac{1}{s}\right) \div \left(\dfrac{1}{s}\right)\right] - 1$ 50. $\left(\dfrac{1}{c} + \dfrac{1}{d}\right) \div \dfrac{1}{cd}$

51. $\left(\dfrac{x}{y} + \dfrac{y}{x}\right) \div \dfrac{2}{xy}$ 52. $\left(\dfrac{1}{x} - \dfrac{1}{y}\right) \div \left(\dfrac{2x - 2y}{x}\right)$

53. $\left(\dfrac{1}{a} - \dfrac{1}{b}\right) \div \left(\dfrac{a - b}{3b}\right)$ 54. $\left(\dfrac{1}{u} + \dfrac{1}{v}\right) \div \left(\dfrac{1}{v} - \dfrac{1}{u}\right)$

55. $\left(1 - \dfrac{1}{r}\right) \div \dfrac{1}{r}$ 56. $\left(3 + \dfrac{1}{a}\right) \div \left(4 - \dfrac{1}{a}\right)$

57. $1 - \dfrac{1}{1 - \dfrac{1}{a}}$ 58. $\dfrac{1 - \dfrac{1}{x} - \dfrac{1}{x^2}}{1 + \dfrac{1}{x} + \dfrac{1}{x^2}}$ 59. $\dfrac{1}{1 + \dfrac{1}{1 + \dfrac{1}{a - 1}}}$ 60. $\dfrac{x + y}{\dfrac{1}{x} + \dfrac{1}{y}}$

Each of the following statements is *false,* as indicated. Prove this by replacing the letters with appropriately chosen numbers in each case to produce an inequality rather than an equation.

61. False. $\dfrac{1}{a} + \dfrac{1}{b} = \dfrac{1}{a + b}$ for all a and b such that the denominators are nonzero.

62. False. $\dfrac{a + b}{a} = 1 + b$ for all a and b with $a \neq 0$.

63. False. $(a/b)/c = a/(b/c)$ for all nonzero a, b, and c.

64. False. $\dfrac{1 + ab}{b} = 1 + a$ for all a and b with $b \neq 0$.

65. False. $a \cdot \dfrac{b}{c} = \dfrac{ab}{ac}$ for all a, b, and c with $a \neq 0$ and $c \neq 0$.

66. False. $\dfrac{a}{a + b} = \dfrac{1}{1 + b}$ for all a and b such that the denominators are nonzero.

SECTION

3

Integral Exponents. Square Roots

A. Positive Integral Exponents

Recall from the last section that if a is a real number and n is a positive integer, then

$$a^n = \overbrace{a \cdot a \cdot \cdots \cdot a}^{n \text{ factors}}.$$

If m and n are both positive integers, then

$$\begin{aligned} a^m a^n &= \overbrace{a \cdot a \cdot \cdots \cdot a}^{m \text{ factors}} \cdot \overbrace{a \cdot a \cdot \cdots \cdot a}^{n \text{ factors}} \\ &= \overbrace{a \cdot a \cdot \cdots \cdot a \cdot a \cdot a \cdot \cdots \cdot a}^{m+n \text{ factors}} \\ &= a^{m+n}. \end{aligned}$$

This justifies the following law of exponents for every real number a and all positive integers m and n.

$$\boxed{a^m a^n = a^{m+n}} \qquad (3.1)$$

EXAMPLE 3.1

$$3^4 \cdot 3^2 = (3 \cdot 3 \cdot 3 \cdot 3)(3 \cdot 3) = 3 \cdot 3 \cdot 3 \cdot 3 \cdot 3 \cdot 3 = 3^6 = 729$$

Think about the following additional laws, (3.2) through (3.6), long enough to convince yourself they are true for all real numbers a and b and all positive integers m and n. The examples, which illustrate each law in a special case, should help.

$$\boxed{(a^m)^n = a^{mn}} \qquad (3.2)$$

EXAMPLE 3.2

$$(x^4)^3 = (x \cdot x \cdot x \cdot x)(x \cdot x \cdot x \cdot x)(x \cdot x \cdot x \cdot x) = x^{4 \cdot 3} = x^{12}$$

$$\boxed{(ab)^m = a^m b^m} \qquad (3.3)$$

EXAMPLE 3.3

$$(ab)^3 = ab \cdot ab \cdot ab = a \cdot a \cdot a \cdot b \cdot b \cdot b = a^3b^3$$

$$\boxed{\frac{a^m}{b^m} = \left(\frac{a}{b}\right)^m \quad (b \neq 0)} \qquad (3.4)$$

EXAMPLE 3.4

$$\frac{y^4}{z^4} = \frac{y \cdot y \cdot y \cdot y}{z \cdot z \cdot z \cdot z} = \frac{y}{z} \cdot \frac{y}{z} \cdot \frac{y}{z} \cdot \frac{y}{z} = \left(\frac{y}{z}\right)^4$$

$$\boxed{(-a)^n = \begin{cases} a^n & \text{if } n \text{ is even} \\ -a^n & \text{if } n \text{ is odd} \end{cases}} \qquad (3.5)$$

EXAMPLE 3.5 Notice that $-a^n$ means $-(a^n)$, which may or may not equal $(-a)^n$. That is, in $-a^n$, the exponent n does not operate on the minus sign.

$$(-2)^4 = 16 = 2^4 \quad \text{but} \quad -2^4 = -16$$
$$(-2)^5 = -32 = -2^5$$

$$\boxed{\frac{a^m}{a^n} = \begin{cases} a^{m-n} & \text{if } m > n \quad (a \neq 0) \\ 1 & \text{if } m = n \quad (a \neq 0) \\ 1/a^{n-m} & \text{if } m < n \quad (a \neq 0) \end{cases}} \qquad (3.6)$$

EXAMPLE 3.6

$$\frac{a^5}{a^3} = \frac{\not{a} \cdot \not{a} \cdot \not{a} \cdot a \cdot a}{\not{a} \cdot \not{a} \cdot \not{a}} = a^{5-3} = a^2$$

$$\frac{a^3}{a^3} = \frac{\not{a} \cdot \not{a} \cdot \not{a}}{\not{a} \cdot \not{a} \cdot \not{a}} = 1$$

$$\frac{a^3}{a^5} = \frac{\not{a} \cdot \not{a} \cdot \not{a}}{\not{a} \cdot \not{a} \cdot \not{a} \cdot a \cdot a} = \frac{1}{a^{5-3}} = \frac{1}{a^2}$$

EXAMPLE 3.7 Laws (3.1)–(3.6) can be extended and combined to handle more complicated cases.

(a) $(3ax^2)(-2ax^3) = -6a^2x^5$
(b) $(-2y^2)^3 = (-2)^3(y^2)^3 = -8y^6$
(c) $a^2a^4a^3 = a^{2+4+3} = a^9$
(d) $(xyz)^3 = x^3y^3z^3$
(e) $(a^2b^3)^5 = (a^2)^5(b^3)^5 = a^{10}b^{15}$
(f) $\left(\frac{x^5}{y^2}\right)^3 = \frac{(x^5)^3}{(y^2)^3} = \frac{x^{15}}{y^6}$

B. Arbitrary Integral Exponents

It is convenient that the notation a^n can be extended naturally to include exponents other than positive integers. Here the notation will be extended to include all integers as exponents; further extensions will be made in Chapters IV and IX.

To motivate the following definition of a^0, notice that if the law $a^m/a^n = a^{m-n}$ [Equation (3.6)] is to apply when $m = n$, then we must have $1 = a^0$, because $a^m/a^m = 1$ and $a^{m-m} = a^0$. [Notice that a is nonzero for (3.7); 0^0 is undefined.]

If a is any nonzero real number, then

$$a^0 = 1. \tag{3.7}$$

EXAMPLE 3.8

$$6^0 = 1 \qquad \left(\frac{2}{3}\right)^0 = 1 \qquad \frac{2}{3^0} = \frac{2}{1} = 2 \qquad (-5)^0 = 1 \qquad -5^0 = -1$$

Now suppose that the law $a^m/a^n = a^{m-n}$ [Equation (3.6)] is to be made to apply when $n > m$ and $m = 0$. Then we must have $a^0/a^n = a^{0-n}$; that is, $1/a^n = a^{-n}$. This leads to the next definition, which shows that a negative exponent corresponds to forming a reciprocal.

If a is any nonzero real number and n is a positive integer, then

$$a^{-n} = \frac{1}{a^n}. \tag{3.8}$$

EXAMPLE 3.9

(a) $a^{-1} = 1/a$
(b) $2^{-3} = 1/2^3 = 1/8$
(c) $(-3)^{-2} = 1/(-3)^2 = 1/9$ but $-3^{-2} = -(1/3^2) = -1/9$
(d) $(-0.2)^{-1} = 1/(-0.2) = -5$
(e) $(a/b)^{-4} = 1/(a/b)^4 = 1/(a^4/b^4) = b^4/a^4 = (b/a)^4$

Notice that Equation (3.8) is valid for $n < 0$ as well as $n > 0$, because if $n < 0$, then

$$a^{-n} = \frac{1}{1/a^{-n}} = \frac{1}{a^n}.$$

For instance,

$$4^2 = 4^{-(-2)} = \frac{1}{4^{-2}}.$$

The next example illustrates that *if the numerator and denominator of a fraction are products,* then a factor of the numerator can be moved to the denominator if we change the sign of the exponent on the factor, and a factor of the denominator can be moved to the numerator if we change the sign of the exponent on the factor.

EXAMPLE 3.10

(a) $\dfrac{a^{-3}b}{c} = \dfrac{b}{a^3c}$ **(b)** $\dfrac{u}{v^{-2}} = uv^2$ **(c)** $\dfrac{xy^{-2}}{x^{-3}y} = \dfrac{xx^3}{yy^2} = \dfrac{x^4}{y^3}$

(d) We *cannot* write $\dfrac{1 + a^{-2}b}{c} = \dfrac{1 + b}{a^2c}$. The reason is that a^{-2} is not a factor of the *whole* numerator. In fact,

$$\frac{1 + a^{-2}b}{c} = \frac{1 + \dfrac{b}{a^2}}{c} = \frac{\dfrac{a^2 + b}{a^2}}{\dfrac{c}{1}} = \frac{a^2 + b}{a^2c}.$$

Laws (3.1)–(3.6) all extend from positive integral exponents to all integral exponents, as shown by the following summary. Law (3.6) has been simplified here because $m - n$ is now defined as an exponent even if it is zero or negative.

Laws of integral exponents. If a and b are nonzero real numbers, and m and n are integers, then

$$a^m a^n = a^{m+n} \tag{3.1}$$

$$(a^m)^n = a^{mn} \tag{3.2}$$

$$(ab)^m = a^m b^m \tag{3.3}$$

$$\frac{a^m}{b^m} = \left(\frac{a}{b}\right)^m \tag{3.4}$$

$$(-a)^n = \begin{cases} a^n & \text{if } n \text{ is even} \\ -a^n & \text{if } n \text{ is odd} \end{cases} \tag{3.5}$$

$$\frac{a^m}{a^n} = a^{m-n}. \tag{3.6}$$

Each of the laws for integral exponents can be proved by using the laws for positive exponents and the definitions in Equations (3.7) and (3.8). Here, for example, is the proof of $(a^m)^n = a^{mn}$ for the case $m > 0$ and $n < 0$.

$(a^m)^n = \dfrac{1}{(a^m)^{-n}}$ by Equation (3.8) since $n < 0$ [$n > 0$ in (3.8) so that $-n$ there corresponds to n here]

$= \dfrac{1}{a^{m(-n)}}$ by Law (3.2) for positive exponents because $m > 0$ and $-n > 0$

$$= \frac{1}{a^{-(mn)}} \qquad \text{because } m(-n) = -(mn)$$

$$= a^{mn} \qquad \text{by Equation (3.8) since } mn < 0.$$

The proofs of the other parts are left as exercises.

EXAMPLE 3.11 The following illustrations show how to simplify expressions so that only positive exponents appear in the final answer. Assume that n is a positive integer in part (e).

(a) $3^{-2} \cdot 3^4 = 3^{-2+4} = 3^2 = 9$

(b) $(a^{-2})^{-5} = a^{(-2)(-5)} = a^{10}$

(c) $\left(\frac{u^2}{v^3}\right)^{-4} = \frac{(u^2)^{-4}}{(v^3)^{-4}} = \frac{u^{-8}}{v^{-12}} = \frac{v^{12}}{u^8}$

(d) $\frac{(x^2y^{-3})^4}{(xy^2)^{-2}} = \frac{(x^2)^4(y^{-3})^4}{x^{-2}(y^2)^{-2}} = \frac{x^8y^{-12}}{x^{-2}y^{-4}}$

$$= x^8y^{-12} \cdot x^2y^4 = x^{10}y^{-8} = \frac{x^{10}}{y^8}$$

(e) $\left(\frac{a^{-1}b^5}{a^3b^{-2}}\right)^{-3n} = \left(\frac{b^5b^2}{a^3a^1}\right)^{-3n} = \left(\frac{b^7}{a^4}\right)^{-3n} = \frac{(b^7)^{-3n}}{(a^4)^{-3n}}$

$$= \frac{b^{-21n}}{a^{-12n}} = \frac{a^{12n}}{b^{21n}}$$

C. Scientific Notation

One useful application of exponents begins from the following observation: To multiply a decimal number by 10^n we simply move the decimal point n places to the right if $n > 0$, and $|n|$ places to the left if $n < 0$. For example,

$$1.47 \times 10^2 = 147 \quad \text{and} \quad 3892 \times 10^{-5} = 0.03892.$$

This means that we can move the decimal point wherever we like provided we compensate by multiplying by an appropriate power of 10. If the decimal point is moved just to the right of the leftmost nonzero digit, then the resulting form is called **scientific notation.** Thus the form of a positive number written in scientific notation is

$$a \times 10^n, \quad \text{where } n \text{ is an integer and } 1 \leq a < 10.$$

EXAMPLE 3.12

(a) To write 147 in scientific notation, first replace 147 by 1.47; then multiply 1.47 by 10^2 to compensate:

$$147 = 1.47 \times 10^2.$$

(b) To write 0.03892 in scientific notation, first replace 0.03892 by 3.892; then multiply by 10^{-2} to compensate:

$$0.03892 = 3.892 \times 10^{-2}.$$

EXAMPLE 3.13 Scientific notation is especially convenient when we are faced with extremely large or extremely small numbers.

The earth is approximately $93{,}000{,}000 = 9.3 \times 10^7$ miles from the sun.

A light-year (the distance light travels in one year) is approximately $5{,}880{,}000{,}000{,}000 = 5.88 \times 10^{12}$ miles.

The mass of a hydrogen atom has been found to be approximately $0.000000000000000000000001675 = 1.675 \times 10^{-24}$ grams.

Calculators generally display extremely large or extremely small numbers in scientific notation like this:

2.5344 11 for 2.5344×10^{11}

and

3.9457 -12 for 3.9457×10^{-12}.

The next example shows how to use scientific notation to keep track of decimal points in computations. Given a product or quotient, we can first write each factor in scientific notation and then determine the proper placement of the decimal point in the final answer merely by working with powers of 10.

EXAMPLE 3.14

(a) $2{,}000{,}000 \times 0.03 \times 160 = (2 \times 10^6) \times (3 \times 10^{-2}) \times (1.6 \times 10^2)$
$= (2 \times 3 \times 1.6) \times 10^{6-2+2} = 9.6 \times 10^6 = 9{,}600{,}000$

(b)
$$\begin{aligned}\frac{70 \times 0.0032}{0.001 \times 800} &= \frac{(7 \times 10) \times (3.2 \times 10^{-3})}{(1 \times 10^{-3}) \times (8 \times 10^2)} \\ &= \frac{7 \times 3.2}{8} \times 10^{1-3+3-2} \\ &= \frac{22.4}{8} \times 10^{-1} \\ &= 2.8 \times 10^{-1} \\ &= 0.28\end{aligned}$$

D. Square Roots

We complete this section with some remarks about square roots, which will arise in applications as early as Chapter III.

If $a \geq 0$, then b is called a **square root** of a provided $b^2 = a$. For example, 3 and -3 are both square roots of 9 because $3^2 = 9$ and $(-3)^2 = 9$. The notation $\sqrt{a}$ is reserved for the *nonnegative* real number b such that $b^2 = a$. That is, for $a \geq 0$:

$$b = \sqrt{a} \text{ means that } b \geq 0 \text{ and } b^2 = a.$$

Thus $\sqrt{9} = 3$. The number $\sqrt{a}$ is called the **principal square root** of a. The notation $\pm\sqrt{a}$ represents the pair, $\sqrt{a}$ and $-\sqrt{a}$. Here is another example for emphasis.

EXAMPLE 3.15 The *principal square root* of 49 is $\sqrt{49} = 7$. The *square roots* of 49 are $\pm\sqrt{49} = \pm 7$—that is, 7 and -7.

EXAMPLE 3.16 To simplify an even power of a square root, write the exponent as 2 times another integer and then use $(\sqrt{a})^2 = a$, as follows:

(a) $(\sqrt{5})^6 = [(\sqrt{5})^2]^3 = 5^3 = 125$
(b) $(\sqrt{0.2})^{-4} = [(\sqrt{0.2})^2]^{-2} = 0.2^{-2} = 1/0.2^2 = 1/0.04 = 25$

Table I in Appendix C gives an approximate value for the principal square root of each of the integers from 1 through 100. Negative numbers do not have real square roots, because $b^2 \geq 0$ for every real number b. For example, $b = \sqrt{-4}$ would imply $b^2 = -4$, which is impossible if b is a real number. Therefore, for the present we'll avoid $\sqrt{a}$ if $a < 0$. (We'll return to the question of $\sqrt{a}$ for $a < 0$ in Chapter VIII, when we extend the number system from the real numbers to the complex numbers.)

Many expressions with square roots can be simplified by using the following law.

$$\sqrt{ab} = \sqrt{a}\sqrt{b} \quad (a \geq 0,\ b \geq 0) \tag{3.9}$$

The law is true because $(\sqrt{a}\sqrt{b})^2 = (\sqrt{a})^2(\sqrt{b})^2 = ab$. A typical example of its use is

$$\sqrt{4x} = \sqrt{4}\sqrt{x} = 2\sqrt{x}.$$

The next example gives further illustrations. In each case we first write the expression under the square root sign as a product, with perfect square factors wherever possible.

EXAMPLE 3.17 Simplify. Assume that each letter represents a positive real number.

(a) $\sqrt{18} = \sqrt{9 \cdot 2} = \sqrt{9}\sqrt{2} = 3\sqrt{2}$
(b) $\sqrt{25xy^2} = \sqrt{25y^2 \cdot x} = \sqrt{25}\sqrt{y^2}\sqrt{x} = 5y\sqrt{x}$
(c) $\sqrt{ab^2c^3} = \sqrt{b^2c^2 \cdot ac} = \sqrt{b^2}\sqrt{c^2}\sqrt{ac} = bc\sqrt{ac}$

The following example exploits the fact that if n is any integer, then

$$\sqrt{10^{2n}} = 10^n \quad [\text{because } (10^n)^2 = 10^{2n}].$$

EXAMPLE 3.18 Use $\sqrt{3.5} \approx 1.8708$ and $\sqrt{35} \approx 5.9161$ to determine approximate decimal values for the following:

(a) $\sqrt{350}$ (b) $\sqrt{0.35}$ (c) $\sqrt{350000}$ (d) $\sqrt{0.00035}$.

(The symbol $\approx$ means *approximately equal.*)

Solution Each of the numbers under a square root sign can be written as either 3.5 or 35 times an even power of 10.

(a) $\sqrt{350} = \sqrt{3.5 \times 10^2} = \sqrt{3.5}\sqrt{10^2} \approx 1.8708 \times 10 = 18.708$
(b) $\sqrt{0.35} = \sqrt{35 \times 10^{-2}} = \sqrt{35}\sqrt{10^{-2}} \approx 5.9161 \times 10^{-1} = 0.59161$
(c) $\sqrt{350000} = \sqrt{35 \times 10^4} = \sqrt{35}\sqrt{10^4} \approx 5.9161 \times 10^2 = 591.61$
(d) $\sqrt{0.00035} = \sqrt{3.5 \times 10^{-4}} = \sqrt{3.5}\sqrt{10^{-4}} \approx 1.8708 \times 10^{-2} = 0.018708$

EXERCISES FOR SECTION 3

Compute each power and simplify.

1. $(-2)^3/9^2$ **2.** $-3^2/(-4)^3$ **3.** $7^2/(-5)^3$ **4.** $(\sqrt{2})^6$ **5.** $(\sqrt{10})^4$
6. $7/(\sqrt{7})^4$ **7.** $(2^{-3})^2$ **8.** $(1/2)^{-5}$ **9.** $(3^{-2})^{-3}$ **10.** $(-0.02)^{-1}$
11. -0.2^{-2} **12.** $(-0.1)^{-3}$

Simplify, using no negative exponents in the final answers. Assume that n is a positive integer.

13. $(a^2)^4a^3$ **14.** $(t^3t)^2$ **15.** $(x^2x^4x^3)^5$ **16.** $(-1.5u^3)^2$ **17.** $(-3v^2)^3$
18. $(-0.5w^4)^3$ **19.** $(3b^{-3})^2$ **20.** $\left(\frac{x^2}{2y}\right)^{-1}$ **21.** $[(2x)^{-2}y]^{-3}$
22. $\frac{(-a^{-2}b)^3}{(ab^{-1})^2}$ **23.** $\frac{x^{-3}y^4}{(-x^2y^2)^{-3}}$ **24.** $-\left(\frac{-ut^{-1}}{3u^{-2}t}\right)^4$ **25.** $\frac{-3a^0b}{(-2ab)^0}$
26. $\frac{(-6x)^0}{(2x^2)^{-1}}$ **27.** $\frac{5t^0}{(-10t)^{-1}}$ **28.** $\left(\frac{\sqrt{5}a^{-1}}{b^2}\right)^2$ **29.** $\frac{(-0.3x)^2}{(10xy)^{-3}}$
30. $\frac{(\sqrt{2}ab)^{-2}}{a^2b^{-1}}$ **31.** $\frac{0.4a}{(5a^3)^{-2}}$ **32.** $ab\left(\frac{-\sqrt{2}a^2}{b^3}\right)^{-4}$ **33.** $\left(\frac{\sqrt{3}a}{b^2}\right)^{-2}\left(\frac{\sqrt{2}b}{a^2}\right)^4$
34. $\frac{a^nb^{2n}}{a^{2n}b^{-n}}$ **35.** $(u^2v)^{-2n}(uv^{-1})^{5n}$ **36.** $\left(\frac{x}{y^2}\right)^{-3n}\left(\frac{y^3}{x}\right)^{-n}$ **37.** $(\sqrt{3}y)^{4n}(x\sqrt{3})^{-2n}$
38. $\left[\frac{\sqrt{2}t^{-3}}{\sqrt{6}t^{-1}}\right]^{4n}$ **39.** $\frac{\left(\frac{\sqrt{a}}{b}\right)^{-6n}}{\left(\frac{\sqrt{b}}{a}\right)^{4n}}$

Write each of the following numbers in scientific notation.

40. 0.00000001 (The length in centimeters of an angstrom, a unit used in expressing the length of light waves.)
41. 1,454 (The height in feet of the Sears Tower in Chicago.)
42. 3,212 (The drop in feet of Angel Falls in Venezuela.)
43. 29,028 (The height in feet of Mount Everest.)
44. 36,000 (The approximate depth in feet of the deepest part of the Pacific Ocean, in the Mariana trench.)

45. 25,200,000,000,000 (The approximate distance in miles to *Proxima Centauri,* the nearest star.)
46. 0.0000000033 (The approximate time in seconds for light to travel one meter in a vacuum.)
47. 0.00067 (The approximate time in seconds for sound to travel one meter in fresh water at room temperature.)
48. 0.0029 (The approximate time in seconds for sound to travel one meter in a vacuum at room temperature.)

Perform the following calculations using scientific notation to handle decimal points, as in Example 3.14.

49. 890×0.0004
50. $400 \times 0.0008 \times 2000$
51. $0.12/0.000006$
52. $14 \times 0.0005/0.007$
53. $(120{,}000 \times 0.0003)/(90 \times 0.02)$
54. $(0.03 \times 5400)/(900 \times 0.0006)$

Give the exact value or an approximate decimal value for each of the following principal square roots. Use Table I if necessary.

55. $\sqrt{81}$
56. $\sqrt{10{,}000}$
57. $\sqrt{121}$
58. $\sqrt{41}$
59. $\sqrt{19}$
60. $\sqrt{74}$
61. $\sqrt{0.38}$
62. $\sqrt{7100}$
63. $\sqrt{130{,}000}$
64. $\sqrt{2500}$
65. $\sqrt{0.0036}$
66. $\sqrt{0.49}$

Simplify each of the following square roots by removing all perfect square factors from under the square root sign, as in Example 3.17. (Do not use decimal approximations.) Assume that each letter represents a positive real number.

67. $\sqrt{12}$
68. $\sqrt{50}$
69. $\sqrt{72}$
70. $\sqrt{8x^2y}$
71. $\sqrt{75a^4b}$
72. $\sqrt{16xy^2z^3}$
73. $\sqrt{98ab^5c}$
74. $\sqrt{200t^2u^3v^4}$
75. $\sqrt{360m^2n^8}$

SECTION

4 Polynomials

A. Terminology

A **variable** (or, more precisely, a **real variable**) is a letter representing an unspecified real number. In the formula $V = x^3$ for the volume of a cube, both V and x are variables. A **monomial** in a variable x is a product of a real number, called the **coefficient,** and a power x^n, where n is a nonnegative integer.* The integer n is called the **degree** of the monomial, except that a monomial with coefficient 0 is not assigned a degree. We write x^n for $1 \cdot x^n$.

EXAMPLE 4.1

	Monomial	Coefficient	Degree
(a)	x^3	1	3
(b)	$-7x^5$	-7	5
(c)	$\frac{1}{2}x^0 = \frac{1}{2}$	$\frac{1}{2}$	0
(d)	$0x = 0$	0	undefined

A **polynomial** in a variable x is a finite sum of monomials in x. Notice that *sum* includes *difference* since we allow negative coefficients: for example, $6x^2 - 3x + 1 = 6x^2 + (-3)x + 1$. The individual monomials are called the **terms** of the polynomial: the terms of $6x^2 - 3x + 1$ are $6x^2$, $-3x$, and 1. The largest degree from all of the terms is the **degree** of the polynomial, and the corresponding coefficient is the **leading coefficient** of the polynomial. The zero polynomial (the polynomial having only zero coefficients) is not assigned a degree. A term of degree 0 is called a **constant term.** A polynomial with no term of degree 0 is said to have constant term 0.

EXAMPLE 4.2

	Polynomial	Degree	Leading coefficient
(a)	$6x^2 - 3x + 1$	2	6
(b)	$x^3 - 5$	3	1
(c)	$-\sqrt{2}x^9 + 4x^{25} - 7x$	25	4
(d)	πx^2	2	π
(e)	-17	0	-17

Observe that we include monomials as polynomials: think of πx^2 and -17 as sums with one term, for instance.

Unless there is a special reason to do otherwise, it is generally best to arrange the

*The phrase "monomial *in* a variable x" may be confusing at first, but it is the accepted terminology. A better phrase would be "monomial *with* x as variable."

terms of a polynomial so that their degrees are either in descending order or in ascending order; we usually choose the former. For example, we would usually write the polynomial in Example 4.2(c) as $4x^{25} - \sqrt{2}x^9 - 7x$.

The variable in a polynomial need not be x. For example, $16t^2 - 32t + 48$ is a polynomial in the variable t.

Polynomials in more than one variable are defined as follows. First, a **monomial** in several variables is a product consisting of a real number, called the **coefficient,** and nonnegative integral powers of the variables. Thus $3xy$ and $-\sqrt{2}x^5y$ are monomials in x and y. The **degree** of a monomial in several variables is the sum of the degrees on its individual variables. For example, $3xy$ has degree 2, and $-\sqrt{2}x^5y$ has degree 6. As in the case of one variable, a **polynomial** in several variables is a finite sum of monomials.

EXAMPLE 4.3

(a) Let x and y denote the lengths of the sides of a rectangle. Then the area of the rectangle is given by xy, which is a monomial (and a polynomial) in x and y.
(b) $u^2 - uv + v^2$ is a polynomial in u and v.

A sum of two monomials (such as $x^2 + y^2$) is called a **binomial.** A sum of three monomials (such as $u^2 - uv + v^2$) is called a **trinomial.**

B. Operations with Polynomials

Since the variables in polynomials represent real numbers, we can add, subtract, multiply, and divide polynomials by using the usual laws of arithmetic (from Section 1). Sums and differences of polynomials can be simplified by adding and subtracting similar terms as in the next two examples.

EXAMPLE 4.4 Add $2x^2 - 3x + 6$ and $x^2 + x - 4$.

$$\begin{array}{r} 2x^2 - 3x + 6 \\ + \quad x^2 + x - 4 \\ \hline 3x^2 - 2x + 2 \end{array}$$

We usually carry out such operations without writing one polynomial below the other. Also, the intermediate steps in examples like the following can often be done mentally.

EXAMPLE 4.5

(a) $(3x - y + 4) + (2x + 2y - 3) = (3 + 2)x + (-1 + 2)y + (4 - 3)$
$= 5x + y + 1$
(b) $(x^2 + 4xy + y^2) - (2x^2 - 3xy + y^2)$
$= (1 - 2)x^2 + (4 + 3)xy + (1 - 1)y^2 = -x^2 + 7xy + 0y^2$
$= -x^2 + 7xy$

EXAMPLE 4.6 To multiply monomials, multiply their coefficients and variables separately.

(a) $(2x)(x^3) = 2x^4$
(b) $\frac{1}{4}(8x^6) = 2x^6$
(c) $(3xy^2z)(-5x^2y^2) = [3 \cdot (-5)](x \cdot x^2)(y^2 \cdot y^2)z = -15x^3y^4z$

EXAMPLE 4.7 To multiply a monomial times a polynomial with more than one term, multiply the monomial times each term of the polynomial and then simplify. This amounts to using one of the distributive laws (Section 1).

(a) $2x(5x + 4) = 2x(5x) + 2x(4) = 10x^2 + 8x$
(b) $uv(u^2 - uv + v^2) = u^3v - u^2v^2 + uv^3$

The next two examples show two ways to compute more complicated products. The point is that we must multiply each term in one of the polynomials by each term in the other polynomial. Then we simplify.

EXAMPLE 4.8 Compute $(3x + 2)(x^2 - 4x + 1)$.

$$\begin{array}{rrrrl} & x^2 & -\ 4x & +\ 1 & \\ \times & & 3x & +\ 2 & \\ \hline 3x^3 & -\ 12x^2 & +\ 3x & & = 3x(x^2 - 4x + 1) \\ & 2x^2 & -\ 8x & +\ 2 & = 2(x^2 - 4x + 1) \quad \text{add} \\ \hline 3x^3 & -\ 10x^2 & -\ 5x & +\ 2 & = (3x + 2)(x^2 - 4x + 1) \end{array}$$

This shows what each line to the left represents, and can be omitted.

EXAMPLE 4.9 In this example, we multiply the second polynomial first by x, then by $-y^2$, and then add and simplify.

$$\begin{aligned} (x - y^2)(x^2 + xy^2 + y^4) &= x(x^2 + xy^2 + y^4) - y^2(x^2 + xy^2 + y^4) \\ &= x^3 + x^2y^2 + xy^4 - x^2y^2 - xy^4 - y^6 \\ &= x^3 - y^6 \end{aligned}$$

Division of polynomials will be discussed in Section 27. Any polynomial, or result obtained from combining polynomials through a finite number of additions, subtractions, multiplications, divisions, or extractions of roots, is called an **algebraic expression** (or simply an **expression**).

The **domain** of an expression containing one variable is the set of all real numbers that can be substituted for the variable so that the expression is also equal to a real number. Numbers that must be excluded from the domain include those that make a denominator zero and those that make the number under a square root sign negative. Domains can be recorded concisely with the **set-builder notation**

$$\{x: \ \cdots\},$$

which means

the set of all x such that $\cdots$

For example,

$$\{x: \ x < 5\}$$

represents

the set of all x such that $x < 5$.

(From the context, x is assumed to represent real numbers in this case. For more examples with set-builder notation, see Appendix B.)

EXAMPLE 4.10

(a) The domain of $\sqrt{x - 2}$ is

$$\{x: \quad x \geq 2\}.$$

Each real number less than 2 is excluded from the domain because $x - 2 < 0$ if $x < 2$, and the square root of a negative number is not a real number.

(b) The domain of $1/(t - 1)$ is

$$\{t: \quad t \neq 1\}.$$

The number 1 is excluded because it would lead to division by 0.

(c) The domain of any polynomial with one variable is the set of all real numbers.

C. Some Special Products

Any product of two binomials has the following form.

$$(p + q)(r + s) = pr + ps + qr + qs \tag{4.1}$$

The four products on the right can be visualized and remembered like this:

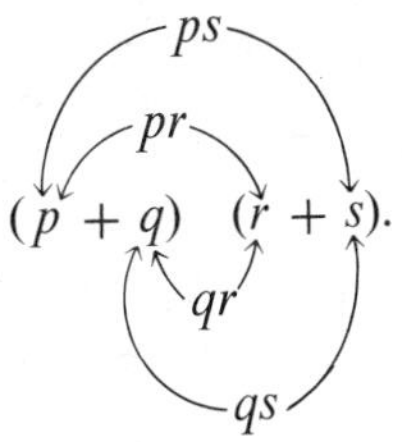

In any special case we simply make appropriate substitutions for p, q, r, and s. For example, with

$$p = u, \quad q = v, \quad r = u, \quad \text{and} \quad s = -v,$$

Equation (4.1) yields

$$(u + v)(u - v) = u^2 - uv + vu - v^2 = u^2 - v^2.$$

This justifies the first of the three products that follow; the others can be justified in the same way. All three products definitely should be memorized.

Difference of two squares

$$(u + v)(u - v) = u^2 - v^2 \tag{4.2}$$

Square of a binomial sum

$$(u + v)^2 = u^2 + 2uv + v^2 \tag{4.3}$$

Square of a binomial difference

$$(u - v)^2 = u^2 - 2uv + v^2 \tag{4.4}$$

EXAMPLE 4.11

(a) $(x + 2)(x - 2) = x^2 - 2^2 = x^2 - 4$

(b) $(t + 3)^2 = t^2 + 2(3)t + 3^2 = t^2 + 6t + 9$

(c) $(5y - \sqrt{2})^2 = (5y)^2 - 2(5y)(\sqrt{2}) + (\sqrt{2})^2 = 25y^2 - 10\sqrt{2}y + 2$

Equation (4.1) leads to the following scheme for the coefficients of $(ax + by)(cx + dy)$.

$(ax + by)(cx + dy)$ ac is the coefficient of x^2

$(ax + by)(cx + dy)$ $ad + bc$ is the coefficient of xy

$(ax + by)(cx + dy)$ bd is the coefficient of y^2

This gives the following general formula. (Suggestion: To compute products like this think in terms of the scheme above; don't memorize the formula.)

$$(ax + by)(cx + dy) = acx^2 + (ad + bc)xy + bdy^2 \tag{4.5}$$

Other variables can replace x and y, of course.

EXAMPLE 4.12

(a) $(x + 2y)(3x + 4y) = 1 \cdot 3x^2 + (1 \cdot 4 + 2 \cdot 3)xy + 2 \cdot 4y^2 = 3x^2 + 10xy + 8y^2$

(b) $(u - 2v)(3u + 4v) = 3u^2 + (4 - 6)uv - 8v^2 = 3u^2 - 2uv - 8v^2$

(c) $(0.1m + 0.3n)(0.2m + 0.4n) = 0.02m^2 + (0.04 + 0.06)mn + 0.12n^2$
$= 0.02m^2 + 0.1mn + 0.12n^2$

D. Introduction to Factoring

To **factor** a polynomial is to write it as a product of other polynomials; these other polynomials are then called **factors** of the original polynomial. For example, $x^2 - 2x = x(x - 2)$, so that x and $x - 2$ are factors of $x^2 - 2x$. Factoring is essential in simplifying algebraic expressions and in solving equations. The general problem of factoring will be considered in Chapter IV; but factoring of second-degree polynomials in one variable will be needed before that, so it will be treated here.

EXAMPLE 4.13 A second-degree polynomial with one variable and no constant term is easy to factor—the variable itself is a factor.

(a) $x^2 - 2x = x(x - 2)$
(b) $3t^2 + 15t = 3t(t + 5)$

Notice that $x^2 - 2x$ could also be factored as $2x(\frac{1}{2}x - 1)$. However, it is better to avoid fractions if possible. On the other hand, integral common factors, such as 3 in Example 4.13(b), generally should be removed from each term.

The next simplest cases are those that fit one of the standard forms in Equations (4.2)–(4.4). To use these equations for factoring, read each equation from right to left rather than from left to right. We can extend the applicability of these forms by substituting freely for u and v. It is the *form* that's important, not the specific variables or numbers. In Equation (4.2), for example, the form is revealed by the name, a *difference of two squares.*

EXAMPLE 4.14 Factor.

(a) $x^2 - 9 = (x + 3)(x - 3)$ [Equation (4.2) with $u = x$, $v = 3$.]
(b) $y^2 + 10y + 25 = (y + 5)^2$ [Equation (4.3) with $u = y$, $v = 5$.]
(c) $z^2 - 14z + 49 = (z - 7)^2$ [Equation (4.4) with $u = z$, $v = 7$.]

The next example shows still more variety in the possibilities from Equations (4.2)–(4.4).

EXAMPLE 4.15 Factor.

(a) $4m^2 - 9n^2 = (2m)^2 - (3n)^2 = (2m + 3n)(2m - 3n)$
(b) $9t^2 + 6t + 1 = (3t)^2 + 2(1)(3t) + 1^2 = (3t + 1)^2$
(c) $0.01x^2 - 0.2xy + y^2 = (0.1x)^2 - 2(0.1x)y + y^2 = (0.1x - y)^2$

The most general form for a factored second-degree polynomial in x is the following special case of Equation (4.5).

$$acx^2 + (ad + bc)x + bd = (ax + b)(cx + d) \tag{4.6}$$

To apply Equation (4.6), keep in mind the scheme preceding Equation (4.5). In this section we use only integral coefficients when working with this most general form.

EXAMPLE 4.16 Factor $x^2 + x - 2$.

Solution We want factors of the form $ax + b$ and $cx + d$ so that

$$(x^2 + x - 2) = (ax + b)(cx + d).$$

In particular, we must have

$$ac = 1 \quad \text{and} \quad bd = -2.$$

The choices $a = 1$, $c = 1$, $b = 2$, and $d = -1$ will work. That is, they will give the required coefficient on x (namely, 1) as well as the required constant term and coefficient on x^2:

$$x^2 + x - 2 = (x + 2)(x - 1).$$

EXAMPLE 4.17 Factor $3x^2 + 7x + 2$.

Solution Here we want

$$3x^2 + 7x + 2 = (ax + b)(cx + d).$$

Thus

$$ac = 3 \quad \text{and} \quad bd = 2.$$

The choices $a = 3$, $c = 1$, $b = 2$, and $d = 1$ will not work:

$$(3x + 2)(x + 1) = 3x^2 + 5x + 2.$$

(The coefficient on x must be 7, not 5.) But the choices $a = 3$, $c = 1$, $b = 1$, and $d = 2$ will work:

$$(3x + 1)(x + 2) = 3x^2 + 7x + 2.$$

EXAMPLE 4.18 To factor $12t^2 - t - 6$, we need

$$12t^2 - t - 6 = (at + b)(ct + d).$$

In particular, then, we must have $ac = 12$ and $bd = -6$. The first condition can be met in several ways: $12 = 12 \cdot 1 = 6 \cdot 2 = 4 \cdot 3 = 3 \cdot 4 = 2 \cdot 6 = 1 \cdot 12$. (Both factors of 12 could also be negative, but it is sufficient to consider positive factors here.) The condition $-6 = bd$ can be met with b and d chosen appropriately form ± 1, ± 2, ± 3, and ± 6. Trial and error leads to $(12t^2 - t - 6) = (4t - 3)(3t + 2)$.

EXERCISES FOR SECTION 4

Carry out the indicated operations and simplify.

1. $(x^2 - x + 3) + (3x^2 - x - 5)$ **2.** $(y^3 + 2y - 5) - (y^3 + y^2 - 1)$

3. $2(t^2 + 1) - (t^2 - 1)$ **4.** $2(z^2 - 1) + (z + 4) - (z^2 - z - 1)$

5. $3(x^3 - x) - (x^3 - x^2 + x) + (x^2 + 2x - 1)$

6. $(v^2 - v + 1) - (2v^2 - v) + 3(v^2 + v - 1)$

7. $(0.5u^2v)(-2uv^3)$ **8.** $(-5a^2b^3)(-7ab^2)$ **9.** $(-\frac{1}{3}xyz)(12xz^2)$

10. $\frac{1}{2}x(2x^2 - 6x + 1)$ **11.** $3u^2(\frac{1}{3}u^2 - \frac{1}{2}u)$ **12.** $\frac{1}{2}x^4(x - \frac{1}{3})$

13. $(3x - 1)(x^2 + x - 2)$ **14.** $(t + 2)(t^2 - 3t + 1)$

15. $(4y - 1)(y^2 - y - 3)$
16. $2t + 3[t - 5(t + 1)]$
17. $0.5x - 2[0.3x - 0.1(x - 2)]$
18. $\frac{2}{3}[3(x - \frac{1}{2}) - 2(\frac{1}{3} + x)]$
19. $xy + y[2(x - 1) - (x + 3)]$
20. $3u[v + \frac{5}{2}(1 - v)] - v[3u + \frac{1}{2}(1 - u)]$
21. $m(1 + n + n^2) - mn(1 + m + n) - n(2 - m - m^2)$
22. $1.5[a + b(u + v)] - 0.5[a + b(u - v)]$
23. $\frac{1}{2}(c - d)m - \frac{1}{2}(c + d)m$
24. $0.2[u - 0.6(t - 1)] + 0.4[t - 0.3(u - 1)]$

Compute these products using the methods of Subsection C.

25. $(x + y)(2x + y)$
26. $(m - n)(2m + n)$
27. $(3u + v)(u + 3v)$
28. $(0.1y + 0.2z)(0.3y - 0.4z)$
29. $(1.5a - 0.5b)(a + b)$
30. $(0.1x + 0.02y)(0.01x + 0.2y)$
31. $(\frac{1}{2}u + \frac{1}{3}v)(\frac{1}{2}u - \frac{1}{3}v)$
32. $(y - \sqrt{2})^2$
33. $(m - \frac{1}{3}n)^2$
34. $(x + \sqrt{5})^2$
35. $(x + 1)(x - 1)$
36. $(z + \sqrt{3})(z - \sqrt{3})$
37. $(t - c)^2$
38. $(u + \frac{1}{2})^2$
39. $(a + 2b)^2$

Factor, using Equations (4.2)–(4.4).

40. $x^2 - 25$
41. $u^2 + 12u + 36$
42. $x^2 - 2x + 1$
43. $t^2 - 6t + 9$
44. $x^2 - 10x + 25$
45. $m^2 + 14m + 49$
46. $y^2 + 8y + 16$
47. $t^2 - 3$
48. $25w^2 + 10w + 1$
49. $4x^2 + 4x + 1$
50. $v^2 + v + \frac{1}{4}$
51. $2y^2 + 2\sqrt{2}y + 1$
52. $u^2 + 2\sqrt{3}u + 3$
53. $9y^2 - 6y + 1$
54. $x^2 - \frac{2}{3}x + \frac{1}{9}$
55. $9z^2 - 0.01$
56. $0.04a^2 - 0.2ab + 0.25b^2$
57. $w^2 + 1.4w + 0.49$

Factor.

58. $x^2 - 4x + 3$
59. $z^2 + 3z + 2$
60. $x^2 + 6x + 5$
61. $2y^2 + y - 1$
62. $5x^2 + 6x + 1$
63. $3w^2 + 4w + 1$
64. $9m^2 + 12m + 4$
65. $4y^2 - 2y - 6$
66. $4t^2 - 20t + 25$
67. $6v^2 + 5v - 6$
68. $4p^2 - 12p + 9$
69. $4n^2 - 13n + 3$

Determine the domain of each expression. (See Example 4.10.)

70. $1/x$
71. $\sqrt{x - 4}$
72. $3x^5 + 2x - 1$
73. $\sqrt{w + 5}$
74. $\dfrac{1}{y + 2}$
75. $\dfrac{\sqrt{z}}{z - 2}$

SECTION

5 Equations

A. Equations with Variables

Stripped to the essentials, algebra consists almost entirely of equations that contain variables. Nearly everything in this book will concern such equations: how to form them, combine them, simplify them, solve them, analyze them. We begin with the most basic facts and terminology about equations with one variable.

Consider these examples.

$$3x + 2 = 4 \tag{5.1}$$

$$\sqrt{x + 4} = x - 2 \tag{5.2}$$

$$y = y + 1 \tag{5.3}$$

$$(t + 1)^2 = t^2 + 2t + 1 \tag{5.4}$$

If the variable in any such equation is replaced by a real number that is in the domain of both sides of the equation, then each side of the equation will also be equal to a real number. If these numbers on the two sides are equal, then the number that replaced the variable is called a **solution,** or **root,** of the equation; we also say in this case that the number **satisfies** the equation. To check whether a number is a solution of an equation, simply substitute the number for the variable in the equation and then simplify to see if the two sides are equal. Here are some examples that involve Equations (5.1)–(5.4). (We are not concerned yet with *finding* solutions; only with *checking* possibilities.)

EXAMPLE 5.1 Is $\frac{2}{3}$ a solution of $3x + 2 = 4$?

Left side *Right side*

$$3\left(\tfrac{2}{3}\right) + 2 \stackrel{?}{=} 4$$

$$4 = 4$$

The answer is Yes.

EXAMPLE 5.2

(a) Is 5 a solution of $\sqrt{x + 4} = x - 2$?

Left side *Right side*

$$\sqrt{5 + 4} \stackrel{?}{=} 5 - 2$$

$$\sqrt{9} \stackrel{?}{=} 3$$

$$3 = 3$$

The answer is Yes.

(b) Is 0 a solution of $\sqrt{x + 4} = x - 2$?

Left side *Right side*

$$\sqrt{0 + 4} \stackrel{?}{=} 0 - 2$$
$$2 \neq -2$$

The answer is No.

EXAMPLE 5.3 The equation $y = y + 1$ has no solution, since the right side will be one more than the left side no matter what number y represents.

EXAMPLE 5.4 Every real number is a solution of

$$(t + 1)^2 = t^2 + 2t + 1$$

since the left side is just the factored form of the right side.

The equation in Example 5.4 is an *identity:* An equation is an **identity** if its two sides have the same domain and if every real number in that domain is a solution of the equation. An equation that contains a variable but is *not* an identity is called a **conditional equation.** To **solve** an equation means to find its solutions. Special techniques are required to solve some types of equations, but others can be solved just by using the following general ideas.

Two equations with the same variable are called **equivalent** if they have the same solutions. The most basic plan for solving an equation is to replace it by simpler and simpler equivalent equations until the solutions (if there are any) become obvious. In doing this we make frequent use of the following laws, which put into words properties of equality that were stated in Section 1.

> If equal numbers, or expressions that represent equal numbers, are added to (or subtracted from) both sides of an equation, then the resulting equation will be equivalent to the original equation. (5.5)

> If both sides of any equation are multiplied (or divided) by equal nonzero numbers, or expressions that represent equal nonzero numbers, then the resulting equation will be equivalent to the original equation. (5.6)

EXAMPLE 5.5 The two equations in each of the following pairs are equivalent.

(a) $2x - 4 = x - 4$, $2x = x$ — Add 4 to both sides.

(b) $15x^2 = 5$, $3x^2 = 1$ — Divide both sides by 5.

(c) $-x^3 + 2x = 5$, $x^3 - 2x = -5$ — Multiply both sides by -1.

(d) $\left.\begin{aligned} 2x - 3 &= 3x \\ 3x &= 2x - 3 \end{aligned}\right\}$ Interchange the two sides. (Remember that if $a = b$, then $b = a$.)

Notice that the law in (5.6) insists that we multiply or divide only by *nonzero* expressions. Whenever we multiply or divide by an expression involving the variable, we run a risk of violating this restriction and creating an inequivalent equation. Here's an example to show why.

EXAMPLE 5.6 We can transform $x^2 = x$ into $x = 1$ by *dividing* both sides by x. Conversely, we can transform $x = 1$ into $x^2 = x$ by *multiplying* both sides by x. But the equations $x^2 = x$ and $x = 1$ are not equivalent, because 0 is a solution of $x^2 = x$ but it is not a solution of $x = 1$.

B. Linear Equations

An equation that is equivalent to an equation of the form

$$ax + b = 0, \tag{5.7}$$

where a and b are real numbers with $a \neq 0$, is called a **linear equation.** The equations in Examples 5.7 through 5.11 are linear. (The reason these are called *linear* equations will become clear in Chapter V.)

The following examples show how to solve linear equations. The idea is to transform each equation by using the Laws (5.5) and (5.6) until the variable is isolated on one side and only numbers remain on the other. Each answer can be checked by substitution.

EXAMPLE 5.7

$$\begin{aligned} x - 3 &= 2 \\ x - 3 + 3 &= 2 + 3 && \text{Add 3 to both sides.} \\ x &= 5 && \text{Simplify.} \end{aligned}$$

Here we have transformed the original equation into the equivalent equation $x = 5$, and the solution to the latter is obviously the number 5. The plan will be similar in the other examples.

EXAMPLE 5.8

$$\begin{aligned} 4y - 5 &= 3y + 2 \\ 4y - 5 + 5 &= 3y + 2 + 5 && \text{Add 5 to both sides.} \\ 4y &= 3y + 7 && \text{Simplify.} \\ 4y - 3y &= 3y + 7 - 3y && \text{Subtract } 3y \text{ from both sides.} \\ y &= 7 && \text{Simplify.} \end{aligned}$$

Hereafter, we frequently combine steps to save space.

EXAMPLE 5.9

$3(2t - 2) = 2(9t + 1)$	
$6t - 6 = 18t + 2$	Remove parentheses.
$6t = 18t + 8$	Add 6 to both sides and simplify.
$-12t = 8$	Subtract $18t$ from both sides and simplify.
$t = -\frac{2}{3}$	Divide by -12 and simplify.

EXAMPLE 5.10

$\dfrac{2x-1}{5} - \dfrac{3x-2}{7} = \dfrac{6}{35}$	
$35 \cdot \left(\dfrac{2x-1}{5} - \dfrac{3x-2}{7}\right) = 35 \cdot \dfrac{6}{35}$	Multiply by 35, the l.c.d. of all the terms.
$7(2x - 1) - 5(3x - 2) = 6$	Cancel the denominators.
$14x - 7 - 15x + 10 = 6$	Remove parentheses.
$-x + 3 = 6$	Simplify.
$-x = 3$	Subtract 3 from both sides and simplify.
$x = -3$	Multiply by -1.

Some equations with the variable in a denominator are equivalent to linear equations, as in the next example. The first step with such an equation is to multiply both sides by an expression that will remove the variable from the denominator. If we do that, however, then we *must* check the final answer. For equations such as those in Examples 5.7 through 5.10 the only reason to check an answer is to catch an error in calculation; in contrast, for equations such as the following we can get an incorrect answer even without such an error. (Remember Example 5.6 and the remark that preceded it.) An answer that arises in this way is called an **extraneous solution** (or **extraneous root**); it is *not* a solution of the original equation. Example 5.12 will give an illustration.

EXAMPLE 5.11

$\dfrac{1}{x} - \dfrac{3}{7x} = \dfrac{8}{21}$	
$21x\left(\dfrac{1}{x} - \dfrac{3}{7x}\right) = 21x \cdot \dfrac{8}{21}$	Multiply by $21x$, the l.c.d. of all the terms.
$21 - 9 = 8x$	Remove parentheses and cancel denominators.

$$8x = 12$$ Simplify and interchange sides.

$$x = \tfrac{3}{2}$$ Divide by 8 and simplify.

Check

$$\frac{1}{\frac{3}{2}} - \frac{3}{7(\frac{3}{2})} \stackrel{?}{=} \frac{8}{21}$$

$$\frac{2}{3} - \frac{2}{7} \stackrel{?}{=} \frac{8}{21}$$

$$\frac{14 - 6}{21} \stackrel{?}{=} \frac{8}{21}$$

$$\frac{8}{21} = \frac{8}{21}$$

Thus $\frac{3}{2}$ is a solution of the original equation.

EXAMPLE 5.12

$$\frac{1}{x} + \frac{1}{x - 1} = \frac{1}{x(x - 1)}$$

$$x(x - 1)\left(\frac{1}{x} + \frac{1}{x - 1}\right) = x(x - 1)\frac{1}{x(x - 1)}$$ Multiply by $x(x - 1)$, the l.c.d. of all the terms.

$$x - 1 + x = 1$$ Remove parentheses and cancel denominators.

$$x = 1$$ Add 1 to both sides, simplify, and divide by 2.

Check The terms $1/(x - 1)$ and $1/[x(x - 1)]$ in the original equation are undefined if $x = 1$, so that 1 is not a solution of the original equation; it is an extraneous solution. The original equation has no solution.

Sometimes we must solve an equation for a selected variable in terms of one or more other variables in the equation. To do that we treat the other variables just like real numbers.

EXAMPLE 5.13 Solve the following equation for z in terms of x and y.

$$\frac{x}{2} + \frac{y}{3} + \frac{z}{4} = 1$$

$$6x + 4y + 3z = 12$$ Multiply by 12.

$$3z = 12 - 6x - 4y$$ Subtract $6x + 4y$ from both sides.

$$z = \tfrac{1}{3}(12 - 6x - 4y)$$ Divide by 3 (multiply by $\frac{1}{3}$).

EXAMPLE 5.14 Let C denote temperature in degrees Celsius and F temperature in degrees Fahrenheit. It can be shown that the formula giving C in terms of F is

$$C = \tfrac{5}{9}(F - 32). \tag{5.8}$$

If the temperature is 50° Fahrenheit, for example, then it is

$$C = \tfrac{5}{9}(50 - 32) = \tfrac{5}{9}(18) = 10° \text{ Celsius.}$$

To convert Equation (5.8) to a formula giving F in terms of C, we must solve for F.

$$\begin{aligned} C &= \tfrac{5}{9}(F - 32) \\ \tfrac{9}{5}C &= \tfrac{9}{5} \cdot \tfrac{5}{9}(F - 32) \\ \tfrac{9}{5}C &= F - 32 \\ F - 32 &= \tfrac{9}{5}C \\ F &= \tfrac{9}{5}C + 32 \end{aligned}$$

If the temperature is $-10°$ Celsius, for example, then it is

$$F = \tfrac{9}{5}(-10) + 32 = 14° \text{ Fahrenheit.}$$

EXERCISES FOR SECTION 5

Determine by substitution whether the given number is a solution of the given equation.

1. $4x + 2 = 2(x + 6)$; 5
2. $\tfrac{1}{2}(4 - 6x) = x - 1$; $\tfrac{1}{2}$
3. $2[x - (1 - x)] = 5(x - 1)$; 3
4. $y^3 - y = 7y + 5$; 3
5. $u^4 - 3u^2 = u^2(3u - 1)$; 4
6. $z^3 - 2z = 0.001(20 - z)$; 0.1
7. $\sqrt{x + 5} = 7 - x$; 11
8. $-\sqrt{t + 1} = 5 - t$; 8
9. $\sqrt{2x - 4} = 2 - x$; 2

Solve each equation.

10. $3x - 5 = 2x + 1$
11. $4y = 5y - 7$
12. $6z - 1 = 7z + 1$
13. $10(2u - 3) = 4(5 - 2u)$
14. $5(x - 1) = 2(4x + 1)$
15. $3(2x - 1) = -2(3x + 1)$
16. $0.2(1 - 2x) = -0.3(x + 1)$
17. $-1.5(x + 2) = 0.6(4 - x)$
18. $0.4(4x + 2) = 0.6(1 + x)$
19. $\dfrac{v}{2} - \dfrac{v}{3} = \dfrac{1}{5}$
20. $\dfrac{2t + 1}{3} - \dfrac{3t + 1}{2} = \dfrac{1}{4}$
21. $\tfrac{1}{2}(y + \tfrac{1}{3}) = \tfrac{1}{3}(y + \tfrac{1}{4})$

Solve each equation. Be sure to check for extraneous solutions.

22. $\dfrac{1}{x} - \dfrac{2}{3x} = \dfrac{1}{12}$
23. $\dfrac{1}{t} + \dfrac{1}{t^2} = \dfrac{1}{2t}$

24. $\dfrac{2}{z-2} + \dfrac{3}{z} = \dfrac{4}{z(z-2)}$

25. $\dfrac{2}{y+1} - \dfrac{3}{y-1} = \dfrac{-4}{(y+1)(y-1)}$

26. $\dfrac{x}{x-1} = -3 + \dfrac{1}{x-1}$

27. $\dfrac{1}{z} - \dfrac{3}{z+1} = \dfrac{1}{2z}$

Solve each equation twice: (a) for y in terms of x; (b) for x in terms of y.

28. $2x + 3y = 1$

29. $4y = 2(x - 1)$

30. $5(x - 1) = 6(y + 2)$

31. $y + 5(x - 1) = 7(1 - y) - x$

32. $2[x - (y - 1)] = 3$

33. $2x + 1 = 3[y + 2(1 - x)]$

Solve each equation for the variable indicated. (Each letter with a subscript, like v_0, r_1, r_2, or r_m, is a separate variable.)

34. $I = Prt$ for r.

35. $R^3 = KT^2$ for K.

36. $P = k/V$ for V.

37. $v = gt + v_0$ for t.

38. $P = 2L + 2W$ for L.

39. $T = 2\pi rh + 2\pi r^2$ for h.

40. $B = V - (xV/n)$ for V.

41. $V = (a + b)h/2$ for h.

42. $A = P(1 + rt)$ for r.

43. $\dfrac{1}{r} = \dfrac{1}{r_1} + \dfrac{1}{r_2}$ for r.

44. $R = \dfrac{ab}{a + b}$ for a.

45. $r = r_m\left(\dfrac{K - N}{K}\right)$ for K.

SECTION 6 Inequalities. Absolute Value

A. Order Properties

Order properties relate to inequality. In Section 1 we looked briefly at inequality in terms of a real line directed to the right:

$$a < b \quad \text{iff} \quad a \text{ is to the left of } b.$$

Also, a number was defined to be *positive* if it is to the right of 0 and *negative* if it is to the left of 0. It is advantageous to restate the condition for $a < b$ in the following equivalent form:

$$\boxed{a < b \quad \text{iff} \quad b - a \text{ is positive.}}$$

EXAMPLE 6.1 (See Figure 6.1.)

(a) $2 < 7$ because $7 - 2 = 5$ is positive.
(b) $-6 < -4.5$ because $-4.5 - (-6) = -4.5 + 6 = 1.5$ is positive.
(c) $-5 < \frac{2}{3}$ because $\frac{2}{3} - (-5) = \frac{2}{3} + 5 = \frac{17}{3}$ is positive.

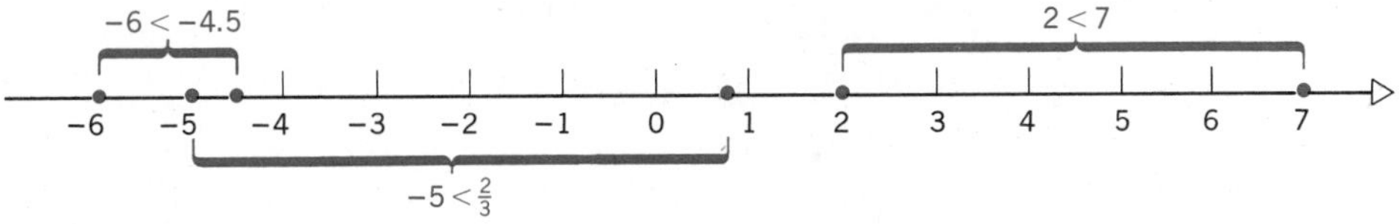

Figure 6.1

(Appendix A indicates how to define *positive* without appealing to a real line. When *positive* has been defined independently of a real line, then the equivalence "$a < b$ iff $b - a$ is positive" makes the idea of *inequality* also independent of a real line.)

The notations $a \leq b$, $b > a$, and $b \geq a$, as well as the terminology *less than* and *greater than,* are all defined in terms of $a = b$ and $a < b$ as in Section 1B. The basic properties of inequalities are the order properties that follow. Properties (6.1) and (6.2) are actually axioms; Properties (6.3) through (6.6) can be proved using (6.1), (6.2), and other properties of the real numbers. (See Examples 6.3 and 6.4.) *Pay special attention to Properties* (6.5) *and* (6.6): the former states that if both sides of an inequality are multiplied by the same *positive* number, then the sense (direction) of the inequality remains *unchanged;* the latter states that if both sides are multiplied by the same *negative* number, then the sense is *reversed.*

Order properties.

If a is any real number, then exactly one of the following is true: (6.1)

$$a > 0, \quad a = 0, \quad \text{or } -a > 0.$$

If $a > 0$ and $b > 0$, then $a + b > 0$ and $ab > 0$. (6.2)

If $a < b$ and $b < c$, then $a < c$. (6.3)

If $a < b$, then

$$a + c < b + c \quad \text{and} \quad a - c < b - c \tag{6.4}$$

for every c.

If $a < b$ and $c > 0$, then $ac < bc$. (6.5)

If $a < b$ and $c < 0$, then $ac > bc$. (6.6)

EXAMPLE 6.2 Because $2 < 7$, Property (6.4) implies each of the following:

$$2 + 11 < 7 + 11$$
$$2 + 20x < 7 + 20x \quad \text{for every } x$$
$$2 - \pi < 7 - \pi$$
$$2 - x^2 < 7 - x^2 \quad \text{for every } x.$$

Also, $2 < 7$ and Property (6.5) imply

$$2 \cdot 9 < 7 \cdot 9$$
$$2\sqrt{2} < 7\sqrt{2}.$$

And $2 < 7$ and Property (6.6) imply

$$2(-12) > 7(-12)$$
$$2(-\pi^2) > 7(-\pi^2).$$

The next two examples give proofs of Properties (6.4) and (6.5), using "$a < b$ iff $b - a$ is positive" as the definition of $a < b$. We use without hesitation the fact that $b > a$ means the same as $a < b$.

EXAMPLE 6.3 Prove this part of Property (6.4):

If $a < b$, then $a + c < b + c$.

Proof.

I. Assume $a < b$.
II. $0 < b - a$. (Step I and the definition of $<$)
III. $0 < (b + c) - (a + c)$. [Step II and $b - a = (b + c) - (a + c)$]
IV. $a + c < b + c$. (Step III and the definition of $<$)

EXAMPLE 6.4 Use Property (6.2) to prove Property (6.5).

Proof.

I. Assume $a < b$ and $c > 0$.
II. $b - a > 0$ and $c > 0$. (Step I and the definition of $a < b$)
III. $(b - a)c > 0$. [Step II and (6.2), which guarantees that the product of two positive numbers is positive]
IV. $bc - ac > 0$. (Step III)
V. $bc > ac$ and $ac < bc$. (Step IV and the definitions of $>$ and $<$)

B. Linear Inequalities

The basic terminology about equations with one variable also applies to inequalities with one variable. A real number is a **solution** of such an inequality (or **satisfies** the inequality) if it leads to a true inequality of numbers when it replaces the variable. Thus 4 is a solution of

$$3x < 15$$

because $3 \cdot 4 = 12$ and $12 < 15.$

The set of all solutions of an inequality is called its **solution set.** To **solve** an inequality is to find its solution set. Two inequalities with the same variable are **equivalent** if they have the same solution set.

As with an equation, the most basic plan for solving an inequality is to replace it by simpler and simpler equivalent inequalities until the solutions (if there are any) become obvious. The most frequently used tools for this are Properties (6.4)–(6.6). For the present we concentrate on **linear inequalities**—that is, inequalities that are equivalent to those of the form

$$ax + b < 0 \quad \text{or} \quad ax + b \leq 0, \quad \text{with } a \neq 0.$$

Linear inequalities are solved just like linear equations, except that the sense of the inequality sign must be reversed whenever the sides are interchanged or both sides are multiplied (or divided) by a negative number.

EXAMPLE 6.5

$$\begin{aligned} 9x - 6 &< 7x + 4 & & \\ 9x &< 7x + 10 & & \text{Add 6 to both sides.} \\ 2x &< 10 & & \text{Subtract } 7x \text{ from both sides.} \\ x &< 5 & & \text{Divide both sides by 2.} \end{aligned}$$

The original inequality has been transformed into the equivalent inequality $x < 5$. Thus, expressed with set-builder notation (Section 4), the solution set of the original inequality is $\{x\colon\ x < 5\}$.

Notice that Example 6.5 has an infinite number of solutions. That is typical of linear inequalities—in contrast with linear equations, which have only one solution each.

The rules for solving inequalities with $\leq$ or $\geq$ are the same as those for $<$ and $>$. Here is an example.

EXAMPLE 6.6

$$\begin{aligned} \frac{3u-2}{5} + 3 &\geq \frac{4u-1}{3} & & \\ 3(3u-2) + 15 \cdot 3 &\geq 5(4u-1) & & \text{Multiply by 15.} \\ 9u - 6 + 45 &\geq 20u - 5 & & \text{Remove parentheses.} \\ 9u &\geq 20u - 44 & & \text{Subtract 39.} \\ -11u &\geq -44 & & \text{Subtract } 20u. \\ u &\leq 4 & & \text{Divide by } -11 \text{ and change } \geq \text{ to } \leq. \end{aligned}$$

Thus the solution set is $\{u\colon\ u \leq 4\}$.

C. Intervals

Assume that a and b are real numbers with $a < b$. The notation $a < x < b$ means that both $a < x$ and $x < b$; that is, x is between a and b. The notations $[a, b]$, $[a, b)$, $(a, b]$, and (a, b) are used to denote the four special sets of real numbers defined in Table 6.1. Each set is called an **interval** with **endpoints** a and b. The keys are that [is used if a is to be *included* in the set, and (is used if a is to be *excluded*. Similarly for] and). In the figures, a closed circle (•) is used to denote an endpoint that is to be *included*, and an open circle (○) is used to denote an endpoint that is to be *excluded*.

TABLE 6.1

Interval	*Definition*	*Figure*
$[a, b]$	$\{x:\ a \le x \le b\}$	
$[a, b)$	$\{x:\ a \le x < b\}$	
$(a, b]$	$\{x:\ a < x \le b\}$	
(a, b)	$\{x:\ a < x < b\}$	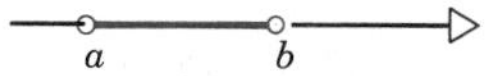

Table 6.2 shows the five types of **unbounded intervals**—that is, intervals without an endpoint on one or both ends. The symbols $-\infty$ and ∞ *do not* denote real numbers: "$(-\infty$" simply denotes the absence of an endpoint on the left, and "$\infty)$" denotes the absence of an endpoint on the right.

TABLE 6.2

Interval	*Definition*	*Figure*
$[a, \infty)$	$\{x:\ x \ge a\}$	
$(-\infty, b]$	$\{x:\ x \le b\}$	
(a, ∞)	$\{x:\ x > a\}$	
$(-\infty, b)$	$\{x:\ x < b\}$	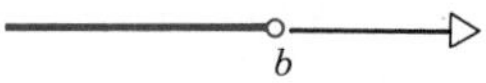
$(-\infty, \infty)$	$\{x:\ x$ is a real number$\}$	

Sometimes it is convenient to use set notation when working with intervals. For this we need the following ideas. (For more examples of these ideas see Appendix B.) Let A and B represent sets (collections of numbers, for example).

$x \in A$ denotes that x is an element of (or is a member of, or belongs to) the set A.

$\{a, b, \ldots\}$ denotes the set whose members are $a, b, \ldots$.

$\emptyset$ denotes the **empty set;** that is, the set that contains no elements.

$A \cup B$ denotes $\{x:\ x \in A$ or $x \in B\}$.

$A \cap B$ denotes $\{x:\ x \in A$ and $x \in B\}$.

The set $A \cup B$ is called the **union** of A and B ("$x \in A$ or $x \in B$" includes the possibility that x belongs to *both* A and B). The set $A \cap B$ is called the **intersection** of A and B. (See Figure 6.2.)

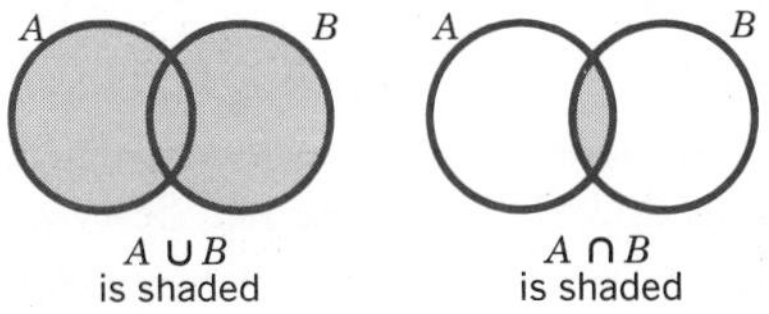

Figure 6.2

EXAMPLE 6.7 Figure 6.3 shows the intervals that appear on the left in these equations.

(a) $[-1, 3] \cup [1, 4) = [-1, 4)$
(b) $[-1, 3] \cap [1, 4) = [1, 3]$
(c) $(-\infty, 1] \cap [1, 4) = \{1\}$
(d) $(-\infty, 1] \cap (-1, \infty) = (-1, 1]$
(e) $(-3, -2) \cap [1, 4) = \varnothing$
(f) The set $(-3, -2) \cup [1, 4)$ is not an interval.

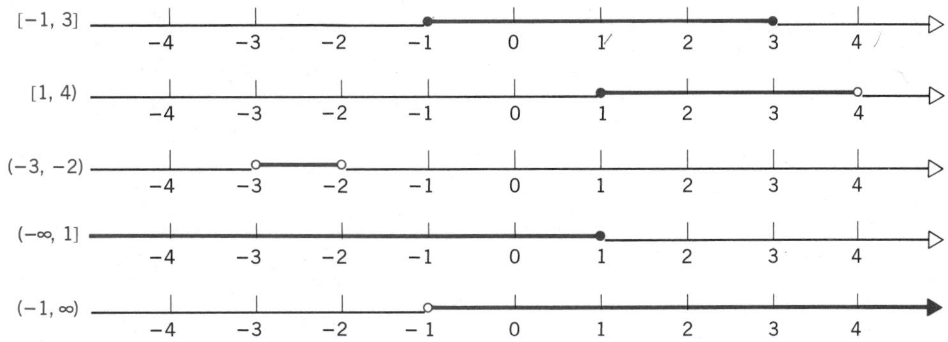

Figure 6.3

D. Absolute Value

The **absolute value** of a real number a, denoted $|a|$, is the distance between the points with coordinates 0 and a on a real line, without regard to direction.

EXAMPLE 6.8

(a) $|5| = 5$ and $|-5| = 5$. (See Figure 6.4.)

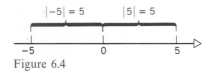

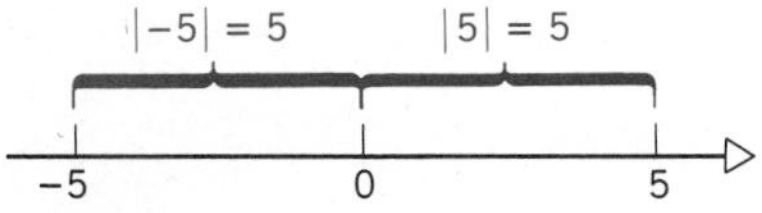

Figure 6.4

(b) $|\pi| = \pi$ and $|-\pi| = \pi$.

Here are three properties of absolute value.

$$|0| = 0.$$
$$\text{If } a \neq 0, \text{ then } |a| > 0.$$
$$|a| = |-a| \text{ for every } a.$$

If a is positive, then $|a| = a$. If a is negative, however, then $|a| = -a$. For instance, $|-5| = 5 = -(-5)$. The point is that if a is negative, then $-a$ is positive. This allows us to make the following equivalent definition of $|a|$, which has the advantage that it does not refer to a real line.

$$|a| = \begin{cases} a & \text{if } a \geq 0 \\ -a & \text{if } a < 0 \end{cases}$$

If a and b are real numbers, then the **distance** between a and b on a real line, without regard to direction, is $|a - b|$. (Strictly speaking, $|a - b|$ is the distance between the *points* with *coordinates* a and b.) Notice that $|a - b| = |b - a|$.

EXAMPLE 6.9 (See Figure 6.5.)

(a) The distance between 8 and 5 is $|8 - 5| = |3| = 3$ or $|5 - 8| = |-3| = 3$.
(b) The distance between 3.5 and -4 is $|3.5 - (-4)| = |7.5| = 7.5$ or $|-4 - 3.5| = |-7.5| = 7.5$.

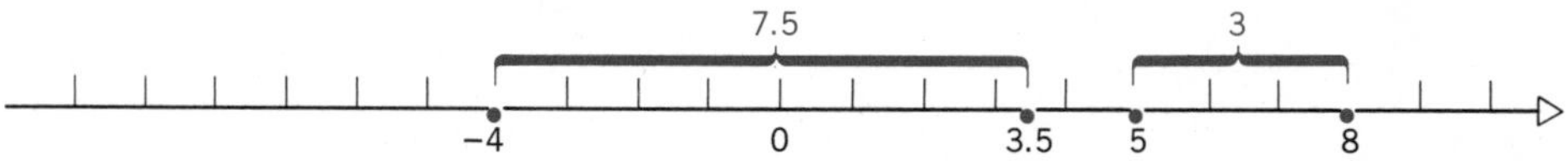

Figure 6.5

To solve an inequality that contains absolute values, we can first convert it to an inequality that does not contain absolute values. To do this we use the facts that, for $c > 0$, $|y| < c$ iff y is *within* c units of 0 (Figure 6.6), and $|y| > c$ iff y is *more* than c units from 0 (Figure 6.7).

For any real numbers y and c such that $c > 0$:

$$|y| < c \text{ iff } -c < y < c \qquad (6.7)$$
$$|y| > c \text{ iff } y < -c \text{ or } y > c \qquad (6.8)$$

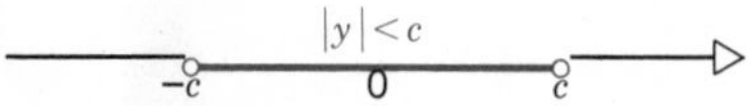

Figure 6.6

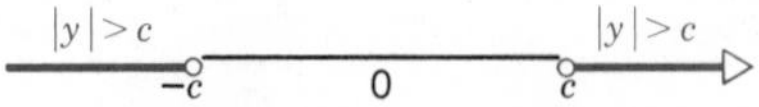

Figure 6.7

EXAMPLE 6.10 Solve $|x| < 3$.

Solution By (6.7), with $y = x$ and $c = 3$, the solution set is $\{x: \quad -3 < x < 3\}$. In interval notation, the solution set is $(-3, 3)$.

EXAMPLE 6.11 Solve $|3x - 2| < 4$.

Solution Using (6.7) with $y = 3x - 2$ and $c = 4$, we see that the given inequality is equivalent to

$$-4 < 3x - 2 < 4.$$

Now proceed as in Subsection B. Add 2 to each part (this is equivalent to adding 2 to both sides of $-4 < 3x - 2$ and to both sides of $3x - 2 < 4$):

$$(-4) + 2 < (3x - 2) + 2 < (4) + 2$$
$$-2 < 3x < 6.$$

Now divide by 3:

$$-\tfrac{2}{3} < x < 2.$$

In interval notation, the solution set is $(-\tfrac{2}{3}, 2)$.

EXAMPLE 6.12 Solve $|2x| \geq 7$.

Solution Use (6.8) with $y = 2x$ and $c = 7$, and the corresponding statement with $=$ in place of $>$. We see that the given inequality is equivalent to

$$2x \leq -7 \text{ or } 2x \geq 7$$
$$x \leq -\tfrac{7}{2} \text{ or } \quad x \geq \tfrac{7}{2}.$$

Thus the solution set is $(-\infty, -3.5] \cup [3.5, \infty)$.

EXAMPLE 6.13 Solve $5 \leq |4 - x|$.

Solution This is equivalent to $|4 - x| \geq 5$. Apply (6.8) with $y = 4 - x$ and $c = 5$, and the corresponding statement with $=$ in place of $>$. This leads to

$$4 - x \leq -5 \text{ or } 4 - x \geq 5$$
$$-x \leq -9 \text{ or } \quad -x \geq 1$$
$$x \geq 9 \quad \text{or} \quad x \leq -1.$$

Thus the solution set is $(-\infty, -1] \cup [9, \infty)$.

EXERCISES FOR SECTION 6

Replace each $\square$ with the appropriate sign, either $<$ or $>$.

1. $-1.4 \square -1.5$

2. $\frac{6}{25} \square \frac{7}{25}$

3. $0 \square -4$

4. $\frac{1}{5} \square \frac{1}{6}$

5. $-0.2 \square -0.19$
6. $1/0.2 \square 1/0.1$
7. $3 - x^3 \square 1 - x^3$
8. $-7\pi \square -5\pi$
9. $-5\sqrt{2} \square -5\sqrt{3}$
10. $1.2 \times 10^6 \square 1.2 \times 10^7$
11. $1.2 \times 10^{-13} \square 1.2 \times 10^{-14}$
12. $-1.2 \times 10^7 \square -1.2 \times 10^8$

Solve each inequality and express the solution set with set-builder notation.

13. $6x + 2 > 3x - 4$
14. $5y - 4 < 7y + 2$
15. $1 + 5z < 2 - z$
16. $2(4 - t) \leq 5(1 + 2t)$
17. $\frac{1}{3}(u + 1) > \frac{1}{4}(u - 1)$
18. $\frac{1}{2}(3 - 2x) \geq \frac{1}{10}(x + 4)$
19. $0.05[v - 3(1 - v)] \leq 0.4v$
20. $1.46 \geq 0.2[1 + 2.1(r - 3)]$
21. $-y \geq 1.4[2(2 - y) + 0.5]$

Graph each interval on a real line. (See Figure 6.3.) Then simplify the set (if possible) as in Example 6.7.

22. $(-2, 1] \cup [-1, 2)$
23. $[0, 3] \cap [2, 4]$
24. $[0, 3] \cup [2, 4]$
25. $(1, 5] \cap [-4, 1)$
26. $[1.5, 2.5] \cap [2.5, \infty)$
27. $(-\infty, 1) \cup (2, \infty)$
28. $[0, \infty) \cup (-\infty, 0]$
29. $(-\infty, 0) \cap (0, \infty)$
30. $(-\infty, -1] \cup [1, \infty)$

Rewrite without absolute value signs and simplify.

31. $|-2| + |3|$
32. $|7| - |-4|$
33. $3 - |5|$
34. $-(|12| - |2|)$
35. $-|-4| + |-7|$
36. $|-0| - |-2|$
37. $|2(-5)|$
38. $-0|-5|$
39. $-2|5|$
40. $-|4|/|-2|$
41. $|15/(-3)|$
42. $-|20/(-8)|$
43. $(|-12| - |0|)/|3|$
44. $(|9| - |5|) - |-4|$
45. $-|-7| - (|2| - |8|)$

Determine the distance on a real line between the points with the given coordinates.

46. 12 and -2
47. -5.1 and -7
48. $16\frac{1}{2}$ and 4.2
49. -3π and $-5\pi/2$
50. $\frac{1}{7}$ and $-\frac{1}{8}$
51. $1/\pi$ and $-3/(2\pi)$

Solve each inequality and express the solution set with interval notation.

52. $|x| \leq 4$
53. $|2x| > 14$
54. $|-x| \leq 5$
55. $|6x| > |-3|$
56. $|-6| \geq |-0.1x|$
57. $|-1| < |\frac{1}{2}x|$
58. $|2x - 1| < 7$
59. $|3x + 1| \geq 5$
60. $|\frac{1}{2}x - 1| \leq \frac{3}{2}$
61. $|0.2 - 0.1x| \geq 0.3$
62. $|1 - 5x| < 1$
63. $|3 - 2x| > 1$

It can be proved that if a and b are any real numbers, then

$$|a + b| \leq |a| + |b| \qquad \textbf{(triangle inequality).}$$

Verify the triangle inequality for each pair of given numbers.

64. 4 and 2
65. -3 and 2
66. 6 and -1
67. 1.2 and -2.3
68. -1 and -5
69. $-\frac{1}{3}$ and $-\frac{1}{4}$
70. Prove Property (6.3). [Use the definition of <. And, $c - a = (c - b) + (b - a)$.]
71. Prove this part of Property (6.4): If $a < b$, then $a - c < b - c$. (See Example 6.3.)
72. Use Property (6.2) to prove Property (6.6). (See Example 6.4. Also, notice that if $c < 0$, then $-c > 0$.)
73. Use Property (6.6) to prove that the product of two negative numbers is positive. (Let $b = 0$.)
74. Use Property (6.5) to prove that the product of a positive number and a negative number is negative. (Let $b = 0$.)

75. Prove that if $a \neq 0$, then $a^2 > 0$. [By Property (6.1), either $a > 0$ or $-a > 0$. For $a > 0$, use Property (6.5) with a in place of b and c, and 0 in place of a. For $-a > 0$, use $(-a)^2 = a^2$ and the case $a > 0$.]
76. Prove that if $a > b > 0$, then $a^2 > b^2$. (Suggestion: First explain why $a^2 > ab$ and $ab > b^2$.)
77. Prove that if $a > b > 0$, then $1/b > 1/a$.
78. Prove that if $a < b$, then $a < (a+b)/2 < b$. [Suggestion: First explain why $a < (a+b)/2$ iff $a/2 < b/2$, and $(a+b)/2 < b$ iff $a/2 < b/2$.]

REVIEW EXERCISES FOR CHAPTER 1*

1. Which of the following numbers are natural numbers? integers? rational numbers? irrational numbers?

$$20.\overline{1},\quad 3\pi,\quad 0/\sqrt{2},\quad \sqrt{5},\quad -12/\sqrt{9},\quad \sqrt{36},\quad -3.14,\quad -\sqrt{2}\sqrt{2}$$

2. Define each of the following sets: natural numbers, integers, rational numbers, irrational numbers, and real numbers.

Simplify.

3. $(-8)(-0.025)$
4. $(0.16)(-5.5)$
5. $(-0.048) \div (0.12)$
6. $(-16.4) \div (-41)$
7. $(9.2 - 6.56) \div 4$
8. $(10.1 - 12.25) \div 5$
9. $4[1 - 2(3 + 4)] - 2(3 - 6)$
10. $5 - 2[1 - 6(7 - 9)]$
11. $-2\{5(b - a) - 2[a + (-3a - b)]\}$
12. $a - c - \{-(a - b) + 2(b - c) - [(a - b) - c]\}$

Write each expression as a single fraction and simplify.

13. $\frac{5}{6} - \frac{3}{10}$
14. $\frac{1}{14} - \frac{1}{6}$
15. $-(2 + \frac{2}{3}) \div (-\frac{2}{3})$
16. $(-\frac{1}{5}) \div (\frac{1}{5} - 1)$
17. $\left(1 + \dfrac{1}{x}\right) \div \left(1 - \dfrac{1}{x}\right)$
18. $2 + \dfrac{1 - \dfrac{2}{y}}{1 + \dfrac{2}{y}}$
19. $\left(\dfrac{2}{a} - \dfrac{1}{a^2}\right) \div \left(\dfrac{1}{a} - 2\right)$
20. $\left(\dfrac{2}{b} + \dfrac{2}{c}\right) \div \left(\dfrac{1}{bc^2} + \dfrac{1}{b^2c}\right)$

Simplify.

21. $(-\sqrt{7})^4$
22. $-(-1/\sqrt{10})^6$
23. -0.3^{-2}
24. $(-0.2)^{-3}$

Simplify, using no negative exponents or square root signs in the final answers. Assume that n is a positive integer.

25. $[x(3y)^{-1}]^3$
26. $(2a^2b^{-1})^5$
27. $\left(\dfrac{(-6a)^2}{(2b)^0}\right)^{-1}$
28. $\left(\dfrac{(15y)^0}{(-y/2)^4}\right)^{-2}$
29. $(\sqrt{100a^4b^2})^3$
30. $(\sqrt{25x^6y^{10}})^3$

---

* Answers to all of the review exercises are at the back of the book.

31. $\left(\frac{a^2}{b}\right)^n\left(\frac{b^2}{a}\right)^{-4n}$

32. $\frac{(\sqrt{2}u)^{6n}}{(\sqrt{6}uv^{-1})^{4n}}$

Write each of these numbers in scientific notation.

33. 772,000,000

34. 0.0000048

Use $\sqrt{3} \approx 1.7321$ and $\sqrt{30} \approx 5.4772$ to determine an approximate decimal value for each of the following square roots.

35. $\sqrt{300{,}000}$

36. $\sqrt{0.0003}$

Carry out the indicated operations and simplify.

37. $(2x - 1)(x^2 + x + 1)$

38. $(y - 3)(y^2 + 2y - 2)$

39. $(a - \frac{1}{3})^2$

40. $(b + 0.2)^2$

Factor.

41. $t^2 - 9$

42. $u^2 - 22u + 121$

43. $v^2 + 12v + 36$

44. $4w^2 - 1$

45. $10x^2 - x - 2$

46. $9y^2 + 3y - 2$

Determine the domain of each expression.

47. $1/(2x + 1)$

48. $\sqrt{x + 6}$

Solve each equation for x.

49. $5(x - 2) = 3(1 - 2x)$

50. $\frac{x + 1}{2} - \frac{2x - 1}{5} = \frac{1}{4}$

51. $4(x - 1) = 3y$

52. $y = 4(x + 2) + 1$

53. $\frac{1}{x + 2} - \frac{2}{x} = 0$

54. $\frac{x}{x + 2} + 1 = \frac{-2}{x + 2}$

Solve each inequality and express the solution set with set-builder notation.

55. $-3(x + 1) < 2(3 - x)$

56. $\frac{1}{2}(y + 1) - \frac{1}{5} \geq \frac{1}{4}(1 - y) + \frac{1}{10}$

Rewrite without absolute value signs and simplify.

57. $-|12| \div (|-2| - |5|)$

58. $(|-2.5\pi| - |\pi/2|) \div (-|1.5\pi|)$

Rewrite each set as a single interval.

59. $[-3, 2) \cap [-1, 4)$

60. $(-\pi, \sqrt{2}] \cup [0, \sqrt{3}]$

61. $(-\infty, 7] \cap (-2, \infty)$

62. $[2, 6] \cup (-\infty, 5)$

Determine the distance on a coordinate line between the points with the given coordinates.

63. 5.1 and 3.11

64. $4.\overline{3}$ and $-5\frac{2}{3}$

Solve each inequality and express the solution set with interval notation.

65. $|-x + 1| < 6$

66. $|4x + 3| \geq 5$

67. $|5x + 2| \geq 8$

68. $|1 - 3x| < 5$

CHAPTER II
INTRODUCTION TO APPLICATIONS

Using algebra requires more than the ability to manipulate symbols and solve equations and inequalities. It also requires the ability to translate problems into algebraic language, to find the right method or formula when it's needed, and to handle such things as units and approximations. This chapter is concerned with some of these extra factors that make algebra more than a mere intellectual pastime. The chapter also reviews percentage, ratio, proportion, and variation—ideas that arise repeatedly in applications.

SECTION

7 Problem Solving

A. Forming Equations

To solve problems with algebra we must frequently form equations based on information given verbally. This step in problem solving is often the most difficult, but some facts stated in words can be translated into equations almost mechanically; by going over those here we can be free later to concentrate on the cases that require more imagination.

Following is a short list of key words with their algebraic equivalents. Example 7.1 gives illustrations.

Word or phrase	Algebraic equivalent
is	$=$
is more than (amount unspecified)	$>$
is more than (amount specified)	$+$
is less than (amount unspecified)	$<$
is less than (amount specified)	$-$
increase	add
decrease	subtract
twice	2 times
half of	$\frac{1}{2}$ times
of	times

EXAMPLE 7.1

Phrase	Algebraic equivalent
y is four times x	$y = 4x$
a is more than b	$a > b$
a is three more than b	$a = b + 3$
L is less than half of M	$L < \frac{1}{2}M$
A does not exceed B	$A \leq B$
x increased by 3 is y	$x + 3 = y$
y is the reciprocal of x	$y = 1/x$

Sometimes we must choose variables before forming equations. It is generally best to choose letters that suggest what they represent, if possible (A for *area*, f for *focal length*, C for *cost*, and so on). Also, notice the use of subscripts in part (d) of the following example, to distinguish variables that are similar.

EXAMPLE 7.2 In each of parts (a) through (d) a statement is translated into an equation or an inequality by first assigning variables.

(a) The area of a triangle is one-half the length of the base times the height (altitude). Let

$$A = \text{area}$$
$$b = \text{length of the base}$$
$$h = \text{height.}$$

Then

$$A = \tfrac{1}{2}bh.$$

(b) The geometric mean of two positive numbers is the square root of their product. Let

$$G = \text{geometric mean}$$
$$a = \text{positive number}$$
$$b = \text{positive number.}$$

Then

$$G = \sqrt{ab}.$$

(c) The absolute value of the sum of two numbers is less than or equal to the sum of their absolute values. Let

$$a = \text{number}$$
$$b = \text{number.}$$

Then

$$|a + b| \leq |a| + |b|.$$

(d) The reciprocal of the focal length of a simple lens is equal to the sum of the reciprocals of the object distance and the image distance. Let

$$f = \text{focal length}$$
$$d_o = \text{object distance}$$
$$d_i = \text{image distance.}$$

Then

$$\frac{1}{f} = \frac{1}{d_o} + \frac{1}{d_i}.$$

EXAMPLE 7.3 Problems about numbers often contain phrases like the following.

Phrase	Algebraic equivalent
x is an even integer	$x = 2n$ for some integer n
x is an odd integer	$x = 2n + 1$ for some integer n
three consecutive integers	$n, n + 1, n + 2$ for some integer n
consecutive even integers	$2n, 2n + 2, 2n + 4, \ldots$ for some integer n

B. A General Plan for Problem Solving

Our chances of solving a problem are much better if we approach it with a plan. Although no single plan will work for every problem, Steps I–V that follow will at least provide some guidelines. Try to keep these steps in mind, and be patient and persistent. Don't worry if the answer to a problem isn't obvious, or if the problem seems confusing or difficult—the whole reason for using algebra is that it can lead us systematically through problems that we cannot see through at a glance.

Step I. Read the problem carefully. Be sure you know what all of the key words mean. If possible, draw a figure. Also make a note of the units in the problem.

Step II. Assign variables to the unknown or unknowns.

Step III. Form all equations that might have something to do with the problem. Make use of what is given in the problem and also of any general formulas that apply.

Step IV. Use algebra to solve the equations in Step III.

Step V. Check to see if the answers found in Step IV really do solve the original problem. Also be sure to include units (if there are any) in the final answer.

EXAMPLE 7.4 A tank one-third full has 20 gallons added; then it is one-half full. What is the capacity of the tank?

Solution

I and II. The units are gallons. Let C denote the capacity.

III. One-third of the capacity is $\frac{1}{3}C$. One-half of the capacity is $\frac{1}{2}C$. Therefore, from what is given,

$$\tfrac{1}{3}C + 20 = \tfrac{1}{2}C. \tag{7.1}$$

IV. Solve Equation (7.1).

$$\begin{aligned} 6(\tfrac{1}{3}C + 20) &= 6 \cdot \tfrac{1}{2}C \\ 2C + 120 &= 3C \\ C &= 120 \end{aligned}$$

That is, the capacity is 120 gallons.

V. Check $\tfrac{1}{3}(120) + 20 \stackrel{?}{=} \tfrac{1}{2}(120)$

$40 + 20 = 60.$

C. More Examples

EXAMPLE 7.5 The perimeter of a rectangle is 20 centimeters and the length is 3 centimeters more than the width. Determine the length and width.

Solution

I. *Perimeter* means *distance around.* The units are centimeters.

II. Let P = perimeter, l = length, and w = width.

III. By the definition of perimeter, $P = 2l + 2w$. Also, from what is given, $P = 20$ and $l = w + 3$ (Figure 7.1).

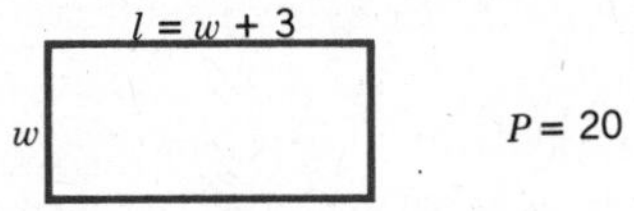

Figure 7.1

IV. From the equations in III,

$$P = 2l + 2w$$
$$20 = 2(w + 3) + 2w \quad \text{(because } P = 20 \text{ and } l = w + 3\text{)}$$
$$20 = 4w + 6$$
$$w = \tfrac{7}{2} = 3.5.$$

That is, the width is 3.5 centimeters. Thus the length is 3.5 + 3 or 6.5 centimeters.

V. Check $20 \overset{?}{=} 2(6.5) + 2(3.5)$
$20 = 13 + 7.$

EXAMPLE 7.6 Suppose that three times the first of three consecutive integers plus twice the second is equal to four times the third. Find the integers.

Solution

I and II. Example 7.3 pointed out that three consecutive integers will have the form n, $n + 1$, and $n + 2$ for some integer n.
III. From what is given,

$3n$	three times the first
$+$	plus
$2(n + 1)$	twice the second
———	equals
$4(n + 2)$	four times the third.

IV. $3n + 2(n + 1) = 4(n + 2)$
$5n + 2 = 4n + 8$
$n = 6$

Therefore, the integers are 6, 7, and 8.

V. Check $3 \cdot 6 + 2 \cdot 7 \overset{?}{=} 4 \cdot 8$
$32 = 32.$

EXAMPLE 7.7 In 1890 the population of the United States was 16 times what it was in 1790. The population increased approximately 69 million between 1890 and 1940, at which time it was approximately 132 million. What was the population in 1790?

Solution

I. A quick sketch, like Figure 7.2, will show all of the essentials. The units are millions.

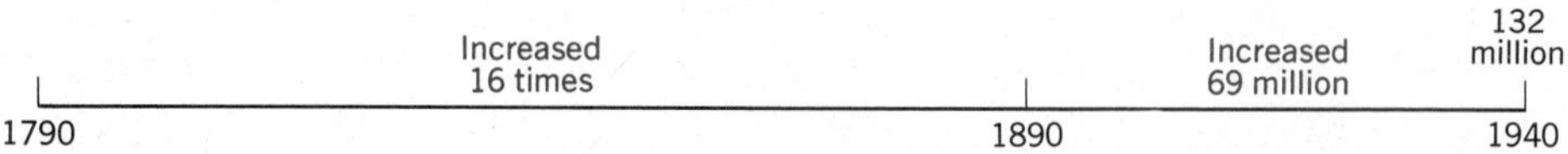

Figure 7.2

II. Let the population in 1790 be x million and the population in 1890 be y million.
III. From the first sentence in the problem,

$$y = 16x. \tag{7.2}$$

From the second sentence in the problem,

$$y + 69 = 132. \tag{7.3}$$

IV. From Equation (7.3), $y = 132 - 69 = 63$. Therefore, from Equation (7.2),

$$63 = 16x$$
$$x \approx 3.9.$$

Thus the population in 1790 was approximately 3.9 million.
V. $16(3.9) + 69 = 131.4 \approx 132$, as required.

EXAMPLE 7.8 The airline distance between Chicago and Washington, D.C., is 600 miles. A plane leaves Chicago for Washington at an average rate of 400 miles per hour. Twenty minutes later, another plane leaves Washington for Chicago at an average rate of 300 miles per hour. How long after the first plane leaves will they meet?

Solution

I. Figure 7.3 summarizes the essential information: the planes are labeled A and B, where B leaves 20 minutes, or $\frac{1}{3}$ hour, after A. We need units for distance, time, and rate. The obvious choices for this problem are *miles, hours,* and *miles per hour.* Thus 20 minutes has been converted to $\frac{1}{3}$ hour.

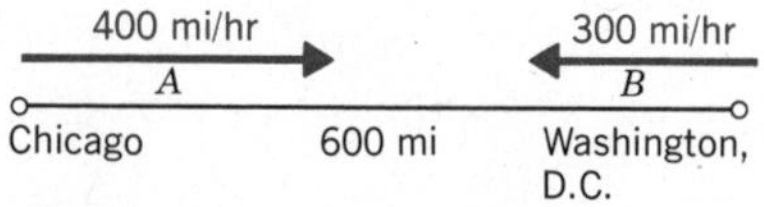

Figure 7.3

II. Let T denote time, measured in hours from when the first plane leaves.
III. The basic formula for motion problems such as this is

$$\text{distance} = \text{rate} \times \text{time}$$
$$D = RT.$$

Let D_A, R_A, and T_A denote the values of distance, rate, and time for plane A. Similarly for plane B. Then $T_A = T$, and $T_B = T - \frac{1}{3}$ because plane B leaves $\frac{1}{3}$ hour after plane A. Thus

$$\begin{aligned} R_A &= 400 \text{ mi/hr} & R_B &= 300 \text{ mi/hr} \\ T_A &= T & T_B &= T - \tfrac{1}{3} \\ D_A &= 400T & D_B &= 300(T - \tfrac{1}{3}). \end{aligned}$$

The problem requires that we determine when

$$D_A + D_B = 600,$$

that is,

$$400T + 300(T - \tfrac{1}{3}) = 600. \tag{7.4}$$

IV. Solve Equation (7.4):

$$\begin{aligned} 400T + 300T - 100 &= 600 \\ 700T &= 700 \\ T &= 1. \end{aligned}$$

Therefore, the planes will meet one hour after the first plane leaves.

V. Check At the end of one hour plane A will have traveled

$$D_A = 400 \cdot 1 = 400 \text{ miles.}$$

At the end of one hour plane B will have traveled

$$D_B = 300(1 - \tfrac{1}{3}) = 300 \cdot \tfrac{2}{3} = 200 \text{ miles.}$$

Thus the two planes will indeed have traveled a total of 600 miles at the end of one hour, and the answer checks. Our calculations also show that the meeting place will be 400 miles from Chicago and 200 miles from Washington.

EXAMPLE 7.9 A candy company has decreased the weight of its most popular-selling chocolate bar by one-fifth, keeping the price constant. Faced with bad publicity and a public outcry, the company announces that it will increase the net weight by one-fifth, again keeping the price constant. How (exactly) would this increased weight compare with the original weight?

Solution

I. Notice that we must concentrate on *weight;* the information about constant price makes the problem more interesting, perhaps, but it is irrelevant to the question at hand. No units are given, so we have a choice about the form of the final answer; we can simply work from what is given and use whatever form results.

II. Let

$$W_1 = \text{original weight}$$
$$W_2 = \text{weight after the original decrease}$$
$$W_3 = \text{weight after the planned increase.}$$

We must compare W_1 and W_3. (See Figure 7.4.)

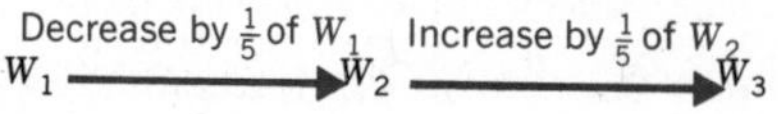

Figure 7.4

III. The decreased weight is the original weight less one-fifth of the original weight:

$$W_2 = W_1 - \tfrac{1}{5}W_1. \tag{7.5}$$

The weight after the planned increase will be more than W_2 by one-fifth of W_2:

$$W_3 = W_2 + \tfrac{1}{5}W_2. \tag{7.6}$$

IV. If we replace W_2 in Equation (7.6) by using Equation (7.5), we get

$$\begin{aligned} W_3 &= (W_1 - \tfrac{1}{5}W_1) + \tfrac{1}{5}(W_1 - \tfrac{1}{5}W_1) \\ &= W_1 - \tfrac{1}{5}W_1 + \tfrac{1}{5}W_1 - \tfrac{1}{25}W_1 \\ &= W_1 - \tfrac{1}{25}W_1. \end{aligned} \tag{7.7}$$

Conclusion: The weight after the planned increase will be $\frac{1}{25}$ less than the original weight. Equivalently, the final weight will be 4% less than the original weight. (Percentage will be reviewed in the next section.) Or, because

$$W_1 - \tfrac{1}{25}W_1 = \tfrac{24}{25}W_1,$$

the final weight will be $\frac{24}{25}$ or 96% of the original weight.

V. To check the answer in this case we must review Steps III and IV to try to ensure that we have made no mistakes in reasoning or calculation. The unit in this problem is actually the original weight; the final answer is expressed as a fraction or percentage of that weight. Had we been given a specific value in grams or ounces for the original weight, then the answer could have been in the same units.

Nearly all of our remarks about applications of equations carry over to applications of inequalities. Here is an example.

EXAMPLE 7.10 What can we conclude about the radius of a circle if we know that the circumference exceeds the perimeter of a square whose edges are each 10 centimeters?

Solution

I and II. *Circumference* and *perimeter* both mean *distance around.* The units are centimeters. Figure 7.5 shows the circle and square, with the unknown radius labeled r.

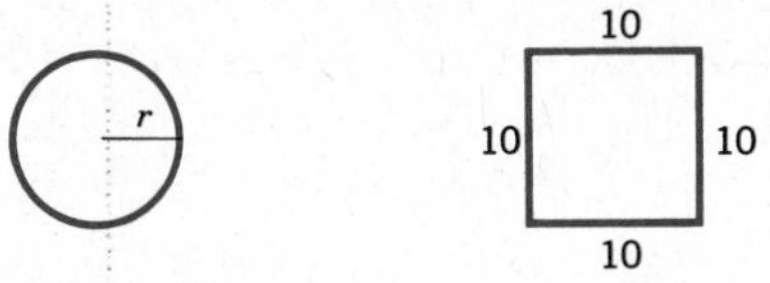

Figure 7.5

III. If C denotes circumference and P denotes perimeter, then, from what is given, $C > P$. Also,

$$C = 2\pi r \quad \text{and} \quad P = 4 \cdot 10 = 40.$$

Therefore,

$$2\pi r > 40.$$

IV. From $2\pi r > 40$ we conclude that $r > 40/2\pi$, or $r > 20/\pi$. Thus the radius exceeds $20/\pi$ centimeters.

V. If $r > 20/\pi$, then $C = 2\pi r > 2\pi(20/\pi) = 40$. That is, $C > 40$, as required.

In the remainder of the book the solutions of applied problems will not be outlined in Steps I–V as they have been here. It will still be a good idea to think in these terms, however, both in reading examples and in working exercises.

EXERCISES FOR SECTION 7

Assign variables as needed, and then translate each statement into an equation or an inequality.

1. The surface area of a sphere is 4π times the square of the radius.
2. The volume of a sphere is four-thirds of π times the cube of the radius (radius raised to the third power).
3. The volume of a pyramid is one-third the area of the base times the height (altitude).
4. Profit is revenue minus cost.
5. Total cost is fixed cost plus variable cost.
6. Average price per unit is the total price divided by the number of units.
7. The arithmetic mean (average) of two numbers is one-half of their sum.
8. The harmonic mean of two numbers is twice the reciprocal of the sum of the reciprocals of the two numbers.
9. The square root of the product of two positive numbers does not exceed one-half of the sum of the two numbers.
10. Force equals mass times acceleration.
11. The frequency of a periodic motion is the reciprocal of the period.
12. Average velocity is the distance traveled divided by the time elapsed.
13. The absolute value of the difference of the absolute values of two numbers does not exceed the absolute value of the difference of the two numbers.

14. Less than three-tenths of the surface of the earth is covered by land.
15. The length of the diagonal of a rectangular parallelopiped (the shape of an ordinary box) is the square root of the sum of the squares of the lengths of the length, width, and height.

In each of the following exercises, indicate clearly the method used to obtain the solution.

16. Ten more than four times a number is 16. What is the number?
17. One-half of a number plus one-third of the number is four less than the number. What is the number?
18. One less than the reciprocal of a number is four. What is the number?
19. The perimeter of a rectangle is 50 feet and the width is 4 feet less than the length. Determine the length and width.
20. The perimeter of a rectangular lot is 360 meters and the length is 10 meters more than the width. Determine the length and width.
21. The perimeter of a rectangular lot is 370 meters and the width is 20 meters more than half the length. Determine the length and width.
22. Assume that the sum of three consecutive integers is 87. What are the integers?
23. Assume that nine times the third of three consecutive integers minus seven times the first is equal to three times the second. What are the integers?
24. Assume that twice the first of three consecutive integers plus twice the second is seven more than three times the third. What are the integers?
25. The population of the American Colonies doubled between 1660 and 1680, and then increased by 100,000 between 1680 and 1700. The population in 1700 was approximately 250,000. What was the population in 1660?
26. An apple contains twice as many calories as a peach. Two apples and one peach contain as many calories as two bananas. If a banana contains 100 calories, how many calories are in a peach?
27. The total world production of wheat in 1975 was approximately 356 million tons. The U.S. and the U.S.S.R. together produced approximately one-third of the total, and the U.S.S.R. produced 8 million tons more than the U.S. Based on these figures, what was the U.S. production?
28. What is the average rate in miles per hour of a four-minute miler?
29. In April of 1978 Henry Rono of Kenya ran 5000 meters in 13 minutes 8.4 seconds. What was his average rate in meters per second?
30. A marathon is approximately $26\frac{1}{5}$ miles. What is the average rate in miles per hour for a runner who completes the course in 2 hours 10 minutes? (The record for the Boston Marathon is slightly faster than this.)
31. The airline distance between Chicago and Washington, D.C., is 600 miles. A plane leaves Chicago for Washington at an average rate of 400 miles per hour. At the same time, another plane leaves Washington for Chicago. If the second plane arrives in Chicago 10 minutes after the first plane arrives in Washington, what was the average rate for the second plane?
32. A plane leaves San Francisco for Chicago at an average rate of 465 miles per hour. At the same time, another plane leaves Chicago for San Francisco at an average rate of 430 miles per hour. The planes meet 2 hours and 5 minutes later. What is the airline distance between the two cities?
33. I step on an escalator to leave a subway. Three seconds later someone else steps

on a parallel escalator to enter the subway, and we pass 7 seconds after that. I know the rate of each escalator to be 0.5 meters per second. What is the length of the escalators?

34. Solve the problem in Example 7.9 under the assumption that the decrease and increase are both one-third rather than one-fifth.

35. Solve the problem in Example 7.9 under the assumption that the decrease and increase are both one-fourth rather than one-fifth.

36. Solve the problem in Example 7.9 under the assumption that the decrease and increase are both one-tenth rather than one-fifth.

37. For which sets of three consecutive positive integers will four times the sum of the first two integers exceed seven times the third of the integers?

38. For which sets of four consecutive positive integers is twice the sum of the first three integers less than five times the fourth of the integers?

39. If you are told that the perimeter of a square is less than the circumference of a circle whose radius is 100 meters, what can you conclude about the length of each edge of the square?

SECTION 8 More about Applications

A. Approximate Numbers

In applying mathematics it is important to distinguish between *exact* numbers (numbers that measure something exactly) and *approximate* numbers (numbers that measure something approximately). Here are two examples.

> There are 100 centimeters in a meter. (*100 is exact.*)
>
> The world population in 1975 was 4 billion. (*4 billion is approximate, although accurate at least to the nearest billion.*)

Approximate numbers generally arise either from measurement or from decimal approximations of irrational numbers like π and $\sqrt{2}$. In particular, no measurement can be more accurate than the instrument used to obtain it.

For dealing with approximate numbers it is convenient to use the notion of significant figures, which can be thought of as the digits in a number about whose accuracy we feel reasonably certain. The significant figures in a number sometimes must simply be inferred. If the airline distance from New York to San Francisco is given as 2570 miles, for example, the digits 2, 5, and 7 should be thought of as significant. Unless we are certain that a zero on the end of a number is a significant figure, we should assume that it is not. If a number in scientific notation (Section 3) is $a \times 10^n$, then the digits in a will be the significant figures.

EXAMPLE 8.1

(a) There are four significant figures in 9.721×10^{-6}—namely, 9, 7, 2, and 1.
(b) There are three significant figures in 2.46×10^5. But there are four significant figures in 2.460×10^5. The assumption is that the 0 in 2.460 would not be written if we weren't reasonably certain about it.

To **round** a number means to replace it by the nearest approximation having a specified degree of accuracy. In rounding numbers we use the following three rules; each rule applies for any placement of the decimal point in the number. Example 8.2 illustrates each rule.

Rule 1. If the part to be discarded is less than . . .5000. . . , simply discard it (*round down*).
Rule 2. If the part to be discarded is more than . . .5000. . . , increase the last digit retained by one (*round up*).
Rule 3. If the part to be discarded is exactly . . .5000. . . , round either down or up, whichever will make the last digit retained an even digit.

EXAMPLE 8.2

(a) By Rule 1, 875.4650 rounded to three significant figures is 875.
(b) By Rule 2, 875.4650 rounded to two significant figures is 880.
(c) By Rule 3, 875.4650 rounded to five significant figures is 875.46.

The guiding principle in working with approximate numbers is that we must not assume a final answer to be more accurate than the numbers with which we begin. Here are two rules that result from that principle. Examples follow the rules.

Rule. The result of *adding* or *subracting* approximate numbers should not be given to more *decimal places* than any of the numbers with which we begin.
Rule. The result of *multiplying* or *dividing* approximate numbers should not be given with more *significant figures* than any of the numbers with which we begin.

These rules are really only thinly disguised common sense. If one edge of a rectangular field is measured only to the nearest meter, for instance, but an adjacent edge is measured to the nearest centimeter, it would be foolish to carry out the computation of the perimeter beyond the nearest meter.

EXAMPLE 8.3

(a) Give the sum of the approximate numbers 4.321 and 91.3 as 95.6, not 95.621.
(b) Give the product of the approximate numbers 43.21 (four significant figures) and 123 (three significant figures) as 5310 (three significant figures), not 5314.83 (six significant figures), or even 5315 (four significant figures).

Exact numbers put no restriction on the accuracy of an answer. For example, the accuracy of an answer from using the formula $P = 4s$ for the perimeter of a square is restricted only by the accuracy of s (the length of a side); the number 4 is exact in this formula. In Example 8.3(b), if the number 123 were exact rather than approximate, then the product would be given as 5315 (four significant figures).

In computations using π, it is reasonable to begin with an approximation for π having one more significant figure than the greatest number of significant figures in the other numbers being used. (Similarly for numbers like $\sqrt{2}$ or $\sqrt{3}$ in place of π.) If the radius of a circle is given as 13 centimeters, for example, compute the circumference ($C = 2\pi r$) by using 3.14; then give the answer with two significant figures. Thus $2 \times 3.14 \times 13 = 81.64$, which rounds to 82 centimeters. (Using 3.1 for π would give $2 \times 3.1 \times 13 = 80.6$, which rounds to 81 centimeters.)

Reminder. $\approx$ means *approximately equal.*

B. Units

Many problems, especially those involving a change of units, can be solved most easily by using the convenient fact that units can be manipulated much like numbers and variables.

EXAMPLE 8.4 The basic formula for motion problems is

$$\text{distance} = \text{rate} \times \text{time}$$
$$D = RT. \tag{8.1}$$

If D is measured in miles and T is measured in hours, then R will be given in miles per hour, which we write miles/hour. Compare what happens if we replace the symbols in Equation (8.1) by the units with which they are measured:

$$\text{miles} = \frac{\text{miles}}{\text{hour}} \times \text{hours}.$$

The *hours* cancel on the right to leave *miles,* which is what is on the left. Units can always be canceled and multiplied in this way.

More examples will follow these lists of frequently used abbreviations, prefixes, and conversion factors.

ABBREVIATIONS

cm	centimeter	in.	inch	mi	mile
cu	cubic	kg	kilogram	qt	quart
ft	feet	km	kilometer	sec	second
g	gram	*l*	liter	sq	square
gal	gallon	m	meter	yd	yard
hr	hour				

Square and cubic factors are also written with exponents. For example, sq ft = ft^2 and cu cm = cm^3.

PREFIXES

Definitions	*Examples*
milli means multiply by 10^{-3}	1 milliliter = 0.001 liter
centi means multiply by 10^{-2}	1 centimeter = 0.01 meter
deci means multiply by 10^{-1}	1 decigram = 0.1 gram

deka means multiply by 10   1 dekaliter = 10 liters
hecto means multiply by 10^2 1 hectometer = 100 meters
kilo means multiply by 10^3 1 kilometer = 1000 meters

CONVERSION FACTORS

Length

1 in. = 2.54 cm
1 yd = 0.9144 m
1 m = 39.37 in.
1 mi = 1760 yd = 5280 ft
1 km = 0.621 mi
1 mi = 1.61 km

Volume

1 qt (liquid, U.S.) = 0.946 *l*
4 qt = 1 gal

Weight

454 g = 1 lb
1 kg = 2.205 lb

Here's a hint for problems that involve units. After recording the units and assigning variables, look for equations and conversion factors that will lead *from* the units of quantities that are given *to* the units of quantities that are to be found. In the process of finding the correct units, you will often be led automatically to the solution of the problem.

EXAMPLE 8.5 Convert 440 yards to meters. Treat 440 as an exact number.

Solution To convert yards to meters, we look for a conversion factor with yards in the denominator and meters in the numerator; then yards will cancel, leaving us with meters. The conversion factor connecting yards and meters is 1 yd = 0.9144 m. This can be rewritten as either

$$\frac{1 \text{ yd}}{0.9144 \text{ m}} = 1 \quad \text{or} \quad \frac{0.9144 \text{ m}}{1 \text{ yd}} = 1.$$

In this case we need the second choice:

$$440 \text{ yd} = 440 \text{ yd} \times \frac{0.9144 \text{ m}}{1 \text{ yd}} = 440 \times 0.9144 \text{ m} \approx 402.3 \text{ m}.$$ [C]

(This shows that a 440 yard dash is approximately $2\frac{1}{3}$ meters longer than a 400 meter dash.)

EXAMPLE 8.6 Express 60 mi/hr in ft/sec.

Solution Here we must convert both length and time. To convert length, we need a factor for ft/mi, so that *mi* will cancel and *ft* will remain. From 1 mi = 5280 ft, we can use 5280 ft/mi = 1. Thus

$$60 \text{ mi/hr} = (60 \text{ mi/hr}) \times (5280 \text{ ft/mi}) = 60 \times 5280 \text{ ft/hr}.$$

To convert time, we need a factor for hr/sec. From 1 hr = 60 min and 1 min = 60 sec, we have 1 hr = 3600 sec, or (1/3600) hr/sec = 1. Thus

$$
\begin{aligned}
60 \text{ mi/hr} &= (60 \times 5280) \text{ ft/hr} \times (1/3600) \text{ hr/sec} \\
&= (60 \times 5280/3600) \text{ ft/sec} \\
&= 88 \text{ ft/sec.}
\end{aligned}
$$

Problems involving density can also be solved by forcing the units to turn out right. Here are two examples.

EXAMPLE 8.7 In 1973 the population density of the earth was 72 persons per square mile (of land area). How many persons were there per square kilometer?

Solution To get from persons/mi^2 to persons/km^2 we need a factor connecting mi^2 and km^2. From 1 km = 0.621 mi we have

$$1 \text{ km}^2 = 0.621^2 \text{ mi}^2 \approx 0.386 \text{ mi}^2. \qquad \boxed{\text{C}}$$

Therefore, 0.386 $\text{mi}^2/\text{km}^2 \approx 1$, and

$$
\begin{aligned}
72 \text{ persons/mi}^2 &\approx (72 \text{ persons/mi}^2)(0.386 \text{ mi}^2/\text{km}^2) \\
&\approx 28 \text{ persons/km}^2. \qquad \boxed{\text{C}}
\end{aligned}
$$

EXAMPLE 8.8 The density of gold (at 20°C) is 19.3 g/cm^3. How many cubic centimeters are in 100 grams of gold?

Solution We are given a quantity in g and a density in g/cm^3, and we need an answer in cm^3. To obtain an answer with cm^3, working from g and g/cm^3, we must divide:

$$\text{g} \div (\text{g/cm}^3) = \text{g} \times (\text{cm}^3/\text{g}) = \text{cm}^3.$$

Therefore, the answer is

$$(100 \text{ g}) \div (19.3 \text{ g/cm}^3) = (100/19.3) \text{ cm}^3 \approx 5.18 \text{ cm}^3. \qquad \boxed{\text{C}}$$

C. Percentage

Recall that **percentage** (or **percent**) means hundredths. For example, $25\% = \frac{25}{100} = 0.25$, $100\% = \frac{100}{100} = 1$, $110\% = \frac{110}{100} = 1.1$, and $\frac{1}{2}\% = 0.5\% = 0.005$. (To remove the % sign from a decimal number, move the decimal point two places to the left.) Percentage is useful because it offers a standard scale of comparison.

EXAMPLE 8.9

(a) 5% of 80 is 0.05×80, which is 4. (Remember that *of* is a signal to multiply.)
(b) If 18 is 75% of x, then $0.75x = 18$, and $x = 18/0.75 = 24$.
(c) If 300 is increased by 15%, the result is

$$300 + 0.15(300) = 300 + 45 = 345.$$

EXAMPLE 8.10 The population of the U.S. in six selected years was as follows (with populations rounded to the nearest 100,000).

Year	Population	Increase
1890	62,900,000	
		13,100,000
1900	76,000,000	
1920	105,700,000	
		17,100,000
1930	122,800,000	
1950	150,700,000	
		28,600.000
1960	179,300,000	

The absolute increases over three 10-year periods are shown at the right in the table. The *percentage* increases are more valuable for most comparison purposes, however. To get these we divide the population at the beginning of each period into the increase over that period.

$$\text{1890 to 1900:} \quad \frac{13{,}100{,}000}{62{,}900{,}000} \approx 20.8\% \qquad \boxed{\text{C}}$$

$$\text{1920 to 1930:} \quad \frac{17{,}100{,}000}{105{,}700{,}000} \approx 16.2\% \qquad \boxed{\text{C}}$$

$$\text{1950 to 1960:} \quad \frac{28{,}600{,}000}{150{,}700{,}000} \approx 19.0\% \qquad \boxed{\text{C}}$$

Thus, for example, the absolute increase over the third 10-year period was more than twice that over the first 10-year period; but as a percentage the increase over the third period was actually less than that over the first period.

EXAMPLE 8.11 A store buys a radio for \$55.00 and sells it for \$79.00. What is the store's profit (ignoring fixed costs or overhead) as a percentage of the cost of the radio to the store?

Solution The profit is \$79.00 − \$55.00 = \$24.00. The percentage profit is

$$\frac{\text{profit}}{\text{cost}} = \frac{\$24.00}{\$55.00} = 0.4\overline{36} \approx 44\%. \qquad \boxed{\text{C}}$$

EXAMPLE 8.12 Suppose that a salesman works on an 8% rate of commission. What must his sales be if he is to earn \$1200.

Solution Let A denote the total amount of sales. We must determine A such that 8% of A is \$1200:

$$8\% \text{ of } A \text{ is } 1200$$

gives

$$0.08A = 1200$$

and

$$A = 1200/0.08 = \$15{,}000.$$

The formula for **simplest interest** is

$$I = Prt, \tag{8.2}$$

where P is the **principal** (amount invested), r is the **rate** per year (as a percentage expressed in decimal form), and t is the **time** of the investment in years. (In practice, most interest is not *simple* interest but *compound* interest, which is discussed in Section 18.)

EXAMPLE 8.13 What is the interest on \$500 invested at 6% per year simple interest for 4 years?

Solution Use $P = \$500$, $r = 0.06$, and $t = 4$ in Equation (8.2):

$$I = 500(0.06)(4) = \$120.$$

EXAMPLE 8.14 What amount must be invested at 8% simple interest so that \$50 interest is earned at the end of 6 months?

Solution Use Equation (8.2) with $I = 50$, $r = 0.08$, $t = \frac{1}{2}$ (6 months $= \frac{1}{2}$ year), and P unknown.

$$P = I/rt$$
$$P = 50/[(0.08)(\tfrac{1}{2})]$$
$$P = 50/0.04$$
$$P = \$1250.$$

Mixture problems such as the following are prime examples of problems that may look difficult at first but become much easier by using equations and moving one step at a time, as outlined in Section 7.

EXAMPLE 8.15 We require 4 liters of a solution that is 15% acohol. We can draw from two solutions: one that is 12% alcohol and another that is 20% alcohol. How many liters of each should we use?

Solution Let

$$x = \text{number of liters of 12\% solution to be used.}$$

Then

$$4 - x = \text{number of liters of 20\% solution to be used.}$$

The 12% solution will contribute $0.12x$ liters of alcohol to the final solution. The 20% solution will contribute $0.20(4 - x)$ liters of alcohol to the final solution. The total amount of alcohol in the final solution will be $0.15(4) = 0.6$ liters. Therefore, for the volume of alcohol,

$$0.12x + 0.20(4 - x) = 0.6$$
$$0.12x + 0.8 - 0.20x = 0.6$$

$$-0.08x = -0.2$$
$$x = 2.5\ l.$$

Thus we should use 2.5 liters of 12% solution and 4 − 2.5 or 1.5 liters of 20% solution.

EXAMPLE 8.16 How many gallons of water must be added to 50 gallons of liquid fertilizer that is 30% nitrogen to produce a solution that is 12% nitrogen?

Solution Let x denote the unknown number of gallons of water. The final amount of solution will be $50 + x$ gallons. Therefore,

$$\begin{array}{lll} & \text{original amount of nitrogen} & = (0.30)50 \text{ gal} \\ + & \text{nitrogen added} & = \quad 0.0x \text{ gal} \\ \hline & \text{final amount of nitrogen} & = 0.12(50 + x) \text{ gal.} \end{array}$$

That is,

$$\begin{aligned} 0.30(50) + 0 &= 0.12(50 + x) \\ 15 &= 6 + 0.12x \\ 0.12x &= 9 \\ x &= 9/0.12 \\ x &= 75 \text{ gal.} \end{aligned}$$

EXERCISES FOR SECTION 8

1. Round each number to two significant figures.
   (a) 81.451 (b) 244.98 (c) 2.35297×10^5 (d) 0.155
2. Round each number to three significant figures.
   (a) 0.09426 (b) 19.35 (c) 41,748 (d) 1.404×10^{-3}
3. Round each number to four significant figures.
   (a) 273.46 (b) 4.7235×10^{-7} (c) 0.013579 (d) 628,849
4. Convert each length to meters and then arrange in order, beginning with the smallest.
   10 meters, 0.09 kilometers, 0.009 kilometers, 200 centimeters, 11 yards.
5. Convert each area to square meters and then arrange in order, beginning with the smallest.
   10 square meters, 500,000 square centimeters, 500 square decimeters, 105 square feet, 11.5 square yards.
6. Convert each (liquid) volume to liters and then arrange in order, beginning with the smallest.
   10 liters, 5000 milliliters, 10 quarts, 11 quarts, 2.6 gallons.
7. Assume that 1 dollar = 800 lira.
   (a) Convert 20 dollars to lira.
   (b) Convert 1000 lira to dollars.
8. Assume that 1 dollar = 2.5 marks.

(a) Convert 20 dollars to marks.
(b) Convert 1000 marks to dollars.

9. Assume that 1 dollar = 5 francs.
(a) Convert 20 dollars to francs.
(b) Convert 1000 francs to dollars.

10. Convert 200 acres to square miles given that 1 square mile = 640 acres.

11. Convert 18 drams to tablespoons given that 1 tablespoon = 4 fluid drams.

12. Convert $9\frac{1}{2}$ furlongs to miles given that 1 mile = 8 furlongs.

13. Express 5 cm/sec in m/min.

14. Express 20 ft/min in yd/hr.

15. Express 10 yd/min in m/sec.

16. In 1850 the population density of the U.S. was 7.9 persons per square mile (of land area).
(a) The land area was 2,940,042 square miles. What was the population?
(b) What was the population density per square kilometer? [C]

17. In 1950 the population density of the U.S. was 50.7 persons per square mile (of land area).
(a) The land area was 2,974,726 square miles. What was the population?
(b) What was the population density per square kilometer? [C]

18. In 1970 the population density of the U.S. was 57.4 persons per square mile (of land area).
(a) The land area was 3,540,023 square miles. What was the population?
(b) What was the population density per square kilometer? [C]

19. The density of lead (at 20°C) is 11.34 g/cm^3. How many cubic centimeters are in 20 grams of lead?

20. The density of aluminum (at 20°C) is 2.7 g/cm^3. How many grams are in 50 cubic centimeters of aluminum?

21. The density of mercury (at 20°C) is 13.55 g/cm^3. How many grams are in 20 cubic centimeters of mercury?

22. What is 2.5% of 70?

23. What is 175% of 20?

24. What is $\frac{1}{2}$% of 40?

25. If 12 is 9% of x, what is x?

26. If 720 is 0.6% of y, what is y?

27. If 15 is $\frac{1}{4}$% of z, what is z?

28. The population of the U.S. in 1800 was 5,300,000 (to the nearest 100,000). The population increased by 35% between 1800 and 1810. What was the population in 1810?

29. The population of the People's Republic of China in 1970 was 772 million. The population increased by 8.7% between 1970 and 1975. What was the population in 1975? (Source: United Nations *Concise Report*.)

30. The population of the U.S.S.R. in 1970 was 243 million. The population increased by 4.5% between 1970 and 1975. What was the population in 1975? (Source: United Nations *Concise Report*.)

31. What is the percentage profit on an item that is bought for \$20.00 and sold for \$22.50?

32. What is the percentage profit on an item that is bought for \$30.00 and sold for \$39.00?

33. What is the percentage profit on an item that is bought for \$80.00 and sold for \$84.00?
34. What total sales will produce a \$500 commission if the commission rate is 20%?
35. What total sales will produce a \$300 commission if the commission rate is 6%?
36. What total sales will produce a \$1800 commission if the commission rate is 9%?
37. What is the interest on \$200 invested at 8% per year simple interest for 2 years?
38. What is the interest on \$1000 invested at 9% per year simple interest for 3 years?
39. What is the interest on \$500 invested at 12% per year simple interest for 6 months?
40. What amount must be invested at 6% per year simple interest so that \$40 interest is earned at the end of 18 months?
41. What amount must be invested at 10% per year simple interest so that \$250 interest is earned at the end of 30 months?
42. What amount must be invested at 8% per year simple interest so that \$200 interest is earned at the end of 5 years?
43. How many liters of solution that is 10% alcohol must be added to 2 liters of solution that is 60% alcohol to produce a solution that is 40% alcohol?
44. Container A holds a solution that is 12% salt and container B holds a solution that is 20% salt. How much solution should we use from each container to form 4 liters of solution that is 15% salt?
45. Part of \$2500 is invested in an account that pays 6% simple annual interest, and the remainder is invested in an account that pays 8% simple annual interest. The total interest earned from the two accounts in one year is \$185. How much is invested in each account?
46. If x is increased by 20% to produce y, and then y is increased by 30% to produce z, and $z = 52$, what is x?
47. One precinct has 400 registered voters of whom 40% are Democrats. Another precinct has 600 registered voters of whom 50% are Democrats. What percent of the registered voters in the two precincts taken together are Democrats?
48. One precinct has 600 registered Republicans, which is 40% of the total registered voters. Another precinct has 500 registered Republicans, which is 50% of the total registered voters. What percent of the registered voters in the two precincts taken together are Republicans?
49. The average price per ounce of gold in New York increased 1.1% from 1974 to 1975, decreased 22.4% from 1975 to 1976, and then increased 18.4% from 1976 to 1977. The average price in 1977 was \$148.30. What was the average price in 1974? [C]
50. In 1975, 25% of the population of Northern America (which consists primarily of the U.S. and Canada) was less than 15 years old, and 42% of the population of Latin America was less than 15 years old. The populations of Northern America and Latin America were approximately 237 million and 324 million, respectively. What percent of the total population of Northern and Latin America was less than 15 years old? [C]
51. In 1970, 73.5% of the U.S. population of 203 million was urban and the remainder was rural. Of the total urban population, 6% lived in New England. Of the total rural population, 5.2% lived in New England. Determine the 1970 population of New England to the nearest 100,000. [C]

SECTION

9 Ratio. Proportion. Variation

A. Ratio and Proportion

The **ratio** of a number a to a nonzero number b is the fraction a/b. The ratio of a to b is often written $a : b$, which is read "a is to b." Thus $a : b = a/b$.

An equality of two ratios is called a **proportion.** Thus a proportion has the form $a : b = c : d$, which is read "a is to b as c is to d." The numbers a and d are called the **extremes** of $a : b = c : d$. The numbers b and c are called the **means** of $a : b = c : d$.

EXAMPLE 9.1

(a) $2 : 6 = 1 : 3$ because $\frac{2}{6} = \frac{1}{3}$.
(b) $\frac{1}{2} : \frac{3}{4} = \frac{1}{2}/\frac{3}{4} = (\frac{1}{2})(\frac{4}{3}) = \frac{2}{3} = 2 : 3$.
(c) $3 : 2 = 3/2 = 1.5/1 = 1.5 : 1$. That is, 3 is to 2 as 1.5 is to 1.

EXAMPLE 9.2 If the two means of a proportion are interchanged, the result will again be a proportion; that is,

$$a : b = c : d \quad \text{iff} \quad a : c = b : d.$$

Here is why:

$$\begin{aligned} a : b = c : d \quad &\text{iff} \quad a/b = c/d \\ &\text{iff} \quad ad = bc \\ &\text{iff} \quad a/c = b/d \\ &\text{iff} \quad a : c = b : d. \end{aligned}$$

EXAMPLE 9.3 The scale of a map is 1 : 750. Determine the distance on the ground corresponding to 3.5 centimeters on the map.

Solution The scale

$$1 : 750$$

represents

$$\text{map distance : ground distance,}$$

where the same units are used for both distances. That is,

1 ft on the map = 750 ft on the ground
1 cm on the map = 750 cm on the ground
and so on,

We must determine x such that $1 : 750 = 3.5 : x$. (map: 1 and 3.5; ground: 750 and x)

$$\begin{aligned} 1 : 750 &= 3.5 : x \\ 1/750 &= 3.5/x \\ x &= (750)(3.5) \\ &= 2625 \text{ cm} \\ &= 2625 \text{ cm} \times (1/100) \text{ m/cm} \\ &\approx 26 \text{ m.} \end{aligned}$$

EXAMPLE 9.4 A mixture is 3 parts alcohol and 5 parts water. How many liters of each are in 20 liters of the mixture?

Solution Let

$$\begin{aligned} A &= \text{number of liters of alcohol} \\ W &= \text{number of liters of water.} \end{aligned}$$

Of each 8 parts of the mixture, 3 are alcohol and 5 are water. Therefore, for any amount of the mixture we have the ratio

$$\frac{\text{amount of alcohol}}{\text{total amount}} = \frac{3}{8}.$$

Thus, when the total amount is 20 liters,

$$\frac{A}{20} = \frac{3}{8}$$

$$A = 20(\tfrac{3}{8}) = \tfrac{60}{8} = 7.5 \text{ liters.}$$

This leaves

$$W = 20 - 7.5 = 12.5 \text{ liters.}$$

EXAMPLE 9.5 One of the many ratios used in business and finance is the *price/earnings* (*P/E*) *ratio* of a stock, which is defined as the ratio of the price per share to the company's annual earnings per share. (P/E ratios provide a good way to compare stocks that have different prices and earnings.) The P/E ratio is also called the *P/E multiple* and is usually reported simply as the multiple of the price to the annual earnings per share. If a stock is selling at \$100 per share with annual earnings of \$5 per share, then its P/E multiple is 20 (= 20/1 = 20 : 1).

EXAMPLE 9.6 Demographers define the *sex ratio* of a population as 100 times the ratio of males to females:

$$\frac{\text{number of males}}{\text{number of females}} \times 100.$$

Denote the sex ratio by r. Then

$$\frac{\text{number of males}}{\text{number of females}} \times 100 = r$$

$$\frac{\text{number of males}}{\text{number of females}} = \frac{r}{100}.$$

The last equation shows that the sex ratio is also equal to the number of males per 100 females. Table 9.1 shows the sex ratios by age of the U.S. population in 1950 and 1975.

Problem Use Table 9.1 to determine the percentage of males among those over 64 in 1975.

TABLE 9.1

	Sex ratio	
Age	*1950*	*1975*
All ages	98.6	94.9
Under 14	103.7	104.2
14–24	98.2	101.3
25–44	96.4	96.5
45–64	100.1	91.7
Over 64	89.6	69.4

SOURCE: *Statistical Abstract of the United States,* 1978.

Solution Because the sex ratio in this case is 69.4, there were 69.4 males for every 100 females. This means that 69.4 of every 169.4 persons over 64 were males, giving a percentage of

$$\frac{69.4}{169.4} \approx 41.0\%.$$

Ⓒ

EXAMPLE 9.7 Two triangles are *similar* if their angles can be paired in such a way that corresponding angles are equal. For example, triangle ABC is similar to triangle $A'B'C'$ in Figure 9.1 at the top of the next page. In geometry it is proved that the ratios between corresponding sides of similar triangles are equal. Thus, in Figure 9.1, $a : a' = b : b' = c : c'$. Also, $a : b = a' : b'$, and so on.

EXAMPLE 9.8 A man 6 feet tall casts a 10-foot shadow when standing 20 feet from the base of a streetlight. How high is the light above ground level?

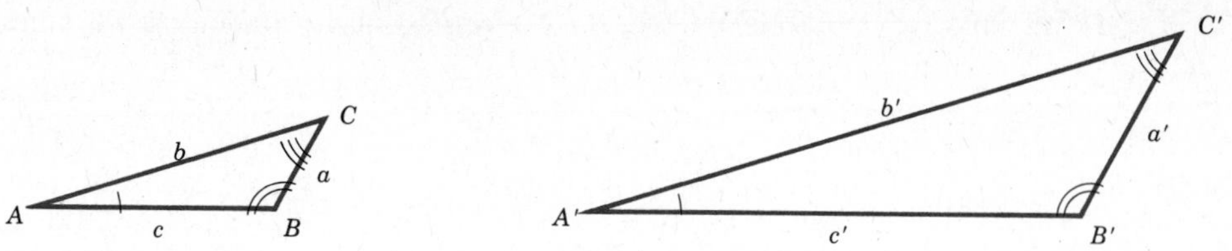

Figure 9.1

Solution Figure 9.2 shows the essentials. Triangles ABC and ADE are similar (Example 9.7). Therefore,

$$\frac{BC}{AC} = \frac{DE}{AE}$$

$$\frac{x}{30} = \frac{6}{10}$$

$$x = \frac{30 \cdot 6}{10} = 18 \text{ ft.}$$

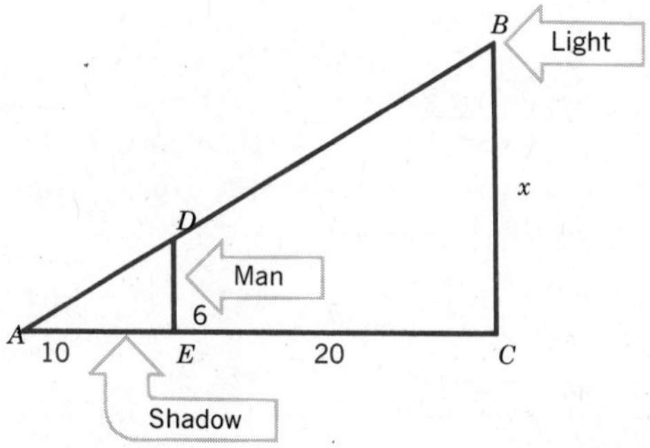

Figure 9.2

B. Variation

If two variables are related in such a way that their ratio is a nonzero constant, then one is said to **vary directly** as the other. If y and x are such variables, then $y/x = k$ for some nonzero real number k, or

$$y = kx. \tag{9.1}$$

The number k is called the **constant of proportionality.** We also say that y is **directly proportional** to x in this case. The following example will emphasize that if a single pair of corresponding values of x and y is known, then any other pair can be computed.

EXAMPLE 9.9 Suppose that y varies directly as x and that $y = 20$ when $x = 6$. What is x when $y = 9.5$?

Solution First, $y = kx$. Therefore, since $y = 20$ when $x = 6$, we must have

$$20 = k \cdot 6 \quad \text{or} \quad k = \tfrac{20}{6} = \tfrac{10}{3}.$$

Thus

$$y = \tfrac{10}{3}x.$$

If $y = 9.5$, this gives

$$9.5 = \tfrac{10}{3}x$$
$$x = 9.5(\tfrac{3}{10}) = 2.85.$$

EXAMPLE 9.10 If the speed (rate) of a moving object is held constant, then the distance traveled varies directly as the time, because $D = RT$. If, for example, a ship travels at a constant rate of 30 knots, then the distance it travels in T hours is

$$D = 30T \text{ nautical miles.}$$

(1 nautical mile $\approx$ 1.15 land miles and 1 knot = 1 nautical mile per hour.)

EXAMPLE 9.11 Figure 9.3 (on the next page) shows a coil spring hanging freely (a), and the same spring (b) with an object attached that weighs W pounds. The scale on the right, directed downward, measures the length L that the spring stretches because of the weight. A simple law of physics, *Hooke's law,* states that L varies directly as W:

$$L = cW \text{ for some constant } c.$$

If W is doubled, then L will double; if W is tripled, then L will triple; and so on: that is precisely what distinguishes direct variation.

The value of c is determined by the spring—each spring has its own value of c. If we know the value of L corresponding to just a single weight W, then we can find c.

Problem Suppose that a weight of 5 pounds stretches a spring 2 inches.

(a) How far will a 20-pound weight stretch the spring?
(b) If a weight of W pounds stretches the spring 3/4 inch, what is W?

Solution With $L = 2$ and $W = 5$, $L = cW$ gives $2 = 5c$, so that $c = \tfrac{2}{5}$. Therefore, for this spring,

$$L = \tfrac{2}{5}W. \tag{9.2}$$

(a) If $W = 20$, then $L = \tfrac{2}{5}(20) = 8$ inches.
(b) If $L = \tfrac{3}{4}$, then $\tfrac{3}{4} = \tfrac{2}{5}W$, and $W = (\tfrac{5}{2})(\tfrac{3}{4}) = \tfrac{15}{8} = 1.875$ pounds.

If n is an integer and

$$y = kx^n \quad (k \neq 0), \tag{9.3}$$

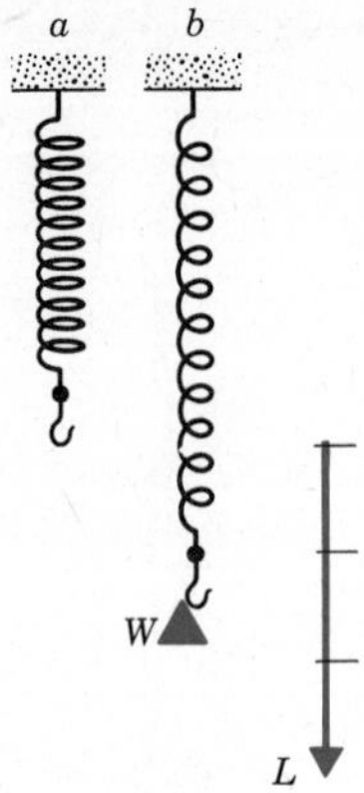

Figure 9.3

we say that y **varies as the nth power of** x, or that y is **proportional to the nth power of** x. Again, k is the **constant of proportionality.** The relation in Equation (9.3) is also represented by

$$y \propto x^n.$$

Here are some special cases.

$n = 1$	$y = kx$	y **varies directly as** x
$n = -1$	$y = k/x$	y **varies inversely as** x
$n = 2$	$y = kx^2$	y **varies as the square of** x

EXAMPLE 9.12 *Boyle's law* states that if the temperature is constant, then the pressure P of a confined gas varies inversely as the volume V of the gas:

$$P = k/V. \tag{9.4}$$

Problem Suppose the volume of a confined gas is 100 cubic inches when it is subjected to a pressure of 40 pounds per square inch. Determine the volume when the pressure is increased to 45 pounds per square inch.

Solution First we determine k, using Equation (9.4) with $P = 40$ and $V = 100$:

$$40 = k/100$$
$$k = 4000.$$

Thus

$$P = 4000/V,$$

so that

$$V = 4000/P.$$

When $P = 45$, this gives

$$V = 4000/45 \approx 89 \text{ cubic inches.}$$

Ⓒ

EXAMPLE 9.13 The intensity I of a sound wave is inversely proportional to the square of the distance r from the source:

$$I \propto 1/r^2.$$

This means that if you double your distance from a noise, the intensity will decrease by a factor of 4, if you triple your distance, the intensity will decrease by a factor of 9, and so on. (This assumes ideal conditions, including the absence of nearby reflecting or absorbing surfaces.) If r is measured in meters, then I is measured in watts/m^2.

Problem Suppose the intensity of a sound is 0.05 watts/m^2 when measured 30 meters from the source. What is the intensity 20 meters from the source?

Solution Use $I = 0.05$ and $r = 30$ in $I = k/r^2$:

$$0.05 = k/30^2$$
$$k = (0.05)(30^2) = 45.$$

Thus $I = 45/r^2$. If $r = 20$, then

$$I = 45/20^2 \approx 0.11 \text{ watts/m}^2.$$

Often the value of one variable depends on the values of several other variables. In special cases this dependence can be described in terms of **joint variation,** as in the following important example.

EXAMPLE 9.14 Newton's *law of universal gravitation* states that any two objects are attracted by a force that is directly proportional to the product of their masses and inversely proportional to the square of the distance between them:

$$F = G\frac{m_1 m_2}{r^2}, \tag{9.5}$$

where G, the constant of proportionality, is the *universal constant of gravitation.*

Problem If the masses of two objects are doubled and the distance between them is tripled, how will the force between them change?

Solution Let F denote the original force and F' the force after the change. Then F is given by Equation (9.5). To compute F', we must replace

$$m_1 \text{ by } 2m_1$$
$$m_2 \text{ by } 2m_2$$

and

$$r \text{ by } 3r.$$

Thus

$$F' = G\frac{(2m_1)(2m_2)}{(3r)^2} = \frac{4}{9}G\frac{m_1 m_2}{r^2} = \frac{4}{9}F.$$

That is, the new force will be $\frac{4}{9}$ of the original force.

EXERCISES FOR SECTION 9

1. What is x if $15 : 12 = x : 1$?
2. What is y if $7 : y = 4 : 1$?
3. What is z if $z : 3 = 4.5 : 1$?
4. If two cities are represented by points 10 centimeters apart on a map having scale 1 : 4,000,000, how far apart are the cities in kilometers?
5. Two cities that are 300 kilometers apart are represented on a map by points 15 centimeters apart. What is the scale of the map?
6. If two cities are 200 kilometers apart, how far apart will the points that represent them be on a map having the scale 1 : 500,000?
7. The width and length of an 8×10 photograph are reduced proportionally so that the new length is 6 inches. What is the new width?
8. A basketball court is 94 feet by 50 feet. In a scale drawing the length is 20 centimeters. What is the width?
9. A scale drawing of a soccer field is 23 centimeters by 15 centimeters. The width of the corresponding field is 75 yards. What is the length?
10. What is the P/E multiple for a stock that is selling at $50 per share if the company's annual earnings are $4 per share?
11. What is the selling price for a stock if its P/E multiple is 15 and the company's annual earnings are $2.60 per share?
12. On September 3, 1929, the composite P/E multiple for Dow Jones Industrial stocks was 19.1. What would the annual earnings be for a stock having that P/E multiple and a price of $50 per share?
13. If a population has sex ratio 80, what is the percentage of males?
14. If a population has sex ratio 90, what is the percentage of females?
15. If a population is 60% females, what is its sex ratio?
16. The sides of a triangle have lengths 6 inches, 8 inches, and 11 inches. The shortest side of a similar triangle has length 10 centimeters. What are the lengths of the other two sides?
17. The sides of a triangle have lengths 5 meters, 6 meters, and 7 meters. The longest side of a similar triangle is 10 feet. What are the lengths of the other two sides?
18. The two legs (the sides containing the right angle) of a right triangle have lengths 10 inches and 18 inches. The shorter leg of a similar right triangle has length 6 centimeters. What is the length of the longer leg?
19. A man 6 feet tall is standing 35 feet from the base of a streetlight that is 15 feet high. How long is his shadow?
20. A boy 4 feet tall casts a 6-foot shadow when standing near a streetlight that is 15 feet high. How far is he from the base of the light?
21. A man casts a 12-meter shadow when standing 20 meters from the base of a streetlight that is 4.8 meters high. How tall is the man?
22. Suppose that y varies directly as x and that $y = 5$ when $x = 20$. What is y when $x = 13$?
23. Suppose that y varies inversely as x and that $y = 70$ when $x = 20$. What is x when $y = 28$?
24. Suppose that y varies as the square of x and that $y = 10$ when $x = 5$. What is y when $x = 7$?

25. (a) Convert 6 nautical miles to land miles.
 (b) Convert 10 land miles to nautical miles.
26. Convert 30 mi/hr to knots, where *mi* represents *land* miles.
27. Convert 20 knots to mi/hr, where *mi* represents *land* miles.
28. Assume that a weight of 10 pounds will stretch a spring 3 inches. How far will a weight of 7 pounds stretch the spring?
29. Assume that a weight of 2 kilograms will stretch a spring 5 centimeters. If a weight of W kilograms stretches the spring 3.5 centimeters, what is W?
30. Assume that a weight of 3 kilograms will stretch a spring 2 centimeters. How far will a weight of 4.2 kilograms stretch the spring?
31. Suppose the volume of a confined gas is 50 cubic inches when it is subjected to a pressure of 20 pounds per square inch. Determine the volume when the pressure is decreased to 18 pounds per square inch.
32. Suppose the volume of a confined gas is 100 cubic inches when it is subjected to a pressure of 60 pounds per square inch. Determine the pressure that will yield a volume of 70 cubic inches.
33. Suppose the volume of a confined gas is 40 cubic inches when it is subjected to a pressure of 15 pounds per square inch. How much must the pressure be decreased or increased to change the volume to 25 cubic inches?
34. Suppose the intensity of a sound is 0.05 watts/m^2 when measured 30 meters from the source. What is the intensity 50 meters from the source?
35. Suppose the intensity of a sound is 0.001 watts/m^2 when measured 15 meters from the source. What is the intensity 20 meters from the source?
36. Suppose the intensity of a sound is 10^{-5} watts/m^2 when measured 10 meters from the source. What is the intensity 15 meters from the source?
37. If the masses of two objects are tripled and the distance between them is doubled, how will the force between them change? (See Example 9.14.)
38. If the masses of two objects are each increased by 10% and the distance between them is increased by 20%, how will the force between them change? (See Example 9.14.)
39. If the mass of one object is doubled and the mass of another object is increased by 50%, and the distance between them is cut in half, how will the force between them change? (See Example 9.14.)
40. If $a : b = b : c$, then b is called a *mean proportional* between a and c. Find a mean proportional between 12 and 75.
41. Show that if the two extremes of a proportion are interchanged, the result will again be a proportion. (Compare Example 9.2.)
42. A constant of proportionality must have units so that the units on the two sides of an equation in which it appears will be the same. What are the units on the constant 2/5 in Equation (9.2) (Example 9.11)?
43. What was the percentage of females in the U.S. population over 64 in 1950? (Use Table 9.1.) [C]
44. What was the percentage of males in the total U.S. population in 1975? (Use Table 9.1.) [C]
45. What was the percentage of females in the total U.S. population in 1950? (Use Table 9.1.) [C]

REVIEW EXERCISES FOR CHAPTER II

1. If one-half is added to twice the reciprocal of a number, the result is three. What is the number?
2. Ten times the first of three consecutive integers plus nine times the second exceeds seventeen times the third by fifteen. What are the integers?
3. If the price of gold increases 15% one day and then decreases by 10% the next day, how does the price at the end of the second day compare with the price at the start of the first day?
4. The United States public debt was $43 billion in 1940. It increased 498% between 1940 and 1950, and 11% between 1950 and 1960. What was the public debt in 1960?
5. The distance of the Kentucky derby is $1\frac{1}{4}$ miles. What is the average rate in miles per hour for a horse that runs the race in 2 minutes? (In 1973 Secretariat won the race in 1 minute 59.2 seconds.)
6. What is the average rate in miles per hour for a 25,000 meter race run in 1 hour 15 minutes? (The world record is slightly faster than this.)
7. Express 1000 yd/hr in ft/min.
8. Express 5000 gal/hr in *l*/min.
9. The density of gold (at 20°C) is 19.3 g/cm^3. How many grams are in 4 cubic centimeters of gold?
10. The density of mercury (at 20°C) is 13.55 g/cm^3. How many cubic centimeters are in 50 grams of mercury?
11. What is 1.8% of 75?
12. What is 210% of 70?
13. What percent of 60 is 125?
14. If 16 is 0.25% of x, what is x?
15. What is the interest on $2000 invested at 7% per year simple interest for 18 months?
16. What amount must be invested at 10% per year simple interest so that $60 interest is earned at the end of 4 years?
17. How many liters of solution that is 20% alcohol must be added to 4 liters of solution that is 60% alcohol to produce a solution that is 50% alcohol?
18. Part of $1200 is invested in an account that pays 8% per year simple interest, and the remainder is invested in an account that pays 10% per year simple interest. The total interest earned from the two accounts in one year is $113. How much was invested in each account?
19. If two cities are represented by points 15 centimeters apart on a map having a scale 1 : 2,000,000, how far apart are the cities in kilometers?
20. A football field is 360 feet (between goal posts) by 160 feet. In a scale drawing the length is 10 centimeters. What is the width?
21. The sides of a triangle have lengths 10 inches, 12 inches, and 16 inches. The longest side of a similar triangle has length 20 centimeters. What are the lengths of the other sides?
22. A man 6 feet tall casts an 8-foot shadow when standing near a streetlight that is 15 feet high. How far is he from the base of the light?
23. The purity of a gold alloy is measured in *karats*. Pure (100%) gold is 24 karats; other degrees of purity are expressed as proportional parts of 24. Thus a 12-karat

gold alloy is 50% gold and a 15-karat gold alloy is 62.5% gold. How many ounces of gold are in 4 ounces of a 14-karat gold alloy?

24. Suppose that y varies inversely as the square of x and that $y = \frac{1}{4}$ when $x = 6$. What is y when $x = 12$?

25. Assume that a weight of 4 kilograms will stretch a spring 1.6 centimeters? How far will a weight of 3.5 kilograms stretch the spring? ($L = cW$)

26. Suppose the volume of a confined gas is 40 cubic centimeters when it is subjected to a pressure of 25 pounds per square inch. Determine the pressure that will yield a volume of 45 cubic centimeters. ($P = k/V$)

CHAPTER III
QUADRATIC EQUATIONS

The equations considered thus far have been linear equations—that is, equations that involve only first-degree polynomials in the unknown variable. In this chapter we consider equations that contain second-degree polynomials. We also consider some typical applications of such equations.

SECTION

10 Solving Quadratic Equations. I

A. Introduction

DEFINITION. A **quadratic equation** in a variable x is one that is equivalent to an equation of the form

$$ax^2 + bx + c = 0, \tag{10.1}$$

where a, b, and c denote real numbers with $a \neq 0$.

Quadratic equations in other variables are defined similarly. Each of the following is a quadratic equation in its variable:

$$x^2 + 2x - 2 = 0$$
$$-3z^2 + z = 0$$
$$\pi t^2 = 0$$
$$(y - 2)(y + 4) = -2y^2 - 7.$$

The last equation is quadratic because it can be rewritten as

$$y^2 + 2y - 8 = -2y^2 - 7$$
$$3y^2 + 2y - 1 = 0.$$

For an equation to be quadratic in a variable the equation *must* contain a second-degree term in that variable after it has been simplified; it may or may not contain a first-degree term or a constant term.

This section and the next will present the methods used to solve quadratic equations and will show which method is best for each different type of equation. We'll see that among the real numbers a quadratic equation may have no solution, one solution, or two solutions.

B. Pure Equations

A quadratic equation with no first-degree term is called a **pure quadratic equation.** A pure quadratic equation with variable x should first be solved for x^2. Then use the fact that, for each nonnegative real number p,

$$x^2 = p \quad \text{iff} \quad x = \pm\sqrt{p}. \tag{10.2}$$

(Remember from Section 3D that $\sqrt{p}$ denotes the principal or nonnegative square root of p.)

EXAMPLE 10.1 To solve the pure equation $3x^2 - 75 = 0$, write

$$3x^2 = 75$$
$$x^2 = 25$$
$$x = \pm 5.$$

Each of the equations above is equivalent to the one that precedes it, so that $3x^2 - 75 = 0$ has *two* solutions: 5 and -5.

Both solutions in Example 10.1 can be checked by direct substitution. However, by the equivalence of the successive equations in the solution process, checking is necessary only to catch mistakes in computation. The same remarks about checking will apply to all of the solutions in this section and the next.

Note well that $p \geq 0$ in Statement (10.2). If $p < 0$, then $x^2 = p$ has no solution in the real numbers, since $x^2 \geq 0$ for every real number x. For example, $x^2 = -3$ has no real solutions. (Chapter VIII will consider a system of numbers that contains the real numbers and that has solutions to equations like $x^2 = -3$.)

Sometimes, as in the next example, the circumstances dictate that only one of the solutions in Statement (10.2) is to be used.

EXAMPLE 10.2 If the area of a circle is 20 square centimeters, what is the radius?

Solution The relevant formula is $A = \pi r^2$. For $A = 20$ we must have

$$20 = \pi r^2$$
$$r^2 = 20/\pi$$
$$r = \sqrt{20/\pi} \approx 2.52 \text{ cm.}$$

[C]

(The negative solution has no significance here.)

C. Solution by Factoring

Any pure quadratic equation should be solved as in the preceding examples. The first method to try on any other quadratic equation is *factoring*, which depends on the following fact about real numbers (Section 1C):

$$ab = 0 \quad \text{iff} \quad a = 0 \quad \text{or} \quad b = 0. \tag{10.3}$$

EXAMPLE 10.3 Solve $(t - 2)(t + 3) = 0$.

Solution The statements in the successive lines that follow are equivalent.

$$(t - 2)(t + 3) = 0$$
$$t - 2 = 0 \quad \text{or} \quad t + 3 = 0$$
$$t = 2 \quad \text{or} \quad t = -3.$$

Thus the solutions are 2 and -3.

To use this factoring method we must be able to factor the expression that results after the equation has been written so that only 0 remains on one side of the equality sign. (To review factoring, see Section 4D.) The next example illustrates how to use factoring to solve a quadratic equation with no constant term.

EXAMPLE 10.4 Solve $x^2 = 3x$.

Solution

$$x^2 = 3x$$
$$x^2 - 3x = 0$$
$$x(x - 3) = 0$$
$$x = 0 \quad \text{or} \quad x = 3.$$

That is, the solutions of $x^2 = 3x$ are 0 and 3.

Notice that if we were to divide both sides of $x^2 = 3x$ by x, we would be left with $x = 3$. Thus we would have lost the solution $x = 0$ of the original equation. Remember that a solution may be lost whenever both sides of an equation are divided by either the variable or an expression that involves the variable.

EXAMPLE 10.5 At the first step in this example we divide both sides of the equation by 2, which is a common factor of the coefficients.

$$6y^2 + 4y - 2 = 0$$
$$3y^2 + 2y - 1 = 0$$
$$(3y - 1)(y + 1) = 0$$
$$3y - 1 = 0 \text{ or } y + 1 = 0$$
$$y = \tfrac{1}{3} \text{ or } y = -1.$$

EXAMPLE 10.6 The sum of a number and its square is 20. What is the number?

Solution Let x denote the unknown number. Then $x^2 + x = 20$.

$$x^2 + x = 20$$
$$x^2 + x - 20 = 0$$
$$(x - 4)(x + 5) = 0$$
$$x = 4 \quad \text{or} \quad x = -5.$$

It can be verified that both 4 and -5 are indeed solutions of the problem.

By the method we adopted for pure quadratic equations (Subsection B), an equation like $x^2 - p = 0$ would be solved by writing

$$x^2 = p \quad \text{iff} \quad x = \pm\sqrt{p}.$$

Such equations can also be solved by factoring:

$$x^2 - p = 0 \quad \text{iff} \quad (x + \sqrt{p})(x - \sqrt{p}) = 0 \quad \text{iff} \quad x = \pm\sqrt{p}.$$

The following example will serve as a bridge between pure equations and the method to be discussed next.

EXAMPLE 10.7 Solve $y^2 - 2y + 1 = 9$.

Solution

$$y^2 - 2y + 1 = 9$$

$$(y-1)^2 = 9 \qquad \text{Because } y^2 - 2y + 1 = (y-1)^2.$$

$$y - 1 = \pm 3 \qquad \text{Use Statement (10.2) with } x = y - 1 \text{ and } p = 9.$$

$$y = 1 \pm 3 \qquad \text{Add 1 to both sides.}$$

$$y = 4 \quad \text{or} \quad y = -2.$$

Both 4 and -2 are solutions of $y^2 - 2y + 1 = 9$ since the statements in successive lines of the solution process are equivalent.

In Example 10.7 we exploited the fact that $y^2 - 2y + 1$ is a *perfect square,* namely, $(y-1)^2$. The method of completing the square depends on the fact that if there is not a perfect square, then we can create one. The method will be outlined in the following five steps, and each step will be illustrated with the equation

$$3x^2 + x - 2 = 0.$$

I. Arrange the equation with the constant term on the right of the equality sign and the other terms on the left.

$$3x^2 + x = 2 \qquad \text{Add 2 to both sides of the original equation.}$$

II. Divide both sides by the coefficient of the second-degree term.

$$x^2 + \tfrac{1}{3}x = \tfrac{2}{3} \qquad \text{Divide by 3.}$$

III. Add the square of half the coefficient of the first-degree term to both sides.*

$$x^2 + \tfrac{1}{3}x + \tfrac{1}{36} = \tfrac{2}{3} + \tfrac{1}{36} \qquad \text{Add } [\tfrac{1}{2} \cdot \tfrac{1}{3}]^2 = \tfrac{1}{36} \text{ to both sides.}$$

IV. The previous step made the left side a perfect square; write it as such, and simplify the right side.

$$x^2 + 2(\tfrac{1}{6})x + (\tfrac{1}{6})^2 = \tfrac{24}{36} + \tfrac{1}{36}$$

$$(x + \tfrac{1}{6})^2 = \tfrac{25}{36}$$

---

*Compare $x^2 + \frac{1}{3}x$ with $x^2 + 2ax$ in $x^2 + 2ax + a^2 = (x + a)^2$. Step III amounts to adding the missing term a^2 to create $x^2 + 2ax + a^2$ on the left side.

V. Use the idea in Statement (10.2), and simplify.

$x + \frac{1}{6} = \pm\sqrt{\frac{25}{36}}$ Use Statement (10.2) with x replaced by $x + \frac{1}{6}$ and $p = \frac{25}{36}$.

$$x + \frac{1}{6} = \pm\frac{5}{6}$$
$$x = -\frac{1}{6} \pm \frac{5}{6}$$
$$x = \frac{-1+5}{6} \quad \text{or } x = \frac{-1-5}{6}$$
$$x = \frac{2}{3} \text{ or } x = -1.$$

EXAMPLE 10.8 Solve $2x^2 = 5x - 1$ by completing the square.

Solution Compare each step with the preceding outline.

$$2x^2 = 5x - 1$$

I. $2x^2 - 5x = -1$ Isolate the constant on the right.

II. $x^2 - \frac{5}{2}x = -\frac{1}{2}$ Divide by the coefficient of x^2.

III. $x^2 - \frac{5}{2}x + (\frac{-5}{4})^2 = -\frac{1}{2} + (\frac{-5}{4})^2$ Complete the square.

IV. $x^2 - 2(\frac{5}{4})x + (\frac{-5}{4})^2 = -\frac{1}{2}\cdot\frac{8}{8} + \frac{25}{16}$ Simplify.

$$(x - \frac{5}{4})^2 = \frac{17}{16}$$

V. $x - \frac{5}{4} = \pm\sqrt{\frac{17}{16}}$ Solve for x.

$$x - \frac{5}{4} = \pm\frac{\sqrt{17}}{4}$$
$$x = \frac{5 \pm \sqrt{17}}{4}.$$

The two solutions are $\dfrac{5 + \sqrt{17}}{4}$ and $\dfrac{5 - \sqrt{17}}{4}$.

Check Let's check $(5 + \sqrt{17})/4$ as a reminder of how checking is done. We begin by substituting in the original equation.

Left side Right side

$$2\left(\frac{5 + \sqrt{17}}{4}\right)^2 \stackrel{?}{=} 5\left(\frac{5 + \sqrt{17}}{4}\right) - 1$$
$$2\left(\frac{25 + 10\sqrt{17} + 17}{16}\right) \stackrel{?}{=} \frac{5(5 + \sqrt{17}) - 4}{4}$$
$$\frac{42 + 10\sqrt{17}}{8} \stackrel{?}{=} \frac{21 + 5\sqrt{17}}{4}$$
$$\frac{21 + 5\sqrt{17}}{4} = \frac{21 + 5\sqrt{17}}{4}$$

EXERCISES FOR SECTION 10

Solve for x, y, or t, or indicate that there is no real solution. Assume $A > 0$ and $B > 0$ in Exercises 11–13.

1. $x^2 = 36$	**2.** $y^2 = 144$	**3.** $t^2 - 100 = 0$
4. $16y^2 = 9$	**5.** $0.8t^2 - 1.8 = 0$	**6.** $98x^2 - 2 = 0$
7. $5t^2 + 20 = 0$	**8.** $10x^2 = 0$	**9.** $9y^2 + 25 = 0$
10. $0.01x^2 - 8 = 0$	**11.** $A^2y^2 = 27$	**12.** $0 = ABt^2$
13. $By^2 = 0$	**14.** $3t^2 + 12 = 0$	**15.** $0.49x^2 = 0.04$

Solve by factoring.

16. $x^2 + 4x = 0$	**17.** $2x^2 - 6x = 0$	**18.** $5x^2 = -25x$
19. $y^2 - 2y + 1 = 0$	**20.** $y^2 + 4y = -4$	**21.** $y^2 = 6y - 9$
22. $2t^2 + 2t = 4$	**23.** $3t^2 - 9t - 12 = 0$	**24.** $t^2 - t = 12$
25. $0.1z^2 = 0.3z$	**26.** $z^2 + 0.2z + 0.01 = 0$	**27.** $z^2 = z - 0.25$
28. $2y^2 + 14y + 20 = 0$	**29.** $3y^2 - 33y + 90 = 0$	**30.** $2y^2 - 2y - 4 = 0$
31. $5x^2 + 4x - 1 = 0$	**32.** $3x^2 - x - 2 = 0$	**33.** $5x^2 - 8x + 3 = 0$

Solve by completing the square. Check at least one solution (by substitution) in each exercise.

34. $x^2 + 2x = 2$	**35.** $x^2 = 1 - 6x$	**36.** $x^2 + 4x = 1$
37. $3y^2 + 2y = 1$	**38.** $2y^2 + 9y = 5$	**39.** $5y^2 = 2y + 1$
40. $2x^2 + 2x - 1 = 0$	**41.** $3x^2 + x - 3 = 0$	**42.** $4x^2 + 2x - 1 = 0$

43. If the area of a square is 30 square meters, what is the length of a side?

44. If the area of a circle is 8π square inches, what is the radius?

45. If the surface area S of a sphere is 20 square centimeters, what is the radius? $(S = 4\pi r^2)$

46. The sum of a number and two times its square is 55. What is the number? (Find all solutions.)

47. If twice the square of a positive number is equal to the number, what is the number?

48. If half the square of a positive number is equal to three times the number, what is the number?

49. If the sum of the squares of the first two of three consecutive positive integers is equal to the square of the third of the integers, what are the integers?

50. If the square of a number is decreased by 15, the result is equal to the number increased by 15. What is the number? (Find all solutions.)

51. If the surface area of a cube is 150 square inches, what is the length of each edge of the cube?

Solve with a calculator having a square root key. [C]

52. $t^2 - 12.31 = 0$ **53.** $509s^2 = 36$ **54.** $4 \cdot 3.14r^2 = 605$

SECTION

11 Solving Quadratic Equations. II

A. The Quadratic Formula

The solutions of any quadratic equation can be found by completing the square, as in Section 10. It is more efficient, however, to complete the square once for the most general quadratic equation,

$$ax^2 + bx + c = 0 \quad (a \neq 0), \tag{11.1}$$

and then use the formula that results. Here is the calculation, with the steps numbered as in Section 10.

I. $$ax^2 + bx = -c$$ *Isolate the constant term on the right.*

II. $$x^2 + \left(\frac{b}{a}\right)x = -\left(\frac{c}{a}\right)$$ *Divide by the coefficient of x^2.*

III. $$x^2 + \left(\frac{b}{a}\right)x + \left(\frac{b}{2a}\right)^2 = -\left(\frac{c}{a}\right) + \left(\frac{b}{2a}\right)^2$$ *Complete the square.*

IV. $$x^2 + 2\left(\frac{b}{2a}\right)x + \left(\frac{b}{2a}\right)^2 = -\left(\frac{c}{a}\right)\left(\frac{4a}{4a}\right) + \frac{b^2}{4a^2}$$ *Simplify.*

$$\left[x + \left(\frac{b}{2a}\right)\right]^2 = \frac{b^2 - 4ac}{4a^2}$$

V. $$x + \frac{b}{2a} = \pm\sqrt{\frac{b^2 - 4ac}{4a^2}}$$ *Solve for x.*

$$x = -\frac{b}{2a} \pm \frac{\sqrt{b^2 - 4ac}}{2a}$$

Exercises 46 and 47 ask you to show by substitution that both the + and − options here do give solutions of Equation (11.1). Thus we arrive at the following conclusion, which definitely should be memorized.

> **Quadratic formula.** The solutions of $ax^2 + bx + c = 0$, with $a \neq 0$, are
>
> $$x = \frac{-b \pm \sqrt{b^2 - 4ac}}{2a}.$$

EXAMPLE 11.1 Solve $6x^2 + x - 2 = 0$ using the quadratic formula.

Solution Use $a = 6$, $b = 1$, and $c = -2$.

$$\begin{aligned} x &= \frac{-1 \pm \sqrt{1^2 - 4(6)(-2)}}{2(6)} \\ &= \frac{-1 \pm \sqrt{1 + 48}}{12} \\ &= \frac{-1 \pm \sqrt{49}}{12} \\ &= \frac{-1 \pm 7}{12} \\ &= \frac{-1 + 7}{12} \quad \text{or} \quad \frac{-1 - 7}{12} \\ &= \tfrac{1}{2} \quad \text{or} \quad -\tfrac{2}{3} \end{aligned}$$

That is, the solutions are $\frac{1}{2}$ and $-\frac{2}{3}$.

The next example emphasizes that the numbers a, b, and c in the quadratic formula are the coefficients after a quadratic equation has been written so that only zero remains on one side of the equality sign.

EXAMPLE 11.2 Solve $x^2 + 8x = 2$ using the quadratic formula.

Solution First rewrite the equation with only zero on the right of the equality sign:

$$x^2 + 8x - 2 = 0.$$

This gives $a = 1$, $b = 8$, and $c = -2$. Therefore,

$$\begin{aligned} x &= \frac{-8 \pm \sqrt{8^2 - 4(1)(-2)}}{2(1)} \\ &= \frac{-8 \pm \sqrt{72}}{2} \\ &= \frac{-8 \pm \sqrt{36 \cdot 2}}{2} \\ &= \frac{-8 \pm 6\sqrt{2}}{2} \\ &= -4 \pm 3\sqrt{2}. \end{aligned}$$

The solutions are $-4 + 3\sqrt{2}$ and $-4 - 3\sqrt{2}$.

EXAMPLE 11.3 Solve $4x^2 - 0.4x + 0.01 = 0$ using the quadratic formula.

Solution Here $a = 4$, $b = -0.4$, and $c = 0.01$.

$$x = \frac{0.4 \pm \sqrt{(-0.4)^2 - 4(4)(0.01)}}{2(4)}$$

$$= \frac{0.4 \pm \sqrt{0.16 - 0.16}}{8}$$

$$= \frac{0.4 \pm 0}{8}$$

$$= 0.05$$

Thus there is one solution, 0.05.

EXAMPLE 11.4 The equation

$$s = \tfrac{1}{2}gt^2 + v_0 t$$

arises in the study of falling objects. (Section 13B will explain its significance.) Solve the equation for t.

Solution To solve the equation for t using the quadratic formula, use t in place of x and treat the other letters in the equation as constants. First, rewrite the equation as

$$\tfrac{1}{2}gt^2 + v_0 t - s = 0.$$

This gives $a = \frac{1}{2}g$, $b = v_0$, and $c = -s$. Therefore,

$$t = \frac{-v_0 \pm \sqrt{v_0^2 - 4(\frac{1}{2}g)(-s)}}{2(\frac{1}{2}g)}$$

$$= \frac{-v_0 \pm \sqrt{v_0^2 + 2gs}}{g}.$$

Here are suggestions for solving quadratic equations.

- If there is no first-degree term, solve the equation as a pure equation (Section 10).
- If the equation is not pure, try to solve it by factoring (Section 10). (This applies, in particular, if there is no constant term.)
- If the first suggestion does not apply and the second is difficult, use the quadratic formula.

B. The Character of the Solutions

EXAMPLE 11.5 If we apply the quadratic formula to

$$x^2 + x + 1 = 0,$$

we obtain

$$x = \frac{-1 \pm \sqrt{1^2 - 4 \cdot 1 \cdot 1}}{2 \cdot 1}$$
$$= \frac{-1 \pm \sqrt{-3}}{2}.$$

There is no real number whose square is -3, so the equation has no real solution. The following analysis categorizes this and all other quadratic equations.

The expression $b^2 - 4ac$ in the quadratic formula is called the **discriminant** of $ax^2 + bx + c$.

If the discriminant is negative, then the quadratic formula involves the square root of a negative number (as in Example 11.5) so that there is no real solution.

If the discriminant is positive, the quadratic formula gives two (unequal) real solutions:

$$\frac{-b + \sqrt{b^2 - 4ac}}{2a} \quad \text{and} \quad \frac{-b - \sqrt{b^2 - 4ac}}{2a}.$$

If the discriminant is zero, then the two numbers given by the quadratic formula are the same:

$$\frac{-b + \sqrt{0}}{2a} = -\frac{b}{2a} \quad \text{and} \quad \frac{-b - \sqrt{0}}{2a} = -\frac{b}{2a}.$$

Table 11.1 summarizes the preceding remarks.

TABLE 11.1

Discriminant	*Character of the solutions*
$b^2 - 4ac < 0$	No real solution
$b^2 - 4ac = 0$	One real solution
$b^2 - 4ac > 0$	Two real solutions

EXAMPLE 11.6 Determine the character of the solutions of $2x^2 = 3x - 5$.

Solution To compute the discriminant, first rewrite the equation as $2x^2 - 3x + 5 = 0$. This gives $a = 2$, $b = -3$, and $c = 5$, so that

$$b^2 - 4ac = (-3)^2 - 4(2)(5) = 9 - 40 = -31.$$

Because $-31 < 0$, there is no real solution. (Quadratic equations with negative discriminants will be discussed again in Chapter VIII.)

EXAMPLE 11.7 The discriminant of $x^2 + 8x - 2 = 0$ is

$$8^2 - 4(1)(-2) = 64 + 8 = 72.$$

Because $72 > 0$, there are two real solutions. They were computed in Example 11.2.

EXAMPLE 11.8 The discriminant of $4x^2 - 0.4x + 0.01 = 0$ is

$$(-0.4)^2 - 4(4)(0.01) = 0.16 - 0.16 = 0.$$

Thus there is one real solution. It was computed in Example 11.3.

EXERCISES FOR SECTION 11

Without solving these equations, use the discriminant to determine whether there is no real solution, one real solution, or two real solutions.

1. $x^2 - 3x + 2 = 0$
2. $x^2 - 12x = -9$
3. $3x^2 + 2x + 1 = 0$
4. $2x^2 = 4 - x$
5. $9x^2 = 6x - 1$
6. $4x^2 - 5x + 1 = 0$
7. $2x^2 + 2x + 1 = 0$
8. $x^2 + 2x + 3 = 0$
9. $-x^2 - 6x = 9$
10. $25x^2 + 20x + 4 = 0$
11. $x^2 + 3x - 1 = 0$
12. $x^2 - x + 1 = 0$

Use the quadratic formula to solve for x.

13. $2x^2 - x - 2 = 0$
14. $-3x^2 + x + 1 = 0$
15. $5x^2 - x - 4 = 0$
16. $8x^2 = 3 - 10x$
17. $-3x^2 + 13x = 10$
18. $x^2 = 2x + 4$
19. $x^2 - 0.3x + 0.02 = 0$
20. $x^2 - 0.1x - 0.06 = 0$
21. $x^2 - 0.3x - 0.04 = 0$
22. $qx^2 = qx + r$
23. $x^2 + qx = q$
24. $rx^2 + qx + p = 0$

Solve for x, y, or t. Use any method.

25. $x^2 - 7x = 0$
26. $y^2 = 5y$
27. $4t^2 + t = 0$
28. $4x^2 - 1 = 0$
29. $5y^2 - 15 = 0$
30. $4t^2 = 12$
31. $x^2 - 2x = 2$
32. $y^2 + 3y + 1 = 0$
33. $t^2 + t = 1$
34. $x^2 - 0.09 = 0$
35. $0.2 = 0.8y^2$
36. $0 = 0.1t^2 - 0.001$
37. $1 = \frac{1}{2}(x + x^2)$
38. $2y^2 + 3y = 0$
39. $\frac{1}{3}t^2 + \frac{5}{3}t + 2 = 0$
40. Solve $h^2 + (s/2)^2 = s^2$ for h.
41. Solve $V = k(R^2 - r^2)$ for r.
42. Solve $S = 2\pi rh + 2\pi r^2$ for r.
43. The product of two consecutive even integers is 168. Find the integers. (There are two sets of solutions; find them both. Example 7.3 may help.)
44. If the square of a positive number is four more than twice the number, what is the number?
45. If a positive number plus its square is three, what is the number?
46. Verify by substitution that

$$\frac{-b + \sqrt{b^2 - 4ac}}{2a}$$

is a solution of Equation (11.1).

47. Verify by substitution that

$$\frac{-b - \sqrt{b^2 - 4ac}}{2a}$$

is a solution of Equation (11.1).

Use the quadratic formula to verify each of the following facts about the solutions of $ax^2 + bx + c = 0$ $(a \neq 0)$.

48. Assume $a > 0$ and $c > 0$. Then there is only one real solution iff $b = \pm 2\sqrt{ac}$.
49. The sum of the solutions is $-b/a$.
50. The product of the solutions is c/a.
51. The square of the difference of the solutions (in either order—take your choice) is $|b^2 - 4ac|/a^2$.

SECTION

12 Miscellaneous Equations

A. Radical Equations

Frequently, an equation that is not strictly linear (first degree) or quadratic (second degree) can be solved by changing it into an equation that *is* linear or quadratic. The first examples illustrating this will be **radical equations,** so-called because the unknown variable appears under a radical sign (such as a square root sign).

EXAMPLE 12.1 Solve $\sqrt{2x + 3} = 7$.

Solution To solve this equation we square both sides to remove the radical. If $\sqrt{2x + 3} = 7$, then

$$\begin{aligned}(\sqrt{2x + 3})^2 &= 7^2 \\ 2x + 3 &= 49 \\ 2x &= 46 \\ x &= 23.\end{aligned}$$

Check $\sqrt{2 \cdot 23 + 3} \stackrel{?}{=} 7$

$\sqrt{49} = 7.$

Thus the equation has one solution, 23.

Two points must be made about the method just used. First, if the original equation has solutions, then the method will produce them. This is because if p and q denote expressions such that $p = q$, then $p^n = q^n$ for each positive integer n. Thus, in Example 12.1, $\sqrt{2x + 3} = 7$ implies $(\sqrt{2x + 3})^2 = 7^2$, so that any solution of the first equation will also be a solution of the second; by solving the second equation we find the solution of the first equation.

The second point about the method in Example 12.1 is that as well as producing the solutions of the original equation, the method can also produce answers that are *not* solutions of the original equation. These are called **extraneous solutions,** and they were discussed in Example 5.6 and preceding Example 5.11. Extraneous solutions can arise, in particular, whenever both sides of an equation are squared or whenever both sides of an equation are multiplied by an expression that involves the variable. For this reason we must always check the answers given by the method in Example 12.1. Here's an example to illustrate this point.

EXAMPLE 12.2 Solve $\sqrt{x+4} = x - 2$.

Solution If $\sqrt{x+4} = x - 2$, then

$$
\begin{aligned}
(\sqrt{x+4})^2 &= (x-2)^2 \\
x + 4 &= x^2 - 4x + 4 \\
x^2 - 5x &= 0 \\
x(x-5) &= 0 \\
x = 0 &\text{ or } x - 5 = 0 \\
x = 0 &\quad\text{or}\quad x = 5.
\end{aligned}
$$

Check For $x = 0$:

$$
\begin{aligned}
\sqrt{0+4} &\stackrel{?}{=} 0 - 2 \\
2 &\neq -2.
\end{aligned}
$$

For $x = 5$:

$$
\begin{aligned}
\sqrt{5+4} &\stackrel{?}{=} 5 - 2 \\
3 &= 3.
\end{aligned}
$$

Thus 0 is not a solution of the original equation; it is an extraneous solution. But 5 is a solution of the original equation.

In the next example we must square, simplify, and then square again.

EXAMPLE 12.3 Solve $\sqrt{3x-3} - \sqrt{x} = 1$.

Solution Before squaring, isolate one of the radicals on one side of the equation; this will make the calculations simpler. If $\sqrt{3x-3} - \sqrt{x} = 1$, then

$$
\begin{aligned}
\sqrt{3x-3} &= 1 + \sqrt{x} \\
(\sqrt{3x-3})^2 &= (1+\sqrt{x})^2 \\
3x - 3 &= 1 + 2\sqrt{x} + (\sqrt{x})^2 \\
3x - 3 &= 1 + 2\sqrt{x} + x \\
2x - 4 &= 2\sqrt{x} \\
x - 2 &= \sqrt{x}.
\end{aligned}
$$

Notice that the remaining radical has been isolated on one side of the equation. This was done so there will be no radicals left after we square again. Thus, squaring again,

$$
\begin{aligned}
(x-2)^2 &= (\sqrt{x})^2 \\
x^2 - 4x + 4 &= x \\
x^2 - 5x + 4 &= 0 \\
(x-1)(x-4) &= 0 \\
x = 1 \quad \text{or} \quad x &= 4.
\end{aligned}
$$

Check For $x = 1$:

$$
\begin{aligned}
\sqrt{3 \cdot 1 - 3} - \sqrt{1} &\stackrel{?}{=} 1 \\
-1 &\neq 1.
\end{aligned}
$$

For $x = 4$:

$$
\begin{aligned}
\sqrt{3 \cdot 4 - 3} - \sqrt{4} &\stackrel{?}{=} 1 \\
\sqrt{9} - \sqrt{4} &\stackrel{?}{=} 1 \\
1 &= 1.
\end{aligned}
$$

Thus 1 is not a solution; it is an extraneous solution. But 4 is a solution. Remember: To check an answer, always substitute in the *original* equation.

B. The Unknown in a Denominator

To solve an equation with the unknown variable in a denominator, we can first multiply both sides of the equation by a common multiple of the denominators in the equation. We did this in Section 5, where the result was always a linear equation. Here are two examples that produce quadratic equations. Notice that we must check for extraneous solutions in these examples.

EXAMPLE 12.4 Solve $\dfrac{1}{t+2} = \dfrac{4}{t} + \dfrac{1}{2}$.

Solution If we multiply both sides of the original equation by the common multiple $2t(t+2)$ of the denominators, we obtain

$$
\begin{aligned}
2t(t+2)\left(\frac{1}{t+2}\right) &= 2t(t+2)\left(\frac{4}{t} + \frac{1}{2}\right) \\
2t &= 8(t+2) + t(t+2) \\
2t &= 8t + 16 + t^2 + 2t \\
t^2 + 8t + 16 &= 0 \\
(t+4)^2 &= 0 \\
t &= -4.
\end{aligned}
$$

Check $$\frac{1}{-4+2} \stackrel{?}{=} \frac{4}{-4} + \frac{1}{2}$$
$$\frac{-1}{2} = \frac{-1}{2}.$$

Thus the equation has one solution, -4.

EXAMPLE 12.5 Solve $\frac{1}{x} + \frac{1}{x+1} = \frac{-1}{x(x-1)}$.

Solution A common multiple of x, $x + 1$, and $x(x - 1)$ is $x(x + 1)(x - 1)$.

$$x(x+1)(x-1)\left(\frac{1}{x} + \frac{1}{x+1}\right) = x(x+1)(x-1)\left(\frac{-1}{x(x-1)}\right)$$
$$(x+1)(x-1) + x(x-1) = -(x+1)$$
$$x^2 - 1 + x^2 - x = -x - 1$$
$$2x^2 = 0$$
$$x^2 = 0$$
$$x = 0$$

Check Both sides of the original equation are undefined for $x = 0$; that is, 0 is not in the domain of either side of the equation (Section 4B). Therefore, 0 is extraneous and the original equation has no solution.

Equations with an unknown variable in a denominator sometimes arise from rate problems, as in the following example.

EXAMPLE 12.6 A water tank has two drain pipes. With only the first pipe open the tank will empty in 2 hours. With only the second pipe open the tank will empty in 3 hours. How long will be required for the tank to empty with both pipes open?

Solution Just as the basic formula for motion problems is

$$\text{distance} = \text{rate} \times \text{time},$$

the basic formula for the problem at hand is

$$\text{volume} = \text{rate} \times \text{time}.$$

Write this as

$$V = RT. \tag{12.1}$$

The problem asks for a *time* ("how long"), which suggests that we rewrite Equation (12.1) as

$$T = \frac{V}{R}. \tag{12.2}$$

Here V denotes the volume of the tank and R denotes the rate at which water will leave the tank when *both* pipes are open. Let's call the first pipe A and the second pipe B. Then

$$R = R_A + R_B,$$

where R_A and R_B denote the rates at which water leaves from pipes A and B, respectively. Because pipe A will empty the tank in 2 hours,

$$R_A = \frac{V}{2}.$$

Because pipe B will empty the tank in 3 hours,

$$R_B = \frac{V}{3}.$$

Therefore,

$$R = \frac{V}{2} + \frac{V}{3}. \tag{12.3}$$

If we substitute from Equation (12.3) into Equation (12.2), we obtain

$$\begin{aligned} T &= \frac{V}{\dfrac{V}{2} + \dfrac{V}{3}} \\ &= \frac{V}{\dfrac{3V + 2V}{6}} \\ &= V \cdot \frac{6}{3V + 2V} \\ &= \tfrac{6}{5}. \end{aligned}$$

We conclude that with both pipes open the tank will drain in $\frac{6}{5}$ hours, or 1 hour and 12 minutes. (Notice that the volume of the tank was not given in this problem, but that didn't matter because V canceled at the last step.)

C. Substitutions

The equations on the left in the following list are not quadratic equations. If, however, we replace the variable by a new variable through the indicated substitution in each case, then we do get a quadratic equation, as shown on the right. We can solve the equation on the left if we first solve its related quadratic equation and then use each solution to obtain a corresponding value of the original variable. Examples 12.7 and 12.8 will show the details for the first two examples.

Equation	Substitution	New (quadratic) equation
$y^4 - 3y^2 - 4 = 0$	$x = y^2$	$x^2 - 3x - 4 = 0$
$x + 5\sqrt{x} - 24 = 0$	$t = \sqrt{x}$	$t^2 + 5t - 24 = 0$
$z^{-2} + z^{-1} - 20 = 0$	$x = z^{-1}$	$x^2 + x - 20 = 0$
$\left(y + \frac{1}{y}\right)^2 + 5\left(1 + \frac{1}{y}\right) + 6 = 0$	$x = y + \frac{1}{y}$	$x^2 + 5x + 6 = 0$

EXAMPLE 12.7 Solve $y^4 - 3y^2 - 4 = 0$.

Solution Substitute x for y^2. This gives the quadratic equation

$$x^2 - 3x - 4 = 0.$$

Solve for x:

$$(x + 1)(x - 4) = 0$$
$$x = -1 \quad \text{or} \quad x = 4.$$

Now return to the original variable y by using $x = y^2$:

$$x = 4 \text{ implies } y^2 = 4$$
$$x = -1 \text{ implies } y^2 = -1.$$

There is no real number y such that $y^2 = -1$. But $y^2 = 4$ if $y = 2$ or $y = -2$. You can verify by substitution that both 2 and -2 are solutions of the original equation.

EXAMPLE 12.8 Solve $x + 5\sqrt{x} - 24 = 0$.

Solution Substitute t for $\sqrt{x}$. Then $t^2 = x$, and the original equation becomes

$$t^2 + 5t - 24 = 0$$
$$(t - 3)(t + 8) = 0$$
$$t = 3 \quad \text{or} \quad t = -8.$$

Therefore, since $t = \sqrt{x}$, x will be a solution of the original equation if

$$\sqrt{x} = 3 \quad \text{or} \quad \sqrt{x} = -8.$$

There is no real number x such that $\sqrt{x} = -8$. But $\sqrt{x} = 3$ if $x = 9$. You can verify by substitution that 9 is indeed a solution of the original equation.

EXERCISES FOR SECTION 12

Solve each equation. Be sure to check for extraneous solutions.

1. $\sqrt{3x + 4} = 5$

2. $\sqrt{10 - 4x} = 2$

3. $3 = \sqrt{3x + 11}$

4. $\sqrt{2x + 2} = x - 3$

5. $\sqrt{5x - 4} = x$

6. $\sqrt{4x + 1} = x - 1$

7. $\sqrt{3u + 1} - \sqrt{u + 4} = 1$

8. $\sqrt{3v} - 2\sqrt{v + 4} + 2 = 0$

9. $\sqrt{w} - \sqrt{2w + 7} = -2$

10. $\sqrt{x}\sqrt{10 - 9x} = 1$

11. $\sqrt{y}\sqrt{4y + 3} = 1$

12. $\sqrt{8z} - 2\sqrt{z} = 6$

13. $\sqrt{p - 1} = \dfrac{\sqrt{3}}{\sqrt{p + 1}}$

14. $\sqrt{q} - \dfrac{2}{\sqrt{q}} = 1$

15. $\sqrt{r^2+1} - r = 2$

16. $\dfrac{x}{x-1} = \dfrac{8}{x+2}$

17. $\dfrac{4}{x} = \dfrac{x}{x-1}$

18. $x + \left(\dfrac{0.01}{x+0.2}\right) = 0$

19. $\dfrac{2}{k^2-1} - \dfrac{k}{k-1} = \dfrac{1}{k+1}$

20. $\dfrac{m}{m+1} + \dfrac{2}{m} = \dfrac{5}{m(m+1)}$

21. $\dfrac{1}{n^2-n} + \dfrac{1}{n} = \dfrac{n}{n-1}$

Solve 22 through 30 by first making a substitution to obtain a quadratic equation.

22. $x^4 - 9 = 0$

23. $y + 10\sqrt{y} + 25 = 0$

24. $z^{-2} - 8z^{-1} + 16 = 0$

25. $\left(1+\dfrac{1}{x}\right)^2 + 5\left(1+\dfrac{1}{x}\right) + 4 = 0$

26. $0.0016 - y^4 = 0$

27. $(z+\frac{1}{5})^2 - 4(z+\frac{1}{5}) - 5 = 0$

28. $x^{-4} - 4x^{-2} = 0$

29. $(y^2-1)^2 + (y^2-1) - 6 = 0$

30. $z - 7\sqrt{z} + 10 = 0$

31. Solve $r = \sqrt{A/\pi}$ for A.

32. Solve $t = \sqrt{2s/g}$ for s.

33. Solve $T = 2\pi\sqrt{l/g}$ for g.

34. Solve $d = \sqrt{x^2+y^2}$ for y if $y > 0$.

35. Solve $S = \pi r\sqrt{r^2+h^2}$ for h if $h > 0$.

36. Solve $y = \sqrt{(1-x)/2}$ for x.

37. If one minus the square root of a positive number is equal to twice the number, what is the number?

38. If the sum of a positive number and its reciprocal is 5, what is the number?

39. There is exactly one positive number L such that $1 : L = L : (L+1)$. What is it?

40. A water tank has two drain pipes. With only the first pipe open the tank will empty in 3 hours. With only the second pipe open the tank will empty in 5 hours. How long will be required for the tank to empty with both pipes open?

41. A water tank has two drain pipes. With both pipes open the tank will empty in 3 hours. With only the first pipe open the tank will empty in 5 hours. How long will it take for the tank to empty with only the second pipe open?

42. A water tank has two drain pipes. The tank will empty in one hour if both pipes are open. The tank will empty four times as fast with only the first pipe open as it will with only the second pipe open. How long will it take for the tank to empty with only the first pipe open?

SECTION

13 Applications

In this section we'll add to the list of applications of quadratic equations given in Sections 10, 11, and 12. We'll see, in particular, how to answer each of the following questions:

How can we compute the height of a pyramid?

How far is the horizon from an astronaut 100 miles above the earth's surface?

How far can one see from the top of the Sears Tower?

If a baseball is dropped from the top of the Washington Monument, when will it reach the ground?

How can we estimate the depth of a well with a stopwatch and a rock?

A. Geometric Problems

Each example in this subsection will rely on the following theorem.

Pythagorean Theorem
If a and b denote the lengths of the legs of a right triangle and c denotes the length of the hypotenuse (all expressed in the same units, such as meters), then

$$a^2 + b^2 = c^2.$$

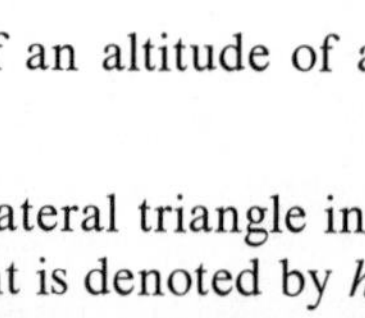

Figure 13.1

EXAMPLE 13.1 What is the length of an altitude of an equilateral triangle in terms of the length of a side?

Solution Figure 13.2 shows an equilateral triangle in which the lengths of the equal sides are denoted by s and the height is denoted by h. The goal is to express h in terms of s. By the Pythagorean Theorem applied to triangle ADC,

$$h^2 + \left(\frac{s}{2}\right)^2 = s^2.$$

Then

$$h^2 = s^2 - \frac{s^2}{4}$$

$$= \frac{4s^2 - s^2}{4}$$

$$= \frac{3s^2}{4}.$$

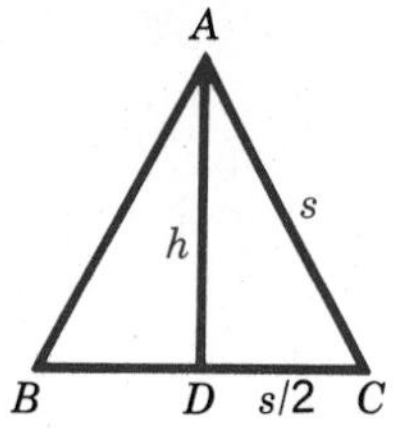

Figure 13.2

Therefore,

$$h = \sqrt{\frac{3s^2}{4}} = \frac{s\sqrt{3}}{2}.$$

For an approximate answer we can use $\sqrt{3} \approx 1.732$ (from either Table I or a calculator). If, for example, each side is 5 meters, then the altitude is

$$h = \frac{5\sqrt{3}}{2} \approx 4.33 \text{ m.}$$ Ⓒ

EXAMPLE 13.2 The Great Pyramid of Khufu, near Cairo, was built with a square base 230 meters on each side. The distance from the top to the midpoint of a side of the base was 186 meters. How high was the pyramid?

Solution The relevant parts of the pyramid are labeled in Figure 13.3. From what is given,

$$\overline{BC} = 115 \text{ m} \quad (\text{because } \overline{BD} = 230 \text{ m}),$$
$$\overline{AB} = 186 \text{ m}, \quad \text{and}$$
$$\overline{AC} = h = \text{the unknown height.}$$

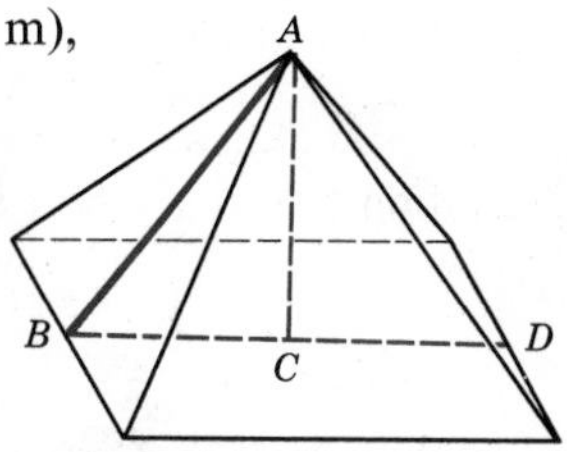

Figure 13.3

By the Pythagorean Theorem applied to triangle ABC,

$$\overline{AC}^2 + \overline{BC}^2 = \overline{AB}^2$$
$$h^2 + 115^2 = 186^2$$
$$h^2 = 186^2 - 115^2 = 34{,}596 - 13{,}225 = 21{,}371$$
$$h = \sqrt{21371} \approx 146 \text{ m.}$$ Ⓒ

That is, the original height was approximately 146 meters. (The present height is approximately 137 meters.)

EXAMPLE 13.3 An astronaut is in orbit 100 miles above the earth's surface. How far away is the horizon?

Solution For efficiency we answer the following, more general, problem; then we can apply the answer to other examples as well as this one.

What is the distance to the horizon from a point h miles above the earth's surface? (13.1)

Figure 13.4 shows a cross section of the earth, thought of as a sphere with O at the center and with radius r. The problem asks for the length of PA, which has been labeled x. Angle PAO is a right angle because PA is tangent to the circle. Therefore, the Pythagorean Theorem can be applied to triangle PAO (with hypotenuse PO) to give

$$x^2 + r^2 = (r + h)^2.$$

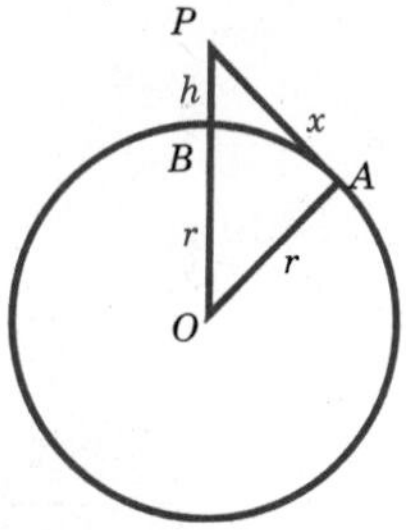

Figure 13.4

Solve this for x:

$$x^2 + r^2 = r^2 + 2rh + h^2$$
$$x^2 = 2rh + h^2$$
$$x = \sqrt{2rh + h^2}.$$

To summarize: The distance to the horizon from a point h miles above the earth's surface is

$$x = \sqrt{2rh + h^2} \text{ mi.} \tag{13.2}$$

Since distance is measured in miles, $r = 3960$, the radius of the earth in miles. In the original problem the astronaut is 100 miles above the earth's surface, so $h = 100$. With these values for r and h, Equation (13.2) gives

$$x = \sqrt{2 \cdot 3960 \cdot 100 + 100^2} \approx 896 \text{ mi.}$$ C

That is, the straight-line distance from the astronaut to the horizon is approximately 896 miles.

Continuation. Notice that if we use 3960 miles for r and 1 mile for h, then Equation (13.2) gives

$$x = \sqrt{2 \cdot 3960 \cdot 1 + 1^2} = \sqrt{7920 + 1} \approx \sqrt{7920}. \tag{13.3}$$

Dropping 1 from $7920 + 1$ in Equation (13.3) is equivalent to dropping h^2 from Equation (13.2). The point here is that if h is small compared to r, then in place of Equation (13.2) we can use the simplified approximation

$$x \approx \sqrt{2rh}. \tag{13.4}$$

If r is in miles and h is in feet, and we want x in miles, then in place of h we must use $h/5280$ (to convert the height to miles). Then Equation (13.4) becomes

$$x \approx \sqrt{2 \cdot 3960 \cdot h/5280}$$
$$x \approx \sqrt{3h/2} \text{ mi.}$$

To summarize: The distance to the horizon from a point h feet above the earth's surface is approximately

$$\sqrt{3h/2} \text{ mi.} \tag{13.5}$$

EXAMPLE 13.4 The tallest office building in the world is the Sears Tower in Chicago, which is 1454 feet tall. How far can one see from the top?

Solution Apply (13.5) with $h = 1454$:

$$h \approx \sqrt{3 \cdot 1454/2} \approx 47 \text{ mi.}$$

C

This is straight-line distance, but it will vary little from distance measured along the earth's surface in this case. The solution assumes ideal visibility, of course. Expression (13.5) also ignores such things as valleys and mountains.

B. Falling Objects

Assume that an object is dropped near the earth's surface, and let s denote the distance it falls in the first t seconds. A law of physics tells us that if we ignore all forces (such as air resistance) except gravity, then

$$s = \tfrac{1}{2}gt^2. \tag{13.6}$$

Here g denotes a constant whose value depends on the units used to measure s. If s is measured in feet, then $g \approx 32$; if s is measured in meters, then $g \approx 9.8$. Therefore, approximately,

$$s = \begin{cases} 16t^2 \text{ feet} \\ 4.9t^2 \text{ meters.} \end{cases} \tag{13.7}$$

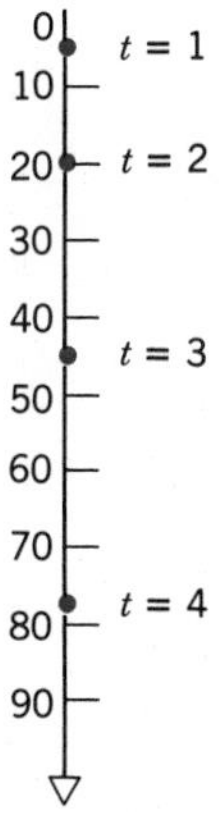

Figure 13.5

EXAMPLE 13.5 Figure 13.5 shows the distance, in meters, that an object will have dropped at the end of 1, 2, 3, and 4 seconds. At the end of 3 seconds, for example, Equation (13.7) gives

$$s = 4.9(3)^2 \approx 44 \text{ m.}$$

EXAMPLE 13.6 A baseball is dropped from the top of the Washington Monument, which is 555 feet tall. Estimate when it will reach the ground.

Solution We must determine the value of t that will yield $s = 555$ feet in Equation (13.7):

$$555 = 16t^2$$
$$t = \sqrt{555/16} = \sqrt{555}/4 \approx 5.9 \text{ sec.}$$

If an object is thrown down near the earth's surface rather than just dropped, then the distance the object travels in the first t seconds is approximately

$$s = \tfrac{1}{2}gt^2 + v_0 t, \tag{13.8}$$

where v_0 denotes the *initial velocity* of the object, that is, the velocity when $t = 0$. If we use feet to measure s, feet per second to measure v_0, and $g = 32$, then

$$s = 16t^2 + v_0t. \tag{13.9}$$

EXAMPLE 13.7 A rock is thrown directly down from the top of a building 336 feet tall with an initial velocity of 64 ft/sec. When will it reach the ground?

Solution From Equation (13.9),

$$s = 16t^2 + 64t. \tag{13.10}$$

If the building is 336 feet tall, then the object will reach the ground when $s = 336$. Thus we use Equation (13.10) with $s = 336$ and solve for t:

$$\begin{aligned} 336 &= 16t^2 + 64t \\ 16t^2 + 64t - 336 &= 0 \\ t^2 + 4t - 21 &= 0 \\ (t + 7)(t - 3) &= 0 \\ t &= -7, 3. \end{aligned}$$

The circumstances dictate a positive value for t, so the required time is $t = 3$ seconds.

EXAMPLE 13.8 A rock is dropped into a water well and the sound of the splash is heard 5 seconds later. How deep is the well?

Solution We can solve this with Equation (13.7) and the fact that the speed of sound in air is approximately 1100 feet per second.

For part of the 5 seconds the rock will be falling, and for the other part the sound from the splash will be traveling to the top. The law governing the falling rock is Equation (13.7),

$$s = 16t^2.$$

(We use s in feet since we're using ft/sec rather than m/sec to measure the speed of sound.) The law governing the sound is

$$s = 1100t \tag{13.11}$$

(distance = rate × time).

If the time for the rock to fall is t_1, and the time for the sound to reach the top is t_2, then

$$t_1 + t_2 = 5. \tag{13.12}$$

If s_1 denotes the distance the rock must travel, then

$$s_1 = 16t_1^2. \tag{13.13}$$

The sound must travel the same distance, so, in addition,

$$s_1 = 1100t_2. \tag{13.14}$$

If we equate the expressions for s_1 from Equations (13.13) and (13.14), and also use $t_2 = 5 - t_1$ from Equation (13.12), the result is

$$16t_1^2 = 1100(5 - t_1).$$

We solve this for t_1, choosing the positive solution:

$$16t_1^2 + 1100t_1 - 5(1100) = 0$$
$$4t_1^2 + 275t_1 - 1375 = 0$$
$$t_1 = \frac{-275 + \sqrt{(275)^2 - 4(4)(-1375)}}{2(4)}$$
$$t_1 \approx 4.7 \text{ sec.}$$

C

If we substitute this value in Equation (13.13), we get the approximate depth of the well:

$$s_1 \approx 16(4.7)^2$$
$$s_1 \approx 350 \text{ ft.}$$

C

C. Rate Problems

The ideas in the following example are similar to those in Example 12.6, but this example involves distance = rate × time ($D = RT$) rather than volume = rate × time ($V = RT$).

EXAMPLE 13.9 I can leave a subway by either an escalator or a parallel stairway. I know from previous trips that if I stand still on the escalator the trip to the top will take 50 seconds, but if I walk at my standard pace on the escalator the trip will take 20 seconds. How long would the trip take if I chose the stairway and walked at my standard pace?

Solution The basic formula here is $D = RT$. Also, if

$$R_W = \text{walking rate ("my standard pace")},$$
$$R_E = \text{rate of the escalator, and}$$
$$R_C = \text{combined rate,}$$

then

$$R_C = R_E + R_W.$$

Let L denote the length of the escalator (and stairway). Then $D = RT$ or $R = D/T$ gives

$$R_E = \frac{L}{50} \quad \text{(escalator requires 50 seconds)}$$

and

$$R_C = \frac{L}{20} \quad \text{(combined method requires 20 seconds).}$$

Therefore,

$$\begin{aligned} R_W &= R_C - R_E \\ &= \frac{L}{20} - \frac{L}{50}. \end{aligned} \tag{13.15}$$

If T_W denotes the time required to walk to the top using the stairway, then

$$\begin{aligned} T_W &= \frac{L}{R_W} \\ &= \frac{L}{\frac{L}{20} - \frac{L}{50}} \\ &= \frac{L}{\frac{5L - 2L}{100}} \\ &= \frac{100L}{3L} \\ &= 33\tfrac{1}{3} \text{ sec.} \end{aligned}$$

Notice that we solved this problem without knowing the length of the escalator and the stairway.

EXERCISES FOR SECTION 13

1. There are 4840 square yards in one acre. How long is each side of a square one-fourth acre lot?
2. There are 10,000 square meters in one hectare. How long is a diagonal of a square hectare lot?
3. What is the length of a side of an equilateral triangle in terms of the length of an altitude?
4. Consider an isosceles triangle in which the equal sides each have length s and the third side has length $s/2$. How long is the altitude from the third side to the opposite vertex?
5. Let s denote the length of each edge of a square and let d denote the length of a diagonal. Write an equation that expresses d in terms of s.
6. The diagonal of a square is 10 inches. How long is each side of the square?
7. Khafre's Pyramid was built with a square base 215 meters on each side. The distance from the midpoint of a side of a base to the top was 180 meters. How high was the pyramid?

8. If the height of a pyramid is 100 meters and the base is a square, 150 meters on each side, what is the distance from the top to the midpoint of a side of the base?

9. Compute the distance from the top of Khufu's Pyramid to a corner of the base (Example 13.2).

10. Solve Equation (13.2) for h. Interpret your solution by stating a question similar to Statement (13.1) for which your solution provides the answer (with $r = 3960$).

11. A camera is 30 miles above the moon's surface. How far away is the (moon's) horizon? (Assume a smooth surface and use 1080 miles for the radius of the moon.)

12. An astronaut is 150 miles above Kansas City, which is 1300 miles from the nearest point in the Pacific Ocean. Can the astronaut see the Pacific?

13. The world's tallest self-supporting tower is the CN Tower in Toronto, which is 1820 feet tall. You're on top, it's a clear day, and a ship is 65 miles away on Lake Erie. Can you see it?

14. Solve Equation (13.4) for h. Assume $r = 3960$ and then interpret your answer with a statement similar to that accompanying (13.5).

15. You're on the 20th story of the World Trade Center in New York City, which has 110 stories. You want to double the distance to the horizon. To which story should you move? Why? [Assume the stories to be equal in height; you do not need to know what that height is. Think about (13.5).]

16. Consider Equation (13.6), with t measured in seconds. What are the units for g: (a) if s is measured in feet? (b) if s is measured in meters? (Units were discussed in Section 8.)

17. Consider Equation (13.7) with s measured in feet. Solve for t.

18. Write the version of Equation (13.7) with s measured in centimeters.

19. If an object is dropped near the earth's surface, how many feet will it fall in the first 5 seconds? How long will it take to fall 225 feet?

20. If an object is dropped near the earth's surface, how many meters will it drop in the first 3 seconds? How long will it take to fall 19.6 meters?

21. A baseball is dropped from the top of the Gateway Arch in St. Louis, which is 630 feet tall. Estimate when it will reach the ground.

22. An object is thrown down near the earth's surface with an initial velocity of 24 feet per second.
(a) Write the equation giving the position after t seconds. [Use Equation (13.8).]
(b) Where is the object 1 second after it is thrown?
(c) How long will it take for the object to travel the first 16 feet?

23. An object is thrown down from the top of a building 400 feet tall, and 4 seconds later it hits the ground. What was the initial velocity? [Use Equation (13.9).]

24. A rock is dropped from the top of a building, and 4 seconds later the sound of the rock hitting the pavement is heard at the top of the building. How tall is the building? (See Example 13.8.)

25. I can leave a subway by either an escalator or a parallel stairway. I know from previous trips that if I stand still on the escalator the trip to the top will take 40 seconds, while if I walk at my standard pace on the escalator the trip will take 15 seconds. How long would the trip take if I chose the stairway and walked at my standard pace?

26. An escalator has known length 100 feet. One person travels from the bottom to the top while walking 5 feet per second on the escalator in the same time another person on the escalator travels 60 feet toward the top while standing still. What is the speed of the escalator? (Suggestion: The time for each person can be expressed in terms of distance and rate. Equate these times.)

27. If I stand still on an escalator I can move from one floor of a building to another in 20 seconds. If I walk at my standard pace on a parallel set of stairs the trip takes 15 seconds. How long would the trip take if I walked on the escalator at my standard pace?

REVIEW EXERCISES FOR CHAPTER III

Solve by factoring.

1. $x^2 + 3x - 10 = 0$ **2.** $2y^2 + 5y = -3$

Solve by completing the square.

3. $2t^2 = 1 - 3t$ **4.** $3u^2 - u = 1$

Solve by any method.

5. $0.09w^2 - 1.21 = 0$ **6.** $x^2 + px + 1 = 0$
7. $7y^2 - 6y + 1 = 0$ **8.** $25z^2 = 49$

Use the discriminant to determine whether each equation has no real solution, one real solution, or two real solutions.

9. $r^2 = 4 + r$ **10.** $0.1s^2 - 0.25s + 0.1 = 0$
11. $\frac{1}{2}t = \frac{1}{4}t^2 + \frac{5}{6}$ **12.** $u^2 = 4u - 4$

Solve each equation. Be sure to check for extraneous solutions.

13. $\sqrt{4x + 6} = 3 + 2x$ **14.** $\sqrt{x + 7} + x + 1 = 0$
15. $\sqrt{4x + 7} - \sqrt{2 + 4x} = 1$ **16.** $\sqrt{2x - 3} - \sqrt{x - 2} = 1$
17. $\dfrac{x}{3} + \dfrac{5 - x}{x - 1} = 0$ **18.** $\dfrac{x - 1}{x - 2} = \dfrac{2x - 3}{x - 2}$
19. Solve $f = (1/2L)\sqrt{T/d}$ for T. **20.** Solve $r + 1 = (1 + k)^2$ for k.

21. If one-half of a positive number is three more than the reciprocal of the number, what is the number?

22. If twice the reciprocal of the square root of a real number equals one-fifth of the square root of the number, what is the number?

23. The top of an equilateral trapezoid has length x, the base has length $2x$, and the height is h (Figure 13.6). Express s, the length of a slant height, in terms of x and h.

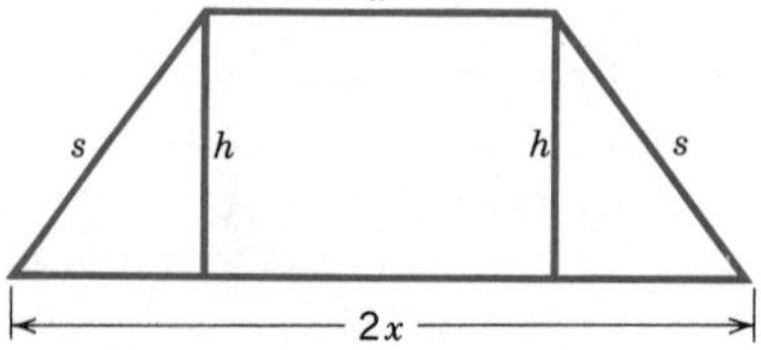

Figure 13.6

24. The radius of the circle in Figure 13.7 is r, and the length of AB is s. Show that the length of CD is $r - \sqrt{r^2 - (s/2)^2}$.

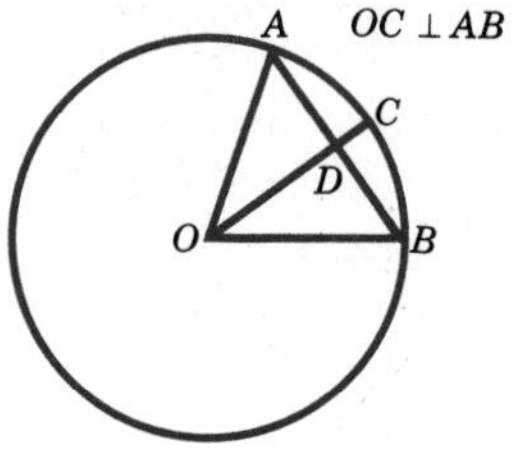

Figure 13.7

25. If a basketball is dropped from the rim of a goal, which is 10 feet above the floor, when will it reach the floor?

26. A rock is thrown directly down from the top of a building 73.5 meters tall with an initial velocity of 9.8 m/sec. When will it reach the ground?

27. A water tank has two drain pipes. With only the first pipe open the tank will empty in one hour. With only the second pipe open the tank will empty in 30 minutes. How long will it take the tank to empty with both pipes open?

28. If I stand still on an escalator I can move from one floor of a building to another in 30 seconds. If I walk at my standard pace on a parallel set of stairs the trip will take 25 seconds. How long would the trip take if I walked on the escalator at my standard pace?

CHAPTER IV
MORE ABOUT ALGEBRAIC EXPRESSIONS

This chapter, like Chapter I, concerns basic algebra. Here we'll look more thoroughly at factoring, exponents, and the simplification of algebraic expressions.

SECTION

14 Factoring

A. Factoring Quadratics: Review

For convenience, let us recall three important formulas for factoring **quadratics** (second-degree polynomials).

$$u^2 - v^2 = (u + v)(u - v) \tag{14.1}$$
$$u^2 + 2uv + v^2 = (u + v)^2 \tag{14.2}$$
$$u^2 - 2uv + v^2 = (u - v)^2 \tag{14.3}$$

These formulas were illustrated in Section 4. The following additional examples will emphasize that squares can appear in disguise. For instance,

$$7 = (\sqrt{7})^2, \quad x^4 = (x^2)^2, \quad \text{and} \quad a^{2n} = (a^n)^2.$$

EXAMPLE 14.1 Factor.

(a) $4x^2 - 7 = (2x)^2 - (\sqrt{7})^2 = (2x + \sqrt{7})(2x - \sqrt{7})$
(b) $x^4 + 2x^2y^2 + y^4 = (x^2)^2 + 2x^2y^2 + (y^2)^2 = (x^2 + y^2)^2$
(c) $a^{2n} - 2a^n + 1 = (a^n)^2 - 2a^n + 1^2 = (a^n - 1)^2$

Notice that in Example 14.1(a) we used the irrational coefficient $\sqrt{7}$ in two of the factors. That brings out the point that the coefficients in our polynomials can be any real numbers. If only integral coefficients were allowed, then $4x^2 - 7$ could not be factored. Although any real coefficients will be allowed, most of the polynomials in our examples and exercises can be factored with integral coefficients.

Factoring like the following was also done in Section 4. (For a review, see Equation (4.6) and the examples that follow it.)

EXAMPLE 14.2 Factor.

(a) $m^2 + 8m + 15 = (m + 3)(m + 5)$
(b) $2t^2 + 7t + 6 = (2t + 3)(t + 2)$
(c) $12z^2 - 13z - 4 = (3z - 4)(4z + 1)$

B. Cubes

Here are the first two of four special products involving cubes (third powers). The exercises ask you to verify that these are correct. Make a special note of the signs on both sides of these equations.

Sum of two cubes

$$u^3 + v^3 = (u + v)(u^2 - uv + v^2) \quad (14.4)$$

Difference of two cubes

$$u^3 - v^3 = (u - v)(u^2 + uv + v^2) \quad (14.5)$$

EXAMPLE 14.3 Factor.

(a) $u^3 + 1$ **(b)** $a^3 + 8b^3$ **(c)** $27x^6 - 8y^6$ **(d)** $x^{3n} - y^{3n}$

Solution

(a) Use Equation (14.4) with 1 in place of v.

$$u^3 + 1 = u^3 + 1^3 = (u + 1)(u^2 - u + 1)$$

(b) Use Equation (14.4) with a in place of u, and $2b$ in place of v.

$$\begin{aligned} a^3 + 8b^3 &= a^3 + (2b)^3 \\ &= (a + 2b)[a^2 - a(2b) + (2b)^2] \\ &= (a + 2b)(a^2 - 2ab + 4b^2) \end{aligned}$$

(c) Use Equation (14.5) with $3x^2$ in place of u, and $2y^2$ in place of v.

$$\begin{aligned} 27x^6 - 8y^6 &= (3x^2)^3 - (2y^2)^3 \\ &= (3x^2 - 2y^2)[(3x^2)^2 + (3x^2)(2y^2) + (2y^2)^2] \\ &= (3x^2 - 2y^2)(9x^4 + 6x^2y^2 + 4y^4) \end{aligned}$$

(d) Use Equation (14.5) with x^n in place of u, and y^n in place of v.

$$\begin{aligned} x^{3n} - y^{3n} &= (x^n)^3 - (y^n)^3 \\ &= (x^n - y^n)[(x^n)^2 + x^ny^n + (y^n)^2] \\ &= (x^n - y^n)(x^{2n} + x^ny^n + y^{2n}) \end{aligned}$$

Equations (14.4) and (14.5) concern sums and differences of cubes; the next equations concern cubes of sums and differences. Contrast $u^3 + v^3$ with $(u + v)^3$, for example. Equations (14.4)–(14.7) should all be memorized.

Cube of a binomial sum

$$(u + v)^3 = u^3 + 3u^2v + 3uv^2 + v^3 \quad (14.6)$$

Cube of a binomial difference

$$(u - v)^3 = u^3 - 3u^2v + 3uv^2 - v^3 \quad (14.7)$$

Although the preceding two identities are included in this section on factoring, we are more likely to need them for expanding each left side than for factoring each right side. The next two examples illustrate both uses.

EXAMPLE 14.4 Expand and simplify.

(a) $(a + 2b)^3 = a^3 + 3a^2(2b) + 3a(2b)^2 + (2b)^3$
$= a^3 + 6a^2b + 12ab^2 + 8b^3$

(b) $(u - \frac{1}{2}t)^3 = u^3 - 3u^2(\frac{1}{2}t) + 3u(\frac{1}{2}t)^2 - (\frac{1}{2}t)^3$
$= u^3 - \frac{3}{2}u^2t + \frac{3}{4}ut^2 - \frac{1}{8}t^3$

EXAMPLE 14.5 Factor.

(a) $m^6 + 3m^4n^2 + 3m^2n^4 + n^6$
$= (m^2)^3 + 3(m^2)^2(n^2) + 3(m^2)(n^2)^2 + (n^2)^3$
$= (m^2 + n^2)^3$

(b) $8 - 12z + 6z^2 - z^3 = 2^3 - 3 \cdot 2^2z + 3 \cdot 2z^2 - z^3$
$= (2 - z)^3$

C. Common Factors. Regrouping

The first step in factoring should be to remove all *common factors*. These include the highest power of each variable that appears in every term. If the coefficients are all integers, also remove the greatest integral common factor.

EXAMPLE 14.6 Factor.

(a) $10x^3 - 15x^2 = 5x^2(2x - 3)$
(b) $2at^2 + 4at + 2a = 2a(t^2 + 2t + 1) = 2a(t + 1)^2$
(c) $q^4 + 8q = q(q^3 + 8) = q(q + 2)(q^2 - 2q + 4)$

Some expressions can be factored more easily after the terms have been regrouped.

EXAMPLE 14.7 Factor.

(a) $2x + 2y + ax + ay = 2(x + y) + a(x + y)$ [$(x + y)$ is a common factor]
$= (2 + a)(x + y)$

(b) $a^2 - 2ab + b^2 - 1 = (a^2 - 2ab + b^2) - 1$
$= (a - b)^2 - 1$ [difference of two squares]
$= [(a - b) + 1][(a - b) - 1]$
$= (a - b + 1)(a - b - 1)$

D. Complete Factorization

Recall that each positive integer can be written as a product of prime powers (Section 2). For example, $60 = 2^2 \cdot 3 \cdot 5$. We say that 60 has been *factored completely* when we write $60 = 2^2 \cdot 3 \cdot 5$, because primes cannot be factored using positive integral factors (except trivially, like $5 = 1 \cdot 5$). We now look at the corresponding idea for polynomials.

The analogue of a *prime number* is an *irreducible polynomial:* A polynomial is **reducible** if it can be written as a product of two other polynomials that are both of positive degree. A polynomial that is not reducible is said to be **irreducible.**

EXAMPLE 14.8

(a) $x^2 + x$ is reducible because $x^2 + x = x(x + 1)$ and both x and $x + 1$ have degree one.

(b) $y^3 + .001$ is reducible because

$$y^3 + 0.001 = (y + 0.1)(y^2 - 0.1y + 0.01).$$

(c) $2x^2 + 2$ is irreducible. Although we can write

$$x^2 + 2 = 2(x^2 + 1),$$

we cannot write $2x^2 + 2$ as a product of two polynomials of degree one, which would be the case if $2x^2 + 2$ were reducible. (In Chapter VII we'll see that a quadratic polynomial is irreducible iff its discriminant is negative.)

Any first-degree polynomial is irreducible. Example 14.8 shows that some second-degree polynomials are irreducible and some are not. Each polynomial of degree greater than two is reducible, but it may not easily yield factors.

In analogy with the case of integers, we say that a polynomial has been **factored completely** when it has been written as a product of irreducible polynomials. Although factoring can be tricky, many polynomials can be handled by applying the following suggestions. (Practice helps, too.)

- Remove all common factors.
- Look for the standard forms in Equations (14.1), (14.2), and (14.3).
- If the polynomial is quadratic and (14.1), (14.2), and (14.3) do not apply, try to factor it as a general quadratic (Example 14.2).
- Look for the standard forms in Equations (14.4) and (14.5), and then in (14.6) and (14.7).
- Attempt to regroup the terms so that one of the previous suggestions applies.

EXAMPLE 14.9 Factor completely.

(a) $x^3 - 6x^2 + 9x = x(x^2 - 6x + 9) = x(x - 3)^2$

(b) $15 - 3t^2 = 3(5 - t^2) = 3(\sqrt{5} + t)(\sqrt{5} - t)$

(c) $x^4 - 1 = (x^2)^2 - 1 = (x^2 + 1)(x^2 - 1)$
$= (x^2 + 1)(x + 1)(x - 1)$

(d) $y^6 + y^5 + \frac{1}{4}y^4 = y^4(y^2 + y + \frac{1}{4}) = y^4(y + \frac{1}{2})^2$

(e) $ac^2 - bd^2 - ad^2 + bc^2 = ac^2 - ad^2 + bc^2 - bd^2$
$= a(c^2 - d^2) + b(c^2 - d^2)$
$= (a + b)(c^2 - d^2)$
$= (a + b)(c + d)(c - d)$

EXERCISES FOR SECTION 14

Factor completely.

1. $x^2 + 12x + 36$
2. $y^2 - 6y + 9$
3. $z^2 - 1$
4. $y^3 - y$
5. $z^4 + 2z^3 + z^2$
6. $x^3 - 4x^2 + 4x$
7. $4z^2 - 4z + 1$
8. $9x^2 - 25$
9. $9y^2 + 6y + 1$
10. $u^3 - 27$
11. $v^3 - 1$
12. $w^3 - 8$
13. $v^4 - 4v^3 + 3v^2$
14. $2w - 3w^2 + w^3$
15. $u^5 + 7u^4 + 6u^3$
16. $w^3 + u^3$
17. $u^3 - 1000v^3$
18. $8v^3 + w^3$
19. $r^4 - 125r$
20. $27s^5 + s^2$
21. $64t - t^4$
22. $s^4 - t^4$
23. $t^4 - 2t^2 + 1$
24. $1 - r^4$
25. $t^3 - 0.001$
26. $0.027 + r^3$
27. $s^3 + 0.008$
28. $a^{2n} + 2a^n + 1$
29. $b^{2n} - 1$
30. $4 + 4c^n + c^{2n}$
31. $b^3 + 3b^2 + 3b + 1$
32. $c^3 - 3c^2 + 3c - 1$
33. $a^3 - 6a^2 + 12a - 8$
34. $c^3 - 3c^2a + 3ca^2 - a^3$
35. $a^3 + 9a^2b + 27ab^2 + 27b^3$
36. $b^3 + 0.3b^2 + 0.03b + 0.001$
37. $x^3y - 10x^2y^2 + 25xy^3$
38. $4u^3v - 9uv^3$
39. $r^3s - 6r^2s^2 + 9rs^3$
40. $m^3 + \frac{3}{2}m^2 + \frac{3}{4}m + \frac{1}{8}$
41. $\frac{1}{27} - \frac{1}{3}n + n^2 - n^3$
42. $p^3 + \frac{3}{2}p^2q + \frac{3}{4}pq^2 + \frac{1}{8}q^3$
43. $ax + ay - bx - by$
44. $ax - ay + bx - by$
45. $ax + ay + bx + by$
46. $ax^3 - bx^3 + ay^3 - by^3$
47. $cu^2 + du^2 - cv^2 - dv^2$
48. $ap^3 + aq^3 - bp^3 - bq^3$

Expand and simplify.

49. $(t + 1)^3$
50. $(u - 2)^3$
51. $(v + 5)^3$
52. $(y - 0.5)^3$
53. $(x + \frac{1}{2})^3$
54. $(a^2 - b^2)^3$
55. $(\frac{1}{2}x^2 + 1)^3$
56. $(z^2 + 0.1)^3$
57. $(y - \frac{2}{3})^3$
58. Verify Equation (14.4) by showing that when the two factors on the right side are multiplied, the result simplifies to the left side.
59. Same as Exercise 58, with Equation (14.5) in place of Equation (14.4).
60. Verify Equation (14.6) by multiplying the right side of $(u + v)^2 = u^2 + 2uv + v^2$ by $u + v$ and showing that the result simplifies to the right side of Equation (14.6).
61. Verify Equation (14.7) by multiplying the right side of $(u - v)^2 = u^2 - 2uv + v^2$ by $u - v$ and showing that the result simplifies to the right side of Equation (14.7).
62. Verify that $(u + v)^4 = u^4 + 4u^3v + 6u^2v^2 + 4uv^3 + v^4$.
63. Verify that $(u - v)^4 = u^4 - 4u^3v + 6u^2v^2 - 4uv^3 + v^4$.

SECTION

15 Rational Expressions

A. Introduction

Recall that a number is *rational* if it can be written as a fraction in which both the numerator and nonzero denominator are integers. Similarly, an expression with one or more variables is called a **rational expression** if it can be written as a fraction in which both the numerator and nonzero denominator are polynomials.

EXAMPLE 15.1 Each of the following is a rational expression.

(a) $\dfrac{x^3 - 3x + 2}{x - 2}$

(b) $\dfrac{x^2y^2}{x^2 + y^2}$

(c) $\dfrac{x}{2} + \dfrac{2}{x} = \dfrac{x^2 + 4}{2x}$

Also, any polynomial is a rational expression. For example, $x^2 + 1 = (x^2 + 1)/1$ is a rational expression, just as $3 = \frac{3}{1}$ is a rational number. The expression $\sqrt{x + 1}$ is not rational. In later sections we'll meet other examples of expressions that are not rational.

This section concerns how to simplify rational expressions. In general, any expression that results from combining polynomials through addition, subtraction, multiplication, or division is a rational expression; by simplification, if necessary, it can be written as a single fraction in which both the numerator and denominator are polynomials.

Since the variables in rational expressions represent real numbers, we can manipulate rational expressions just like ordinary fractions. Remember, however, that a fraction is undefined when its denominator is zero. Therefore, we must always exclude values that make a denominator zero. When we work with $1/(x^2 - 4)$, for example, we must exclude both $x = 2$ and $x = -2$. Such restrictions will always be assumed, whether explicitly stated or not.

Following is a summary of the basic laws about fractions, which were stated in Section 2. The first law concerns equality.

$$\frac{a}{b} = \frac{c}{d} \text{ iff } ad = bc. \tag{15.1}$$

This leads to the following *law of cancellation*.

$$\text{If } d \neq 0, \text{ then } \frac{ad}{bd} = \frac{a}{b}. \tag{15.2}$$

Property (15.1) implies (15.2) because $adb = bda$.

EXAMPLE 15.2 Simplify.

(a) $\dfrac{3x^2y^2}{9y^2z} = \dfrac{x^2}{3z}$ (cancel $3y^2$)

(b) $\dfrac{4 - 2x}{x - 2} = \dfrac{-2(x - 2)}{x - 2} = -2$ (cancel $x - 2$)

(c) $\dfrac{x^2 - 1}{x - 1} = \dfrac{(x + 1)(x - 1)}{x - 1} = x + 1$ (cancel $x - 1$)

Notice that the last equality is valid only for $x \neq 1$, because if $x = 1$, then $(x^2 - 1)/(x - 1)$ is undefined. A similar remark applies to the other examples.

Here are the rules for multiplication and division.

$$\frac{a}{b} \cdot \frac{c}{d} = \frac{ac}{bd} \tag{15.3}$$

$$\frac{\dfrac{a}{b}}{\dfrac{c}{d}} = \frac{ad}{bc} \tag{15.4}$$

EXAMPLE 15.3 Simplify.

(a) $\dfrac{3x}{y} \cdot \dfrac{x}{2z} = \dfrac{3x^2}{2yz}$

(b) $\dfrac{p^2q}{r^2} \cdot \dfrac{r^3}{pq^2} = \dfrac{p^2qr^3}{r^2pq^2} = \dfrac{pr}{q}$

(c) $\dfrac{x^2}{x + 1} \cdot \dfrac{x^2 + 2x + 1}{x} = \dfrac{x^2(x + 1)^2}{(x + 1)x} = x(x + 1)$

EXAMPLE 15.4 Simplify.

(a) $\dfrac{6x/y^2}{10x^2/y} = \dfrac{6x}{y^2} \cdot \dfrac{y}{10x^2} = \dfrac{3}{5xy}$

(b) $\dfrac{(ab^2)/c}{(a^2b)/c^2} = \dfrac{ab^2}{c} \cdot \dfrac{c^2}{a^2b} = \dfrac{ab^2c^2}{a^2bc} = \dfrac{bc}{a}$

(c) $\dfrac{t/(t + 2)}{t^2/(t^2 - 4)} = \dfrac{t(t^2 - 4)}{(t + 2)t^2} = \dfrac{t(t + 2)(t - 2)}{t^2(t + 2)} = \dfrac{t - 2}{t}$

(d) $\dfrac{(u^3 - v^3)/u^2v^2}{(v - u)/uv} = \dfrac{(u^3 - v^3)uv}{u^2v^2(v - u)} = \dfrac{(u - v)(u^2 + uv + v^2)uv}{-u^2v^2(u - v)} = \dfrac{-(u^2 + uv + v^2)}{uv}$

Fractions with common (that is, equal) denominators are easy to add or subtract:

$$\frac{a}{d} + \frac{b}{d} = \frac{a + b}{d}, \qquad \frac{a}{d} - \frac{b}{d} = \frac{a - b}{d}. \tag{15.5}$$

EXAMPLE 15.5 Simplify.

(a) $\dfrac{x - 1}{x} + \dfrac{1}{x} = \dfrac{(x - 1) + 1}{x} = \dfrac{x}{x} = 1$

(b) $\dfrac{u - v}{t} - \dfrac{2u - v}{t} + \dfrac{u}{t} = \dfrac{(u - v) - (2u - v) + u}{t} = \dfrac{0}{t} = 0$

EXAMPLE 15.6 In these two cases we add after converting to common denominators.

(a) $\dfrac{1}{u} + \dfrac{1}{v} = \dfrac{1}{u} \cdot \dfrac{v}{v} + \dfrac{1}{v} \cdot \dfrac{u}{u} = \dfrac{v + u}{uv}$

(b) $\dfrac{a}{x} + \dfrac{b}{y} + \dfrac{c}{z} = \dfrac{a}{x} \cdot \dfrac{yz}{yz} + \dfrac{b}{y} \cdot \dfrac{xz}{xz} + \dfrac{c}{z} \cdot \dfrac{xy}{xy} = \dfrac{ayz + bxz + cxy}{xyz}$

B. Least Common Denominators

The simplest denominator to use when adding two or more rational expressions is usually a *least common denominator* (*l.c.d.*). As with fractions of integers (Section 2), l.c.d.'s for rational expressions are defined in terms of least common multiples. One polynomial is a **multiple** of a second if the second is a factor of the first. For example, $x^2 + x$ is a multiple of $x + 1$ because $x^2 + x = x(x + 1)$. A **least common multiple (l.c.m.)** of a set of polynomials is a polynomial of smallest degree that is a multiple of each polynomial in the set.

The method for finding an l.c.m. of polynomials is the same as that for finding the l.c.m. of integers, except that in place of prime integers as the building blocks (factors) we use irreducible polynomials. Compare the outline for finding l.c.m.'s in Section 2B with the example that follows.

EXAMPLE 15.7 Find an l.c.m. of $x^2 - y^2$, $x^2 + 2xy + y^2$, and $x^2 - xy$.

Solution First, factor each polynomial completely.

$$x^2 - y^2 = (x + y)(x - y)$$
$$x^2 + 2xy + y^2 = (x + y)^2$$
$$x^2 - xy = x(x - y)$$

Now, form the product of the irreducible polynomials that appear in these factorizations, and on each irreducible factor place the greatest exponent that appears on that factor in any of the above factorizations. In this case, we use x, $(x + y)^2$, and $x - y$. That is, an l.c.m. is $x(x + y)^2(x - y)$.

A **least common denominator (l.c.d.)** of a set of rational expressions is an l.c.m. of the set of denominators of the expressions.

EXAMPLE 15.8 Perform the indicated operations and simplify.

(a) $\dfrac{1}{2xy^2} + \dfrac{1}{6x^2y}$ **(b)** $\dfrac{x}{x^2 - y^2} - \dfrac{y}{x^2 + 2xy + y^2}$

Solution

(a) An l.c.m. of $2xy^2$ and $6x^2y$ is $6x^2y^2$. To use this l.c.m. as a denominator, write

$$\frac{1}{2xy^2} + \frac{1}{6x^2y} = \frac{1}{2xy^2} \cdot \frac{3x}{3x} + \frac{1}{6x^2y} \cdot \frac{y}{y}$$
$$= \frac{3x + y}{6x^2y^2}.$$

(b) An l.c.d. from $x^2 - y^2$ and $x^2 + 2xy + y^2$ is $(x + y)^2(x - y)$. Therefore, we have

$$\frac{x}{x^2 - y^2} - \frac{y}{x^2 + 2xy + y^2}$$

$$= \frac{x}{(x + y)(x - y)} - \frac{y}{(x + y)^2}$$ Factor denominators to determine an l.c.d.

$$= \frac{x}{(x + y)(x - y)} \cdot \frac{x + y}{x + y}$$
$$- \frac{y}{(x + y)^2} \cdot \frac{x - y}{x - y}$$ Convert to common denominators.

$$= \frac{x(x + y) - y(x - y)}{(x - y)(x + y)^2}$$ Subtract.

$$= \frac{x^2 + xy - xy + y^2}{(x - y)(x + y)^2}$$ Simplify.

$$= \frac{x^2 + y^2}{(x - y)(x + y)^2}.$$

Although a least common denominator is usually the simplest denominator to use when adding rational expressions, the *product* of the denominators (which may or may not be a *least* common denominator) can also be used. The main idea is to use *some* common denominator and then simplify in the end.

C. Complex Fractions

A fraction or rational expression in which the numerator or denominator (or both) are also fractions or rational expressions is called a **complex fraction.** Complex fractions can be simplified by first simplifying the numerator and denominator separately, and then inverting the denominator and multiplying [Equation (15.4)].

EXAMPLE 15.9 Simplify.

$$\frac{x - \dfrac{x}{x+2}}{x + \dfrac{x}{x+1}} = \frac{\dfrac{x(x+2)}{1(x+2)} - \dfrac{x}{x+2}}{\dfrac{x(x+1)}{1(x+1)} + \dfrac{x}{x+1}}$$

$$= \frac{\dfrac{x(x+2)-x}{x+2}}{\dfrac{x(x+1)+x}{x+1}} = \frac{\dfrac{x^2+x}{x+2}}{\dfrac{x^2+2x}{x+1}}$$

$$= \frac{x^2+x}{x+2} \cdot \frac{x+1}{x^2+2x}$$

$$= \frac{x(x+1)^2}{x(x+2)^2}$$

$$= \frac{(x+1)^2}{(x+2)^2}$$

EXAMPLE 15.10 Simplify.

$$\frac{\dfrac{1}{x} - \dfrac{1}{y}}{x - y} = \frac{\dfrac{1}{x} \cdot \dfrac{y}{y} - \dfrac{1}{y} \cdot \dfrac{x}{x}}{x - y} = \frac{\dfrac{y-x}{xy}}{\dfrac{x-y}{1}}$$

$$= \frac{y-x}{xy} \cdot \frac{1}{x-y} = \frac{-(x-y)}{xy(x-y)} = -\frac{1}{xy}$$

As with fractions of integers (Examples 2.15 and 2.16), another method for simplifying complex fractions is to multiply both the numerator and denominator by an l.c.d. of all the fractions in both the numerator and denominator.

EXAMPLE 15.11 An l.c.m. of $x + 1$ and $x - 1$ is $(x + 1)(x - 1)$. Therefore, in the following simplification we multiply by $\dfrac{(x+1)(x-1)}{(x+1)(x-1)}$.

$$\frac{\dfrac{1}{x+1} - \dfrac{2}{x-1}}{\dfrac{3}{x+1} + \dfrac{4}{x-1}} = \frac{(x+1)(x-1)}{(x+1)(x-1)} \cdot \frac{\dfrac{1}{x+1} - \dfrac{2}{x-1}}{\dfrac{3}{x+1} + \dfrac{4}{x-1}}$$

$$= \frac{(x-1) - 2(x+1)}{3(x-1) + 4(x+1)} = \frac{-x-3}{7x+1}$$

EXERCISES FOR SECTION 15

Simplify.

1. $\dfrac{x+2}{5x+10}$
2. $\dfrac{6x-3}{2x-1}$
3. $\dfrac{8x^2+4}{2x^2+1}$
4. $\dfrac{y^2+2y+1}{y^2-1}$
5. $\dfrac{2y-6}{y^2-6y+9}$
6. $\dfrac{y^2+2y}{y^2-4}$
7. $\dfrac{z^3-1}{z^2-1}$
8. $\dfrac{z^3+8}{z+2}$
9. $\dfrac{(z+1)^3}{z^3+1}$

Perform the indicated operations and simplify.

10. $\dfrac{x+5}{x}\cdot\dfrac{x^2}{(x+5)^2}$
11. $\dfrac{4y}{(y+1)^2}\cdot\dfrac{y+1}{2y^2}$
12. $\dfrac{x}{y}\cdot\dfrac{xy+y^2}{x^2+xy}$
13. $\dfrac{a^2/(a-1)}{a/(a-1)^2}$
14. $\dfrac{1/b}{2/(b-1)}$
15. $\dfrac{(c-1)/(c+1)}{1/(c+1)}$
16. $\dfrac{(r^2-10r+25)/r}{r-5}$
17. $\dfrac{(1-s^3)/(1+s)}{(1-s)/(1+s^3)}$
18. $\dfrac{(t^2+t+1)/(t+1)}{t^3-1}$
19. $\dfrac{(2u+v)/u^2v}{(8u^3+v^3)/uv^2}$
20. $\dfrac{w^3-5\sqrt{5}}{(w-\sqrt{5})/w}$
21. $\dfrac{(9v^2-4w^2)/(3v-2w)}{(6v+4w)/10}$
22. $\dfrac{y-2}{y}-\dfrac{y+2}{y}$
23. $\dfrac{2y+1}{yz}+\dfrac{y-1}{yz}$
24. $\dfrac{z^3+1}{z^2}-\dfrac{z^2+1}{z^2}$
25. $\dfrac{1}{a+1}+\dfrac{a^3}{a+1}$
26. $\dfrac{2b^2}{b+1}-\dfrac{2}{b+1}$
27. $\dfrac{1000}{100-c^2}-\dfrac{c^3}{100-c^2}$
28. $\dfrac{1}{x+h}-\dfrac{1}{x}$
29. $\dfrac{1}{(x+h)^2}-\dfrac{1}{x^2}$
30. $\dfrac{1}{(x+h)^3}-\dfrac{1}{x^3}$

Find an l.c.m. for each set of polynomials. Write the answers in factored form.

31. $x^2,\ x^2+x,\ x-1$
32. $x^3-1,\ x+1$
33. $x^2+9x+20,\ (x+5)^2$
34. $a^2-b^2,\ (a-b)^2$
35. $a^3-ab^2,\ a^2b^2,\ b^2-ab$
36. $a-b,\ a^2+b^2,\ a^4-b^4$

Perform the indicated operations and simplify.

37. $\dfrac{1}{a-b}-\dfrac{b}{a^2-b^2}$
38. $\dfrac{1}{x+y}+\dfrac{y}{x^2+xy}$
39. $\dfrac{c}{c-d}+\dfrac{d}{d-c}$
40. $\dfrac{2}{x^2+x}+\dfrac{1}{x-1}-\dfrac{1}{x+1}$
41. $\dfrac{1}{c-d}-\dfrac{c}{c^2-d^2}+\dfrac{1}{2c+2d}$
42. $\dfrac{1}{a+5}-\dfrac{5}{a^2+10a+25}$
43. $\dfrac{1+\dfrac{1}{x-1}}{1-\dfrac{1}{x-1}}$
44. $\dfrac{1+\dfrac{1}{y}}{y-\dfrac{1}{y}}$
45. $\dfrac{2+\dfrac{2}{z+1}}{\dfrac{2z}{z+1}-2}$

46. $\dfrac{y+1}{y-\dfrac{1}{y}}$

47. $\dfrac{z^2-\dfrac{8}{z}}{z^2-4}$

48. $\dfrac{25+10y+y^2}{1+\dfrac{5}{y}}$

49. $\dfrac{z-\dfrac{1}{z^2}}{1-\dfrac{1}{z^2}}$

50. $\dfrac{x+2}{\dfrac{x}{x+1}+\dfrac{2}{x+1}}$

51. $\dfrac{\dfrac{1}{y-1}-\dfrac{1}{y+1}}{\dfrac{1}{y-1}-\dfrac{1}{y}}$

52. $\dfrac{\dfrac{1}{a}+\dfrac{1}{b}}{a+b}$

53. $\dfrac{\dfrac{2}{c}-\dfrac{2}{d}}{c-d}$

54. $\dfrac{m+n}{\dfrac{1}{m^2}-\dfrac{1}{n^2}}$

55. $\dfrac{\dfrac{1}{x+h}-\dfrac{1}{x}}{h}$

56. $\dfrac{\dfrac{1}{(x+h)^2}-\dfrac{1}{x^2}}{h}$

57. $\dfrac{\dfrac{1}{(x+h)^3}-\dfrac{1}{x^3}}{h}$

SECTION 16 Radicals

A. Definitions

If $a \geq 0$, then $\sqrt{a}$, the principal square root of a, is the unique real number $b \geq 0$ such that $b^2 = a$ (Section 3D). That is:

> For $a \geq 0$ and $b \geq 0$, $b = \sqrt{a}$ iff $b^2 = a$.

Thus $3 = \sqrt{9}$ because $3^2 = 9$. Let's now look at how to extend this idea to integral exponents greater than 2.

Let a denote a real number and n a positive integer. If $a \geq 0$, then there is precisely one real number $b \geq 0$ such that $b^n = a$. This number b is called the **principal *n*th root** of a and is denoted $\sqrt[n]{a}$. That is:

> For $a \geq 0$, $b \geq 0$, and n a positive integer, $b = \sqrt[n]{a}$ iff $b^n = a$. (16.1)

In particular, $\sqrt[2]{a}$ is the same as $\sqrt{a}$.

EXAMPLE 16.1

(a) $10 = \sqrt[3]{1000}$ because $10^3 = 1000$.
(b) $2 = \sqrt[5]{32}$ because $2^5 = 32$.
(c) $\frac{1}{3} = \sqrt[4]{\frac{1}{81}}$ because $(\frac{1}{3})^4 = \frac{1}{81}$.

If $a < 0$, we must consider two cases, depending on whether the integer n is even or odd. If $a < 0$ and n is even, then there is *no* real number b such that $b^n = a$, because if n is even then $b^n \geq 0$ for every real number b. However, if $a < 0$ and n is odd, then there is precisely one real number $b < 0$ such that $b^n = a$. Again, this number b is called the **principal *n*th root** of a and is denoted $\sqrt[n]{a}$. Thus:

For $a < 0$, $b < 0$, and n an odd positive integer, $b = \sqrt[n]{a}$ iff $b^n = a$. (16.2)

EXAMPLE 16.2

(a) $-2 = \sqrt[3]{-8}$ because $(-2)^3 = -8$.
(b) $-0.1 = \sqrt[5]{-0.00001}$ because $(-0.1)^5 = -0.00001$.
(c) $-\frac{3}{4} = \sqrt[3]{-\frac{27}{64}}$ because $(-\frac{3}{4})^3 = -\frac{27}{64}$.

In the notation $\sqrt[n]{a}$, the integer n is called the **index,** a is called the **radicand,** and $\sqrt[n]{\ }$ is called the **radical.** For each positive integer n, $\sqrt[n]{0} = 0$.

Notice that if n is even and $a > 0$, then there are two real numbers b such that $b^n = a$, but $\sqrt[n]{a}$ denotes only the positive one of these. For example, $\sqrt[4]{625} = 5$, even though $(-5)^4 = 625$ as well as $5^4 = 625$. In general, for n even and $a > 0$, both $\sqrt[n]{a}$ and $-\sqrt[n]{a}$ are called ***n*th roots** of a, but the *principal* nth root is $\sqrt[n]{a}$. The notation $\pm\sqrt[n]{a}$ stands for the pair, $\sqrt[n]{a}$ and $-\sqrt[n]{a}$.

B. Properties

Laws of exponents and radicals. Each of the following equations is valid for all values of m, n, a, and b for which the roots in the equation exist.

$$(\sqrt[n]{a})^n = a \qquad (16.3)$$

If n is even, then $\sqrt[n]{a^n} = |a|$. (16.4)

If n is odd, then $\sqrt[n]{a^n} = a$. (16.5)

$$\sqrt[n]{ab} = \sqrt[n]{a}\,\sqrt[n]{b} \qquad (16.6)$$

$$\sqrt[n]{a/b} = (\sqrt[n]{a})/(\sqrt[n]{b}) \qquad (16.7)$$

$$\sqrt[m]{\sqrt[n]{a}} = \sqrt[mn]{a} \qquad (16.8)$$

$$\sqrt[n]{a^m} = (\sqrt[n]{a})^m \qquad (16.9)$$

EXAMPLE 16.3 Here are illustrations, in order, of Laws (16.3)–(16.9).

(a) $(\sqrt{5})^2 = 5$ and $(\sqrt[3]{-8})^3 = (-2)^3 = -8$

(b) $\sqrt[4]{2^4} = \sqrt[4]{16} = 2 = |2|$ and $\sqrt[4]{(-2)^4} = \sqrt[4]{16} = 2 = |-2|$

(c) $\sqrt[3]{2^3} = \sqrt[3]{8} = 2$ and $\sqrt[3]{(-2)^3} = \sqrt[3]{-8} = -2$

(d) $\sqrt{4 \cdot 9} = \sqrt{36} = 6 = 2 \cdot 3 = \sqrt{4}\sqrt{9}$

(e) $\sqrt[3]{(-27/64)} = -3/4 = (\sqrt[3]{-27})/(\sqrt[3]{64})$

(f) $\sqrt[3]{\sqrt[2]{64}} = \sqrt[3]{8} = 2 = \sqrt[6]{64}$

(g) $\sqrt[5]{(-32)^2} = \sqrt[5]{1024} = 4 = (-2)^2 = (\sqrt[5]{-32})^2$

Proofs of Laws (16.3)–(16.9). Following are proofs of a representative set of these laws. The laws are obviously true for $a = 0$, so we assume throughout that $a \neq 0$. Law (16.3) simply restates the meaning of $\sqrt[n]{a}$.

For Law (16.4), first remember that, since n is even, $\sqrt[n]{a^n}$ denotes the unique positive real number b such that $b^n = a^n$. If $a > 0$, it follows from the uniqueness of b that $b = a$. If $a < 0$, then, since $-a > 0$ and $(-a)^n = a^n$, it follows from the uniqueness of b that $b = -a$. So, in either case, $b = |a|$, as required. (Remember that $|a| = -a$ if $a < 0$.) Law (16.5) is similar to (16.4).

For Law (16.6), let $c = \sqrt[n]{a}$ and $d = \sqrt[n]{b}$. Then $c^n = a$ and $d^n = b$. Therefore, $(cd)^n = c^n d^n = ab$, and so $cd = \sqrt[n]{ab}$, that is, $\sqrt[n]{a}\sqrt[n]{b} = \sqrt[n]{ab}$. Law (16.7) is similar.

Here is a proof of the easiest case of Law (16.8). Assume that $a > 0$, and let $b = \sqrt[n]{a}$ and $c = \sqrt[m]{b}$. Then $c = \sqrt[m]{\sqrt[n]{a}}$, and so we must show that $c = \sqrt[mn]{a}$. This is the same as showing that $c^{mn} = a$; but that is true because $c^{mn} = (\sqrt[m]{b})^{mn} = [(\sqrt[m]{b})^m]^n = b^n = a$. The proofs of the other cases are similar.

If $m > 0$, Law (16.9) follows from repeated application of (16.6). If $m = 3$, for example, we have

$$\sqrt[n]{a^3} = \sqrt[n]{a \cdot a \cdot a} = \sqrt[n]{a}\,\sqrt[n]{a}\,\sqrt[n]{a} = (\sqrt[n]{a})^3.$$

The case $m < 0$ follows from other laws and the case $m > 0$ (remember that $-m > 0$ if $m < 0$):

$$\sqrt[n]{a^m} = \sqrt[n]{\left(\frac{1}{a}\right)^{-m}} = \frac{\sqrt[n]{1}}{\sqrt[n]{a^{-m}}} = \frac{1}{(\sqrt[n]{a})^{-m}} = (\sqrt[n]{a})^m. \qquad \square$$

C. Rationalizing Denominators

We sometimes encounter fractions with one or more radicals in their denominators. To **rationalize the denominator** of such a fraction is to rewrite the fraction in a form without a radical in the denominator. Most cases can be handled by one of the two techniques that follow. Each technique depends on multiplying the fraction by 1 in an appropriately chosen form.

> If the denominator has the form $a\sqrt{b}$, multiply the fraction by $\sqrt{b}/\sqrt{b}$ and simplify.

EXAMPLE 16.4 Rationalize each denominator.

(a) $\dfrac{1}{\sqrt{2}} = \dfrac{1}{\sqrt{2}} \cdot \dfrac{\sqrt{2}}{\sqrt{2}} = \dfrac{\sqrt{2}}{2}$

(b) $\dfrac{\sqrt{3}}{2\sqrt{5}} = \dfrac{\sqrt{3}}{2\sqrt{5}} \cdot \dfrac{\sqrt{5}}{\sqrt{5}} = \dfrac{\sqrt{15}}{10}$

> If the denominator has the form $a + \sqrt{b}$, multiply the fraction by $(a - \sqrt{b})/(a - \sqrt{b})$ and simplify. The expression $a - \sqrt{b}$ is called the **conjugate** of $a + \sqrt{b}$. Similarly, for $\sqrt{a} + \sqrt{b}$ use the conjugate $\sqrt{a} - \sqrt{b}$, and for $c\sqrt{a} + d\sqrt{b}$ use the conjugate $c\sqrt{a} - d\sqrt{b}$.

These ideas work because $(u + v)(u - v) = u^2 - v^2$. In particular,

$$(a + \sqrt{b})(a - \sqrt{b}) = a^2 - (\sqrt{b})^2 = a^2 - b$$

and

$$(\sqrt{a} + \sqrt{b})(\sqrt{a} - \sqrt{b}) = (\sqrt{a})^2 - (\sqrt{b})^2 = a - b.$$

EXAMPLE 16.5 Rationalize each denominator.

(a) $\dfrac{1}{2 + \sqrt{3}} = \dfrac{1}{2 + \sqrt{3}} \cdot \dfrac{2 - \sqrt{3}}{2 - \sqrt{3}} = \dfrac{2 - \sqrt{3}}{4 - 3} = 2 - \sqrt{3}$

(b) $\dfrac{\sqrt{r}}{\sqrt{r} - 2\sqrt{s}} = \dfrac{\sqrt{r}}{\sqrt{r} - 2\sqrt{s}} \cdot \dfrac{\sqrt{r} + 2\sqrt{s}}{\sqrt{r} + 2\sqrt{s}} = \dfrac{r + 2\sqrt{rs}}{r - 4s}$

(c) $\dfrac{1}{1 - \sqrt{x + 1}} = \dfrac{1}{1 - \sqrt{x + 1}} \cdot \dfrac{1 + \sqrt{x + 1}}{1 + \sqrt{x + 1}} = \dfrac{1 + \sqrt{x + 1}}{1 - (x + 1)}$

$= -\dfrac{1 + \sqrt{x + 1}}{x}$

Sometimes there is a good reason to rationalize the *numerator* of a fraction. The following example arises in calculus; it uses an obvious variation on one of the techniques for rationalizing denominators.

EXAMPLE 16.6 Rationalize the numerator.

$$\frac{\sqrt{x + h} - \sqrt{x}}{h} = \frac{\sqrt{x + h} - \sqrt{x}}{h} \cdot \frac{\sqrt{x + h} + \sqrt{x}}{\sqrt{x + h} + \sqrt{x}}$$

$$= \frac{(x + h) - x}{h(\sqrt{x + h} + \sqrt{x})}$$

$$= \frac{1}{\sqrt{x + h} + \sqrt{x}}$$

D. More Examples

Here are some things to keep in mind when simplifying expressions that involve radicals.

- A power can be removed from under a radical by changing its exponent if the index of the radical is a factor of the exponent. For example, 5 is a factor of 10, and $\sqrt[5]{a^{10}} = a^2$ because $a^{10} = (a^2)^5$.
- Multiples of radicals with the same index and the same radicand can be combined through addition and subtraction. Thus, for example, $6\sqrt{5} - 4\sqrt{5} = (6 - 4)\sqrt{5} = 2\sqrt{5}$.
- If $a \neq 0$, $b \neq 0$, and $n > 1$, then $\sqrt[n]{a + b} \neq \sqrt[n]{a} + \sqrt[n]{b}$. For example, $\sqrt{4 + 9} = \sqrt{13}$, but $\sqrt{4} + \sqrt{9} = 2 + 3 = 5$; thus $\sqrt{4 + 9} \neq \sqrt{4} + \sqrt{9}$.
- Try to use $\sqrt[n]{a^m} = (\sqrt[n]{a})^m$ when faced with computing $\sqrt[n]{a^m}$, because roots are easier to recognize for small numbers than for large ones. For example, $\sqrt[3]{27^4} = \sqrt[3]{531{,}441}$, and we aren't likely to recognize 531,441 as a perfect cube. However, $\sqrt[3]{27^4} = (\sqrt[3]{27})^4 = 3^4 = 81$.

EXAMPLE 16.7 Simplify. (This means, in particular, to remove all possible powers from under the radical signs.) Assume that each variable represents a positive real number.

(a) $\sqrt{a^3} = \sqrt{a^2 \cdot a} = \sqrt{a^2}\sqrt{a} = a\sqrt{a}$

(b) $(\sqrt[3]{x})^{12} = (\sqrt[3]{x^3})^4 = x^4$

(c) $\sqrt[5]{x^{11}} = \sqrt[5]{x^{10}x} = \sqrt[5]{x^{10}}\sqrt[5]{x} = x^2\sqrt[5]{x}$

(d) $\sqrt{8} - \sqrt{18} = \sqrt{4 \cdot 2} - \sqrt{9 \cdot 2} = \sqrt{4}\sqrt{2} - \sqrt{9}\sqrt{2} = 2\sqrt{2} - 3\sqrt{2} = -\sqrt{2}$

(e) $6\sqrt{3a} - \sqrt{12a} + 2\sqrt{75a^3} = 6\sqrt{3a} - \sqrt{4}\sqrt{3a} + 2\sqrt{25a^2}\sqrt{3a}$
$= 6\sqrt{3a} - 2\sqrt{3a} + 10a\sqrt{3a} = (4 + 10a)\sqrt{3a}$

EXERCISES FOR SECTION 16

Assume that the variables in these exercises represent positive real numbers. Find each root.

1. $\sqrt[4]{16}$ **2.** $\sqrt[4]{10{,}000}$ **3.** $\sqrt[3]{64}$ **4.** $\sqrt[4]{81}$
5. $\sqrt[5]{-32}$ **6.** $\sqrt[5]{100{,}000}$ **7.** $\sqrt[3]{-125}$ **8.** $\sqrt[3]{8000}$
9. $\sqrt[3]{-1000}$ **10.** $\sqrt[4]{0.0001}$ **11.** $\sqrt[3]{-0.008}$ **12.** $\sqrt[3]{0.125}$
13. $\sqrt[5]{-1/32}$ **14.** $\sqrt[4]{81/16}$ **15.** $\sqrt[3]{-8/27}$

Simplify. Use Laws (16.3)–(16.9) as much as possible. Lengthy calculations should be unnecessary.

16. $\sqrt[4]{(-7)^4}$ **17.** $(\sqrt[4]{20})^4$ **18.** $(\sqrt[5]{100})^5$
19. $\sqrt[3]{(-13)^3}$ **20.** $\sqrt[10]{(-1/5)^{10}}$ **21.** $\sqrt[7]{(-1/12)^7}$
22. $(\sqrt[6]{25})^6$ **23.** $\sqrt[5]{(-1.7)^5}$ **24.** $\sqrt[6]{(-3)^6}$
25. $(\sqrt{0.09})^3$ **26.** $(\sqrt[3]{0.064})^2$ **27.** $(\sqrt{0.81})^3$

28. $\sqrt{5} + \sqrt{20}$
29. $\sqrt{12} - \sqrt{3}$
30. $\sqrt{50} + \sqrt{18}$
31. $3\sqrt{200} - 5\sqrt{72}$
32. $2\sqrt{40} + \sqrt{90}$
33. $2\sqrt{48} - 5\sqrt{27}$
34. $\sqrt{12\sqrt[3]{27}}$
35. $\sqrt[3]{270\sqrt{10000}}$
36. $\sqrt{5\sqrt[3]{8000}}$
37. $(\sqrt[3]{a})^6$
38. $(\sqrt{b})^4$
39. $(\sqrt[5]{c})^{10}$
40. $\sqrt[6]{x^{19}}$
41. $\sqrt[3]{y^{10}}$
42. $\sqrt{z^7}$
43. $2\sqrt{9u} - \sqrt{25u}$
44. $6\sqrt{v} - \sqrt{100v}$
45. $\sqrt{w} + \sqrt{49w}$
46. $2\sqrt{r^3} - r(\sqrt{r} - \sqrt{9r})$
47. $s\sqrt{2s} - \sqrt{18s^3} + s\sqrt{50s}$
48. $\sqrt[3]{t^4} - 2t(\sqrt[3]{t} - \sqrt[3]{8t})$

Rationalize each denominator and simplify.

49. $2/\sqrt{7}$
50. $10/\sqrt{10}$
51. $1/\sqrt{3}$
52. $1/(3 + \sqrt{2})$
53. $2/(1 - \sqrt{3})$
54. $5/(1 - \sqrt{6})$
55. $1/(\sqrt{x} - \sqrt{y})$
56. $1/(\sqrt{x+1} - \sqrt{x})$
57. $1/(\sqrt{a} + 1)$

Rationalize each numerator and simplify.

58. $(1 - \sqrt{2})/3$
59. $(2 - \sqrt{3})/(2 + \sqrt{3})$
60. $(\sqrt{5} + \sqrt{2})/3$
61. $(\sqrt{x-h} - \sqrt{x})/h$
62. $(\sqrt{x+1+h} - \sqrt{x+1})/h$
63. $(\sqrt{1-x+h} - \sqrt{1-x})/h$

SECTION 17 Rational Exponents

A. Definitions

Positive integral exponents relate to repeated multiplication, such as $a^2 = a \cdot a$. *Negative integral* exponents relate to multiplication and inversion, such as $a^{-2} = 1/(a \cdot a)$. We'll now see that *rational* exponents relate to radicals. We'll also see that it is often much easier to manipulate radicals if we use such exponents. The link between radicals and exponents can be described in three steps.

Step I. If a is a real number and n is a positive integer, then we define

$$a^{1/n} = \sqrt[n]{a}, \tag{17.1}$$

provided $\sqrt[n]{a}$ exists (that is, provided $a \geq 0$, or $a < 0$ and n is odd).

EXAMPLE 17.1 Compute.

(a) $25^{1/2} = \sqrt{25} = 5$

(b) $(-27)^{1/3} = \sqrt[3]{-27} = -3$

Notice that the definition of $a^{1/n}$ in Equation (17.1) is just what it must be to make the law $(a^m)^n = a^{mn}$ for integral exponents carry over for $1/n$ in place of m: If we are to have

$$(a^{1/n})^n = a^{(1/n)n} = a^1 = a,$$

then necessarily

$$a^{1/n} = \sqrt[n]{a}.$$

In the same way, the next step is just what it must be to make the laws for integral exponents carry over to all positive rational exponents: If we are to have

$$(a^{1/n})^m = a^{m/n} \quad \text{and} \quad a^{1/n} = \sqrt[n]{a},$$

then necessarily

$$a^{m/n} = (a^{1/n})^m = (\sqrt[n]{a})^m.$$

Step II. If a is a real number, and m and n are positive integers such that $\sqrt[n]{a}$ exists, then we define

$$\boxed{a^{m/n} = (\sqrt[n]{a})^m.} \tag{17.2}$$

Because $\sqrt[n]{a^m} = (\sqrt[n]{a})^m$, by Law (16.9), it is also true that $a^{m/n} = \sqrt[n]{a^m}$.

EXAMPLE 17.2 Compute.

(a) $25^{3/2} = (\sqrt{25})^3 = 5^3 = 125$

(b) $\sqrt[5]{x^{11}} = x^{11/5} = x^{2+(1/5)} = x^2x^{1/5} = x^2\sqrt[5]{x}$. [Compare Example 16.7(c).]

To extend the meaning of $a^{m/n}$ to include negative as well as positive exponents, we simply observe that if m/n is negative, then we can assume $m < 0$ and $n > 0$ [for example, $-(2/3) = (-2)/3$]. Then we use Equation (17.2) again, as follows.

Step III. If a is a real number, and m and n are integers such that $n > 0$ and $\sqrt[n]{a}$ exists, then $a^{m/n}$ is defined by Equation (17.2).

EXAMPLE 17.3 Compute.

(a) $8^{-2/3} = (\sqrt[3]{8})^{-2} = 2^{-2} = 1/2^2 = 1/4$

(b) $0.04^{-3/2} = (\sqrt{0.04})^{-3} = (0.2)^{-3} = 1/0.2^3 = 1/0.008 = 125$

B. Properties and Examples

The laws for integral exponents (Section 3B) carry over to rational exponents as follows.

Laws for rational exponents. Each of the following equations is valid for all real numbers a and b and all rational numbers r and s for which both sides of the equation exist.

$$a^r a^s = a^{r+s} \tag{17.3}$$

$$(a^r)^s = a^{rs} \tag{17.4}$$

$$(ab)^r = a^r b^r \tag{17.5}$$

$$a^r / b^r = (a/b)^r \tag{17.6}$$

$$a^r / a^s = a^{r-s} \tag{17.7}$$

Here is a proof of Law (17.3); the proofs of the other parts are similar. Assume that $r = m/n$ and $s = u/v$ (m, n, u, and v integers). Then

$$\begin{aligned}
a^r a^s &= a^{m/n} a^{u/v} \\
&= a^{(mv)/(nv)} a^{(nu)/(nv)} \\
&= (\sqrt[nv]{a})^{mv} (\sqrt[nv]{a})^{nu} && \text{by (17.2)} \\
&= (\sqrt[nv]{a})^{mv+nu} && \text{by (3.1)} \\
&= a^{(mv+nu)/nv} && \text{by (17.2)} \\
&= a^{(m/n)+(u/v)} \\
&= a^{r+s}.
\end{aligned}$$

Many expressions that contain radicals can be simplified most easily by converting to rational exponents, as in parts of the following example. In examples like this it is important to understand the reason behind every step. Here is a good test: Could you explain it to someone else?

EXAMPLE 17.4 Simplify. The variables represent positive real numbers.

(a) $\sqrt{t}/\sqrt[3]{t} = t^{1/2}/t^{1/3} = t^{(1/2)-(1/3)} = t^{1/6} = \sqrt[6]{t}$

(b) $\sqrt[3]{25}\,\sqrt[3]{5^4} = (5^2)^{1/3} \cdot 5^{4/3} = 5^{2/3} \cdot 5^{4/3} = 5^{(2/3)+(4/3)} = 5^2 = 25$

(c) $(27a^6)^{2/3} = 27^{2/3}(a^6)^{2/3} = (\sqrt[3]{27})^2 a^{6(2/3)} = 3^2 a^{12/3} = 9a^4$

(d) $t^{-3}\sqrt[2]{\sqrt[3]{t^{12}}} = t^{-3}[(t^{12})^{1/3}]^{1/2} = t^{-3}(t^4)^{1/2} = t^{-3}t^2 = t^{-1} = 1/t$

(e) $\dfrac{(x^{-1/2}y)^3 x^{-5/2} y^4}{(xy^2)^{-2}} = \dfrac{x^{-3/2}y^3x^{-5/2}y^4}{x^{-2}(y^2)^{-2}} = \dfrac{x^{-4}y^7}{x^{-2}y^{-4}} = \dfrac{y^{7+4}}{x^{4-2}} = \dfrac{y^{11}}{x^2}$

(f) $\sqrt{\dfrac{x^{m-2}}{x^{4-m}}} = [x^{(m-2)-(4-m)}]^{1/2} = [x^{2m-6}]^{1/2} = x^{(2m-6)/2} = x^{m-3}$

Here is one more example of how to simplify an expression; this one arises in calculus.

EXAMPLE 17.5 Simplify.

$$\frac{(x^2+1)^{1/2} \cdot 1 - x \cdot \frac{1}{2}(x^2+1)^{-1/2} \cdot 2x}{[(x^2+1)^{1/2}]^2}$$

$$= \frac{(x^2+1)^{1/2} - \dfrac{x^2}{(x^2+1)^{1/2}}}{x^2+1} = \frac{\dfrac{(x^2+1)-x^2}{(x^2+1)^{1/2}}}{x^2+1}$$

$$= \frac{1}{(x^2+1)^{1/2}} \cdot \frac{1}{(x^2+1)} = \frac{1}{(x^2+1)^{3/2}}$$

C. More Numerical Examples

EXAMPLE 17.6 Compute $81^{0.75}$.

Solution Because $0.75 = \frac{3}{4}$, we can apply Equation (17.2):

$$81^{0.75} = 81^{3/4} = (\sqrt[4]{81})^3 = 3^3 = 27.$$

The idea here can be used for any exponent that is a decimal number provided it is also rational. (Irrational exponents will be discussed in Section 35.)

Powers of numbers can be computed quickly on a calculator having a $\boxed{y^x}$ key. Typically, we simply *enter y, press* $\boxed{y^x}$, *enter x, press* $\boxed{=}$, and then read the answer in the display. This does not lessen the importance of knowing how to use the laws of exponents. In manipulating algebraic expressions, for example, we are often faced not with specific numbers but with letters that represent numbers.

EXAMPLE 17.7 Compute $8^{2/3}$.

Solution without a calculator. $8^{2/3} = (\sqrt[3]{8})^2 = 2^2 = 4.$

Solution with a calculator. (One method: for a calculator having a $\boxed{y^x}$ key.) In place of 2/3 we use the nearest possible decimal approximation. On a calculator with an eight-digit display, for example, we use .66666667. We can *enter* 8, *press* $\boxed{y^x}$, *enter* .66666667, and then *press* $\boxed{=}$. The answer, 4, should be displayed. (The answer may vary slightly because of rounding.) $\boxed{C}$

Solution with a calculator. (Another method: for a calculator having $\boxed{y^x}$, $\boxed{(}$, and $\boxed{)}$ keys.) *Enter* 8, *press* $\boxed{y^x}$, *press* $\boxed{(}$, *enter* 2 , *press* $\boxed{\div}$, *enter* 3, *press* $\boxed{)}$, *press* $\boxed{=}$. The answer should be displayed. $\boxed{C}$

EXERCISES FOR SECTION 17

Compute. (Do not use a calculator.)

1. $4^{5/2}$
2. $8^{2/3}$
3. $9^{3/2}$
4. $9^{-1/2}$
5. $4^{-7/2}$
6. $25^{-3/2}$
7. $0.001^{4/3}$
8. $0.25^{5/2}$
9. $0.027^{2/3}$
10. $(-1000)^{-1/3}$
11. $(-32)^{-3/5}$
12. $(-64)^{-2/3}$
13. $(-27000)^{2/3}$
14. $(-125)^{5/3}$
15. $81^{-3/4}$
16. $25^{2.5}$
17. $49^{1.5}$
18. $(-32)^{1.4}$
19. $(16^{0.75})^{-3}$
20. $(10^8)^{-0.25}$
21. $(10^{-9})^{0.\overline{3}}$

Simplify. Assume that the variables represent positive real numbers.

22. $\dfrac{x^{1/3}x^{7/3}}{x^{2/3}}$ **23.** $\dfrac{y^{-1/3}y^{2/3}}{y^{-7/3}}$ **24.** $\dfrac{z^2z^{1/2}}{z^{-1/2}}$

25. $\sqrt[3]{a}/\sqrt[4]{a}$ **26.** $\sqrt[4]{b}/\sqrt[3]{b}$ **27.** $\sqrt[5]{c}/\sqrt[4]{c}$

28. $\sqrt[3]{4}\,\sqrt[3]{2^4}$ **29.** $\sqrt{27}\,\sqrt{3^5}$ **30.** $\sqrt[4]{8}\,\sqrt{2^5}$

31. $(125x^3)^{2/3}$ **32.** $(16y^6)^{1/2}$ **33.** $(32z^{10})^{3/5}$

34. $\sqrt[3]{\sqrt{r^{-12}}}$ **35.** $\sqrt{\sqrt{s^{20}}}$ **36.** $(\sqrt{\sqrt[3]{t^2}})^9$

37. $\dfrac{a^2(a^3b^{1/2})^{-4}}{(a^3b)^2}$ **38.** $\dfrac{(c^{-3}d^{2/3})^{-1}d}{cd^{4/3}}$ **39.** $\dfrac{(mn^2)^{-1/2}(m^{1/2}n)^{-1}}{(mn)^{-2}}$

40. $\sqrt{\dfrac{a^{2m+1}}{a^{-3}}}$ **41.** $\sqrt{\dfrac{b^{4n+1}}{b^{2n-1}}}$ **42.** $\sqrt[3]{\dfrac{c^{2n}}{c^{6-n}}}$

43. $(x^{1/2} - y^{1/2})(x^{1/2} + y^{1/2})$ **44.** $(a^{1/3} - b^{1/3})(a^{2/3} + a^{1/3}b^{1/3} + b^{2/3})$

45. $(c^{1/3} + d^{1/3})(c^{2/3} - c^{1/3}d^{1/3} + d^{2/3})$

46. $\dfrac{x \cdot \frac{1}{2}(x^2+1)^{-1/2} \cdot 2x - (x^2+1)^{1/2} \cdot 1}{x^2}$

47. $\dfrac{(x+1)^{3/2} \cdot 2x - x^2 \cdot \frac{3}{2}(x+1)^{1/2}}{[(x+1)^{3/2}]^2}$

48. $\dfrac{(x^2+1)^{1/3} \cdot 2x - x^2 \cdot \frac{1}{3}(x^2+1)^{-2/3} \cdot 2x}{[(x^2+1)^{1/3}]^2}$

49. $\left(\dfrac{a^{n^2+1}}{a^{1-n}}\right)^{1/(n+1)}$ **50.** $\left(\dfrac{b^{n-1}}{b^{n-n^2}}\right)^{1/(n-1)}$

For the next four exercises, compare the proof of Law (17.3) that follows Laws (17.3)–(17.7).

51. Prove Law (17.4). **52.** Prove Law (17.5).

53. Prove Law (17.6). **54.** Prove Law (17.7).

Compute with a calculator having a y^x key. C

55. $10^{1/5}$ **56.** $5^{3/10}$ **57.** $6^{-2/3}$ **58.** $(\sqrt[5]{7.06})^{-12}$ **59.** $\sqrt[4]{12.8^{-3}}$ **60.** $(\sqrt[5]{6.1413})^7$

SECTION

18 Applications

A. Volume and Surface Area

EXAMPLE 18.1 Express the surface area of a cube in terms of the volume of the cube. (See Figure 18.1.)

Solution Let S denote the surface area and V the volume. The first key to this problem is to work through an intermediate quantity, the length of a side, which we denote by x. The volume is given by

$$V = x^3. \tag{18.1}$$

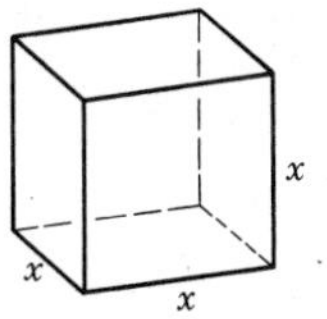

Figure 18.1

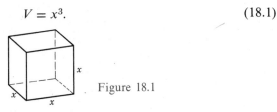

The area of each face of the cube is x^2, and there are six faces, so that

$$S = 6x^2. \tag{18.2}$$

To express S in terms of V we can solve Equation (18.1) for x and substitute the answer in Equation (18.2). Equation (18.1) implies

$$x = V^{1/3}.$$

When x is replaced by $V^{1/3}$ in Equation (18.2), the result is

$$S = 6(V^{1/3})^2$$

$$S = 6V^{2/3}. \tag{18.3}$$

If the volume is 125 m^3, for example, then the surface area is

$$S = 6(125)^{2/3} = 6(\sqrt[3]{125})^2 = 6 \cdot 5^2 = 150 \text{ m}^2.$$

EXAMPLE 18.2 The amount of heat loss from an animal is determined in part by the surface area of the animal. The exact surface area of an animal is hard to determine, but physiologists have shown that it is approximately proportional to the two-thirds power of the weight of the animal:

$$S = KW^{2/3}, \tag{18.4}$$

where

$$S = \text{surface area in square meters } (m^2)$$
$$W = \text{weight in kilograms (kg)}$$
$$K = \text{constant of proportionality.}$$

The constant K here is approximately 0.1 for most animals. Some typical values are shown in Table 18.1.

TABLE 18.1

Animal	K
Cow	0.090
Cat	0.100
Dog (over 4 kg)	0.112
(under 4 kg)	0.101
Horse	0.100
Man	0.11
Swine	0.090

SOURCE: A. C. Guyton, C. E. Jones, and T. G. Coleman, *Circulatory Physiology: Cardiac Output and Its Regulation*, Second Edition, W. B. Saunders Company, Philadelphia, 1973.

Problem Compute the approximate surface area of an 8 kg dog.

Solution Use Equation (18.4) with $W = 8$ and $K = 0.112$ (from Table 18.1):

$$S = 0.112 \times 8^{2/3} = 0.112 \times 4 = 0.448 \text{ m}^2.$$

Notice that Equation (18.3) makes Equation (18.4) seem entirely reasonable, because (18.3) shows that (for a cube, at least) surface area is proportional to the two-thirds power of volume, and weight should be (approximately) proportional to volume.

(Example 18.2 is based on a discussion in R. DeSapio, *Calculus for the Life Sciences,* W. H. Freeman and Company, San Francisco, 1979.)

B. Kepler's Third Law

Kepler's first two laws of planetary motion were published in 1609, and the third was published in 1619. The first law describes the shape of planetary orbits; the second describes the speed of the planets. These first two laws will be discussed in Section 21B.

Kepler's third law establishes a uniformity among the motions of the different planets. Let R denote the average distance between the sun and a planet, measured in millions of miles. Let T denote the *period* of the planet—that is, the time required for one complete orbital revolution about the sun; measure T in Earth years, so that $T = 1$ for the Earth. Then Kepler's third law states that the ratio R^3/T^2 is the same for all planets. If this constant ratio is denoted by K, then

$$R^3/T^2 = K. \tag{18.5}$$

If R and T are known for a single planet, then K can be computed from Equation (18.5). Then, for example, if T can be determined for a second planet, the value of R for the second planet can be computed by solving Equation (18.5) for R:

$$R^3 = KT^2$$
$$R = \sqrt[3]{KT^2}. \tag{18.6}$$

EXAMPLE 18.3 The Earth is approximately 93 million miles from the sun. Determine K in Equation (18.5).

Solution With $R = 93$ and $T = 1$, Equation (18.5) becomes

$$K = 93^3/1^2 \approx 8.04 \times 10^5. \quad \boxed{C} \tag{18.7}$$

EXAMPLE 18.4 The period of Mars is approximately 1.88 Earth years. What is the average distance between Mars and the sun?

Solution Use Equation (18.6) with $T = 1.88$ and $K = 8.04 \times 10^5$ (from the previous example). Then

$$R = \sqrt[3]{8.04 \times 10^5 \times 1.88^2} \approx 142. \quad \boxed{C}$$

That is, Mars is approximately 142 million miles from the sun.

C. Compound Interest

The two basic kinds of interest on invested money are *simple* interest and *compound* interest. In each case an amount of interest is paid at the end of regular time periods. The difference is that for simple interest the amount of interest paid is the same for each period, being computed solely on the original principal (amount invested); for compound interest the amount of interest paid increases from period to period, because the interest paid during each period is added to the principal to create a new principal, and interest for the following period is computed on this new (increased) principal. Simple interest was discussed in Section 8. Here we'll look at compound interest. We'll be led naturally from positive integral exponents to negative integral exponents to rational exponents (radicals).

Compound amount formula. The time periods over which compound interest is computed are called **conversion periods.** Interest rates are conventionally given as annual rates and then converted to rates per conversion period. For example, a rate of 10% compounded quarterly yields a rate of $\frac{1}{4} \cdot 10\% = 2.5\%$ per quarter (the conversion period).

EXAMPLE 18.5 Suppose that \$1000 is invested at an annual rate of 6%, compounded semiannually. How much (principal plus interest) will have accumulated at the end of one year?

Solution The rate per conversion period is $\frac{1}{2} \cdot 6\% = 3\%$. The amount of interest earned during the first period is

$$1000 \times 0.03 = \$30.00.$$

Therefore, the principal for the second period is \$1030 (Figure 18.2).

The interest earned during the second period is

$$1030 \times 0.03 = \$30.90.$$

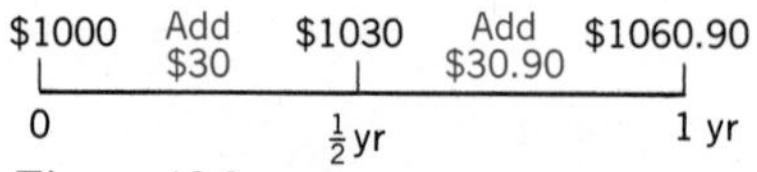

Figure 18.2

The amount accumulated at the end of one year is, therefore,

$$\$1030.00 + \$30.90 = \$1060.90.$$

(At 6% per year *simple* interest, \$1060.00 would have accumulated.)

By generalizing the preceding example we can derive a formula for **compound amount**—that is, the amount accumulated (principal plus interest) in any compound interest problem. Let

$$\begin{aligned} P &= \text{original principal} \\ r &= \text{annual interest rate} \\ n &= \text{number of conversion periods per year} \\ S_k &= \text{compound amount at the end of } k \text{ periods.} \end{aligned}$$

The interest rate per conversion period will be r/n. Therefore, the compound amount at the end of the first period will be

$$S_1 = P + P\frac{r}{n} = P\left(1 + \frac{r}{n}\right).$$

The principal for the second period will be S_1, so that

$$\begin{aligned} S_2 &= S_1 + S_1\left(\frac{r}{n}\right) = P\left(1 + \frac{r}{n}\right) + P\left(1 + \frac{r}{n}\right)\frac{r}{n} \\ &= P\left(1 + \frac{r}{n}\right)\left(1 + \frac{r}{n}\right) \\ &= P\left(1 + \frac{r}{n}\right)^2. \end{aligned}$$

In the same way,

$$S_3 = S_2 + S_2\left(\frac{r}{n}\right) = P\left(1 + \frac{r}{n}\right)^2 + P\left(1 + \frac{r}{n}\right)^2\left(\frac{r}{n}\right) = P\left(1 + \frac{r}{n}\right)^3.$$

Continuing, we see that in general

$$S_k = P\left(1 + \frac{r}{n}\right)^k. \qquad (18.8)$$

Over t years the number k of conversion periods will be nt (the number of conversion periods per year times the number of years). Therefore, Equation (18.8) justifies the following conclusion.

Compound amount. If a principal amount P is invested at an annual rate of $r\%$ compounded n times per year for t years, then the compound amount S is given by

$$S = P\left(1 + \frac{r}{n}\right)^{nt}. \qquad (18.9)$$

EXAMPLE 18.6 What is the compound amount on \$500 invested for two years at 12% compounded quarterly?

Solution Here $P = \$500$, $n = 4$ (quarterly compounding), $r/n = 0.12/4 = 0.03$, and $t = 2$. Equation (18.9) gives

$$S = 500(1.03)^8 = \$633.39.$$ Ⓒ

Throughout this section and its exercises, the rates are annual rates, like r in Equation (18.9).

Present value formula. Suppose we want to know how much to invest at a specified rate and compounding period, and for a specified time, in order to accumulate a specified amount. Then r, n, t, and S are known, and we can find P by solving Equation (18.9). This amount P is called the **present value** of S (for the given r, n, and t).

Present value

$$P = S\left(1 + \frac{r}{n}\right)^{-nt} \qquad (18.10)$$

EXAMPLE 18.7 How much must be invested at 10% compounded semiannually so that \$1000 will be accumulated at the end of two years?

Solution The problem asks for the present value of \$1000 under the stated interest conditions. In this case $r = 0.10$, $n = 2$, $r/n = 0.05$, $t = 2$, and $S = 1000$. Equation (18.10) gives

$$P = 1000(1.05)^{-4} = 1000/1.05^4 = \$822.70.$$ Ⓒ

Effective rate formula. It is natural to ask for the rate of simple annual interest that will produce the same accumulated amount at the end of one year as a given compound rate. This simple rate is called the **effective rate,** and the compound rate to which it corresponds is called the **nominal rate.** Savings institutions frequently publicize the effective rates that are equivalent to the nominal rates for different savings plans—this gives a convenient method of comparison. Table 18.2 gives examples of the effective rate (%) for different nominal rates and conversion periods (rounded to the nearest 0.01%). For instance, the table shows that a rate of 9% compounded quarterly is equivalent to a rate of 9.31% without compounding.

TABLE 18.2

Conversion periods (no. per yr)	*Nominal rate (%)*				
	6	7	8	9	10
1	6.00	7.00	8.00	9.00	10.00
2	6.09	7.12	8.16	9.20	10.25
4	6.14	7.19	8.24	9.31	10.38
12	6.17	7.23	8.30	9.38	10.47

We next look at some examples of how the entries in such a table can be calculated.

EXAMPLE 18.8 What is the effective rate equivalent to 6% compounded semiannually?

Solution The calculations needed here were carried out in Example 18.5. The nominal rate there was also 6% compounded semiannually, and at the end of one year \$1000 accumulated to \$1060.90. It follows that the effective rate is 6.09% (the interest earned, \$60.90, is 6.09% of \$1000).

To derive a general formula, let

$$r_N = \text{nominal rate}$$
$$r_E = \text{effective rate}$$
$$n = \text{number of conversion periods per year.}$$

The amount accumulated at the end of one year from r_E is

$$P + Pr_E = P(1 + r_E). \tag{18.11}$$

The amount accumulated at the end of one year for a rate r_N with n conversion periods is given by Equation (18.9) with $r = r_N$ and $t = 1$:

$$P\left(1 + \frac{r_N}{n}\right)^n \tag{18.12}$$

We seek r_E so that the amounts in Equations (18.11) and (18.12) will be equal.

$$P(1 + r_E) = P\left(1 + \frac{r_N}{n}\right)^n$$

$$1 + r_E = \left(1 + \frac{r_N}{n}\right)^n$$

This gives the following formula, which was used to compute the entries in Table 18.2.

Effective rate

$$r_E = \left(1 + \frac{r_N}{n}\right)^n - 1 \tag{18.13}$$

EXAMPLE 18.9 What effective rate is equivalent to a nominal rate of 6% compounded monthly

Solution Use Equation (18.13) with $n = 12$ and $r_N = 0.06$. The result is

$$r_E = \left(1 + \frac{0.06}{12}\right)^{12} - 1 = (1.005)^{12} - 1 \approx 0.0617, \qquad \boxed{C}$$

or 6.17%. (A closer value is 6.16778%.)

For a fixed value of n (number of conversion periods per year), Equation (18.13) can be used to find the nominal rate that will produce a desired effective rate:

$$r_E + 1 = \left(1 + \frac{r_N}{n}\right)^n$$

$$\sqrt[n]{1 + r_E} = 1 + \frac{r_N}{n}$$

$$\frac{r_N}{n} = \sqrt[n]{1 + r_E} - 1$$

Nominal rate

$$r_N = n(\sqrt[n]{1 + r_E} - 1) \tag{18.14}$$

EXAMPLE 18.10 What nominal rate compounded quarterly will yield an effective rate of 8%?

Solution Use Equation (18.14) with $n = 4$ and $r_E = 0.08$. The result is

$$r_N = 4(\sqrt[4]{1.08} - 1) \approx 0.0777, \qquad \boxed{C}$$

or 7.77%. (A closer value is 7.77062%.)

EXERCISES FOR SECTION 18

1. Express the volume ($V = x^3$) of a cube in terms of the surface area ($S = 6x^2$) of the cube.
2. Express the area ($A = s^2$) of a square in terms of the perimeter ($P = 4s$) of the square.
3. Express the perimeter ($P = 4s$) of a square in terms of the area ($A = s^2$) of the square.
4. Express the area of a circle ($A = \pi r^2$) in terms of the circumference ($C = 2\pi r$) of the circle.
5. Express the surface area ($S = 4\pi r^2$) of a sphere in terms of the volume ($V = \frac{4}{3}\pi r^3$) of the sphere.
6. Express the volume ($V = \frac{4}{3}\pi r^3$) of a sphere in terms of the surface area ($S = 4\pi r^2$) of the sphere.
7. Solve Equation (18.4) for W.
8. What are the units on the constant K in Equation (18.4)? (Suggestion: The units on the two sides of the equation must be the same. See Section 8B.)
9. What are the units on the constant K in Equation (18.5)? (See the suggestion for Exercise 8.)

The next three exercises are based on Example 18.2.

10. Estimate the surface area of a 343 kg cow.
11. Estimate the surface area of a 27 kg boy.
12. Estimate the surface area of a 27 kg swine.

The next six exercises relate to Subsection B and blanks in Table 18.3. (C)

13. What is the average distance between Mercury and the sun?
14. What is the average distance between Venus and the sun?
15. What is the average distance between Jupiter and the sun?
16. What is the period for Saturn?
17. What is the period for Uranus?
18. What is the period for Neptune?

TABLE 18.3

Planet	*Average distance from sun*[a]	*Period*[b]
Mercury		0.24
Venus		0.62
Earth	93	1.00
Mars	142	1.88
Jupiter		11.86
Saturn	886	
Uranus	1782	
Neptune	2793	
Pluto	3672	248.43

[a] *In millions of miles.*
[b] *As multiples of the Earth's period.*

In the next exercises determine the compound amount for the given principal and investment conditions.*

19. \$100 for two years at 12% compounded quarterly.
20. \$200 for one year at 8% compounded quarterly.
21. \$500 for three years at 10% compounded semiannually.

In the next exercises determine the amount (present value) that must be invested to produce the specified compound amount under the given investment conditions.

22. \$1000; two years at 8% compounded quarterly.
23. \$500; one year at 6% compounded monthly.
24. \$2000; three years at 10% compounded semiannually.

In the next exercises determine the effective rate equivalent to each nominal rate.

25. 8% compounded quarterly.
26. 10% compounded quarterly.
27. 12% compounded quarterly.

In the next exercises determine the nominal rate equivalent to each effective rate under the given compounding conditions.

28. 9%; quarterly compounding.
29. 11%; quarterly compounding.
30. 13%; quarterly compounding.
31. What is the compound amount on \$1000 invested for $2\frac{1}{2}$ years at 6% compounded monthly?
32. How much must be invested at 6% compounded monthly so that \$500 accumulates at the end of six months?
33. What is the compound amount on \$500 invested for two years at 9% compounded monthly?
34. I know that in 18 months I'll need \$1000. To take care of this, how much should I invest now in an account that pays 10% compounded quarterly?
35. You place \$1000 in an account that pays 12% compounded quarterly. How much will be in the account after 18 months?
36. You place \$1000 for one year in an account that pays 8% compounded monthly. You place \$1000 for one year in another account that pays r% with no compounding. At the end of the year the amounts in the two accounts are equal. What is r?
37. One investment pays 8% compounded annually and another pays r% compounded monthly. Determine r so that the two investments earn equal amounts each year.
38. One investment pays 11% compounded annually and another investment pays r% compounded monthly. Determine r so that the two investments earn equal amounts each year.
39. In $3\frac{1}{2}$ years you will need \$3000 and you want to take care of this by investing now in an account that pays 10% compounded monthly. How much should you invest?

---

*The integral and fractional powers needed for Exercises 19–39 can be computed with a calculator or taken from the list at the end of the exercises.

The next three exercises can be solved with appropriate interpretations of the ideas and formulas in Subsection C.

40. Suppose the cost of living increases 10% per year over the next 20 years. Compute the cost at the end of that period of an item that costs \$1.00 now. C

41. The Consumer Price Index rose from 100.0 to 104.2 from 1967 to 1968. What would the index have been in 1977 if it had increased at that same annual rate from 1967 to 1977? (The 1977 index was actually 181.5.) C

42. (a) Solve Equation (18.9) for r.

(b) The Consumer Price Index (with 100.0 for 1967) rose from 94.5 to 161.2 over the 10-year period 1965 to 1975. What annual rate of increase, compounded annually, does that represent? C

The following are five-place approximations for the powers needed in Exercises 19–39. (The order here is more or less random.)

$1.02^4 = 1.08243$	$1.03^6 = 1.19405$
$1.005^{30} = 1.16140$	$1.05^6 = 1.34010$
$1.0075^{24} = 1.19641$	$1.03^8 = 1.26677$
$1.01^{12} = 1.12682$	$1.03^4 = 1.12551$
$1.00\overline{6}^{12} = 1.08300$	$1.008\overline{3}^{-42} = 0.70571$
$1.025^4 = 1.10381$	$1.02^{-8} = 0.85349$
$1.025^{-6} = 0.86230$	$1.05^{-6} = 0.74622$
$1.005^{-6} = 0.97052$	$1.09^{1/4} = 1.02178$
$1.11^{1/12} = 1.00873$	$1.11^{1/4} = 1.02643$
$1.13^{1/4} = 1.03103$	$1.00\overline{6}^{1/12} = 1.00055$
	$1.005^{-12} = 0.94191$

REVIEW EXERCISES FOR CHAPTER IV

Factor completely.

1. $9x^3 - 6x^2 + x$

2. $y^5 + y^2$

3. $15a^2 + 4ab - 4b^2$

4. $c^4 - d^4$

5. $m^3 - 3m^2n + 3mn^2 - n^3$

6. $u^2 + 2uv + v^2 - 4$

7. $x^{2n+1} - xy^{2n} + x^{2n}y - y^{2n+1}$

8. $x^{3n} - y^{3n}$

Expand and simplify.

9. $(a + 2b)^3$

10. $(c^2 - d^2)^3$

Perform the indicated operations and simplify.

11. $\dfrac{m^2 + 2m + 1}{3} \cdot \dfrac{6}{m + 1}$

12. $\dfrac{(u - v)}{uv} \cdot \dfrac{v^2}{u^2 + uv - 2v^2}$

13. $\dfrac{x^2 + 3x + 2}{2x} \div \dfrac{2x + 4}{x}$

14. $\dfrac{2x + 2}{x} \div \dfrac{x^2 - 1}{2x^2}$

15. $\dfrac{x}{x^2 - 1} - \dfrac{1}{x + 1}$

16. $\dfrac{1}{x^2 - x - 2} - \dfrac{1}{x^2 + x}$

17. $\dfrac{a-b}{\dfrac{1}{a}-\dfrac{1}{b}}$

18. $\dfrac{1-\dfrac{c^2}{d^2}}{1-\dfrac{c}{d}}$

Simplify. Assume that the variables represent positive real numbers.

19. $\sqrt[3]{8a^6}$

20. $\sqrt[4]{16a^8b^{12}}$

21. $3\sqrt{x^4y}-\sqrt{x^4y}+x^2\sqrt{y}$

22. $(\sqrt{x}-2\sqrt{y})(\sqrt{x}+\sqrt{4y})$

23. $\dfrac{32^{3/5}}{32^{-3/5}}$

24. $\left(\dfrac{-27}{8}\right)^{2/3}$

25. $\left(\dfrac{4a^4b^2}{25a^{-2}b^6}\right)^{-3/2}$

26. $\left(\dfrac{c^{3/2}}{c^{2/3}}\right)^{12}$

Rationalize each denominator and simplify.

27. $\dfrac{5}{\sqrt{3}}$

28. $\dfrac{1}{a^{1/2}-b^{1/2}}$

29. Solve $y^2=2x^3$ for x. (Assume $x>0$ and $y>0$.)

30. Solve $\sqrt[3]{a^2}=b$ for a. (Assume $a>0$ and $b>0$.)

For Exercises 31–36, if you do not have a calculator to evaluate the powers that arise, simply write an expression that will give the answer but do not carry out the computations.

31. What is the compound amount on \$2000 invested for three years at 8% compounded quarterly?
32. What is the compound amount on \$5000 invested for $2\frac{1}{2}$ years at 6% compounded monthly?
33. How much must be invested at 8% compounded quarterly so that \$10,000 will be accumulated at the end of ten years?
34. How much must be invested at 10% compounded quarterly so that \$2000 will be accumulated at the end of four years?
35. What effective rate is equivalent to a nominal rate of 6% compounded quarterly?
36. What nominal rate compounded monthly will yield an effective rate of 10%?

CHAPTER V
INTRODUCTION TO GRAPHS

The idea behind this chapter is so simple that its importance can easily be missed. This idea—to use pairs of numbers to represent points in a plane—is the primary link between algebra and geometry. It allows us to restate geometric problems as problems about numbers, which can often be solved using algebra. Conversely, it allows us to visualize many algebraic problems geometrically. The ability to solve a mathematical problem often rests on little more than the skill to move freely between algebra and geometry in this way.

Sections 19 and 20 are essential for the remainder of the book. Section 21 is optional, as explained at the beginning of the section.

SECTION

19 Cartesian Coordinates. Circles

A. Coordinate Systems

The basis for Figures 19.1 and 19.2 is two perpendicular lines: an **x-axis,** directed to the right, and a **y-axis,** directed upward. The point of intersection of the axes, called

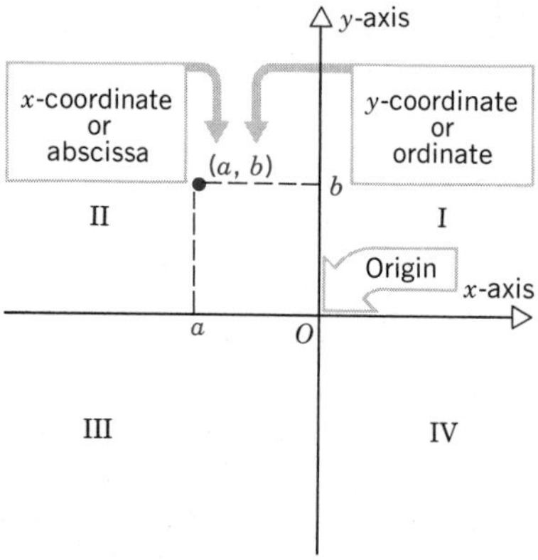

Figure 19.1

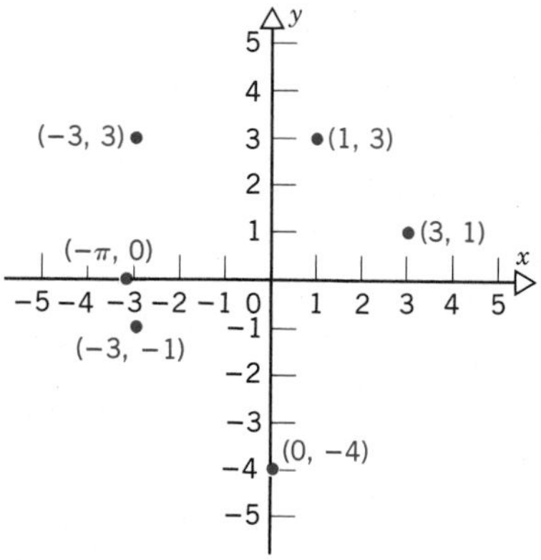

Figure 19.2

the **origin,** corresponds to the zero point on each axis. The origin is denoted by the letter O.

Each point in the plane of the axes is represented by an ordered pair* of real numbers, called its **coordinates:**

> The first coordinate is the directed distance of the point from the y-axis. We say *directed* distance because direction is taken into account—the coordinate is positive if the point is to the right of the y-axis, and negative if the point is to the left.
>
> The second coordinate is the directed distance of the point from the x-axis. This coordinate is positive if the point is above the x-axis, and negative if the point is below.

The first coordinate is called the **x-coordinate** (or the **abscissa**). The second coordinate is called the **y-coordinate** (or the **ordinate**).

By representing points by coordinates in this way we achieve a one-to-one correspondence between the set of points in the plane and the set of ordered pairs of real

*When we say that (a, b) is an **ordered pair,** we mean that it is to be distinguished from (b, a); that is, it is important which symbol appears first (reading from left to right). Thus $(3, 1) \neq (1, 3)$. Coordinate pairs (a, b) and (c, d) are equal iff $a = c$ and $b = d$.

numbers: each point is assigned to precisely one pair, and each pair is assigned to precisely one point. This one-to-one correspondence is called a **Cartesian**† (or **rectangular) coordinate system** for the plane. A plane with a Cartesian coordinate system is called a **Cartesian plane,** and the coordinates of its points are called **Cartesian coordinates.** When there is no chance of confusion we often refer to a point P with coordinates (x, y) as "the point (x, y)." The notation $P(x, y)$ will also denote the point P with coordinates (x, y).

The axes divide their plane into four **quadrants,** always labeled I, II, III, and IV as shown in Figure 19.1. The same unit of distance is generally used on the two axes unless the circumstances make some other choice more convenient. Also, the letters x and y are the standard labels for the axes, but circumstances will sometimes lead to other choices.

B. Graphs

With each equation involving one or both of the variables x and y we can associate a set of points in the plane, as follows. First, an ordered pair (a, b) of numbers is said to **satisfy** the equation if equality results when x is replaced by a and y is replaced by b.

DEFINITION. The **graph** of an equation involving one or both of the variables x and y is the set of all points in the Cartesian plane whose coordinates satisfy the equation.

EXAMPLE 19.1 The pair $(1, 2)$ satisfies $y = 2x$ because $2 = 2 \cdot 1$. The pairs $(-2, -4)$ and $(3, 6)$ also satisfy $y = 2x$. These points have been plotted in Figure 19.3, and the graph of $y = 2x$ is the straight line that passes through them. (We'll see later why the complete graph is a straight line.)

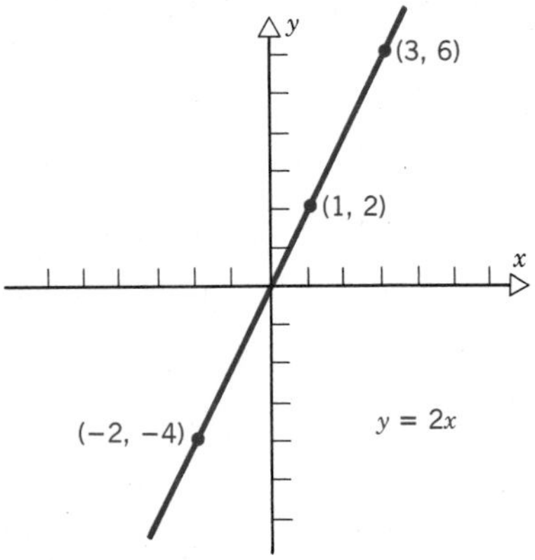

Figure 19.3

The next two examples concern equations with only one of x and y.

EXAMPLE 19.2 Draw the graph (in a Cartesian plane) of the equation $x = 3$.

---

†In honor of René Descartes (1596–1650), who stressed the value of treating algebraic problems geometrically.

Solution A point will be on the graph of $x = 3$ if its x-coordinate is 3, regardless of its y-coordinate. Therefore, the graph is the line shown in Figure 19.4. Lines such as this, parallel to the y-axis, are called **vertical lines.** If c is any real number, the graph of $x = c$ is a vertical line.

EXAMPLE 19.3 Draw the graph of the equation $y = -2$.

Solution A point will be on the graph of $y = -2$ if its y-coordinate is -2, regardless of its x-coordinate. Therefore, the graph of $y = -2$ is the line shown in Figure 19.5. Lines such as this, parallel to the x-axis, are called **horizontal lines.** If c is any real number, the graph of $y = c$ is a horizontal line.

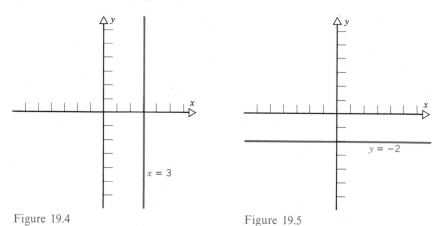

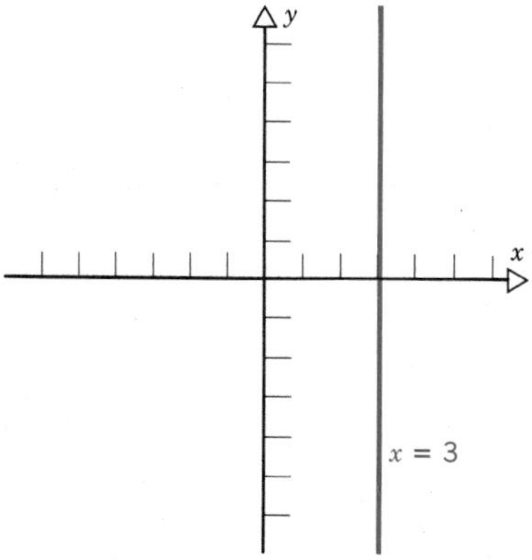

Figure 19.4

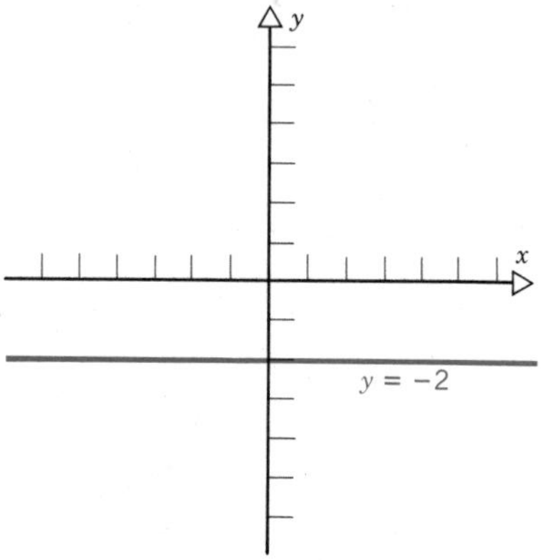

Figure 19.5

To determine the graphs of some equations we must simply plot a number of points and then look for a pattern. In many other cases, however, we can determine what the graphs will be just by looking at the equation. For example, the distance formula that follows will help us determine all of the equations whose graphs are circles. And in the next section we'll determine the equations whose graphs are lines.

C. The Distance Formula

Recall that the distance between points with coordinates a and b on a real line is $|a - b|$ (Section 6D). We now derive the corresponding formula for points in a Cartesian plane.

Two points $P_1(x_1, y_1)$ and $P_2(x_2, y_2)$ are on the same vertical line iff $x_1 = x_2$ (first coordinates equal). The distance between such points is

$$|y_1 - y_2|. \tag{19.1}$$

EXAMPLE 19.4 The points (7, 4) and (7, 1) are on the same vertical line (Figure 19.6). The distance between them is $|4 - 1| = 3$.

Two points $P_1(x_1, y_1)$ and $P_2(x_2, y_2)$ are on the same horizontal line iff $y_1 = y_2$ (second coordinates equal). The distance between such points is

$$|x_1 - x_2|. \tag{19.2}$$

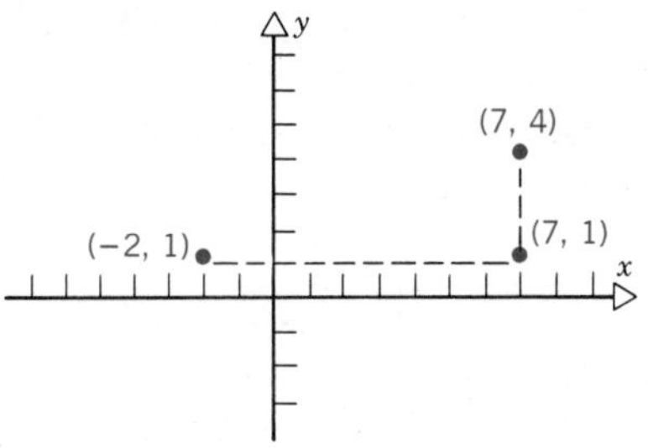

Figure 19.6

EXAMPLE 19.5 The points $(-2, 1)$ and $(7, 1)$ are on the same horizontal line (Figure 19.6). The distance between them is $|-2 - 7| = 9$.

Distance formula. The distance d between points $P_1(x_1, y_1)$ and $P_2(x_2, y_2)$ is

$$d = \sqrt{(x_1 - x_2)^2 + (y_1 - y_2)^2}. \tag{19.3}$$

Proof. Figure 19.7 shows typical points P_1 and P_2. The vertical and horizontal lines through these points, as shown, intersect at the point with coordinates (x_1, y_2), which has been labeled Q. Triangle P_1QP_2 is a right triangle with hypotenuse P_1P_2. The legs P_1Q and QP_2 have lengths $|y_1 - y_2|$ and $|x_1 - x_2|$, respectively. Therefore, using the Pythagorean Theorem and the fact that $|a|^2 = a^2$, we have

$$\begin{aligned} d^2 &= |x_1 - x_2|^2 + |y_1 - y_2|^2 \\ &= (x_1 - x_2)^2 + (y_1 - y_2)^2. \end{aligned}$$

Equation (19.3) follows by taking a square root. □

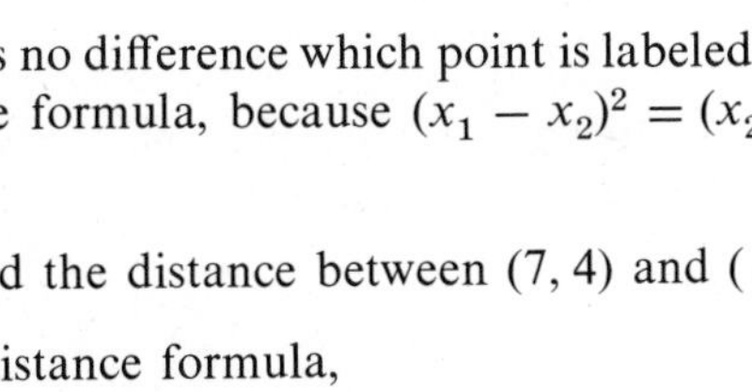

Figure 19.7

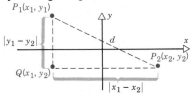

Notice that it makes no difference which point is labeled P_1 and which is labeled P_2 in using the distance formula, because $(x_1 - x_2)^2 = (x_2 - x_1)^2$ and $(y_1 - y_2)^2 = (y_2 - y_1)^2$.

EXAMPLE 19.6 Find the distance between $(7, 4)$ and $(-2, 1)$ (Figure 19.6).

Solution By the distance formula,

$$\begin{aligned} d &= \sqrt{[7 - (-2)]^2 + [4 - 1]^2} = \sqrt{9^2 + 3^2} = \sqrt{90} \\ &= \sqrt{9}\sqrt{10} = 3\sqrt{10}. \end{aligned}$$

D. Circles

Figure 19.8 shows the graph consisting of all points on the circle with center at $P(h, k)$ and radius r. A point $Q(x, y)$ will be on this circle iff its distance from $P(h, k)$ is r. By the distance formula, this will be true iff

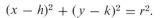

$$\sqrt{(x - h)^2 + (y - k)^2} = r$$

or

$$(x - h)^2 + (y - k)^2 = r^2.$$

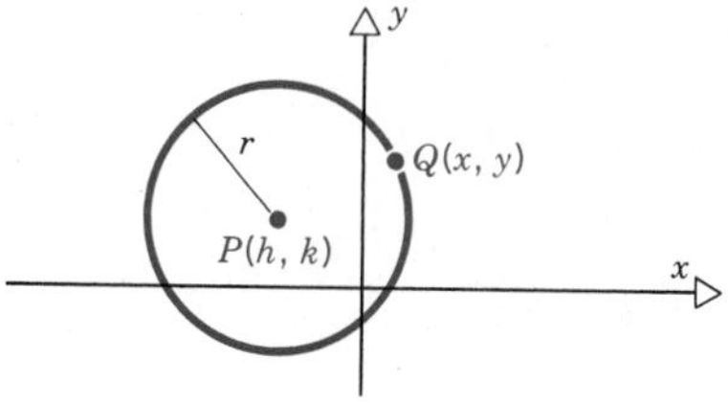

Figure 19.8

This proves the following result.

> If r is a positive real number, then the graph of
>
> $$(x - h)^2 + (y - k)^2 = r^2 \tag{19.4}$$
>
> is a circle with center at $P(h, k)$ and radius r.

For a circle with center at the origin and radius r, Equation (19.4) becomes $x^2 + y^2 = r^2$.

EXAMPLE 19.7 Determine an equation of the circle with center at $(3, -2)$ and radius 5.

Solution By Equation (19.4), the answer is

$$[x - 3]^2 + [y - (-2)]^2 = 5^2$$

or

$$(x - 3)^2 + (y + 2)^2 = 25.$$

EXAMPLE 19.8 Determine the equation of the circle with center at $(-4, 6)$ and passing through $(-1, 2)$.

Solution The circle is shown in Figure 19.9. The radius must be the distance between $(-4, 6)$ and $(-1, 2)$, which is

$$\sqrt{[-4 - (-1)]^2 + [6 - 2]^2} = \sqrt{9 + 16} = 5.$$

Therefore, the answer is

$$(x + 4)^2 + (y - 6)^2 = 25.$$

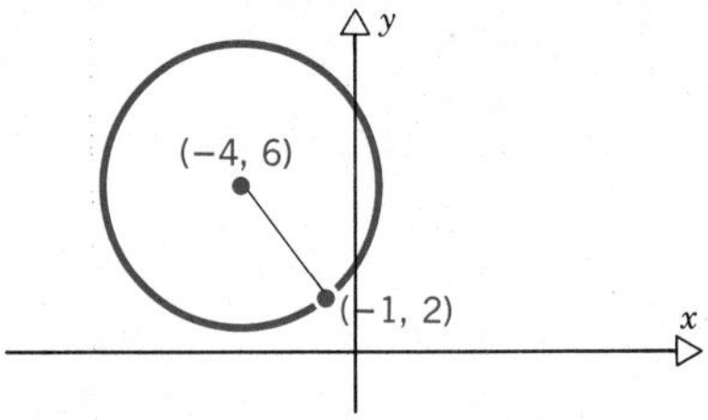

Figure 19.9

The equation of the circle in the preceding example can be rewritten as

$$x^2 + 8x + 16 + y^2 - 12y + 36 = 25$$

or

$$x^2 + y^2 + 8x - 12y + 27 = 0.$$

The last equation has the form

$$x^2 + y^2 + ax + by + c = 0, \tag{19.5}$$

where a, b, and c are real numbers. To determine the graph of such an equation we first complete the squares on the terms involving x and the terms involving y, as in the following examples. (Completing the square is discussed in Section 10.)

EXAMPLE 19.9 Characterize the graph of

$$x^2 + y^2 - 10x + 6y + 30 = 0.$$

Solution To complete the square on the x terms add 25 to both sides of the equation. To complete the square on y add 9 to both sides.

$$(x^2 - 10x + 25) + (y^2 + 6y + 9) + 30 = 25 + 9$$
$$(x - 5)^2 + (y + 3)^2 = 4$$

Comparison of this equation with Equation (19.4) shows that the graph is the circle with center at $(5, -3)$ and radius 2.

EXAMPLE 19.10 Characterize the graph of

$$x^2 + y^2 + 10x - 2y + 26 = 0.$$

Solution If we complete the squares on the x terms and the y terms, the result is

$$(x + 5)^2 + (y - 1)^2 = 0.$$

The only solution of this equation is $(-5, 1)$. Thus the graph consists of a single point.

EXAMPLE 19.11 Characterize the graph of

$$x^2 - 20x + y^2 + 6y + 113 = 0.$$

Solution If we complete the squares we obtain

$$(x - 10)^2 + (y + 3)^2 = -4.$$

Since $(x - 10)^2 \geq 0$ for every x and $(y + 3)^2 \geq 0$ for every y, but $-4 < 0$, the equation has no solution. Thus the graph is the empty set (it contains no points).

The method in the last three examples can be used to reduce any equation of the form (19.5) to the form

$$(x - h)^2 + (y - k)^2 = t.$$

Depending on whether the number t on the right is positive, zero, or negative, we have, respectively, the outcomes in the following statement.

> The graph of an equation that can be written in the form (19.5) is either a circle, a single point, or the empty set.

EXERCISES FOR SECTION 19

For each exercise, draw one set of coordinate axes and plot the given points.

1. $(2, 4), (-2, 4), (0, -3), (3.5, 0), (\sqrt{2}, -\pi), (-1, -2)$
2. $(5, 3), (3, -5), (\pi, 0), (0, -\sqrt{2}), (-1.5, -2.5), (-4, 2)$
3. $(1, 4), (-1, 0), (\sqrt{2}, -1), (0, \pi), (-4, 3), (1.5, 3.5)$

Give the Cartesian coordinates of each point.

4. 3 units above the x-axis and 2 units to the left of the y-axis.
5. 2.5 units to the right of the y-axis and 4 units below the x-axis.
6. 1.7 units below the x-axis and 3 units to the left of the y-axis.
7. On the y-axis and 4 units below the x-axis.
8. On the x-axis and 2.5 units to the left of the y-axis.
9. On the y-axis and 3.2 units above the x-axis.

In Exercises 10–15, determine which of the given points are on the graph of the given equation.

10. $y = 3x$; $(2, 6), (0.3, 1), (-\frac{1}{2}, -\frac{3}{2})$
11. $y = 2$; $(0, 2), (2, 0), (4, 2)$
12. $x = -3$; $(0, -3), (-3, 0), (-3, 2)$
13. $y = x^2$; $(1, 1), (2, -4), (-2, 4)$
14. $y^3 = x$; $(1, -1), (-\frac{1}{2}, -\frac{1}{8}), (-\frac{1}{8}, -\frac{1}{2})$
15. $y^2 + 1 = (x + 1)^2$; $(0, -1), (-1, 0), (-1, -1)$

Draw the graph of each equation.

16. $x = 3$
17. $y = -1$
18. $y = 4$
19. $y = -1.5$
20. $x = 7/5$
21. $x = -2$
22. $x^2 + y^2 = 9$
23. $x^2 + y^2 = 16$
24. $x^2 + y^2 = 4$
25. $(x - 1)^2 + y^2 = 10$

26. $(x + 4)^2 + (y - 3)^2 = 15$
27. $(x + 2)^2 + (y + 1)^2 = 12$

Determine the equations of the following lines and circles.

28. The line parallel to and 4 units above the x-axis.
29. The line parallel to and 3 units below the x-axis.
30. The line parallel to and 2 units to the left of the y-axis.
31. The line parallel to and 5 units to the right of the y-axis.
32. The line parallel to and 1.5 units to the left of the y-axis.
33. The line parallel to and 5 units above the x-axis.
34. The circle with center at $(1.5, 0)$ and radius $\sqrt{6}$.
35. The circle with center at $(-2, -3)$ and radius $\sqrt{2}$.
36. The circle with center at $(0, 4.5)$ and radius 10.
37. The circle with center at the origin and passing through $(3, -4)$.
38. The circle with center at $(1, 2)$ and passing through the origin.
39. The circle with center at $(-3, 1)$ and tangent to the y-axis.

Find the distance between each pair of points.

40. $(2, 5), (-1, 4)$
41. $(\frac{3}{2}, -2), (\frac{9}{2}, 3)$
42. $(-1, \frac{8}{3}), (7, \frac{2}{3})$
43. $(0, \frac{1}{4}), (\frac{1}{2}, -\frac{3}{5})$
44. $(-\frac{1}{2}, -2), (0.4, 0)$
45. $(2, 0), (-0.5, -1)$

Characterize the graph of each equation.

46. $x^2 + y^2 - 2x + 4y - 4 = 0$
47. $x^2 + y^2 + 12x - 2y - 37 = 0$
48. $x^2 + y^2 - 2x - 8y - 8 = 0$
49. $x^2 + y^2 - 6y = -10$
50. $x^2 + y^2 - 2y = 3$
51. $x^2 - 20x + y^2 = -100$
52. $x^2 + y^2 = 0$
53. $x^2 + y^2 + 9 = 0$
54. $x^2 + y^2 - x + y + 1 = 0$

Describe the set of all points in the Cartesian plane whose coordinates satisfy the given condition.

55. $x < 0$
56. $xy > 0$
57. $x/y < 0$
58. $x > y$
59. $x^2 + y^2 > 5$
60. $x^2 + y^2 \leq 1$

61. Use similar triangles to prove that the midpoint P of the segment joining $P_1(x_1, y_1)$ and $P_2(x_2, y_2)$ has coordinates

$$\left(\frac{x_1 + x_2}{2}, \frac{y_1 + y_2}{2}\right) \quad \text{(midpoint formula).}$$

(See Examples 9.7 and 9.8.) Where is the midpoint of the segment joining $(-3, 2)$ and $(4, 7)$?

62. Use the distance formula and the first part of Exercise 61 to prove that the midpoint of the hypotenuse of a right triangle is equidistant from the three vertices of the triangle. [First, explain why the vertices of the triangle can be assumed to be at $P_1(0, 0)$, $P_2(a, 0)$, and $P_3(0, b)$ for some real numbers a and b.]

63. Use the first part of Exercise 61 to prove that the diagonals of every parallelogram bisect each other. [First, explain why the vertices of the parallelogram can be assumed to be at $P_1(0, 0)$, $P_2(a, 0)$, $P_3(a + b, c)$ and $P_4(b, c)$ for some real numbers a, b, and c.]

SECTION 20 Lines

A. Introduction

In this section we study lines (straight lines) and the equations whose graphs are lines. We'll see that the equations are precisely those covered by the following definition.

DEFINITION. An equation with variables x and y is called a **linear equation** if it can be written in the form

$$Ax + By + C = 0, \tag{20.1}$$

where A, B, and C are real numbers with A and B not both 0.

EXAMPLE 20.1 These are linear equations:

$$3x - 2y + 5 = 0 \quad (A = 3, B = -2, C = 5)$$
$$y - 4 = 0 \quad (A = 0, B = 1, C = -4).$$

Also, $2x = -9$ and $y = 2 - x$ are linear equations, since they can be written as

$$2x + 9 = 0 \quad (A = 2, B = 0, C = 9)$$
$$x + y - 2 = 0 \quad (A = 1, B = 1, C = -2).$$

Figure 20.1 shows that the graphs of the linear equations in Example 20.1 are all lines. To move from these examples to more general statements we must first examine some basic facts about lines.

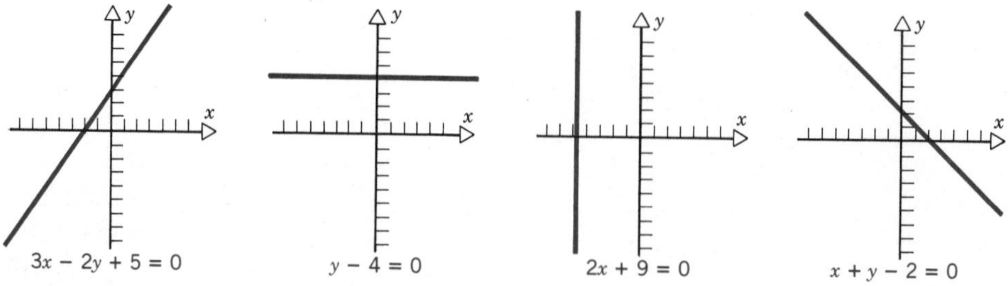

Figure 20.1

B. The Slope of a Line

To study nonvertical lines we begin with the idea of their *slopes*. The slope measures the steepness of a line, and also whether an increase in x causes an increase in y, no change in y, or a decrease in y. In the definition of slope, which follows, we use the capital Greek letter Δ (delta) to denote *change*. In particular, Δx denotes a change in x, and Δy denotes a change in y. In this context Δx and Δy are each to be treated as single symbols: for example, Δx does *not* mean "Δ times x."

DEFINITION. The **slope** of the line through $P_1(x_1, y_1)$ and $P_2(x_2, y_2)$, with $x_1 \neq x_2$, is denoted by m and is defined by

$$m = \frac{\Delta y}{\Delta x} = \frac{y_2 - y_1}{x_2 - x_1}. \tag{20.2}$$

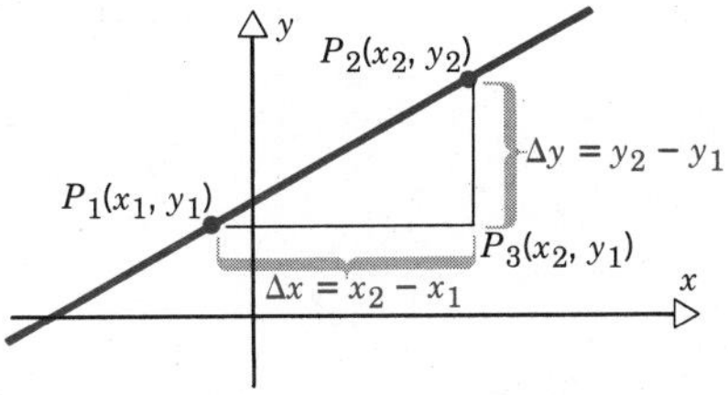

Figure 20.2

From Figure 20.2 we see that the slope is

$$\frac{\text{the change in } y}{\text{the change in } x}$$

as we move from one point to another along the line. Notice, in particular, that if $\Delta x = 1$, then $m = \Delta y$. Thus the slope is the number of units that y increases or decreases for each unit of increase in x; if y increases the slope is positive and if y decreases the slope is negative.

EXAMPLE 20.2 Determine the slope of the line through $(-6, 0)$ and $(2, 4)$.

Solution Use $(x_1, y_1) = (-6, 0)$ and $(x_2, y_2) = (2, 4)$ in Equation (20.2):

$$m = \frac{4 - 0}{2 - (-6)} = \frac{4}{8} = \frac{1}{2}.$$

It makes no difference which point is labeled P_1 and which is labeled P_2 for Equation (20.2), because

$$\frac{y_2 - y_1}{x_2 - x_1} = \frac{y_2 - y_1}{x_2 - x_1} \cdot \frac{(-1)}{(-1)} = \frac{y_1 - y_2}{x_1 - x_2}.$$

Interchanging the order in Example 20.2, for instance, we get

$$m = \frac{0 - 4}{-6 - 2} = \frac{-4}{-8} = \frac{1}{2},$$

as before.

Moreover, we can choose any other pair of points on the same line and we will get the same number for the slope. For Figure 20.3 this can be seen as follows. The triangles P_1PP_2 and P_3QP_4 are similar (that is, corresponding angles are equal) because the angles labeled α are equal and the angles at P and Q are right angles. Therefore, because ratios between corresponding sides of similar triangles are equal,

$$\frac{y_4 - y_3}{x_4 - x_3} = \frac{y_2 - y_1}{x_2 - x_1}. \tag{20.3}$$

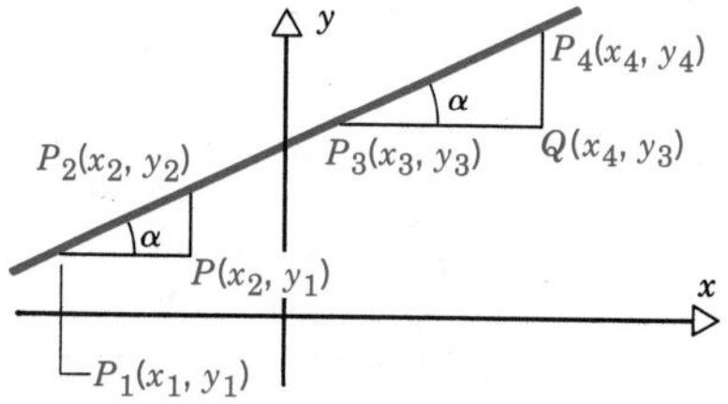

Figure 20.3

If Equation (20.3) were not true, we could not refer to *the* slope of a line, for a line would have more than one slope.

Figure 20.4 shows the lines through the origin having slopes $\pm\frac{1}{2}$, ± 1, ± 2, and ± 4. Also, the x-axis has slope 0. Figure 20.4 illustrates the following facts, which can be proved from equation (20.2).

> $m < 0$ iff y decreases as x increases.
> $m = 0$ iff y is constant as x increases.
> $m > 0$ iff y increases as x increases.
> The larger the number $|m|$, the steeper the line.

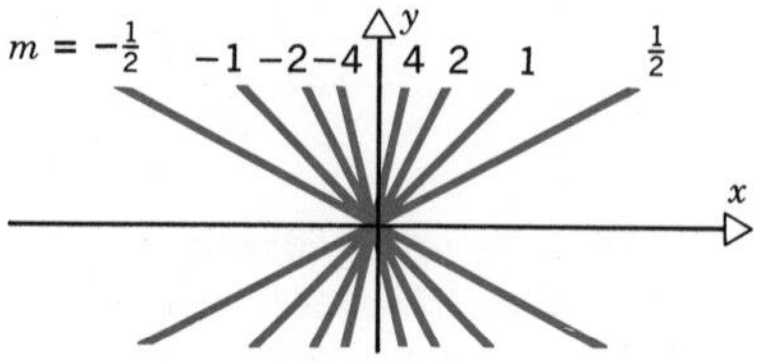

Figure 20.4

Any attempt to compute the slope of a vertical line will lead to division by 0. Therefore, we say that *the slope of a vertical line is undefined.*

C. Point-Slope Form

There is exactly one line through a given point with a given slope. Its equation is determined as follows.

> An equation for the line through the point $P_1(x_1, y_1)$ with slope m is
>
> $$y - y_1 = m(x - x_1). \qquad (20.4)$$
>
> This is called the **point-slope form** for the equation of the line.

Proof. Let L denote the line through $P_1(x_1, y_1)$ with slope m. If $P(x, y)$ is any point on L with $x \neq x_1$ (Figure 20.5), then P and P_1 can be used to compute the slope of L, so that

$$\frac{y - y_1}{x - x_1} = m$$
$$y - y_1 = m(x - x_1).$$

Since (x_1, y_1) also satisfies this equation, which is the same as Equation (20.4), we see that every point on L satisfies Equation (20.4). Conversely, any point $P(x, y)$ whose coordinates satisfy Equation (20.4) must be on L since there is only one line through $P_1(x_1, y_1)$ with slope m. □

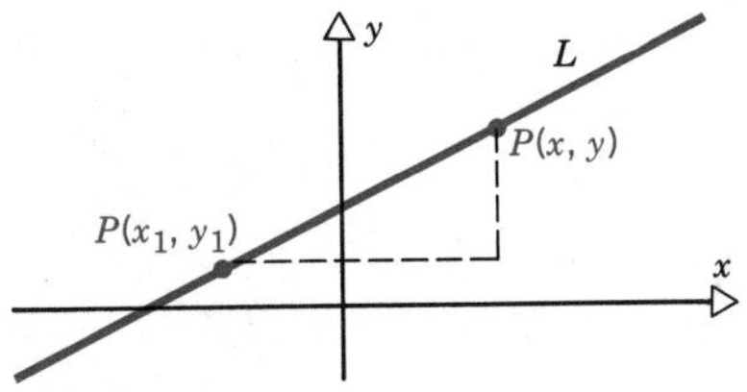

Figure 20.5

EXAMPLE 20.3 Find an equation for the line through $(-1, 5)$ with slope 2.

Solution Use Equation (20.4) with $m = 2$ and $(x_1, y_1) = (-1, 5)$:

$$y - 5 = 2[x - (-1)]$$
$$y - 5 = 2x + 2$$
$$2x - y + 7 = 0.$$

If two points on a line are given, the slope can be found from Equation (20.2), and then the equation of the line can be determined by using the slope and either one of the points in the point-slope form.

D. Lines and Linear Equations

We are now in a position to prove the following statements regarding lines and linear equations.

> Any line is the graph of a linear equation. Conversely, the graph of any linear equation is a line.

Proof. The point-slope form shows that a nonvertical line is the graph of an equation of the form

$$y - y_1 = m(x - x_1),$$

which can be rewritten as

$$y - y_1 = mx - mx_1$$
$$mx - y + (y_1 - mx_1) = 0.$$

This has the form

$$Ax + By + C = 0$$

with

$$A = m, \quad B = -1, \quad \text{and} \quad C = y_1 - mx_1.$$

On the other hand, a vertical line is the graph of a linear equation of the form $x - c = 0$ (Example 19.2). Thus we have verified that *any line is the graph of a linear equation.* Now let's see why, conversely, the graph of any linear equation is a line.

If $B = 0$ in Equation (20.1), $Ax + By + C = 0$, then $A \neq 0$, since we specified that not both A and B are 0. Therefore, the equation can be rewritten as $x = -C/A$, which is the graph of a vertical line.

On the other hand, if $B \neq 0$ in the equation $Ax + By + C = 0$, then the equation can be rewritten as

$$By + C = -Ax$$
$$y + \frac{C}{B} = -\frac{A}{B}x$$
$$y - \left(-\frac{C}{B}\right) = \frac{-A}{B}(x - 0).$$

If we compare this equation with Equation (20.4), we see that this is the equation of the nonvertical line through the point $(0, -C/B)$ with slope $-A/B$. Thus we have shown that *the graph of any linear equation is a line.* □

E. Slope-Intercept Form

Now that we know that the graph of any linear equation is a line, it is easy to draw its graph: plot any two points determined by the equation and then draw the line through those two points. The graph will tend to be more accurate if we choose points that are not too close together. The *intercepts,* whose definition follows, are usually the easiest points to determine and can be used if they are sufficiently far apart. In any case, it's generally a good idea to plot a third point to serve as a check.

The x-coordinate where a graph intersects the x-axis is called an **x-intercept** of the graph. (To compute the x-intercepts set $y = 0$ and solve for x.)

The y-coordinate where a graph intersects the y-axis is called a **y-intercept** of the graph. (To compute the y-intercepts set $x = 0$ and solve for y.)

EXAMPLE 20.4 Draw the graph of $3x - 4y + 12 = 0$.

Solution If $y = 0$, then $3x + 12 = 0$ and $x = -4$. Thus -4 is the x-intercept. If $x = 0$, then $-4y + 12 = 0$ and $y = 3$. Thus 3 is the y-intercept. These intercepts have been plotted to determine the graph in Figure 20.6. Also, if $x = 4$, then $12 - 4y + 12 = 0$ and $y = 6$; the point $(4, 6)$ has been plotted as a check.

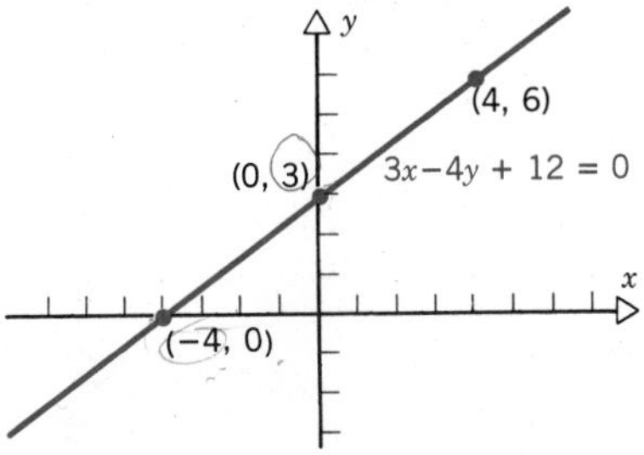

Figure 20.6

The following form for the equation of a line is often more useful than the point-slope form.

> An equation of the line with slope m and y-intercept b is
>
> $$y = mx + b. \tag{20.5}$$
>
> This is called the **slope-intercept form** for the equation of the line.

Proof. To say that the y-intercept is b is the same as to say that the line passes through the point $(0, b)$. Using the point-slope form, Equation (20.4), with $(0, b)$ for the point and m for the slope, we obtain

$$y - b = m(x - 0)$$
$$y = mx + b. \qquad \square$$

EXAMPLE 20.5 An equation of the line with slope -2 and y-intercept 4 is

$$y = -2x + 4.$$

EXAMPLE 20.6 Find the slope and y-intercept of the graph of

$$5x - 2y - 4 = 0.$$

Solution Solve the equation for y and compare the answer with Equation (20.5). First,

$$2y = 5x - 4$$
$$y = \tfrac{5}{2}x - 2.$$

Hence the slope is $\frac{5}{2}$ (the coefficient of x) and the y-intercept is -2 (the constant term).

Figure 20.7 shows the effect of changing m with b held constant. Figure 20.8 shows the effect of changing b with m held constant. Figure 20.8 illustrates the following fact.

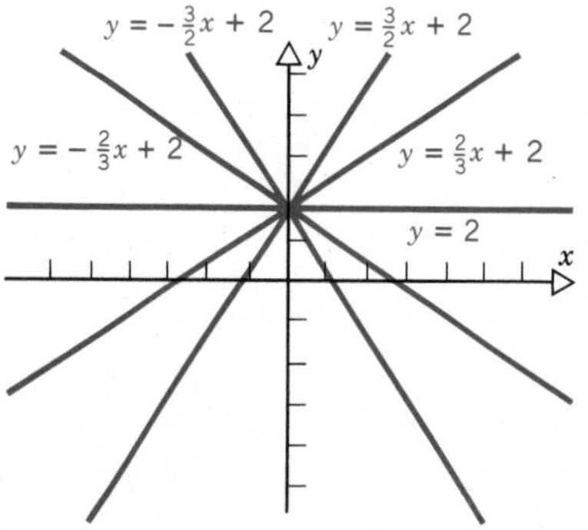

Figure 20.7 $y = mx + 2$ for different values of m.

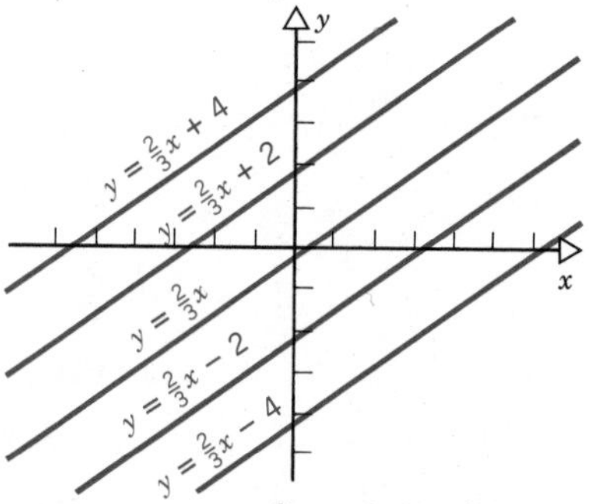

Figure 20.8 $y = \frac{2}{3}x + b$ for different values of b.

> Two lines with slopes m_1 and m_2 are parallel iff $m_1 = m_2$.

Proof. For the case of parallel lines with positive slopes refer to Figure 20.9. The slope of line L_1 can be computed with points P_1 and Q_1. The slope of line L_2 can be

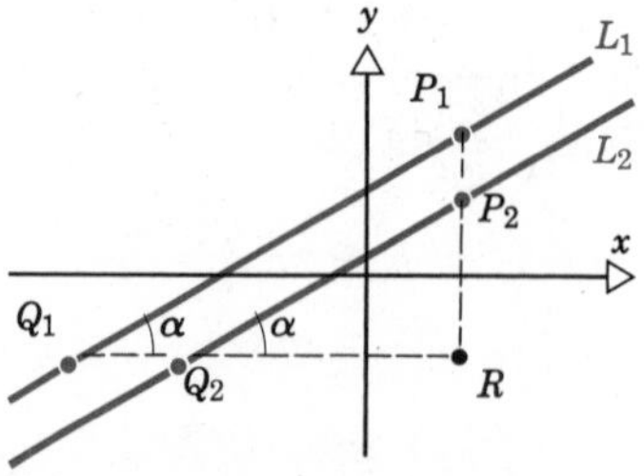

Figure 20.9

computed with points P_2 and Q_2. Using, in turn, facts about parallel lines, similar triangles, and slope, we can write

L_1 and L_2 are parallel

iff

the angles labeled by α are equal

iff

triangles P_1Q_1R and P_2Q_2R are similar

iff

$\overline{P_1R}/\overline{Q_1R} = \overline{P_2R}/\overline{Q_2R}$

iff

$m_1 = m_2$.

The proof for lines with negative slopes is similar. □

EXAMPLE 20.7 The graph of $y = 3x + 5$ is parallel to the graph of $y = 3x - 2$ because each is a line with slope 3, by the slope-intercept form.

EXAMPLE 20.8 Find an equation of the line containing the point $(-1, 3)$ and parallel to the line whose equation is $2x + y - 1 = 0$.

Solution If we rewrite the given equation in the form $y = mx + b$, we have $y = -2x + 1$; thus the slope of the given line, and any parallel line, is -2. Using the point-slope form, Equation (20.4), with the point $(-1, 3)$ and slope -2, we obtain

$$y - 3 = -2(x + 1).$$

It can be proved that two lines with nonzero slopes m_1 and m_2 are *perpendicular* iff $m_1 = -1/m_2$.

EXERCISES FOR SECTION 20

Determine the slope of the line through each pair of points.

1. $(3, 1)$ and $(2, 4)$ **2.** $(1, 3)$ and $(4, 2)$ **3.** $(-7, 1)$ and $(2, 2)$
4. $(-0.5, 4.3)$ and $(0.1, 1.3)$ **5.** $(-2, \frac{7}{3})$ and $(2, \frac{7}{3})$ **6.** $(5.2, 3.1)$ and $(7.6, -0.5)$

Determine an equation of the line through the given point with the given slope.

7. $(-3, 1)$; $m = \frac{1}{2}$ **8.** $(-1, -2)$; $m = -0.4$ **9.** $(4, 1)$; $m = \frac{1}{4}$
10. $(6, -1)$; $m = -3$ **11.** $(3.1, 4)$; $m = 2.5$ **12.** $(-4, -2)$; $m = -3.1$

Determine an equation of the line through each pair of points. Write the answer in the form $ax + by + c = 0$.

13. $(1, 2)$ and $(-4, 0)$ **14.** $(4, 7)$ and $(2, -1)$ **15.** $(-5, -2)$ and $(2, 5)$
16. $(-3, 5)$ and $(-2, -1)$ **17.** $(0, -5)$ and $(1, 1)$ **18.** $(7, 2)$ and $(-1, 0)$

Determine an equation of the line with the given slope (m) and y-intercept (b).

19. $m = -1, b = 2$ **20.** $m = 2, b = -1$ **21.** $m = 0, b = 2$
22. $m = -\frac{1}{3}, b = 0$ **23.** $m = 0, b = 0$ **24.** $m = \frac{1}{2}, b = -1$

Determine the slope and y-intercept of the graph of each equation.

25. $2x + y - 3 = 0$ **26.** $-x + 2y = 5$ **27.** $5x - 10y - 15 = 0$
28. $y - 5 = 0$ **29.** $x = y$ **30.** $2y = -5$

Draw the graph of each equation.

31. $x + y = 0$ **32.** $-2x + y = 0$ **33.** $x - 2y = 0$
34. $x + y + 1 = 0$ **35.** $x - 2y - 2 = 0$ **36.** $2x + 3y - 6 = 0$

In each of Exercises 37–39, use a single Cartesian plane (coordinate system) to draw the three graphs determined by $b = -2$, $b = 0$, and $b = 3$.

37. $y = 2x + b$ **38.** $y = -3x + b$ **39.** $y = \frac{1}{2}x + b$

In each of Exercises 40–42, use a single Cartesian plane (coordinate system) to draw the three graphs determined by $m = -2$, $m = 0$, and $m = 1$.

40. $y = mx + 1$ **41.** $y = mx$ **42.** $y = mx + 3$

43. Determine an equation of the line through the origin and parallel to the line whose equation is $2x - 3y + 5 = 0$.

44. Determine an equation of the line whose y-intercept is 2 and that is parallel to the line $x + y = 0$.

45. Determine an equation of the line whose x-intercept is -2 and that is parallel to the line through $(4, 0)$ and $(0, -4)$.

46. Prove that an equation of the line through $P_1(x_1, y_1)$ and $P_2(x_2, y_2)$, with $x_1 \neq x_2$, is

$$y - y_1 = \frac{y_2 - y_1}{x_2 - x_1}(x - x_1).$$

(This is called the **two-point form** for the equation of the line.)

47. Prove that an equation of the line with x-intercept a and y-intercept b is

$$\frac{x}{a} + \frac{y}{b} = 1 \quad (a \neq 0, b \neq 0).$$

(This is called the **intercept form** for the equation of the line.)

48. Prove that if the x- and y-intercepts of a line are equal, then the slope of the line is -1.

49. Use the slope-intercept form to prove that if a line passes through the point $(1, 1)$, then its slope is one minus its y-intercept.

50. A line cuts the positive x- and y-axes to form an isosceles triangle having area 8 square units. Find its equation. (Exercise 47 may help.)

51. Find an equation of the line that is parallel to the line $y = 2x$ and bisects the circle whose equation is $x^2 - 4x + y^2 + 6y + 9 = 0$.

SECTION

21 Conic Sections (Optional*)

A. Introduction

A **conic section** is a curve formed by the intersection of a plane and a double-napped right circular cone (Figure 21.1). Conic sections are of three basic types: ellipses, hyperbolas, and parabolas. In this section we study conic sections as graphs of quadratic equations in x and y (defined below). The emphasis will be on drawing graphs, not on proofs and derivations.

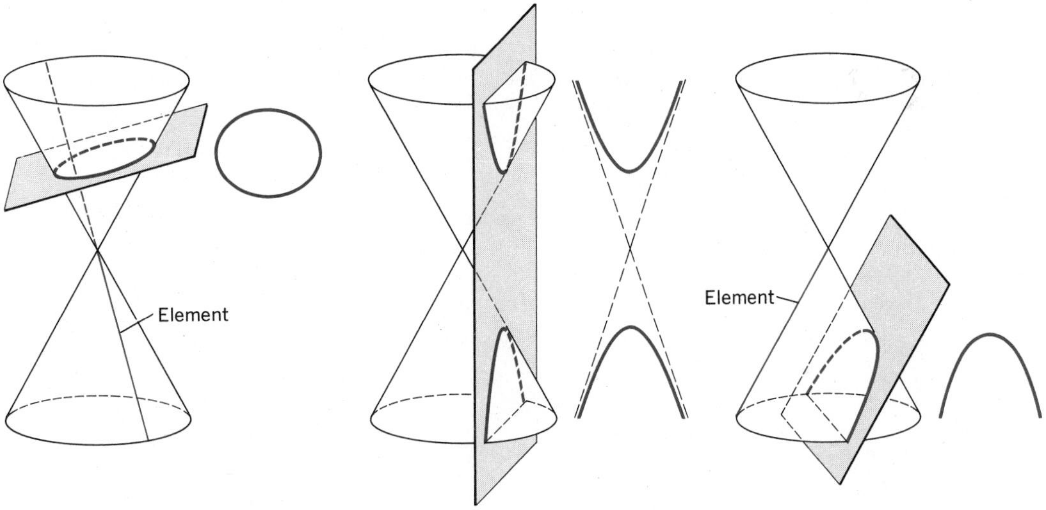

Figure 21.1
(a) **Ellipse.** The intersecting plane cuts all of the elements of the cone.
(b) **Hyperbola.** The intersecting plane cuts both nappes (parts) of the cone.
(c) **Parabola.** The intersecting plane is parallel to an element.

The **general quadratic equation** with variables x and y has the form

$$Ax^2 + Bxy + Cy^2 + Dx + Ey + F = 0, \tag{21.1}$$

where A, B, C, D, E, and F are real numbers with at least one of A, B, and C not zero. This equation allows for all possible terms of degree two or less in x and y. With a few special exceptions that will be mentioned later, the graph of any quadratic equation in x and y is a conic section.

*The only section for which this section is a prerequisite is Section 43 (Nonlinear Systems of Equations), and this section is not essential even there. Section 25 (Quadratic Functions), although related to this section, is independent of it.

In Section 19D we studied the cases of the form

$$x^2 + y^2 + Dx + Ey + F = 0 \tag{21.2}$$

[$A = C = 1$ and $B = 0$ in (21.1)]. We saw that the graph of such an equation is either a circle, a single point, or the empty set. As a conic section, a circle is a special case of an ellipse formed when the intersecting plane is perpendicular to the axis of the cone. A point occurs when the plane intersects the cone only at the point where the two nappes meet. The empty set is not a conic section.

B. Ellipses

The graph of an equation of the form

$$\frac{x^2}{a^2} + \frac{y^2}{b^2} = 1 \tag{21.3}$$

($a \neq 0$, $b \neq 0$) is an ellipse centered at the origin. [Equation (21.3) is Equation (21.1) with $A = 1/a^2$, $C = 1/b^2$, $F = -1$, and $B = D = E = 0$.] If $y = 0$ in Equation (21.3), then $x^2/a^2 = 1$, $x^2 = a^2$, and $x = \pm a$. Thus the x-intercepts are at $(a, 0)$ and $(-a, 0)$. Similarly, the y-intercepts are at $(0, b)$ and $(0, -b)$.

Figure 21.2 shows the three possibilities depending on the relative sizes of a and b. Notice that in each case the graph is symmetric with respect to both the x- and y-axes. (A graph is **symmetric** with respect to the x-axis if for each point on one side of the axis there is a corresponding point on the other side with the same x-coordinate and the same distance from the axis. This means that if the graph is folded along the x-axis, the parts originally above and below the axis will coincide. Symmetry with respect to the y-axis is defined similarly. See Figure 30.5.)

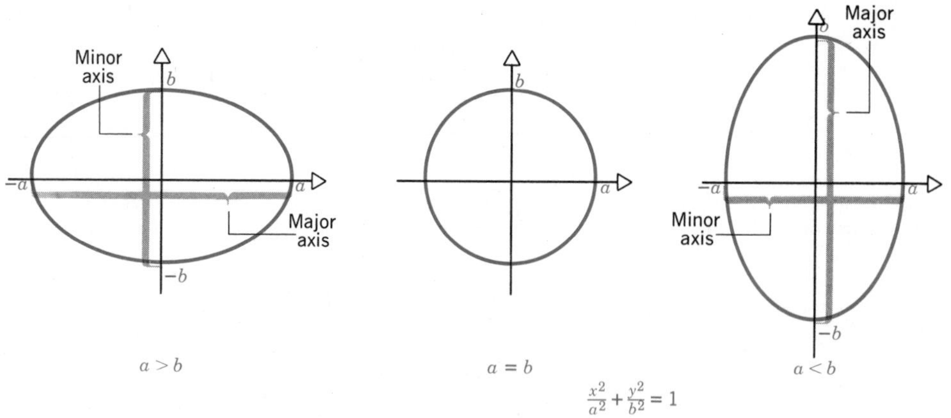

Figure 21.2

EXAMPLE 21.1 Draw the graph of $9x^2 + 4y^2 - 36 = 0$.

Solution Division of both sides by 36 yields

$$\frac{x^2}{4} + \frac{y^2}{9} = 1 \quad \text{or} \quad \frac{x^2}{2^2} + \frac{y^2}{3^2} = 1.$$

The graph is in Figure 21.3. Table 21.1 shows some coordinate pairs. In particular, the graph contains all of the points $(1, 3\sqrt{3}/2)$, $(1, -3\sqrt{3}/2)$, $(-1, 3\sqrt{3}/2)$, and $(-1, -3\sqrt{3}/2)$.

TABLE 21.1

x	0	$\pm\frac{1}{2}$	± 1	± 2
y (exact)	± 3	$\pm 3\sqrt{15}/4$	$\pm 3\sqrt{3}/2$	0
y (approximate)	± 3	± 2.90	± 2.60	0

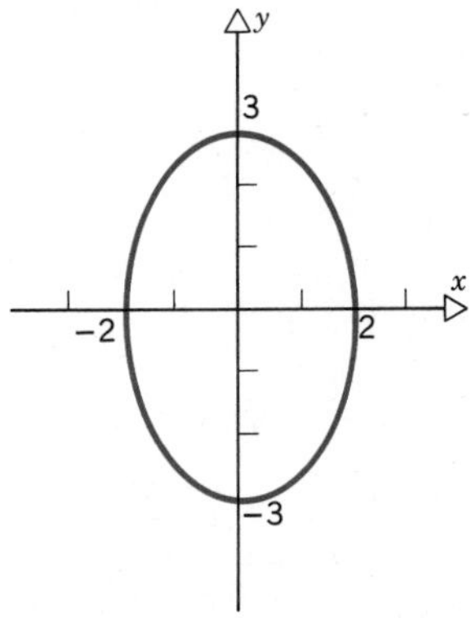

Figure 21.3

An ellipse can also be defined as the set of all points in a plane such that the sum of the distances from two fixed points, called **foci,** is a constant. (*Foci* is the plural of **focus**.) Figure 21.4 shows an example with foci at $F'(-c, 0)$ and $F(c, 0)$; $d_1 + d_2$ is the same for all points P on the ellipse.

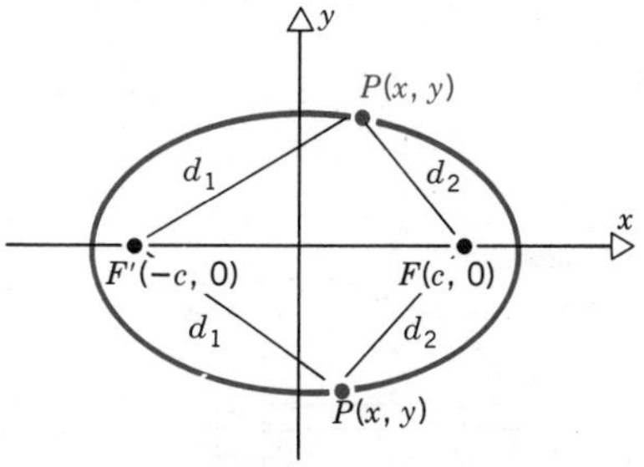

Figure 21.4 $d_1 + d_2$ is the same for all points P on the ellipse.

EXAMPLE 21.2 Suppose that $a > 0$ and $c > 0$, and consider the ellipse shown in Figure 21.4 with foci at $F'(-c, 0)$ and $F(c, 0)$ and with

$$d_1 + d_2 = 2a \tag{21.4}$$

for all $P(x, y)$ on the ellipse. Use the distance formula (Section 19) to show that if $b^2 = a^2 - c^2$, then the equation of the ellipse is Equation (21.3).

Solution The solution will be outlined and the details left as an exercise. If

$$d_1 + d_2 = 2a$$

then

$$\sqrt{(x + c)^2 + y^2} + \sqrt{(x - c)^2 + y^2} = 2a.$$

Therefore,

$$\sqrt{(x + c)^2 + y^2} = 2a - \sqrt{(x - c)^2 + y^2}.$$

Square both sides and simplify, to get

$$a^2 - cx = a\sqrt{(x - c)^2 + y^2}.$$

Now square again and simplify, to get

$$(a^2 - c^2)x^2 + a^2y^2 = a^2(a^2 - c^2).$$

With $b^2 = a^2 - c^2$ (given) this gives

$$b^2x^2 + a^2y^2 = a^2b^2$$

and

$$\frac{x^2}{a^2} + \frac{y^2}{b^2} = 1.$$

One application of ellipses is given by Kepler's three laws of planetary motion (Figure 21.5):

I. The planets move in elliptical orbits with the sun at one focus.
II. The imaginary line connecting the sun to a planet sweeps out equal areas in equal intervals of time.
III. The ratio of the cube of the semimajor axis to the square of the period is a constant, which is the same for all planets.

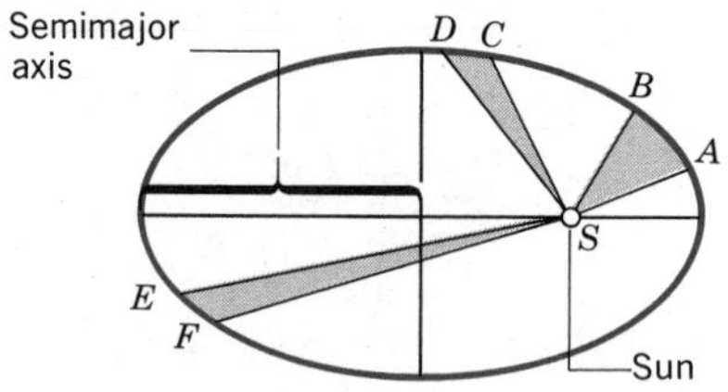

Figure 21.5

The third law (stated in a slightly different form) was discussed in Section 18B. While the orbits described by the first law are elliptical, they do not differ greatly from circular orbits. As to the second law, it implies that if areas such as ASB, CSD, and ESF in Figure 21.5 are all equal, then the planet will move from A to B, from C to D, and from E to F in equal intervals of time (in particular, a planet moves fastest when it is closest to the sun).

C. Hyperbolas

The graph of an equation of the form

$$\frac{x^2}{a^2} - \frac{y^2}{b^2} = 1 \tag{21.5}$$

or

$$\frac{y^2}{a^2} - \frac{x^2}{b^2} = 1 \tag{21.6}$$

($a \neq 0$, $b \neq 0$) is a hyperbola centered at the origin. Figure 21.6 shows both possibilities. The graph of (21.5) has x-intercepts at $x = \pm a$, but it has no y-intercepts. The graph of (21.6) has y-intercepts at $y = \pm a$, but it has no x-intercepts. The two parts of the graph of a hyperbola are called **branches.** The graphs of both (21.5) and (21.6) are symmetric with respect to both the x- and y-axes.

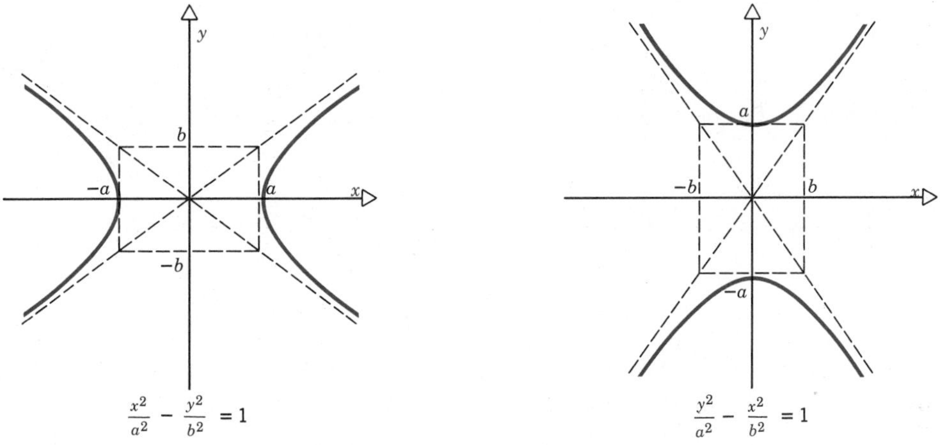

Figure 21.6

Each dashed line in Figure 21.6 is called an **asymptote** of its hyperbola. This means that the distance to the dashed line from a variable point (x, y) on a branch of the hyperbola approaches zero as the point moves outward along the hyperbola. (The asymptotes are an aid in graphing; they are not part of the hyperbola.) It can be shown that the equations of the asymptotes for Equation (21.5) arise from replacing 1 by 0 on the right side of the equation:

$$\frac{x^2}{a^2} - \frac{y^2}{b^2} = 0$$

$$y^2 = \frac{b^2}{a^2}x^2$$

$$y = \pm\frac{b}{a}x. \tag{21.7}$$

(See Exercise 39.) Similarly, the asymptotes for Equation (21.6) arise from replacing

1 by 0 in that equation; the result in this case is

$$y = \pm \frac{a}{b}x. \tag{21.8}$$

The graph of an equation of the form in either (21.5) or (21.6) can be drawn quickly as follows:

- Draw the asymptotes as dashed lines.
- Plot the intercepts.
- Determine several other points by direct calculation. Also plot the additional points that arise from these by symmetry.
- Complete the graph so that it has the general shape of the appropriate form from Figure 21.6.

EXAMPLE 21.3 Draw the graph of

$$25x^2 - 4y^2 + 100 = 0. \tag{21.9}$$

Solution Divide both sides by 100 and then rearrange the terms as shown.

$$\frac{x^2}{4} - \frac{y^2}{25} + 1 = 0$$

$$\frac{y^2}{5^2} - \frac{x^2}{2^2} = 1$$

This is Equation (21.6) with $a = 5$ and $b = 2$. The equations of the asymptotes are

$$\frac{y^2}{5^2} - \frac{x^2}{2^2} = 0 \quad \text{or} \quad y = \pm \frac{5}{2}x.$$

These asymptotes are shown in Figure 21.7, along with the intercepts, (0, 5) and (0, −5). If $x = 1$ in Equation (21.9), then $y^2 = \frac{125}{4}$ so $y = \pm 5\sqrt{5}/2 \approx \pm 5.6$. This

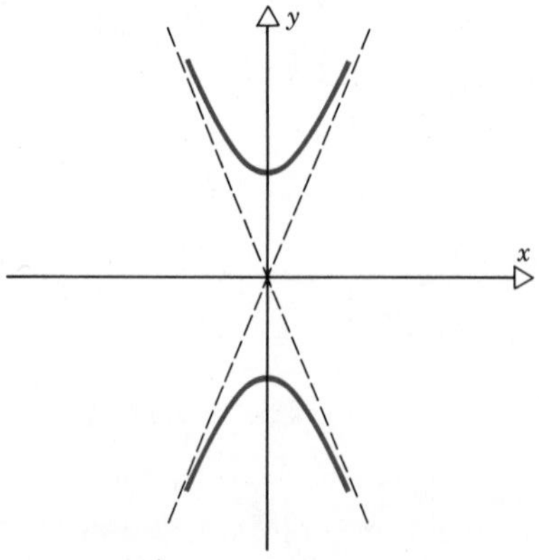

Figure 21.7

gives the points (1, 5.6) and (1, −5.6). Symmetry gives the additional points (−1, 5.6) and (−1, −5.6). Other points can be determined in the same way.

A hyperbola can also be defined as the set of all points in a plane such that the difference of the distances from two fixed points, called **foci,** is a constant. Figure 21.8 shows an example with foci at $F'(-c, 0)$ and $F(c, 0)$; $|d_1 - d_2|$ is the same for all points P on the hyperbola. Exercise 38, which is similar to Example 21.2, asks you to relate this definition to Equation (21.5).

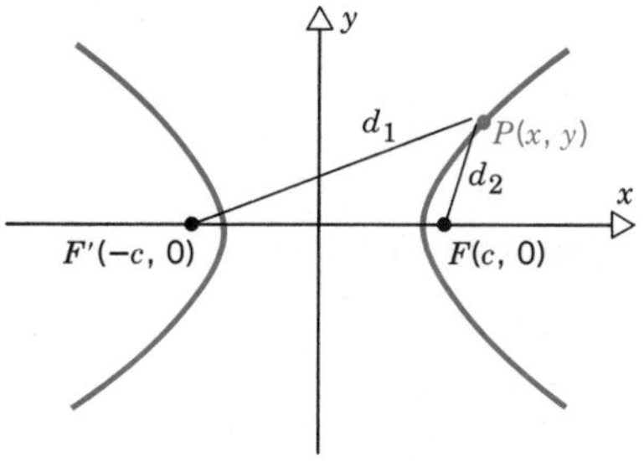

Figure 21.8 $|d_1 - d_2|$ is the same for all points P on the hyperbola.

D. Parabolas

The graph of an equation of the form

$$y = ax^2 \tag{21.10}$$

is a parabola opening upward if $a > 0$ and downward if $a < 0$. The **vertex** (lowest point if $a > 0$, highest point if $a < 0$) is at the origin (Figure 21.9). Whether $a > 0$ or $a < 0$, the graph is symmetric with respect to the y-axis.

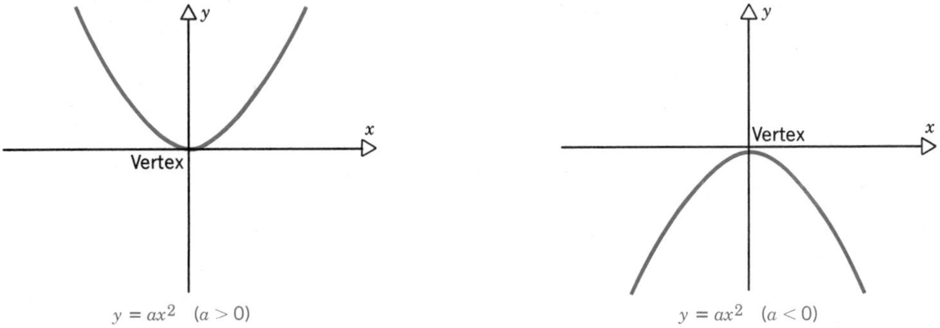

Figure 21.9

The graph of an equation of the form

$$x = ay^2 \tag{21.11}$$

is a parabola opening to the right if $a > 0$ and to the left if $a < 0$. The **vertex** (leftmost point if $a > 0$, rightmost point if $a < 0$) is at the origin in these cases also (Figure 21.10). Whether $a > 0$ or $a < 0$, the graph is symmetric with respect to the x-axis.

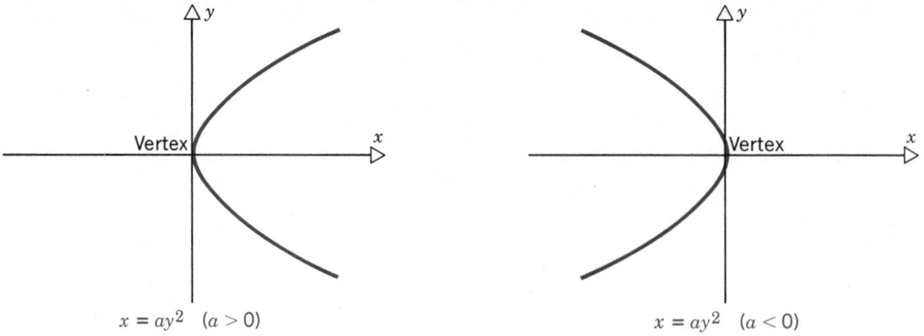

Figure 21.10

EXAMPLE 21.4 Draw the graph of $x = \frac{1}{4}y^2$.

Solution This has the form (21.11) with $a > 0$, so the graph opens to the right. Table 21.2 shows some coordinate pairs and Figure 21.11 shows the graph. In this case it's easier to assign values to y and compute the corresponding values of x than vice versa (unless you have a calculator). Remember to take advantage of symmetry; here, each pair (x, y) with $y > 0$ give rise to another point $(x, -y)$.

TABLE 21.2

y	0	$\pm\frac{1}{2}$	± 1	± 2	± 3	± 4
x	0	$\frac{1}{16}$	$\frac{1}{4}$	1	$\frac{9}{4}$	4

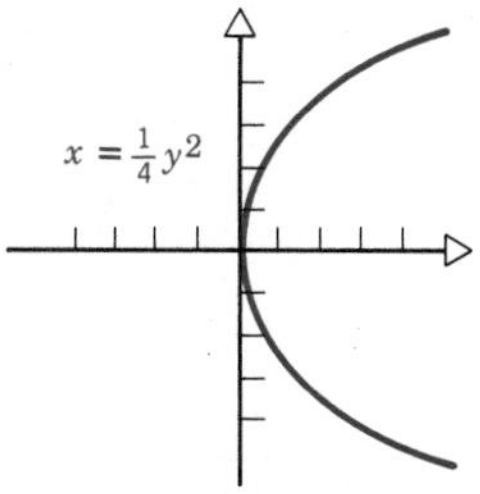

Figure 21.11

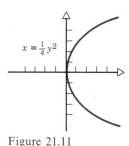

Parabolas will also be discussed in Section 25. Applications will be given there.

E. Other Possibilities

We have seen that the graph of a quadratic equation in x and y can be an ellipse (with a circle and a single point as special cases), a hyperbola, a parabola, or the empty set. The graphs of quadratic equations also include two intersecting lines, two parallel lines, or a single line. Figure 21.12 shows one example of each type.

All of the previous ellipses and hyperbolas have been centered at the origin. Ellipses that have been translated (shifted) so that their centers are not at the origin are covered by the following statement: The graph of

$$\frac{(x-h)^2}{a^2} + \frac{(y-k)^2}{b^2} = 1 \tag{21.12}$$

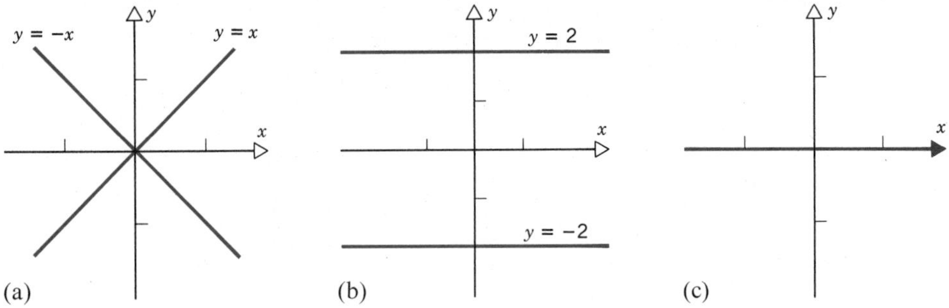

Figure 21.12 (a) **Intersecting lines:** $x^2 - y^2 = 0$ or $y = \pm x$. (b) **Parallel lines:** $y^2 - 4 = 0$ or $y = \pm 2$. (c) **One line:** $y^2 = 0$ or $y = 0$.

$(a \neq 0,\ b \neq 0)$ is the same as the graph of

$$\frac{x^2}{a^2} + \frac{y^2}{b^2} = 1$$

except that the center of the graph of (21.12) is at the point (h, k) rather than at the origin (Figure 21.13). In the next example we transform an equation into the form (21.12) and then draw the graph.

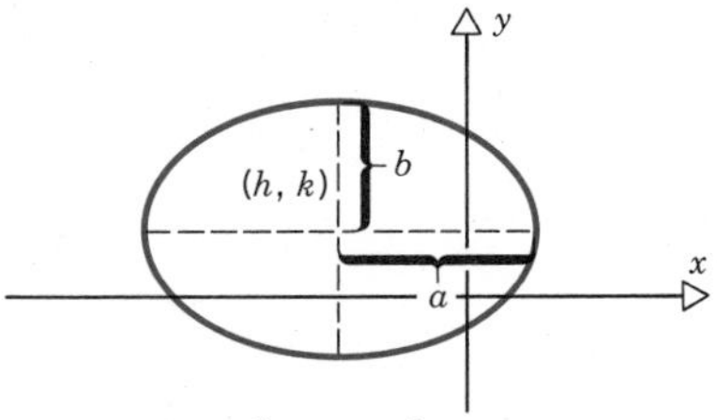

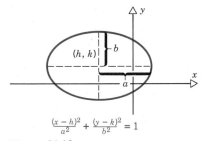

$$\frac{(x-h)^2}{a^2} + \frac{(y-k)^2}{b^2} = 1$$

Figure 21.13

EXAMPLE 21.5 Draw the graph of $9x^2 + 16y^2 - 18x + 64y - 71 = 0$.

Solution

Step 1. Factor the coefficient of x^2 from the x terms, factor the coefficient of y^2 from the y terms, and write the equation with the constant term on the right of the equality sign.

$$9(x^2 - 2x \qquad) + 16(y^2 + 4y \qquad) = 71.$$

Step 2. Complete the squares on the x and y terms, and add appropriate constants to the right side to compensate.

$$\begin{aligned} 9(x^2 - 2x + 1) + 16(y^2 + 4y + 4) &= 71 + 9 + 64 \\ 9(x-1)^2 + 16(y+2)^2 &= 144. \end{aligned}$$

Step 3. Divide both sides by the constant on the right, to put the equation in the form of Equation (21.12).

$$\frac{(x-1)^2}{16}+\frac{(y+2)^2}{9}=1$$

$$\frac{(x-1)^2}{4^2}+\frac{[y-(-2)]^2}{3^2}=1. \qquad (21.13)$$

Step 4. Draw the graph of (21.13) by following the pattern in Figure 21.13. (See Figure 21.14.)

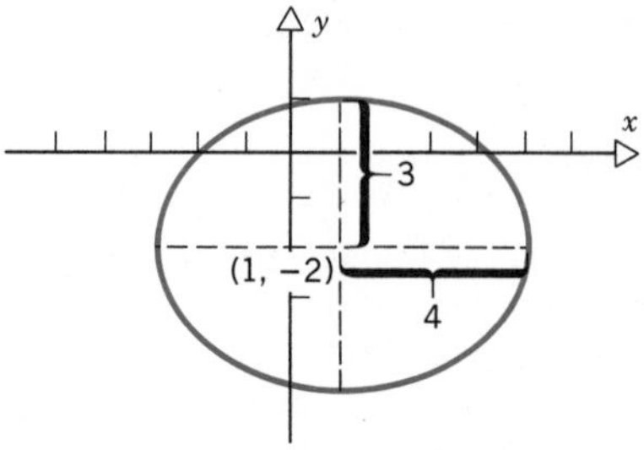

Figure 21.14

The ideas in Example 21.5 also apply to hyperbolas; see Figure 21.15. Similar ideas for parabolas will be given in detail in Section 25.

The effect of introducing an xy term into a quadratic equation is generally to rotate the conic section so that its axes are not parallel to the coordinate axes. An analysis of this and other details about quadratic equations can be found in many books on coordinate (or analytic) geometry.

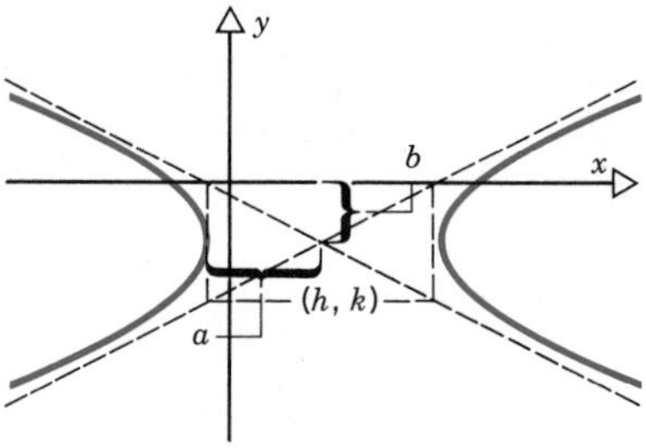

$$\frac{(x-h)^2}{a^2}+\frac{(y-k)^2}{b^2}=1$$

Figure 21.15

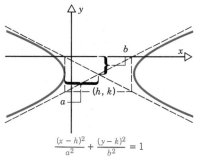

EXERCISES FOR SECTION 21

Draw the graph of each equation.

1. $16x^2+9y^2-144=0$ **2.** $9x^2+4y^2-36=0$ **3.** $4x^2-9y^2-36=0$
4. $4x^2-9y^2+36=0$ **5.** $x^2-y^2-4=0$ **6.** $25x^2+4y^2-100=0$
7. $x^2+25y^2-25=0$ **8.** $2x^2+y=0$ **9.** $x^2-y^2+9=0$
10. $9x+y^2=0$ **11.** $25x^2-9y^2+225=0$ **12.** $3x^2+9y^2-27=0$
13. $5x^2-4y^2-20=0$ **14.** $3x^2+4y^2-12=0$ **15.** $4x^2-y=0$

The graph of each equation in Exercises 16–30 is either the empty set, one point, one line, or two lines. In each case draw the graph or, if appropriate, indicate that it is the empty set.

16. $x^2-9=0$ **17.** $9x^2-y^2=0$ **18.** $y^2=0$

19. $4x^2 - 9y^2 = 0$
20. $2x^2 + 3y^2 + 4 = 0$
21. $x^2 - 4 = 0$
22. $-y^2 = 0$
23. $10x^2 = 0$
24. $x^2 + y^2 + 2 = 0$
25. $3x^2 + 5y^2 = 0$
26. $-x^2 - y^2 = 0$
27. $5x^2 + 3y^2 = 0$
28. $4x^2 + 3y^2 + 2 = 0$
29. $y^2 - 16 = 0$
30. $x^2 - 4y^2 = 0$

Draw the graph of each equation.

31. $3x^2 + y^2 - 12x - 6y + 12 = 0$
32. $16x^2 + 4y^2 + 160x + 256 = 0$
33. $25x^2 + 16y^2 + 50x + 96y - 231 = 0$
34. $9x^2 - 25y^2 + 100y - 325 = 0$
35. $9x^2 - 16y^2 - 18x - 64y - 199 = 0$
36. $9x^2 - 16y^2 - 54x - 64y + 161 = 0$

37. Write out the solution in Example 21.2 supplying all of the missing details.
38. Suppose that $a > 0$ and $c > 0$, and consider the hyperbola shown in Figure 21.8 with $|d_1 - d_2| = 2a$. Use the distance formula to show that if $b^2 = c^2 - a^2$, then the equation of the hyperbola is Equation (21.5).
39. Verify that Equation (21.7) gives asymptotes for the hyperbola defined by Equation (21.5) by carrying out all of the following steps.
 (a) Show that Equation (21.5) can be rewritten as

$$y^2 = \frac{b^2}{a^2}(x^2 - a^2).$$

 (b) Show that the equation in (a) can be rewritten as

$$y^2 = \left(\frac{bx}{a}\right)^2\left(1 - \frac{a^2}{x^2}\right).$$

 (c) Explain why $1 - (a^2/x^2)$ approaches 1 as $|x|$ becomes larger and larger.
 (d) Use the preceding steps to explain why a variable point on the graph of Equation (21.5) approaches the graph of $y = \frac{b}{a}x$ or $y = -\frac{b}{a}$ as $|x|$ becomes larger and larger.

REVIEW EXERCISES FOR CHAPTER V

Determine the distance between each pair of points.

1. $(-3, 4)$ and $(1, 2)$
2. $(6, 0)$ and $(1, -2)$

Determine an equation of the line through each pair of points. Write the answer in the form $ax + by + c = 0$.

3. $(4, -5)$ and $(-1, 2)$
4. $(0, 3)$ and $(-3, 1)$
5. Find an equation of the line with slope -4 and y-intercept 2.
6. Find an equation of the line containing the point $(2, -1)$ and parallel to the line whose equation is $x - y + 5 = 0$.

Draw the graph of each equation.

7. $x^2 + y^2 - 5 = 0$
8. $9x^2 + y^2 = 9$
9. $3x + y = 1$
10. $9x^2 - y^2 = 0$
11. $3x^2 + y = 1$
12. $9x^2 - y^2 = 9$
13. $4x^2 - y^2 = 4$
14. $4x - y = 3$
15. $x^2 + 4y^2 = 4$
16. $2x^2 + 2y^2 = 8$
17. $4x^2 - y^2 = 0$
18. $-3x^2 + y = 1$
19. $9x^2 + 4y^2 + 18x - 24y + 9 = 0$
20. $9x^2 - 4y^2 - 36x - 8y - 4 = 0$

CHAPTER VI
INTRODUCTION TO FUNCTIONS

In using mathematics we repeatedly meet problems in which one quantity is determined by one or more other quantities. For example:

The area of a circle is determined by the radius.

The pressure of a confined gas is determined by the volume and temperature of the gas.

The interest on a loan is determined by the principal, rate, and duration of the loan.

The idea of a *function* encompasses all such examples. This chapter will introduce the basic language and notation for functions, and will thoroughly analyze two of the most important types of functions (those that are *linear* or *quadratic*). This chapter will draw freely on results from Sections 19 and 20 of Chapter V.

SECTION

22 Functions

A. Definition

As stated in the chapter introduction, the idea of a *function* encompasses all examples in which one quantity is determined by one or more other quantities. In most of our examples there will be only two quantities, with one determined by the other.

EXAMPLE 22.1 If x denotes the length of a side of a square in meters and A denotes the area in square meters, then A is determined by x:

$$A = x^2. \tag{22.1}$$

EXAMPLE 22.2 Let C denote temperature in degrees Celsius and F temperature in degrees Fahrenheit. Then C is determined by F:

$$C = \tfrac{5}{9}(F - 32). \tag{22.2}$$

Notice that in each example we can identify one thing as *input* and another thing as *output,* as follows.

Example	Input	Output
22.1	x	$A = x^2$
22.2	F	$C = \frac{5}{9}(F - 32)$

The concept of a function ties together the following observations about these two examples.

- The values that can occur as input form a set (in the first case the possible values for x; in the second case the possible values for F).
- The values that can occur as output also form a set (in the first case the possible values for A; in the second case the possible values for C).
- In each case the output is determined uniquely (unambiguously) by the input.

In the following definitions the set S corresponds to the set of input elements, and the output elements belong to the set T.

DEFINITIONS. Let S and T denote sets. A **function** from S to T is a relationship (formula, rule, correspondence) that assigns a unique element of T to each element of S. The set S is called the **domain** of the function. The set of elements of T to which the elements of S are assigned is called the **range** of the function.

EXAMPLE 22.3

(a) In Example 22.1, x can be any positive real number (length in meters) and A can be any positive real number (area in square meters). Thus the domain and the range are both the set of all positive real numbers.

(b) The domain and range in Example 22.2 are restricted by the laws of nature: F and C both denote real numbers, with $F \geq -459.67$ and $C \geq -273.15$. (These minimum values correspond to absolute zero, the point at which molecules have no heat energy.) In set-builder notation (Section 4),

$$\text{domain} = \{F\colon\ F \geq -459.67\}$$

and

$$\text{range} = \{C\colon\ C \geq -273.15\}.$$

Most of the domains and ranges for functions in this book will be sets of real numbers, as in Example 22.3. The next example illustrates a different possibility.

EXAMPLE 22.4 Let S denote the set of all 50 states and T the set of all cities in the United States. The relationship that assigns a capital to each state is a function: the domain is S and the range is the subset of T consisting of the 50 state capitals.

An alternative but equivalent definition of *function,* in terms of ordered pairs, is given in Appendix B. It will not be used elsewhere in this book.

B. Notation

Example 22.1 ($A = x^2$) will now be used to introduce some extremely important notation for functions.

EXAMPLE 22.5 If $A = x^2$, we say that

$$A \text{ is a function of } x.$$

Let's replace "is" by "=" and abbreviate "a function" to just "f." This gives

$$A = f \text{ of } x.$$

It is convenient and useful to go one step further and replace "f of x" by just "$f(x)$." Then

$$A = f(x) \text{ means } A \text{ is a function of } x.$$

Because $A = x^2$, we can also write

$$A = f(x) = x^2. \qquad (22.3)$$

Notice that in this context $f(x)$ does *not* mean "f times x;" f represents a function, not a number.

Equation (22.3) is concise, but that is not its main advantage. Its main advantage is that if we want to refer to A or x^2 for a specific value of x, we can simply replace x by that specific value in $f(x)$. For example,

$$f(5) = 5^2 = 25$$
$$f(\tfrac{1}{2}) = (\tfrac{1}{2})^2 = \tfrac{1}{4}$$
$$f(t) = t^2.$$

Replacing x by a number or letter in $f(x)$ is a signal to replace x by that same number or letter in x^2. Furthermore, just as we are using x and $f(x)$ to denote *numbers* in this context, we are using f to denote the *function* (relationship) associated with $A = x^2$ in this context.

Moving from the specific example $f(x) = x^2$ to the most general case, we obtain the following convention.

> If f denotes a function, and x denotes an element in the domain of f, then the element that f assigns to x is denoted by $f(x)$.

That is, if the input is x, then the output is $f(x)$ (Figure 22.1). The notation "$f(x)$" is read "f of x."

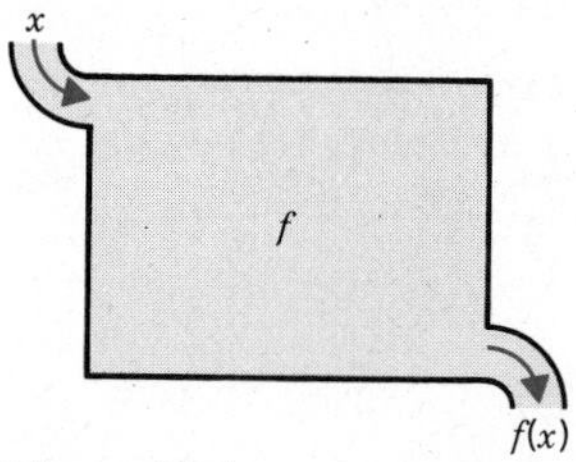

Figure 22.1

EXAMPLE 22.6 If f denotes the function that gives the perimeter P of a square in terms of the length x of a side, then

$$f(x) = 4x \text{ for each positive real number } x.$$

To get the perimeter corresponding to a specific value of x, simply replace x by that value. If, for example, the length of a side is 3 feet, then the perimeter is

$$f(3) = 4 \cdot 3 = 12 \text{ feet.}$$

If the length of a side is 10 centimeters, then the perimeter is

$$f(10) = 4 \cdot 10 = 40 \text{ centimeters.}$$

EXAMPLE 22.7 If f denotes the function that assigns a capital to each state (Example 22.4), then, for example,

$$f(\text{California}) = \text{Sacramento}$$
$$f(\text{New York}) = \text{Albany.}$$

Just as we can use letters other than x to denote variables, we can use letters other than f to denote functions. (Like g in Example 22.8, which follows.) It must be made clear in each case exactly what each letter represents.

Sometimes we must replace the x in $f(x)$ by an expression that is more complicated than a single number or letter. The key to doing this is to remember that if x is replaced by any expression representing an element of the domain of f, then x must be replaced by the same expression throughout the formula or rule giving $f(x)$.

EXAMPLE 22.8 Let $g(x) = x^2 + x$ for each real number x. (That sentence tells us, in particular, that the domain of g is the set of all real numbers.) Then

(a) $g(-2) = (-2)^2 + (-2) = 2$
(b) $g(\sqrt{3}) = (\sqrt{3})^2 + \sqrt{3} = 3 + \sqrt{3}$
(c) $g(t) = t^2 + t$
(d) $g(a + 1) = (a + 1)^2 + (a + 1)$
$= a^2 + 2a + 1 + a + 1$
$= a^2 + 3a + 2$
(e) $g(x/2) = (x/2)^2 + (x/2) = x^2/4 + x/2 = (x^2 + 2x)/4$
(f) $g(\pi^{-2}) = (\pi^{-2})^2 + (\pi^{-2}) = \pi^{-4} + \pi^{-2}$
(g) $g(x^3) = (x^3)^2 + (x^3) = x^6 + x^3$.

Because the domain of g is the set of all real numbers, the letters t, a, and x above can also denote any real numbers.

It is important to realize that the function g in Example 22.8 could be defined equally well by using any other variable in place of x. That is,

$$g(x) = x^2 + x, \quad g(t) = t^2 + t, \quad \text{and} \quad g(u) = u^2 + u$$

all define the same function—they describe the same relationship between the input (a variable) and the output (the sum of that variable and its square). A similar remark applies to other functions.

EXAMPLE 22.9 Let $f(x) = 1 + (1/x)$ for each nonzero real number x. Then

(a) $f(-1) = 1 + (1/-1) = 1 - 1 = 0$
(b) $f(\frac{2}{3}) = 1 + (1/\frac{2}{3}) = 1 + \frac{3}{2} = \frac{5}{2}$
(c) $f(a) = 1 + (1/a) \quad (a \neq 0)$
(d) $f(3/(t + 2)) = 1 + \dfrac{1}{3/(t + 2)} = 1 + [(t + 2)/3] = (3 + t + 2)/3$
$= (t + 5)/3$.

In part (d) we must require $t \neq -2$, for otherwise the input element $3/(t + 2)$ would be undefined.

Expressions that involve one or more functions and algebraic operations are simplified by carrying out whatever steps are indicated.

EXAMPLE 22.10 Let $f(x) = x^2$ and $g(x) = 3x - 2$. Then

(a) $f(x) + g(x) = x^2 + 3x - 2$
(b) $f(2) + g(5) = (2^2) + (3 \cdot 5 - 2) = 17$
(c) $2 \cdot f(1) - 4 \cdot g(-1) = 2(1^2) - 4[3(-1) - 2] = 2(1) - 4(-5) = 22$
(d) $f(t)/g(t) = t^2/(3t - 2)$ for $t \neq \frac{2}{3}$.

C. More about Domains

The domain of the function f in Example 22.9 was the set of all *nonzero* real numbers: $1/x$ is undefined for $x = 0$, so 0 could not be in the domain of f. If a function f is defined by an algebraic expression, like $f(x) = 1 + (1/x)$, and the domain of f is not specified, then the domain will be assumed to be the domain of the algebraic expression—that is, the set of all real numbers x for which $f(x)$ is also a real number (Section 4B). This domain is called the **natural domain** of the function.

EXAMPLE 22.11 Determine the domain (that is, the natural domain) of

$$f(x) = \frac{1}{x^2 - 1}.$$

Solution Both $x = 1$ and $x = -1$ must be excluded from the domain because they lead to division by 0. With no other restriction stated, the domain of f is the set of all real numbers except 1 and -1. In set-builder notation,

$$\text{domain of } f = \{x\colon\ \ x \neq 1 \ \text{ and } \ x \neq -1\}.$$

EXAMPLE 22.12 Determine the domain of

$$g(x) = \frac{\sqrt{x - 2}}{x - 3}.$$

Solution To avoid a square root of a negative number in the numerator we must exclude the real numbers less than 2 from the domain. To avoid zero for the denominator we must exclude $x = 3$. Thus

$$\text{domain of } g = \{x\colon\ \ x \geq 2 \ \text{ and } \ x \neq 3\}.$$

If we write $f(x) = x^2$, with no restriction specified, then the domain is understood to be the natural domain—that is, the set of all real numbers. In Examples 22.1 and 22.5 $[A = f(x) = x^2]$, a restriction was dictated by the context: x represented a length in meters so the domain was the set of positive real numbers. In Example 22.3(b) $[C = f(F) = \frac{5}{9}(F - 32)]$ the domain and range were determined by physical laws. Generally, it is more important to know the domain of a function than the range of a function.

D. Composition

EXAMPLE 22.13 Let $f(x) = x + 1$ and $g(x) = x^2$. The variable x in $g(x) = x^2$ can be replaced by any real number or symbol representing a real number; in particular, we can replace it by $f(x)$. The result is

$$g(f(x)) = g(x + 1) = (x + 1)^2.$$

Figure 22.2 shows what we have done. The function f assigns $x + 1$ to x; then, the function g assigns $(x + 1)^2$ to $x + 1$. As a result of the two steps combined, $(x + 1)^2$

is assigned to x. We can think of this result as a single function made up from the two component functions f and g, as in the following definition.

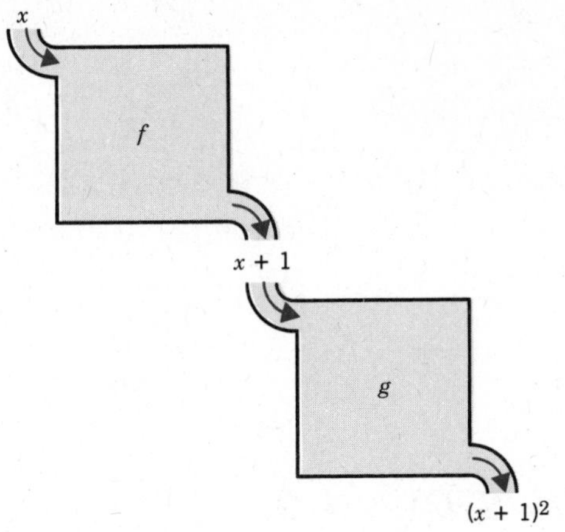

Figure 22.2

DEFINITION. Assume that f and g are functions such that the range of f is contained in the domain of g. Then the **composition** of f and g, denoted $g \circ f$, is the function defined by

$$(g \circ f)(x) = g(f(x)) \tag{22.4}$$

for each x in the domain of f.

We read $g \circ f$ as "g composed with f" or "g circle f." For $g(f(x))$ to be meaningful, $f(x)$ must be in the domain of g whenever x is in the domain of f; that is why the definition requires the range of f to be contained in the domain of g. It is essential that $g \circ f$ be distinguished from $f \circ g$, as the following example makes clear.

EXAMPLE 22.14 Let $f(x) = x - 2$ and $g(x) = x^2 + 1$. Determine: **(a)** $(f \circ g)(x)$; **(b)** $(g \circ f)(x)$; **(c)** $(f \circ g)(3)$; **(d)** $(g \circ f)(3)$.

Solution

(a) $(f \circ g)(x) = f(g(x)) = f(x^2 + 1) = (x^2 + 1) - 2 = x^2 - 1$
(b) $(g \circ f)(x) = g(f(x)) = g(x - 2) = (x - 2)^2 + 1$
$= x^2 - 4x + 4 + 1 = x^2 - 4x + 5$
(c) Use the answer to part (a): $(f \circ g)(3) = 3^2 - 1 = 8$.
(d) Use the answer to part (b): $(g \circ f)(3) = 3^2 - 4 \cdot 3 + 5 = 2$.

Composition can be viewed as a way to construct a new function from two given functions. It can also be viewed as a way to break a given function into simpler components. Before illustrating this we need one more definition.

Suppose f and g are the functions, each with the set of real numbers as domain, such that

$$f(x) = (x + 1)^2 \quad \text{and} \quad g(x) = x^2 + 2x + 1.$$

Because $(x + 1)^2 = x^2 + 2x + 1$ for every real number x, it is reasonable to think of f and g as being equal. This leads us to say that any two functions f and g are **equal** if their domains are equal and if $f(x) = g(x)$ for every x in that common domain; in this case we write $f = g$.

EXAMPLE 22.15 Let $f(x) = (2x - 3)^2$. Find functions g and h such that $h \circ g = f$.

Solution Choose $g(x) = 2x - 3$ and $h(x) = x^2$. Then

$$(h \circ g)(x) = h(g(x)) = h(2x - 3) = (2x - 3)^2 = f(x).$$

Therefore, $h \circ g = f$, as required.

EXERCISES FOR SECTION 22

For each function in Exercises 1–6, determine $f(2)$, $f(0)$, $f(-\sqrt{5})$, $f(t)$, $f(a/2)$, $f(b + 1)$, and $f(x^2)$.

1. $f(x) = 2x - 3$ **2.** $f(x) = 4x + 1$ **3.** $f(x) = -x + 5$
4. $f(x) = 3x^2$ **5.** $f(x) = x^2 + 1$ **6.** $f(x) = 4x^2 - 3$

Determine the domain of each function.

7. $f(x) = 2/(x - 1)$ **8.** $g(x) = -1/(x^2 - 9)$ **9.** $h(x) = -x/(x^2 + 2x + 1)$
10. $g(x) = \sqrt{2x - 1}$ **11.** $h(x) = \sqrt{1 - x}$ **12.** $f(x) = \sqrt{x + 5}$
13. $h(x) = \sqrt{x + 1}/x$ **14.** $f(x) = \sqrt{x}/(x - 1)$ **15.** $g(x) = \sqrt{4 - x}/(x + 1)$

In each of Exercises 16–21, determine:

(a) $f(x) + g(x)$ **(b)** $f(x) - g(x)$ **(c)** $f(x) \cdot g(x)$ **(d)** $f(x)/g(x)$
(e) $f(0) + g(1)$ **(f)** $f(2) - g(-1)$ **(g)** $f(a) \cdot g(\sqrt{3})$ **(h)** $f(t)/g(0)$.

16. $f(x) = 2x - 1$ and $g(x) = x^2 + 1$
17. $f(x) = x + 2$ and $g(x) = x^2 - 2$
18. $f(x) = 3x$ and $g(x) = x^2 - 3$
19. $f(x) = x + b$ and $g(x) = x + a$
20. $f(x) = ax$ and $g(x) = ax + b$
21. $f(x) = ax + b$ and $g(x) = x + b$

In each of Exercises 22–33, determine:

(a) $(f \circ g)(x)$ **(b)** $(g \circ f)(x)$ **(c)** $(f \circ g)(2)$ **(d)** $(g \circ f)(-3)$.

22. $f(x) = 3x$, $g(x) = x - 1$
23. $f(x) = x + 2$, $g(x) = -4x$
24. $f(x) = 1 - x$, $g(x) = 2x$
25. $f(x) = x^3$, $g(x) = -3x$
26. $f(x) = 2x - 1$, $g(x) = x^3$
27. $f(x) = -2x^3$, $g(x) = x + 1$
28. $f(x) = x - 2$, $g(x) = 3$
29. $f(x) = -2$, $g(x) = x^2$
30. $f(x) = 2x + 1$, $g(x) = 0$
31. $f(x) = |x|$, $g(x) = -x$
32. $f(x) = x^2$, $g(x) = -|x|$
33. $f(x) = 1/(x - 1)$, $g(x) = 2/x$

In Exercises 34–39, find functions g and h such that $h \circ g = f$.

34. $f(x) = (x - 4)^2$ **35.** $f(x) = \sqrt{x + 3}$ **36.** $f(x) = \sqrt{2x - 1}$
37. $f(x) = x^2 + 1$ **38.** $f(x) = \sqrt{x} - 1$ **39.** $f(x) = 2\sqrt{x}$

40. Let $g(x) = x^2$ for $x = 0, \pm 1$, and ± 2. (Thus the domain of g is $\{0, \pm 1, \pm 2\}$.) What is the range of g?
41. Let $f(x) = x^2 + 1$ for each real number x. What is the range of f?
42. Let $g(x) = 3x - 1$ for each real number x. Show that for each real number y there is a real number x such that $f(x) = y$. (Solve $3x - 1 = y$ for x.) What do you conclude about the range of f?

SECTION 23 Graphs and More Examples

A. The Graph of a Function

The domains and ranges of the functions in this section will all be sets of real numbers. One of the best ways to study such functions is to look at their graphs, which are determined as follows. First, choose a plane with a Cartesian coordinate system. If x is in the domain of a function f, then $(x, f(x))$ is a pair of real numbers. Therefore, $(x, f(x))$ determines a point in the coordinate plane—the point with x as first coordinate and $f(x)$ as second coordinate.

DEFINITION. The **graph** of a function f is the set of all points having coordinates $(x, f(x))$ for x in the domain of f.

If we use y to represent the second member of a coordinate pair, as usual, then the graph of f is the set of all (x, y) such that $y = f(x)$. Therefore:

> The graph of a *function* f is the same as the graph of the *equation* $y = f(x)$.

EXAMPLE 23.1 Draw the graph of $f(x) = x$. (This function is called the **identity function.**)

Solution The graph is the same as the graph of the equation $y = x$. Typical points on the graph are $(0, 0)$, $(1, 1)$, and $(-5, -5)$. From Section 20 we know that the graph is the line through these points (Figure 23.1).

For each x in the domain of any function f there is only one y such that $y = f(x)$, because each input element determines a unique output element. This means that each vertical line can intersect the graph of a function *at most once:*

If a is in the domain of f, then the vertical line $x = a$ will intersect the graph of f *exactly once*—at the point $(a, f(a))$.

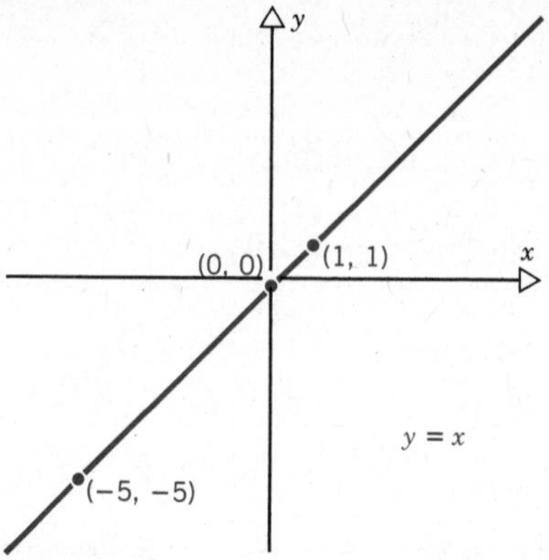

Figure 23.1

If a is not in the domain of f, then the vertical line $x = a$ will *not* intersect the graph of f.

For example, if the curve C_1 in Figure 23.2 were the graph of a function f, then necessarily $f(a) = b$, and $f(a) = c$, an impossibility because $b \neq c$. On the other hand, the curve C_2 in Figure 23.2 *is* the graph of a function; if f denotes this function then, for example, $f(a) = b$. [The curve C_2 is, in fact, the graph of $f(x) = \sqrt{x}$, $x \geq 0$.]

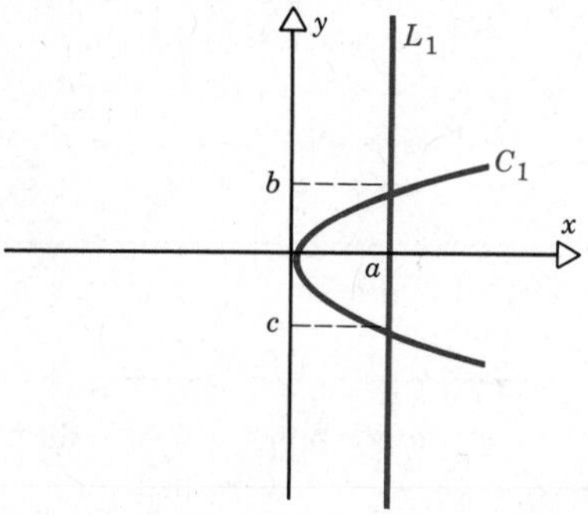

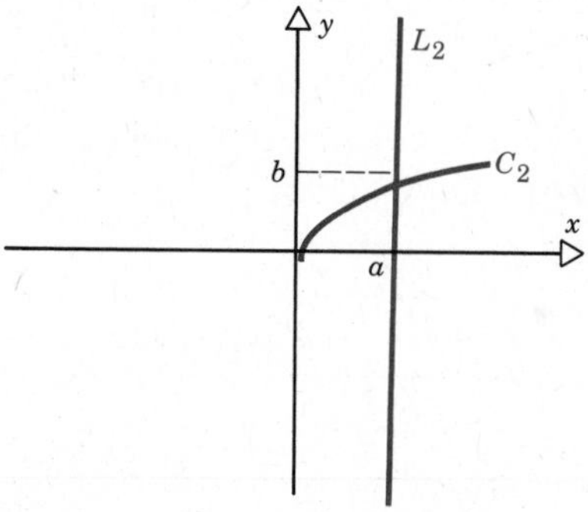

Figure 23.2 (a) The curve C_1 is *not* the graph of a function, because the vertical line L_1 intersects the curve more than once. (b) The curve C_2 is the graph of a function, because each vertical line (such as L_2) intersects the curve at most once.

B. More Examples of Graphs

EXAMPLE 23.2 Draw the graph of $f(x) = -2x + 3$.

Solution The graph is the same as the graph of $y = -2x + 3$, which is shown in Figure 23.3.

EXAMPLE 23.3 Draw the graph of $f(x) = x^3$.

Solution Some typical coordinate pairs are shown in Table 23.1. Figure 23.4 shows the graph of the function.

TABLE 23.1

x	$y = f(x)$
-2	-8
-1	-1
$-\frac{1}{2}$	$-\frac{1}{8}$
0	0
$\frac{1}{2}$	$\frac{1}{8}$
1	1
2	8

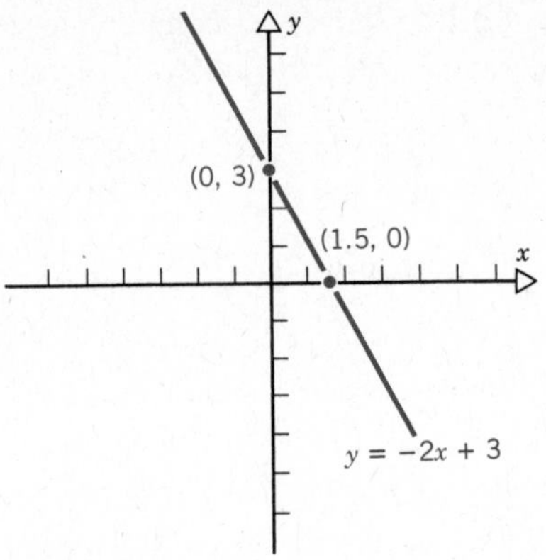

Figure 23.3

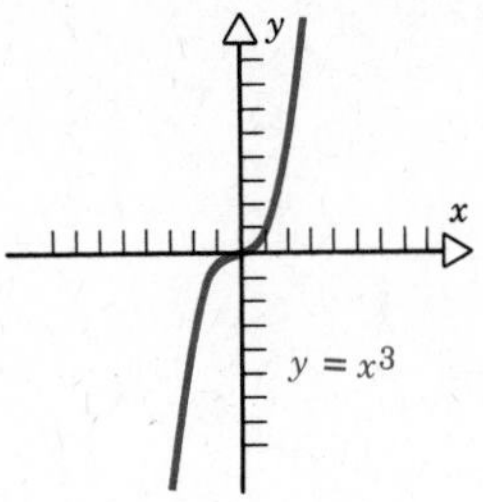

Figure 23.4

EXAMPLE 23.4 Draw the graph of the function f defined by $f(x) = 4$ for each real number x.

Solution For this function the output is 4, no matter what the input is. The graph of f is the horizontal line $y = 4$, which is shown in Figure 23.5. More generally, if b is *any* real number, then the graph of the function $f(x) = b$ will be the horizontal line $y = b$. Any such function is called a **constant function.**

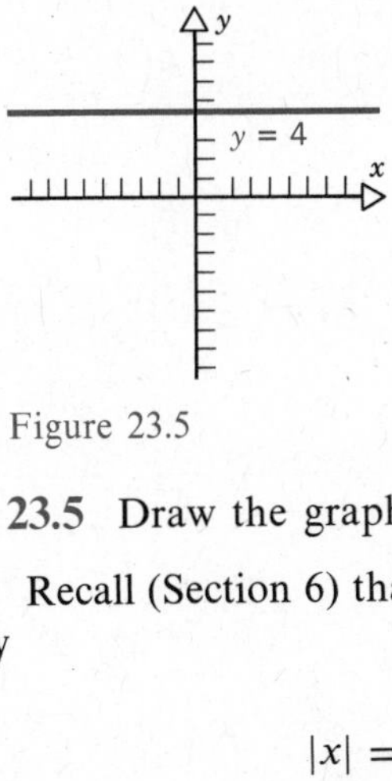

Figure 23.5

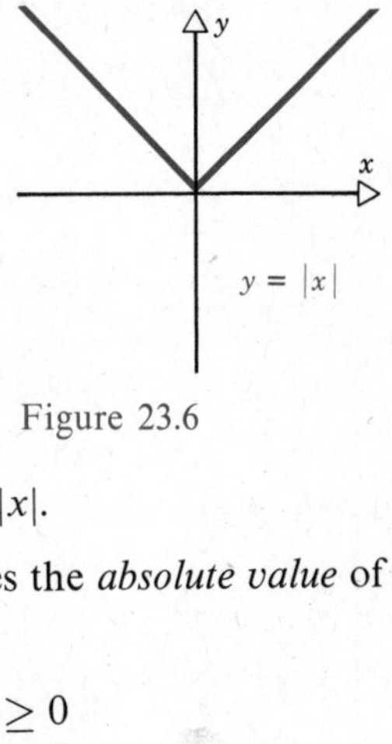

Figure 23.6

EXAMPLE 23.5 Draw the graph of $f(x) = |x|$.

Solution Recall (Section 6) that $|x|$ denotes the *absolute value* of x, which can be defined by

$$|x| = \begin{cases} x & \text{if } x \geq 0 \\ -x & \text{if } x < 0. \end{cases}$$

Therefore, for $x \geq 0$ the graph is the same as the graph of $y = x$. For $x < 0$, however, the graph coincides with the graph of $y = -x$. Typical points are $(-1, 1)$ and $(-2, 2)$. The graph is shown in Figure 23.6.

The function in the next example is defined by one expression for part of its domain and another expression for the remainder of its domain.

EXAMPLE 23.6 The graph of

$$f(x) = \begin{cases} -1 & \text{for } x \leq 0 \\ x + 1 & \text{for } x > 0 \end{cases}$$

is shown in Figure 23.7. The symbol • on the end of a segment means that the endpoint is part of the graph; the symbol o means that the endpoint is not part of the graph.

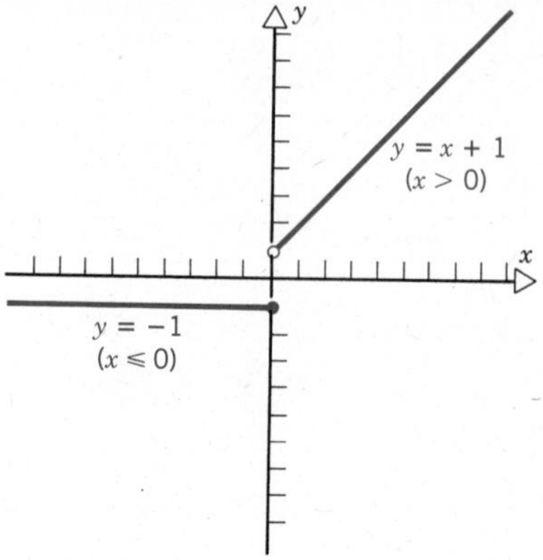

Figure 23.7

EXAMPLE 23.7 The cost of a phone call from one city to another, during a given time period, is usually determined by the duration of the call. For instance, the cost might be \$1.00 for the first 3 minutes (or portion thereof), and then \$0.25 for each additional minute (or portion thereof). The graph of the corresponding function is shown in Figure 23.8, with T (for *time* in minutes) in place of x, and C (for *cost*) in place of y.

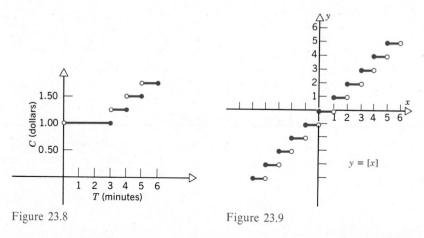

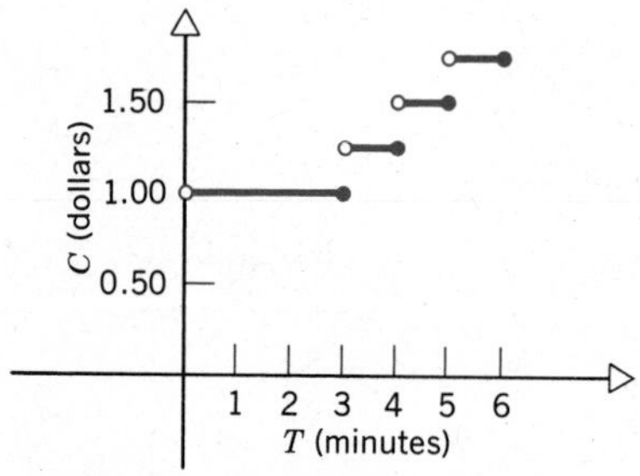

Figure 23.8

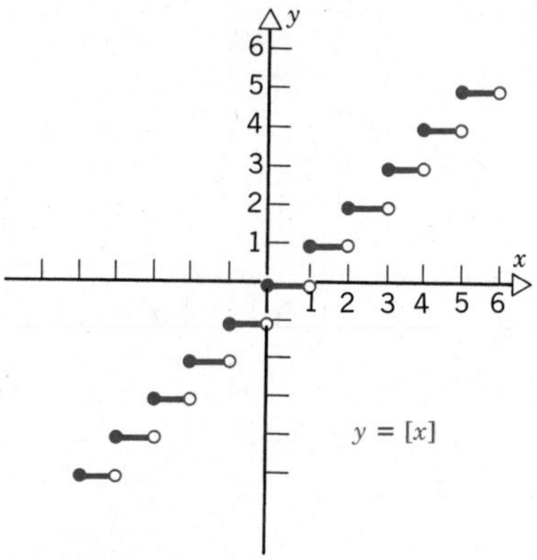

Figure 23.9

EXAMPLE 23.8 If x is any real number, then $[x]$ is often used to denote the greatest integer not exceeding x. In other words, if x is an integer then $[x] = x$; otherwise, x "rounded down" to the next integer is $[x]$. For example,

$$[5] = 5, \quad [1.2] = 1, \quad [\pi] = 3, \quad \text{and} \quad [-17.2] = -18.$$

Figure 23.9 shows the graph of $f(x) = [x]$. This function is called the **greatest integer function.**

C. Functions of Several Variables

The next two examples concern functions of *several variables,* in which one quantity depends on more than one other quantity.

EXAMPLE 23.9 The formula for simple interest was given in Equation (8.2):

$$I = Prt,$$

where I, P, r, and t denote interest, principal, rate, and time, respectively. The variable I is a function of *three* variables: P, r, and t. To handle this we extend the notation $f(x)$ for functions of a single variable by writing

$$I = f(P, r, t) = Prt.$$

If $P = \$100$, $r = 0.05$, and $t = 2$ years, for example, then

$$I = f(100, 0.05, 2) = 100 \times 0.05 \times 2 = \$10.$$

EXAMPLE 23.10 The formula for the volume V of a right circular cylinder of radius r and height h is

$$V = \pi r^2 h. \tag{23.1}$$

(a) Determine a function f such that $h = f(V, r)$.
(b) Determine a function g such that $r = g(V, h)$.

Solution

(a) From (23.1), $h = V/\pi r^2$. Thus the function is

$$f(V, r) = \frac{V}{\pi r^2}.$$

If, for example, the volume is 30 cubic meters and the radius is 2 meters, then the height is

$$h = f(30, 2) = \frac{30}{\pi 2^2} \approx 2.39 \text{ meters.}$$

C

(b) From (23.1), $r^2 = V/\pi h$, so $r = \sqrt{V/\pi h}$. Therefore,

$$g(V, h) = \sqrt{\frac{V}{\pi h}}.$$

D. Change and Difference Quotients

Recall from Section 20 that Δ is used to denote *difference* or *change*. Thus Δx denotes a change in x, and Δy denotes a change in y. Here Δx and Δy are each to be treated as single symbols: Δx does *not* mean "Δ times x."

EXAMPLE 23.11 The area of a square will increase if the length of a side is increased. Let x denote the length of a side and A the area, so that $A = x^2$. Determine the change ΔA in A corresponding to a change Δx in x.

Solution Let $f(x) = x^2$, so that $A = f(x)$. If the original length of a side is x, then the new length will be $x + \Delta x$. The new area will be $f(x + \Delta x)$. The increase in area, ΔA, will be the difference between $f(x + \Delta x)$ and $f(x)$:

$$\begin{aligned}\Delta A &= f(x + \Delta x) - f(x)\\ &= (x + \Delta x)^2 - x^2\\ &= x^2 + 2x(\Delta x) + (\Delta x)^2 - x^2\\ &= 2x(\Delta x) + (\Delta x)^2.\end{aligned}$$

In Figure 23.10 the increase in area is shaded, and it is easily seen to agree with the answer from our calculations.

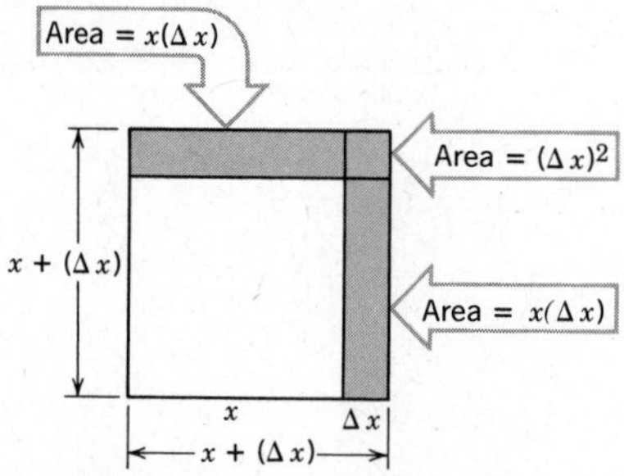

Figure 23.10

The following notation is used to indicate the change in the output of a function corresponding to a change in the input. It is assumed here that both x and $x + \Delta x$ belong to the domain of f.

> If f denotes a function, then
>
> $$\Delta f = f(x + \Delta x) - f(x). \tag{23.2}$$

In Example 23.11, $\Delta A = \Delta f$.

EXAMPLE 23.12 Assume that $f(x) = 3x^2 - 1$.

(a) Determine Δf.
(b) Determine the change in the output of f created by a change in the input from 1 to 4.
(c) Determine the change in the output of f created by a change in the input from 2 to 1.5.

Solution

(a) $$\begin{aligned}\Delta f &= f(x + \Delta x) - f(x)\\ &= [3(x + \Delta x)^2 - 1] - [3x^2 - 1]\\ &= 3x^2 + 6x(\Delta x) + 3(\Delta x)^2 - 1 - 3x^2 + 1\\ &= 6x(\Delta x) + 3(\Delta x)^2 \qquad (23.3)\end{aligned}$$

(b) If the input changes from 1 to 4, we use $x = 1$ and $\Delta x = 3$. Then Equation (23.3) gives

$$f = 6(1)(3) + 3(3)^2 = 18 + 27 = 45.$$

(c) If the input changes from 2 to 1.5, we use $x = 2$ and $\Delta x = -0.5$ (the negative sign is essential). In this case Equation (23.3) gives

$$f = 6(2)(-0.5) + 3(-0.5)^2 = -6 + 0.75 = -5.25.$$

The **difference quotient** of a function f is defined to be

$$\frac{\Delta f}{\Delta x}. \tag{23.4}$$

Difference quotients are important in calculus.

EXAMPLE 23.13 Determine and simplify the difference quotient of $f(x) = 3x^2 - 1$.

Solution The difference Δf for this function was computed in Example 23.12 [Equation (23.3)]:

$$\Delta f = 6x(\Delta x) + 3(\Delta x)^2.$$

Therefore,

$$\frac{\Delta f}{\Delta x} = \frac{6x(\Delta x) + 3(\Delta x)^2}{\Delta x} = 6x + 3(\Delta x).$$

The difference quotient of a function f represents the average rate of change of $f(x)$ with respect to x, over the interval Δx. Thus difference quotients are apt to be of interest whenever one variable changes as a result of a change in another variable.

EXERCISES FOR SECTION 23

Draw the graph of each function.

1. $f(x) = 2x$ **2.** $f(x) = x + 1$ **3.** $f(x) = -x$ **4.** $f(x) = 5$
5. $f(x) = 5 - 2x$ **6.** $f(x) = -2$ **7.** $f(x) = 3 - x$ **8.** $f(x) = 9/2$

9. $f(x) = 1 - x$ **10.** $f(x) = \begin{cases} -2 & \text{for } x \leq 1 \\ x & \text{for } x > 1 \end{cases}$

11. $f(x) = \begin{cases} x & \text{for } x < 0 \\ 2x & \text{for } x \geq 0 \end{cases}$ **12.** $f(x) = \begin{cases} x + 1 & \text{for } x \leq 0 \\ -x + 1 & \text{for } x > 0 \end{cases}$

Evaluate.

13. $[1.99]$ **14.** $[-3.2]$ **15.** $[\pi/2]$ **16.** $\frac{1}{2}[5]$ **17.** $[\frac{1}{2} \cdot 5]$
18. $5[\frac{1}{2}]$ **19.** $|[-3.1]|$ **20.** $[|-3.1|]$ **21.** $|-[3.1]|$

Draw the graph of each function.

22. $f(x) = |x| + 1$ **23.** $f(x) = |x| - 1$ **24.** $f(x) = |2x|$
25. $f(x) = [x] - 1$ **26.** $f(x) = [2x]$ **27.** $f(x) = [x] + 1$

28. The formula for the volume V of a right circular cone having height h and base of radius r is $V = \frac{1}{3}\pi r^2 h$.
(a) Determine a function f such that $h = f(V, r)$.
(b) Determine a function g such that $r = g(V, h)$.

29. The formula for the volume of a pyramid having height h and square base of length b on each side is $V = \frac{1}{3}b^2 h$.
(a) Determine a function f such that $h = f(V, b)$.
(b) Determine a function g such that $b = g(V, h)$.

30. The formula for the weight W of a sphere having radius r and constant density d is $W = \frac{4}{3}\pi\, dr^3$.
(a) Determine a function f such that $d = f(W, r)$.
(b) Determine a function g such that $r = g(W, d)$.

31. Determine a function f such that $S = f(P, r, n)$, where S is the compound amount from a principal amount P invested for one year at an annual rate r (in decimal form) compounded n times annually. [See Equation (18.9).] Compute $f(100, 0.08, 4)$.

32. Determine a function g such that $P = g(S, r, t)$, where P is the amount that must be invested at an annual rate r (in decimal form) compounded quarterly for t years to produce the compound amount S. [See Equation (18.10).] Compute $f(2000, 0.08, 2)$.

33. Determine a function h such that $r_E = h(r_N, n)$, where r_E is the effective rate equivalent to the nominal rate r_N compounded n times annually. [See Equation (18.13)]. Use Table 18.2 to determine $h(0.09, 4)$.

34. If r denotes the radius of a circle and C denotes the circumference, then $C = 2\pi r$. Determine the change ΔC in C corresponding to a change Δr in r.

35. If r denotes the radius of a circle and A denotes the area, then $A = \pi r^2$. Determine the change ΔA in A corresponding to a change Δr in r.

36. If r denotes the radius of a sphere and S denotes the surface area, then $S = 4\pi r^2$. Determine the change ΔS in S corresponding to a change Δr in r.

In Exercises 37–42, determine:

(a) Δf,
(b) the change in the output of f created by a change in the input from 2 to 5,
(c) the change in the output of f created by a change in the input from 1 to -0.5.

37. $f(x) = 4x + 2$ **38.** $f(x) = -2x + 1$ **39.** $f(x) = 1 - x$
40. $f(x) = -x^3$ **41.** $f(x) = x^3 + 1$ **42.** $f(x) = 2x^3 - 1$

Determine the difference quotient of each function.

43. $f(x) = 4x + 2$ **44.** $f(x) = -2x + 1$ **45.** $f(x) = 1 - x$
46. $f(x) = x^2 + 3$ **47.** $f(x) = 2x^2 - 1$ **48.** $f(x) = 4x^2 + 2$

SECTION

24 Linear Functions

A. Functions with Lines as Graphs

In this section we study the functions whose graphs are lines.

DEFINITION. A function f is called a **linear function** if it has the form

$$f(x) = mx + b \tag{24.1}$$

for some pair of real numbers m and b.

> The graph of a function is a straight line
> iff
> the function is linear.

Proof. The graph of a function cannot be a vertical line since the graph of a function f cannot contain two or more points with the same x value; remember that for each x in the domain of f there is just one value of $f(x)$. On the other hand, the slope-intercept form for the equation of a line (Section 20) shows that the nonvertical lines are precisely the graphs of the equations of the form $y = mx + b$. □

If $m = 0$, then Equation (24.1) reduces to $f(x) = b$, so that f is a *constant function* (Example 23.4). Thus every constant function is also a linear function. The graph of a linear function is a horizontal line iff the function is a constant function (like Figure 24.1*a*).

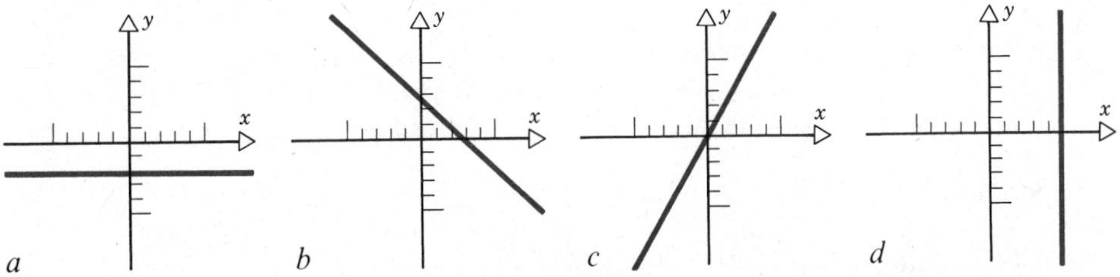

Figure 24.1 (a) Graph of the linear (constant) function $f(x) = -2$. (b) Graph of the linear function $f(x) = -x + 3$. (c) Graph of the linear function $f(x) = 2x$. (d) Graph of $x = 5$, which is not the graph of a function.

Linear functions can be further classified with the following definitions, which will also be useful in studying many other types of functions. Let f denote any function and let S denote any subset of the domain of f.

The function f is **increasing** over S if $x_1 < x_2$ implies $f(x_1) < f(x_2)$ for x_1 and x_2 in S.

The function f is **decreasing** over S if $x_1 < x_2$ implies $f(x_1) > f(x_2)$ for x_1 and x_2 in S.

In terms of a graph, a function is *increasing* if we move *upward* as we move to the right along the graph; and a function is *decreasing* if we move *downward* as we move to the right along the graph. See Figure 24.2.

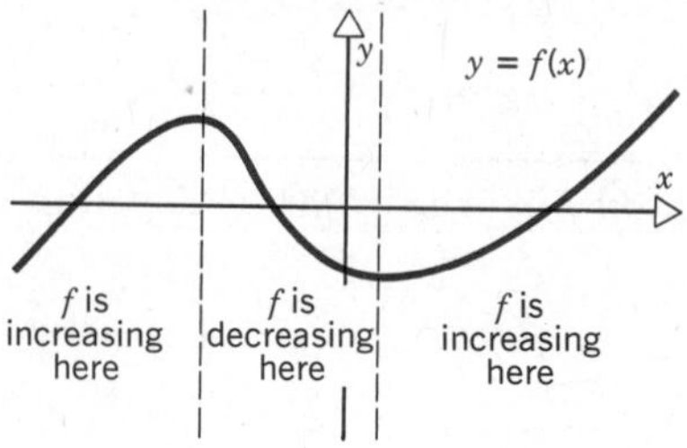

Figure 24.2

When we say simply that a function is *increasing,* without specifying a subset, we mean that it is increasing over its entire domain. Similarly for *decreasing.* From our study of slope in Section 20, we know that a linear function $f(x) = mx + b$ is increasing if $m > 0$, and it is decreasing if $m < 0$. For example, the function in Figure 24.1*c* is increasing, and the function in Figure 24.1*b* is decreasing.

B. Applications

A variable y *varies directly* as a variable x (or is *directly proportional* to x) if

$$y = kx \tag{24.2}$$

for some nonzero real number k, which is called the *constant of proportionality.* (See Section 9.) If $f(x) = kx$, then f is a linear function. Thus any example of direct variation gives an example of a linear function. The slope-intercept form shows that the graph of $f(x) = kx$ passes through the origin and has slope k.

EXAMPLE 24.1 A plane leaves Atlanta at noon and travels at a constant rate of 500 miles per hour for four hours. Express the distance it travels as a function of time, and draw the graph of the function.

Solution This is an application of $D = RT$ (distance = rate × time). In this case $R = 500$ is constant, so that $D = 500T$. Therefore, the required function f is given by

$$f(T) = 500T \quad (0 \leq T \leq 4).$$

Here T is measured in hours with $T = 0$ at noon. The graph is shown in Figure 24.3. The slope is 500. Notice, incidentally, that the units on the D- and T-axes measure different quantities. The scale on these axes is more or less arbitrary; although slope measures "steepness," the connection between slope and steepness also depends on the scale on each axis.

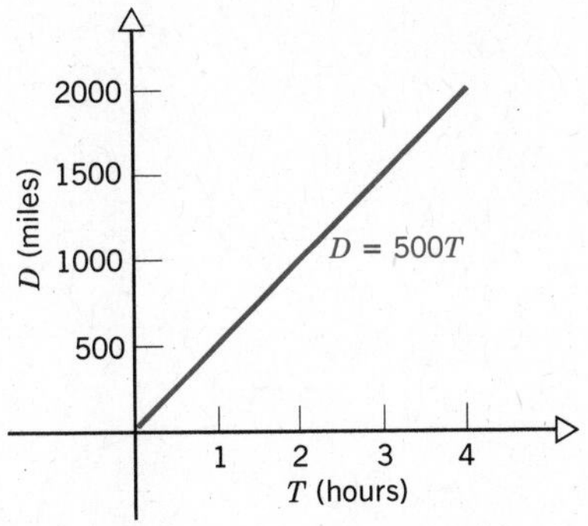

Figure 24.3

If P dollars are invested at a rate of r percent *simple* interest for t years, then the total amount A accumulated will be P (the principal) plus I (the interest):

$$A = P + I$$

$$A = P + Prt. \tag{24.3}$$

(See Section 8 for examples.) If P and r are given, then A will be a linear function of t.

EXAMPLE 24.2 Suppose that \$1000 is invested at 5% simple interest. Express the amount accumulated as a function of time, and draw the graph of the function.

Solution Use Equation (24.3) with $P = 1000$ and $r = 0.05$. Then

$$A = 1000 + (1000)(0.05)t.$$

That is,

$$A = f(t) = 1000 + 50t. \tag{24.4}$$

The graph of this function is shown in Figure 24.4. The A-intercept is $P = 1000$, and the slope is $Pr = 50$. Although here the graph is a straight line, we'll see in Section 38 that the result is quite different for compound interest (that is, when interest is paid on interest as well as principal.)

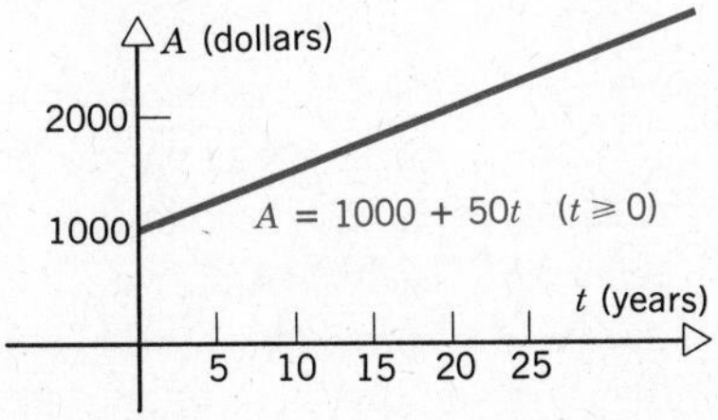

Figure 24.4

C. Determining Linear Functions

Because the graph of a linear function is a line, and two distinct points determine a line, it follows that a linear function is completely determined by any two points on its graph. Said differently, a linear function is determined by the output at any two input elements. To apply this observation we use the following form for the equation of a line.

An equation for the line through $P_1(x_1, y_1)$ and $P_2(x_2, y_2)$, with $x_1 \neq x_2$, is

$$y - y_1 = \frac{y_2 - y_1}{x_2 - x_1}(x - x_1). \tag{24.5}$$

This is called the **two-point form** for the equation of the line.

Proof. Equation (24.5) is simply the point-slope form, Equation (20.4), with m replaced by the slope as defined in Equation (20.2). □

EXAMPLE 24.3 Determine the linear function $f(x) = mx + b$ such that $f(1) = 3$ and $f(2) = -4$.

Solution Since the graph of f is the graph of $y = f(x)$, we need the function whose graph passes through (1, 3) and (2, −4). From Equation (24.5) this is

$$y - 3 = \frac{-4 - 3}{2 - 1}(x - 1)$$
$$y - 3 = -7(x - 1)$$
$$y = -7x + 10.$$

Therefore, the desired function is $f(x) = -7x + 10$.

Check If $f(x) = -7x + 10$, then

$$f(1) = -7 \cdot 1 + 10 = 3 \quad \text{and} \quad f(2) = -7 \cdot 2 + 10 = -4.$$

D. Application: Linear Interpolation

Suppose that y is a function of x, $y = f(x)$, and that x_1, x_2, and a are real numbers such that $x_1 < a < x_2$. Suppose also that $f(x_1)$ and $f(x_2)$ are known. Can we determine $f(a)$? (See Figure 24.5.)

If we know enough about the function f, such as a formula for $f(x)$ that is easy to compute, then the answer is Yes. In many cases we do not know enough about f, however, and then we must settle for a reasonable *estimate* for $f(a)$. One type of estimate is based on *linear interpolation*, in which we assume that the graph of f is nearly a straight line between $(x_1, f(x_1))$ and $(x_2, f(x_2))$. This is an easy idea to apply and it provides estimates that are adequate in many cases where x_1 and x_2 are reasonably close together. Here's how the idea is carried out.

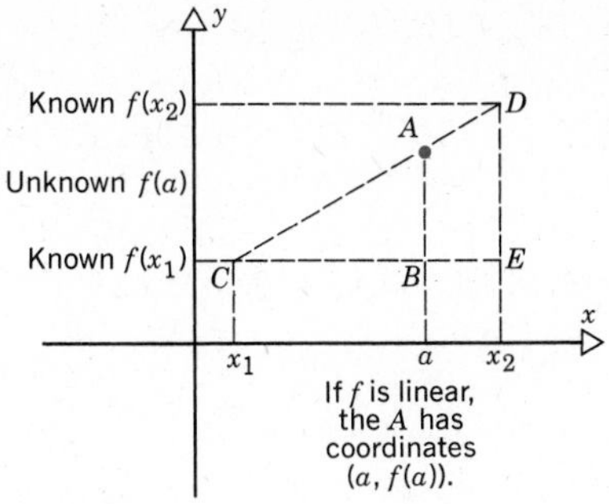

Figure 24.5

Assume that f is linear and that $x_1 < a < x_2$. By the similarity of the triangles ABC and DEC in Figure 24.5,

$$\frac{f(a) - f(x_1)}{a - x_1} = \frac{f(x_2) - f(x_1)}{x_2 - x_1}. \tag{24.6}$$

If $f(x_1)$ and $f(x_2)$ are known, then (24.6) can be used to compute $f(a)$.

EXAMPLE 24.4 Compute an estimate for $\sqrt{70}$ using $\sqrt{64} = 8$, $\sqrt{81} = 9$, and linear interpolation.

Solution Let $f(x) = \sqrt{x}$, so that $f(64) = 8$ and $f(81) = 9$. We want $f(70)$. If f were linear, then $f(70)$ woud be given *exactly* by Equation (24.6) with $x_1 = 64$, $a = 70$, and $x_2 = 81$. As it is, we use (24.6) to get an *estimate* for $f(70)$:

$$\frac{f(70) - f(64)}{70 - 64} \approx \frac{f(81) - f(64)}{81 - 64}$$

$$\frac{f(70) - 8}{6} \approx \frac{9 - 8}{17}$$

$$f(70) - 8 \approx \tfrac{1}{17} \cdot 6$$

$$f(70) \approx \tfrac{142}{17} \approx 8.35.$$

That is, $\sqrt{70} \approx 8.35$. The correct value of $\sqrt{70}$, accurate to the nearest one-hundredth, is 8.37.

EXERCISES FOR SECTION 24

Draw the graph of each function. Also indicate whether the function is constant, increasing, or decreasing.

1. $f(x) = 3x$ **2.** $f(x) = -2x$ **3.** $f(x) = \frac{1}{2}x$
4. $g(x) = -x + 2$ **5.** $g(x) = 6$ **6.** $g(x) = -3x - 2$
7. $h(x) = -\frac{1}{2}$ **8.** $h(x) = 4x - 1$ **9.** $h(x) = -3.5$

Determine the linear function f satisfying the two given conditions.

10. $f(1) = 5$ and $f(3) = 9$ **11.** $f(0) = -1$ and $f(2) = 3$
12. $f(-2) = 1$ and $f(1) = 4$ **13.** $f(-3) = -1$ and $f(0) = -3$
14. $f(2) = 12$ and $f(5) = 12$ **15.** $f(0) = 5$ and $f(4) = -1$

16. Use linear interpolation to estimate $f(-0.2)$, given that $f(-1) = 20$ and $f(0) = 22$.

17. Use linear interpolation to estimate $f(193)$, given that $f(190) = 2.279$ and $f(200) = 2.301$.

18. Use linear interpolation to estimate $f(6.3)$, given that $f(6) = 1.792$ and $f(7) = 1.946$.

19. Compute an estimate for $\sqrt{60}$ using $\sqrt{49} = 7$, $\sqrt{64} = 8$, and linear interpolation.

20. Compute an estimate for $\sqrt{40}$ using $\sqrt{36} = 6$, $\sqrt{49} = 7$, and linear interpolation.

21. Compute an estimate for $\sqrt{30}$ using $\sqrt{25} = 5$, $\sqrt{36} = 6$, and linear interpolation.

22. A plane leaves Dallas at noon and travels at a constant rate of 450 miles per hour for three hours. Determine the function f such that $D = f(T)$, where D denotes the distance traveled in miles and T the time in hours, with $T = 0$ at noon.

23. Assume that y varies directly as x and that $y = 28$ when $x = 5$. Determine the function f such that $y = f(x)$. (See Example 9.9.)

24. Assume that a weight of 3 kilograms will stretch a spring 2 centimeters. Determine the function f such that $L = f(W)$, where L denotes the length that the spring will stretch in centimeters when a weight of W kilograms is attached. (See Example 9.11.)

25. A roll of carpet has width 3 feet. Determine the function f such that $A = f(x)$, where A denotes the area in square feet and x the length in feet of a piece of carpet cut from the roll.

26. The scale of a map is 1 : 2000. Determine the function f such that $D = f(d)$, where D denotes ground distance in kilometers and d denotes map distance in centimeters. (See Section 9.)

27. The scale of a map is 1 : 100,000. Determine the function g such that $d = g(D)$, where d denotes map distance in centimeters and D denotes ground distance in kilometers. (See Section 9.)

28. Suppose that \$5000 is invested at 6% per year simple interest. Express the amount accumulated as a function of time [$A = f(t)$, t in years].

29. Suppose that \$2000 is invested at 8% per year simple interest. Express the amount accumulated as a function of time [$A = f(t)$, t in years].

30. Suppose that \$800 is invested at 6.5% per year simple interest. Express the amount accumulated as a function of time [$A = f(t)$, t in years].

31. You plan to rent a car for the day and find that a rental agency charges \$25 per day plus 20¢ per mile. Find a function f such that $f(x)$ gives the total charge (daily charge plus mileage charge), in dollars, if you drive x miles. Draw the graph of the function. To what cost does the slope correspond?

32. Suppose that electricity costs 3¢ for each of the first 500 kwh (kilowatt-hours) and then 2¢ for each kwh thereafter. Find the function f such that $f(x)$ gives the cost in dollars for x kwh, for $0 \leq x \leq 1000$. (The function will consist of two parts, one part for $0 \le x \le 500$ and another part for $500 < x \le 1000$.)

33. If $y = f(x)$, and x is changed by Δx, then the change in y will be $\Delta y = \Delta f = f(x + \Delta x) - f(x)$. Thus $\Delta y/\Delta x = \Delta f/\Delta x$. Compute the slope of the graph of $f(x) = mx + b$ by computing the difference quotient $\Delta f/\Delta x$. (Difference quotients were discussed in Section 23D.)

SECTION 25 Quadratic Functions

A. Introduction. Intercepts

DEFINITION. A function f is called a **quadratic function** if it has the form

$$f(x) = ax^2 + bx + c, \tag{25.1}$$

where a, b, and c are real numbers with $a \neq 0$.

EXAMPLE 25.1 Draw the graph of the quadratic function $f(x) = x^2$.

Solution Table 25.1 shows some typical coordinate pairs. The graph is shown in Figure 25.1.

TABLE 25.1

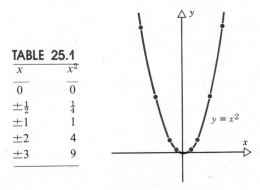

x	x^2
0	0
$\pm\frac{1}{2}$	$\frac{1}{4}$
± 1	1
± 2	4
± 3	9

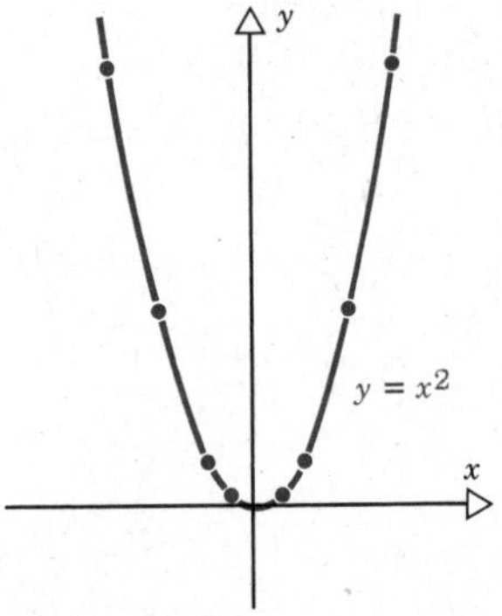

Figure 25.1

The graph of any quadratic function is called a **parabola.** Each figure in this section gives an example. If the parabola opens upward (like Figure 25.1), then the **vertex** is the minimum or lowest point on the graph; if the parabola opens downward (like Figure 25.2*a*), then the **vertex** is the maximum or highest point on the graph. In either case we call the vertex the **extreme point** of the graph. (Parabolas are also discussed in Section 21, but this section is independent of that.)

To analyze a quadratic function we concentrate on three questions, which can be stated in terms of the graph as follows:

Question I. What are the x-intercepts, if any?

Question II. Does the graph open upward? or downward?

Question III. Where is the extreme point on the graph?

We can draw the graph with very little effort once these questions have been answered. Moreover, these questions get to the heart of many other problems besides graphing, as Subsection C will show. We treat Question I here and Questions II and III in Subsection B.

Question I is just that of finding the real numbers that are solutions of

$$ax^2 + bx + c = 0. \tag{25.2}$$

This was disposed of in Chapter III: if possible, solve Equation (25.2) by factoring; otherwise use the quadratic formula.

For convenience, here is the quadratic formula again.

> The solutions of $ax^2 + bx + c = 0$ are
>
> $$x = \frac{-b \pm \sqrt{b^2 - 4ac}}{2a}. \tag{25.3}$$

If the discriminant, $b^2 - 4ac$, is negative, then Equation (25.2) has no real solution; therefore, the graph of (25.1) has no x-intercept.

If the discriminant is 0, then (25.2) has one real solution and the graph of (25.1) has one x-intercept.

If the discriminant is positive, then (25.2) has two unequal real solutions and the graph of (25.1) has two x-intercepts.

EXAMPLE 25.2 Determine the x-intercepts of the graph of each function.

(a) $f(x) = -2x^2 - 6x$
(b) $f(x) = x^2 - 8x + 16$
(c) $f(x) = \frac{1}{4}x^2 + 3$

Solution

(a) Since $-2x^2 - 6x = -2x(x + 3) = 0$ iff $x = 0$ or $x = -3$, there are two x-intercepts, 0 and -3. See Figure 25.2*a*.

(b) Since $x^2 - 8x + 16 = (x - 4)^2 = 0$ iff $x = 4$, there is one x-intercept, 4. See Figure 25.2*b*.

(c) In this case, $b^2 - 4ac = 0^2 - 4(\frac{1}{4})(3) = -3 < 0$. Thus there is no x-intercept. Notice that $f(x) > 0$ for every x because $x^2 \geq 0$ and hence $\frac{1}{4}x^2 + 3 > 0$. See Figure 25.2*c*.

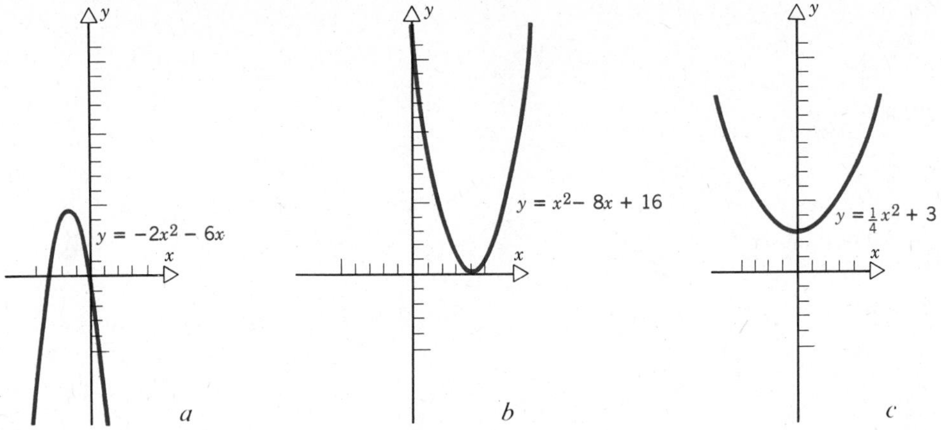

Figure 25.2

B. Maximum and Minimum Values

> The vertex or extreme point of the graph of
>
> $$f(x) = ax^2 + bx + c \quad (a \neq 0)$$
>
> occurs at the point $(-b/2a, f(-b/2a))$.
> If $a > 0$, the graph opens upward and this point is a minimum.
> If $a < 0$, the graph opens downward and this point is a maximum.

(Here is a memory tip: Think of the parabola as a container. If a is *positive,* the container *will* hold water—it opens upward. If a is *negative,* the container *will not* hold water—it opens downward.)

Proof. To begin, we rewrite $ax^2 + bx + c$ by completing the square (Section 10).

$$\begin{aligned} f(x) &= ax^2 + bx + c \\ &= a\left(x^2 + \frac{b}{a}x\right) + c \\ &= a\left(x^2 + \frac{b}{a}x + \frac{b^2}{4a^2}\right) - \frac{b^2}{4a} + c \\ &= a\left(x + \frac{b}{2a}\right)^2 + \frac{4ac - b^2}{4a}. \end{aligned} \tag{25.4}$$

Since the square of every real number is nonnegative,

$$\left(x + \frac{b}{2a}\right)^2 \geq 0 \quad \text{for all real numbers } x.$$

Thus

$$a\left(x + \frac{b}{2a}\right)^2 \geq 0 \quad \text{for all } x \text{ if } a > 0,$$

and

$$a\left(x + \frac{b}{2a}\right)^2 \leq 0 \quad \text{for all } x \text{ if } a < 0.$$

Therefore, from Equation (25.4),

$$f(x) \geq \frac{4ac - b^2}{4a} \quad \text{for all } x \text{ if } a > 0, \tag{25.5}$$

and

$$f(x) \leq \frac{4ac - b^2}{4a} \quad \text{for all } x \text{ if } a < 0. \tag{25.6}$$

Also, (25.4) makes it easy to see that

$$f(x) = \frac{4ac - b^2}{4a} \quad \text{iff} \quad x = \frac{-b}{2a}. \tag{25.7}$$

By (25.5) and (25.7), if $a > 0$, there is a minimum where $x = -b/2a$. By (25.6) and (25.7), if $a < 0$, there is a maximum where $x = -b/2a$. □

EXAMPLE 25.3 Determine the extreme value of $f(x) = -2x^2 + 8x + 3$ and state whether it is a minimum or a maximum value.

Solution Here $a = -2$ and $b = 8$, so that

$$-\frac{b}{2a} = -\frac{8}{2(-2)} = 2.$$

Thus the extreme value is $f(2) = -2 \cdot 2^2 + 8 \cdot 2 + 3 = 11$. Since $a = -2 < 0$, this is a maximum value.

The vertical line $x = -b/2a$, which passes through the vertex, is called the **axis** of the parabola (Figure 25.3). The graph of the parabola is *symmetric* about its axis; that is, for each point of the graph on one side of the axis there is a corresponding point on the other side with the same y-coordinate and the same distance from the axis [compare (2, 4) and (−2, 4) in Figure 25.1]. Figure 25.3 indicates the increasing and decreasing characteristics of quadratic functions.

EXAMPLE 25.4 Draw the graph of $f(x) = 2x^2 + 3x - 2$. Give the coordinates of the vertex, the equation of the axis, the x-intercepts (if there are any), and where the function is increasing and decreasing.

Solution Here $a = 2$, $b = 3$, and $c = -2$. Since $a > 0$, the parabola opens upward. The minimum is where $x = -b/2a = -3/(2 \cdot 2) = -\frac{3}{4}$. The corresponding y-value is

$$f(-\tfrac{3}{4}) = 2(-\tfrac{3}{4})^2 + 3(-\tfrac{3}{4}) - 2 = -\tfrac{25}{8}.$$

Thus the vertex is at $(-\frac{3}{4}, -\frac{25}{8})$. The axis is the line $x = -\frac{3}{4}$.

Because $2x^2 + 3x - 2 = (2x - 1)(x + 2)$, the x-intercepts are at $x = \frac{1}{2}$ and $x = -2$.

The function is decreasing for $x < -\frac{3}{4}$ (to the left of the vertex) and increasing for $x > -\frac{3}{4}$ (to the right of the vertex).

To draw the graph, first plot the vertex and the x-intercepts. Then plot several other points to determine the shape more closely. Some sample points in this case are $(-3, 7)$, $(-1, -3)$, $(0, -2)$, and $(1, 3)$. Remember also that a parabola must be symmetric about its axis. Figure 25.4 shows the graph.

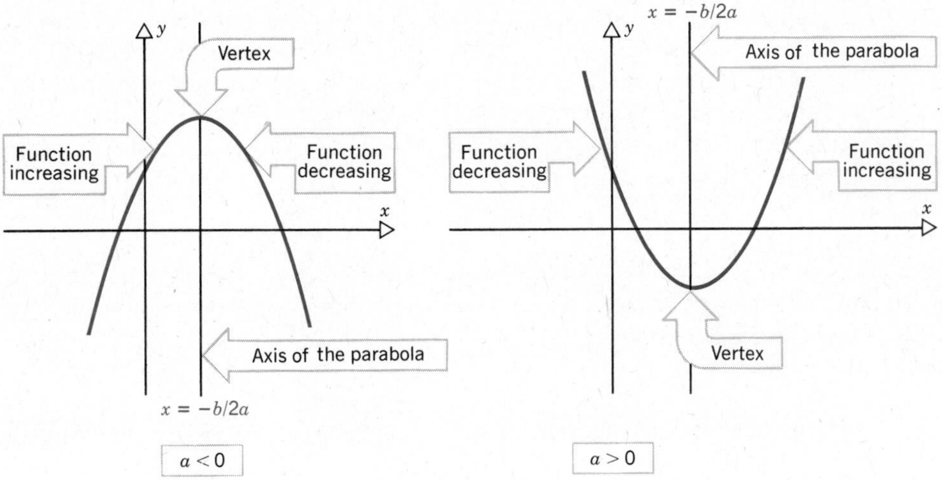

Figure 25.3

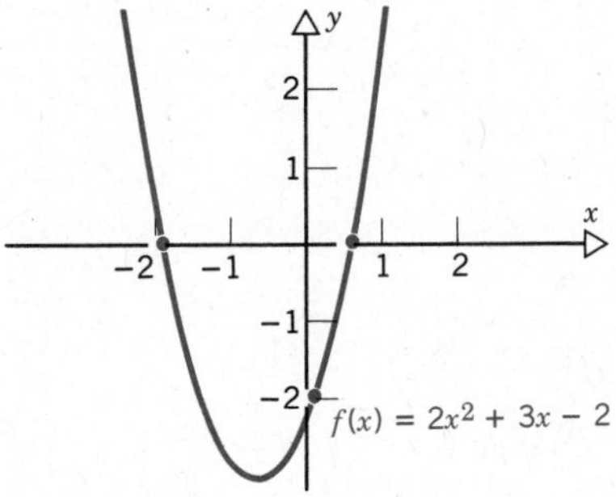

Figure 25.4

C. Applications

EXAMPLE 25.5 Suppose that a small rocket is fired upward near the earth's surface. The distance s of the rocket above the earth's surface after t seconds is given by the quadratic function

$$f(t) = -\tfrac{1}{2}gt^2 + v_0t + s_0 \quad (t \geq 0), \tag{25.8}$$

where

$g \approx 32$ if s is measured in feet,

$g \approx 9.8$ if s is measured in meters,

v_0 = velocity when $t = 0$, and

s_0 = height when $t = 0$.

Remark This type of problem was also considered in Section 13. If you compare Equation (13.8) with Equation (25.8), you'll see that $\frac{1}{2}gt^2$ has been replaced by $-\frac{1}{2}gt^2$ and a new term s_0 has been added. The sign on $\frac{1}{2}gt^2$ has been changed because the positive direction is upward here and it was downward in Section 13. The term s_0 has been added because distance is measured from the ground here, rather than from the initial position, as in Section 13. The number v_0 in Equation (25.8) will be negative if the rocket is fired downward rather than upward. Finally, Equation (25.8) will describe the motion of any other object as well as a rocket, at least within the physical limits that we consider.

Problem A small rocket is fired upward from the top of an 80 foot tall building with an initial velocity of 64 feet per second. Determine its maximum altitude and when it will reach the ground.

Solution The function describing the position s is given by Equation (25.8) with $g = 32$, $v_0 = 64$, and $s_0 = 80$:

$$s = f(t) = -16t^2 + 64t + 80.$$

This has the form of Equation (25.1) with $a = -16$, $b = 64$, $c = 80$, and t in place of x. See Figure 25.5 for the graph.

The rocket will reach its maximum altitude $t = -b/2a = -64/2(-16) = 2$ seconds after it is fired. The maximum altitude will be

$$f(2) = -16(2)^2 + 64(2) + 80 = 144 \text{ feet}$$

above the earth's surface.

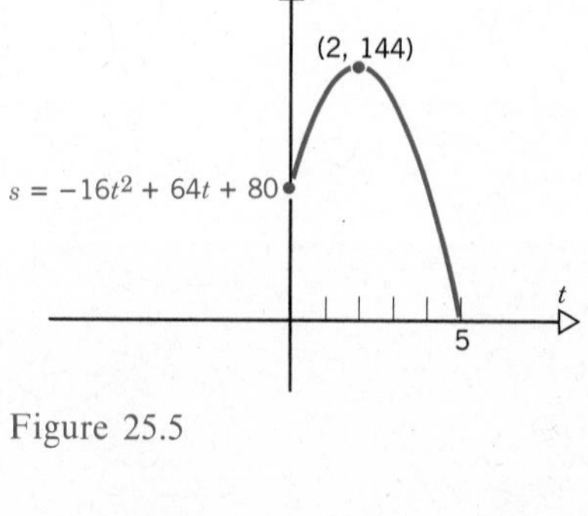

Figure 25.5

The rocket will reach the ground when $s = f(t) = 0$:

$$-16t^2 + 64t + 80 = 0$$
$$-16(t^2 - 4t - 5) = 0$$
$$(t + 1)(t - 5) = 0$$
$$t = -1 \quad \text{or} \quad t = 5.$$

The time $t = -1$ has no physical significance in this problem (remember $t \geq 0$). Therefore, the rocket will reach the ground $t = 5$ seconds after it is fired.

EXAMPLE 25.6 A rectangular field is to be formed with one side along a straight river bank. If the side along the river bank requires no fencing, and if 400 yards of fencing are available for the other three sides, what is the largest possible area for the field?

Solution Let x denote the length of the sides perpendicular to the bank. Then the other length must be $400 - 2x$, as shown in Figure 25.6. Therefore, the total area is

$$A = f(x) = x(400 - 2x) = -2x^2 + 400x.$$

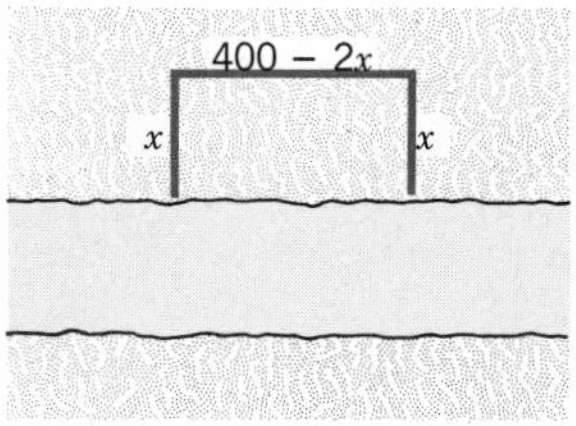

Figure 25.6

This is a quadratic function with $a = -2$, $b = 400$, and $c = 0$. Therefore, the x-value that maximizes the function is

$$-b/2a = -400/[2(-2)] = 100 \text{ yards.}$$

The maximum possible area is

$$f(100) = -2(100)^2 + 400(100)$$
$$= 20{,}000 \text{ square yards.}$$

The graph of f is shown in Figure 25.7.

It is generally harder to determine maximum or minimum values for more complicated functions (such as those involving x^3). That is one of the problems that calculus is designed to solve.

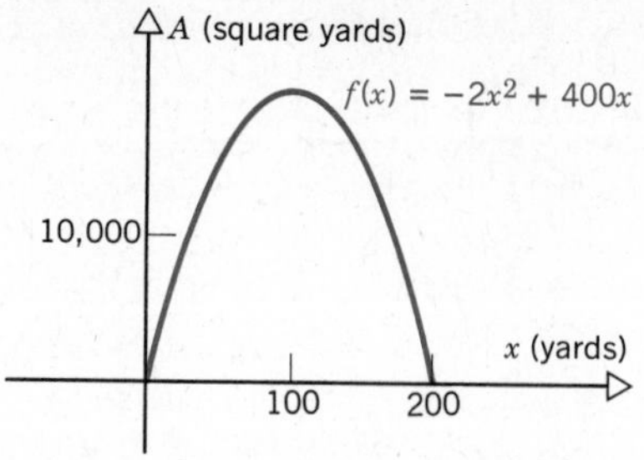

Figure 25.7

Other applications of parabolas are shown in Figures 25.8 and 25.9.

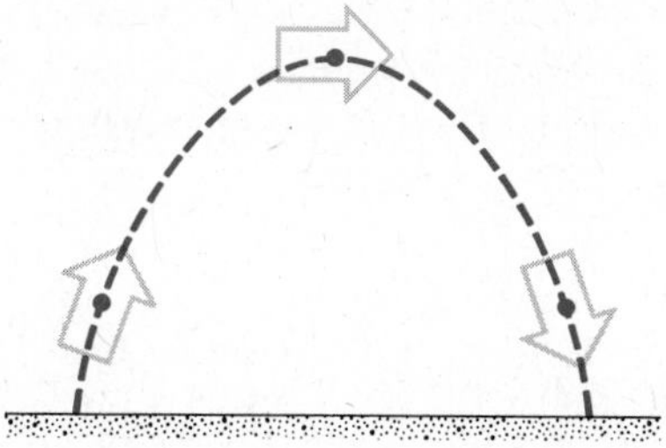

Figure 25.8 If air resistance is ignored, then the path followed by the center of gravity of a leaping leopard or other animal is a parabola. So is the path of a projectile fired from a cannon.

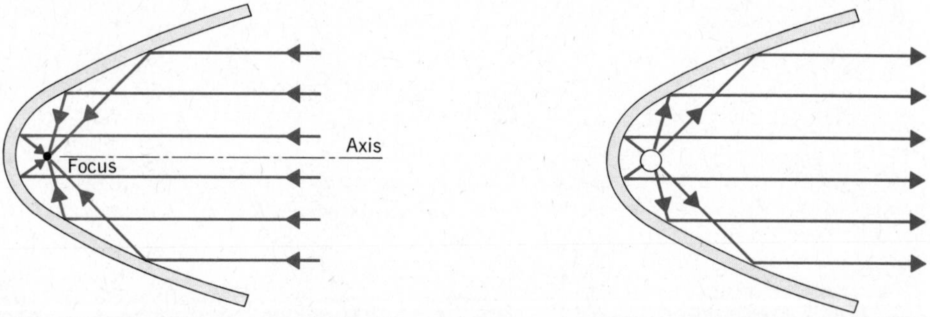

Figure 25.9 A parabola rotated about its axis generates a *parabolic surface.* Such surfaces are used for reflecting telescopes because they concentrate the light from distant objects at a single point (called the *focus*), giving sharp images of stars or other sources of light. Conversely, parabolic surfaces are used for such things as headlights because a light placed at the focus will be reflected outward in rays parallel to the axis.

EXERCISES FOR SECTION 25

1. On a single Cartesian plane (coordinate system) draw the graphs of $y = x^2$, $y = 2x^2$, and $y = -\frac{1}{2}x^2$. How does changing a affect the graph of $y = ax^2$?

2. On a single Cartesian plane (coordinate system) draw the graphs of $y = x^2$, $y = -3x^2$, and $y = \frac{1}{3}x^2$. How does changing a affect the graph of $y = ax^2$?

3. On a single Cartesian plane (coordinate system) draw the graphs of $y = x^2$, $y = 4x^2$, and $y = -\frac{1}{4}x^2$. How does changing a affect the graph of $y = ax^2$?

4. On a single Cartesian plane (coordinate system) draw the graphs of $y = x^2$,

$y = x^2 + 1$, and $y = x^2 - 2$. How does changing c affect the graph of $y = x^2 + c$?

5. On a single Cartesian plane (coordinate system) draw the graphs of $y = x^2$, $y = x^2 + 2$, and $y = x^2 - 1$. How does changing c affect the graph of $y = x^2 + c$?
6. On a single Cartesian plane (coordinate system) draw the graphs of $y = x^2$, $y = x^2 + 3$, and $y = x^2 - 3$. How does changing c affect the graph of $y = x^2 + c$?

Determine the extreme value of each function and state whether it is a minimum or a maximum value.

7. $f(x) = 2x^2 + x + 1$	**8.** $g(x) = x^2 - 2$	**9.** $h(x) = x^2 - 6x$
10. $f(x) = -x^2 - 5$	**11.** $g(x) = -3x^2 + x + 1$	**12.** $h(x) = -2x^2 + 4$

Draw the graph of each function. Give the coordinates of the vertex and the equation of the axis. Also determine the x-intercepts if there are any.

13. $f(x) = -2x^2 - 1$	**14.** $f(x) = -3x^2 + 12$	**15.** $f(x) = 3x^2 + 3$
16. $g(x) = x^2 - 2x + 1$	**17.** $g(x) = -x^2 - x$	**18.** $g(x) = -2x^2 + 12x - 18$
19. $h(x) = (x + 1)^2 - 1$	**20.** $h(x) = -x^2 + 5x$	**21.** $h(x) = x^2 + x + 1$
22. $f(x) = 2x^2 + 3x$	**23.** $f(x) = (3 + x)^2 + 3$	**24.** $f(x) = (x - 2)^2 + 1$

25. An object is thrown upward from the top of a 64 foot tall building with an initial velocity of 48 feet per second. Determine its maximum altitude and when it will reach the ground.
26. An object is thrown upward from the top of a 29.4 meter tall building with an initial velocity of 24.5 meters per second. Determine its maximum altitude and when it will reach the ground. (Remember to use $g = 9.8$.)
27. An object is thrown upward from the top of a 192 foot tall building with an initial velocity of 64 feet per second. Determine its maximum altitude and when it will reach the ground.
28. Repeat Example 25.6 assuming that 1000 meters of fence are available.
29. Repeat Example 25.6 assuming that 500 yards of fence are available.
30. Repeat Example 25.6 assuming that 600 meters of fence are available.
31. Find two numbers whose sum is 7 and whose product is as large as possible. [Suggestion: If x is one number, then the other is $7 - x$. Maximize $x(7 - x)$.]
32. Find two numbers whose sum is 17 and whose product is as large as possible. (Exercise 31 gives a suggestion.)
33. Prove that if x and y are numbers whose sum is s, then the product xy will be maximum when $x = y = s/2$. (Exercise 31 gives a suggestion.)
34. Prove that if a parabola is the graph of a quadratic function and has two (unequal) x-intercepts, then the axis of the parabola bisects the segment connecting the x-intercepts. [Suggestion: The point that bisects the x-intercepts is the average of the x-values in Equation (25.3).]
35. Determine and simplify the difference quotient $\Delta f/\Delta x$ for the general quadratic function $f(x) = ax^2 + bx + c$. (Difference quotients were introduced in Section 23.)
36. The graph of an equation $x = ay^2 + by + c$ will be a parabola with a horizontal axis. Use analogy with the results of this section to answer the following questions.

(a) Under what conditions will the graph have no y-intercept? one y-intercept? two y-intercepts?
(b) Where is the x-intercept?
(c) Under what conditions will the parabola open to the left? to the right?
(d) What are the coordinates of the vertex?

SECTION 26 Inverse Functions

A. Introduction. One-to-One Functions

Consider the following statements, where x represents a real number in each case.

If we choose x, add 5, and then subtract 5 from the result, we get back x.

If we choose x, multiply by 3, and then divide the result by 3, we get back x.

If we choose x positive, square it, and then take the principal square root of the result, we get back x.

In this section we'll see that each of the three statements represents a very special case of a general idea involving functions—the idea of the *inverse* of a function. Before considering inverse functions, however, we need the following facts about *one-to-one* functions.

If f is any function, then for each x in the domain of f there is only *one* y such that $y = f(x)$. In other words, unequal *output* elements of f cannot correspond to the same *input* element. If we also require of f that unequal *input* elements not correspond to the same *output* element, then we get the important class of functions covered by the following definition.

DEFINITION. A function f is **one-to-one** iff

$$x_1 \neq x_2 \text{ implies } f(x_1) \neq f(x_2) \tag{26.1}$$

for all x_1 and x_2 in the domain of f.

Condition (26.1) is equivalent to

$$f(x_1) = f(x_2) \text{ implies } x_1 = x_2 \tag{26.2}$$

for all x_1 and x_2 in the domain of f.

EXAMPLE 26.1 Verify that the function $f(x) = 2x + 5$ is one-to-one.

First Solution Use (26.1): $x_1 \neq x_2$ implies $2x_1 \neq 2x_2$, which implies $2x_1 + 5 \neq 2x_2 + 5$; that is, $f(x_1) \neq f(x_2)$.

Second Solution Use (26.2): $f(x_1) = f(x_2)$ means $2x_1 + 5 = 2x_2 + 5$, which implies $2x_1 = 2x_2$, which in turn implies $x_1 = x_2$.

EXAMPLE 26.2 The function $f(x) = x^2$ is not one-to-one, because (for example) $3 \neq -3$ but $(3)^2 = (-3)^2$. Notice that to show that a function is *not* one-to-one, it suffices to find *one* pair of numbers x_1 and x_2 such that $x_1 \neq x_2$ but $f(x_1) = f(x_2)$.

The definition of one-to-one function can be interpreted geometrically as follows. First, recall that each *vertical* line intersects the graph of a function at most once (Figure 23.2). If a function is one-to-one, then we can also say that each *horizontal* line intersects the graph at most once. The graph on the left in Figure 26.1 represents a function that is not one-to-one: the horizontal line L_1 intersects the graph in points such that $x_1 \neq x_2$ but $f(x_1) = f(x_2)$, violating (26.1).

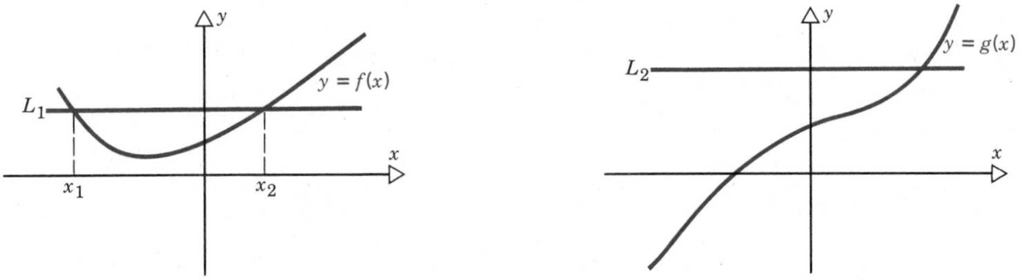

Figure 26.1 (a) The function f is not one-to-one, because the horizontal line L_1 intersects the graph more than once: $x_1 \neq x_2$ but $f(x_1) = f(x_2)$. (b) The function g is one-to-one, because each horizontal line (such as L_2) intersects the graph at most once.

If a function is linear and nonconstant, then its graph is a nonhorizontal line; thus every horizontal line intersects the graph only once, and the function is one-to-one. The same argument applies to any other function that is either increasing (over its entire domain) or decreasing (over its entire domain). This gives us the following statements.

> Every increasing function is one-to-one.
> Every decreasing function is one-to-one.
> Every nonconstant linear function is one-to-one.

B. The Inverse of a One-to-One Function

Assume that f is a one-to-one function and that y is in the range of f. Then there is exactly one element x in the domain of f such that $y = f(x)$; denote this element x by $g(y)$. Thus

$$x = g(y) \text{ iff } y = f(x). \tag{26.3}$$

In this way we obtain a new function g. The domain of g is the range of f, and the range of g is the domain of f.

The relationships in (26.3) can be represented schematically as follows.

$$x \underset{g}{\overset{f}{\rightleftarrows}} y \qquad \begin{array}{l} y = f(x) \text{ so } f \text{ applied to } x \text{ gives } y \\ \\ x = g(y) \text{ so } g \text{ applied to } y \text{ gives } x \end{array}$$

Viewed in this way, it is easy to see what happens if we apply f and then g, or if we apply g and then f; we write the results using composition (Section 22):

$$(g \circ f)(x) = g(f(x)) = g(y) = x \text{ for each } x \text{ in the domain of } f$$

and

$$(f \circ g)(y) = f(g(y)) = f(x) = y \text{ for each } y \text{ in the domain of } g.$$

The conclusion of the last line—that $(f \circ g)(y) = y$ for each y in the domain of g—will remain the same if we replace y throughout by any other variable. In particular, we can use x, which gives "$(f \circ g)(x) = x$ for each x in the domain of g."

The preceding remarks show that g is an *inverse* of f, in the sense of the following definition.

DEFINITION. A function g is an **inverse** of a function f if

$$(g \circ f)(x) = x \text{ for each } x \text{ in the domain of } f$$

and

$$(f \circ g)(x) = x \text{ for each } x \text{ in the domain of } g.$$

EXAMPLE 26.3 Verify that if $f(x) = 2x + 5$ and $g(x) = \frac{1}{2}(x - 5)$, then g is an inverse of f.

Solution

$$(g \circ f)(x) = g(f(x)) = g(2x + 5) = \tfrac{1}{2}[(2x + 5) - 5] = x$$
$$(f \circ g)(x) = f(g(x)) = f(\tfrac{1}{2}(x - 5)) = 2[\tfrac{1}{2}(x - 5)] + 5 = x$$

From the symmetric roles of f and g in the definition of *inverse*, it follows that g is an inverse of f iff f is an inverse of g. It can be proved that f has an inverse iff f is one-to-one. Moreover, if a function does have an inverse, then it has only one; the unique inverse of f, if it exists, is denoted by f^{-1} (read "f inverse"). *Note:* $f^{-1}(x)$ *does not,* in general, mean $\dfrac{1}{f(x)}$. This use of -1 as a superscript is not the same as its use as an exponent.

It follows from the definition of an inverse function that the domain of f^{-1} is the range of f, and the range of f^{-1} is the domain of f.

The three examples at the beginning of this section can be described in the language of functions as follows:

If $f(x) = x + 5$, then $f^{-1}(x) = x - 5$.
If $f(x) = 3x$, then $f^{-1}(x) = x/3$.
If $f(x) = x^2$ for $x \geq 0$, then $f^{-1}(x) = \sqrt{x}$ for $x \geq 0$.

In the last example, we require $x \geq 0$ because otherwise f would not have an inverse; the function defined by $f(x) = x^2$, with no restriction on x, is not one-to-one.

C. Finding Inverse Functions

If the equation $y = f(x)$ can be solved uniquely for x, the result will be $f^{-1}(y)$. If we want an expression for f^{-1} in terms of x, then we can simply replace y by x throughout the equation giving $f^{-1}(y)$.

EXAMPLE 26.4 Find $f^{-1}(x)$ if $f(x) = 4x - 3$.

Solution Let $y = f(x)$ and solve for x:

$$\begin{aligned} y &= 4x - 3 \\ y + 3 &= 4x \\ x &= \tfrac{1}{4}(y + 3). \end{aligned}$$

Therefore,

$$f^{-1}(y) = \tfrac{1}{4}(y + 3). \tag{26.4}$$

To express f^{-1} in terms of x, replace y by x in (26.4):

$$f^{-1}(x) = \tfrac{1}{4}(x + 3).$$

You can check the answer by direct computation, as in Example 26.3.

EXAMPLE 26.5 Find $f^{-1}(x)$ if $f(x) = x^2 + 5$ for $x \geq 0$. (We require $x \geq 0$ so that f will be one-to-one.)

Solution Let $y = f(x)$ and solve for x:

$$\begin{aligned} y &= x^2 + 5 \\ y - 5 &= x^2 \\ x &= \sqrt{y - 5}. \end{aligned}$$

Notice that $y \geq 5$ since $y = x^2 + 5$, so that $\sqrt{y - 5}$ will be a real number. Thus

$$f^{-1}(y) = \sqrt{y - 5} \text{ for } y \geq 5,$$

or

$$f^{-1}(x) = \sqrt{x - 5} \text{ for } x \geq 5.$$

In interval notation (Section 6),

$$\begin{aligned} \text{domain of } f = \text{range of } f^{-1} &= [0, \infty) \\ \text{range of } f = \text{domain of } f^{-1} &= [5, \infty). \end{aligned}$$

D. Graphs

Let f denote a function with an inverse. Then:

$$(a, b) \text{ is on the graph of } y = f(x) \\ \text{iff} \\ (b, a) \text{ is on the graph of } y = f^{-1}(x).$$

Proof. (a, b) is on the graph of $y = f(x)$

$$\text{iff}$$
$$b = f(a)$$
$$\text{iff}$$
$$a = f^{-1}(b)$$
$$\text{iff}$$
$$(b, a) \text{ is on the graph of } y = f^{-1}(x).$$ □

Figure 26.2 shows several pairs of points whose coordinates have this relationship of (a, b) to (b, a). Each pair is located symmetrically about the line $y = x$. This leads to the following conclusion.

> The graphs of $y = f(x)$ and $y = f^{-1}(x)$ are located symmetrically about the line $y = x$.

It follows that if we graph $y = f(x)$ and $y = f^{-1}(x)$ in the same coordinate system, and then turn the figure so that we're looking along the line $y = x$, the figure will be "left-right symmetric."

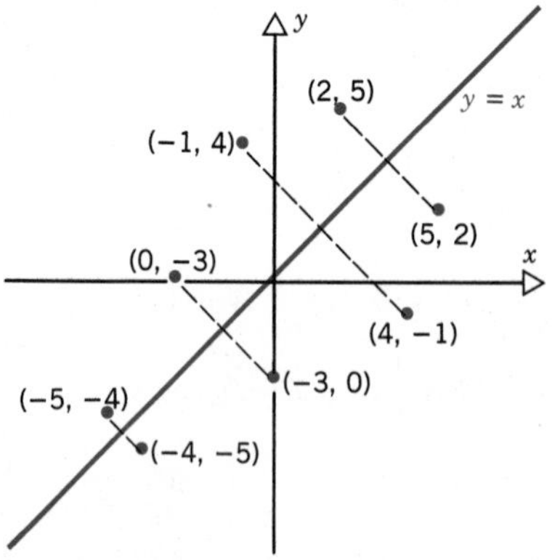

Figure 26.2

EXAMPLE 26.6 If $f(x) = \sqrt{x}$ for $x \geq 0$, then $f^{-1}(x) = x^2$ for $x \geq 0$. Figure 26.3 shows the symmetrical locations of the graphs of f and f^{-1}.

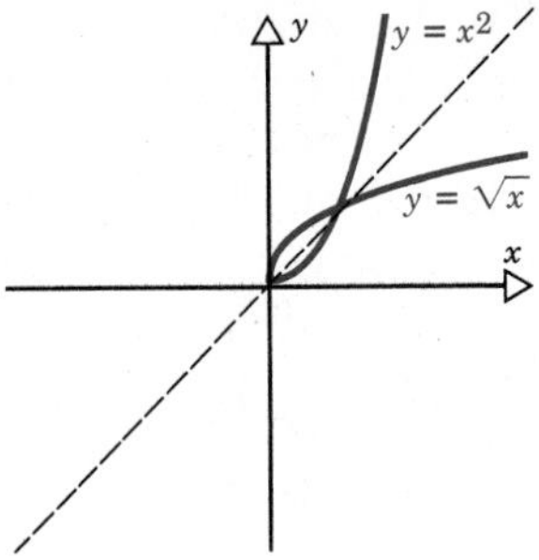

Figure 26.3

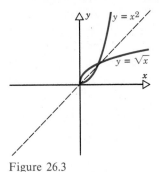

EXAMPLE 26.7 In Example 26.4 we saw that if $f(x) = 4x - 3$, then $f^{-1}(x) = \frac{1}{4}(x + 3)$. Figure 26.4 shows the symmetrical locations of the graphs.

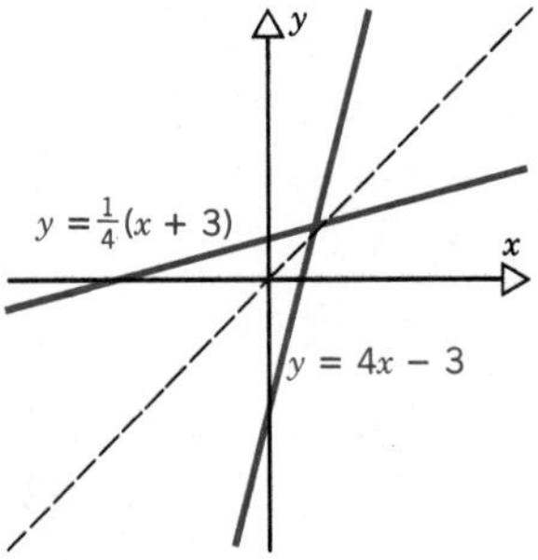

Figure 26.4

EXERCISES FOR SECTION 26

Decide in each case whether the function is one-to-one, and justify your answer. (See Examples 26.1 and 26.2, and the statements at the end of Subsection A.)

1. $f(x) = 10x$
2. $f(x) = -7$
3. $f(x) = 7x + 1$
4. $g(x) = 0$
5. $g(x) = -5x + 7$
6. $g(x) = 2 - x^2$
7. $h(x) = 3x^2$ for $x \geq 0$
8. $h(x) = |x|$
9. $h(x) = 1 + \sqrt{x}$ for $x \geq 0$

For each function f in Exercises 10–24:

(a) Determine $f^{-1}(x)$.
(b) Verify that $(f^{-1} \circ f)(x) = x$ for all x in the domain of f and $(f \circ f^{-1})(x) = x$ for all x in the domain of f^{-1}.
(c) Draw the graphs of f and f^{-1} in the same Cartesian plane.

(Assume $x \geq 0$ in Exercises 19–24.)

10. $f(x) = 2x$
11. $f(x) = 4x$
12. $f(x) = -3x$
13. $f(x) = x + 1$
14. $f(x) = 5 - x$
15. $f(x) = x - 3$
16. $f(x) = 3 - 2x$
17. $f(x) = 3x + 4$
18. $f(x) = 6 - x$
19. $f(x) = \sqrt{x} + 1$
20. $f(x) = \sqrt{2x}$
21. $f(x) = 1 - 2\sqrt{x}$
22. $f(x) = 2x^2$
23. $f(x) = x^2 - 1$
24. $f(x) = 1 - x^2$

25. Determine the inverse of the general linear function $f(x) = ax + b$ $(a \neq 0)$.

26. Determine the inverse of the quadratic function $f(x) = ax^2 + bx + c$, $a \neq 0$, $x \geq -b/2a$. (The condition $x \geq -b/2a$ guarantees that f has an inverse. Why?)
27. Prove that if f and g are functions having inverses, then $f \circ g$ also has an inverse and $(f \circ g)^{-1} = g^{-1} \circ f^{-1}$. [Begin with $((g^{-1} \circ f^{-1}) \circ (f \circ g))(x)$. Along the way, you'll need to use $f^{-1}(f(g(x))) = g(x)$. Why is the last equation true?]

REVIEW EXERCISES FOR CHAPTER VI

Determine the domain of each function.

1. $f(x) = \sqrt{x + 1}/(x^2 - 4)$ 2. $f(x) = \sqrt{2x - 1}/(x + 1)$

In Exercises 3 and 4, determine:

(a) $f(x) + g(x)$ **(b)** $f(x)g(x)$ **(c)** $f(0) - g(2)$
(d) $g(1)/f(-1)$ **(e)** $(f \circ g)(x)$ **(f)** $(g \circ f)(3)$

3. $f(x) = 3x + 5$ and $g(x) = x^2 - 1$ 4. $f(x) = x^2 + 1$ and $g(x) = -2x + 3$

Determine the difference quotient of each function.

5. $f(x) = -3x + 4$ 6. $g(x) = x^2 + 2$

Draw the graph of each function.

7. $f(x) = -x^2 + 2x + 2$ 8. $f(x) = 2x^2 - 6x + 1$

9. $f(x) = \begin{cases} x + 1 & \text{for } x \leq 0 \\ x^2 & \text{for } x > 0 \end{cases}$ 10. $f(x) = \begin{cases} -x^2 + 1 & \text{for } x < 1 \\ x & \text{for } x \geq 1 \end{cases}$

11. Determine the maximum value of $f(x) = x(1 - x)$.
12. Determine the minimum value of $f(x) = (x + 1)(x + 2)$.
13. Find two real numbers such that the first minus the second is 10 and the product of the two numbers is as small as possible.
14. Find two real numbers such that the first plus twice the second is 10 and the product of the two numbers is as large as possible.
15. Compute an estimate for $\sqrt{85}$ using $\sqrt{81} = 9$, $\sqrt{100} = 10$, and linear interpolation.
16. If $f(3) = 10$, $f(5) = 7$, and f is linear, what is $f(-2)$?

Determine $f^{-1}(x)$ and draw the graphs of f and f^{-1} in the same Cartesian plane.

17. $f(x) = 5 - 3x$ 18. $f(x) = x^2 + 2$ $(x \geq 0)$

19. Determine a function f such that $y = f(x)$, if y varies directly as the square of x, and $y = 5$ when $x = 2$.
20. Determine a function g such that $u = g(v)$, if u varies inversely as v, and $u = 10$ when $v = 3$.
21. Suppose that the total production costs of a company per day consist of a fixed cost of \$1000 (independent of the number of items produced) and a variable cost of \$50 per item produced. Determine the function f such that $T = f(n)$, where T denotes the total cost per day for the production of n items. Your answer should be a linear function of n. What is the significance of the variable cost per item in terms of the graph of f?
22. Assume that a state's income tax is computed at the rate of 3% for the first \$2000 of taxable income, 4% for the next \$2000 of taxable income, and 5% for any additional taxable income beyond \$4000. Determine a function f such that

$T = f(I)$ for $0 \leq I \leq 10{,}000$, where I denotes taxable income and T denotes the amount of tax, both in dollars. (The function will consist of three parts.)

23. Newton's law of universal gravitation is given by

$$F = G\frac{m_1 m_2}{r^2} \quad \text{(Example 9.14).}$$

(a) Determine a function f such that $G = f(m_1, m_2, r, F)$.
(b) Determine a function h such that $r = h(m_1, m_2, G, F)$.

24. The formula $\frac{1}{f} = \frac{1}{d_o} + \frac{1}{d_i}$ relates the focal length (f), object distance (d_o), and image distance (d_i) of a simple lens. (Notice that f represents a number, not a function, in this exercise.)
(a) Determine a function g such that $f = g(d_o, d_i)$.
(b) Determine a function h such that $d_o = h(f, d_i)$.

CHAPTER VII
POLYNOMIAL AND RATIONAL FUNCTIONS

In Chapter VI we analyzed the polynomial functions of degrees one and two—the linear and quadratic functions. In the first four sections of this chapter we consider polynomial functions of degree greater than two. In the last two sections we consider quotients of polynomials, which are called *rational functions*. Most of the chapter will draw heavily on the ideas about division that are covered in the first section of the chapter. You can review the most basic facts about polynomials by reading the first half of Section 4.

SECTION

27 Division of Polynomials

A. The Division Algorithm

If we divide any integer (the *dividend*) by a positive integer (the *divisor*), we obtain a *quotient* and a *remainder*. For example,

dividend → 38; remainder → 3; divisor → 5 (under 38); divisor → 5 (under 3); quotient → 7

$$\frac{38}{5} = 7 + \frac{3}{5}. \tag{27.1}$$

This can be rewritten as

$$38 = 5 \cdot 7 + 3.$$

The remainder will always be nonnegative and less than the divisor (in the example, $0 \leq 3 < 5$). The general form of Equation (27.1) is

$$\frac{a}{b} = q + \frac{r}{b}.$$

This can be rewritten as

dividend → a; remainder → r; divisor → b; quotient → q

$$a = bq + r, \quad \text{with} \quad 0 \leq r < b. \tag{27.2}$$

Equation (27.2) summarizes a fact about integers known as the *Division Algorithm*. Our concern now is with the corresponding fact about polynomials. In the statement of this fact, which follows, $b(x) \neq 0$ means that the polynomial $b(x)$ has at least one nonzero coefficient. And $r(x) = 0$ means that $r(x)$ is identically zero; that is, $r(x)$ has *no* nonzero coefficient.

Division Algorithm

If $a(x)$ and $b(x)$ are polynomials such that $b(x) \neq 0$, then there are unique polynomials $q(x)$ and $r(x)$ such that

$$a(x) = b(x)q(x) + r(x)$$
$$\text{with either } r(x) = 0 \text{ or } \deg r(x) < \deg b(x). \tag{27.3}$$

The polynomials $q(x)$ and $r(x)$ are called the **quotient** and the **remainder,** respectively.

Here is an illustration of ordinary long division of integers. The examples that follow the illustration will show that division of polynomials is similar; they will also show how to compute $q(x)$ and $r(x)$ in (27.3).

$$
\begin{array}{rrl}
 & 345 & \leftarrow \text{quotient} \\
\text{divisor} \rightarrow 12 & \overline{)\,4151} & \leftarrow \text{dividend} \\
\text{subtract} & \underline{36} & \\
 & 55 & \\
\text{subtract} & \underline{48} & \\
 & 71 & \\
\text{subtract} & \underline{60} & \\
 & 11 & \leftarrow \text{remainder}
\end{array}
$$

The division process actually consists of subtracting multiples of the divisor from the dividend, until we arrive at a remainder less than the dividend. If you look carefully at the preceding calculation, for example, you will see that we successively subtracted 300×12, 40×12, and 5×12. The same idea applies to division of polynomials.

EXAMPLE 27.1 Determine the quotient and remainder when $6x^4 + 5x^3 - 3x^2 + x + 3$ (the dividend) is divided by $2x^2 - x + 1$ (the divisor).

Solution Write the dividend and divisor as shown, and concentrate first on the terms that are circled.

$$
\boxed{2x^2} - x + 1 \;\overline{)\;\boxed{6x^4} + 5x^3 - 3x^2 + x + 3}
$$

Divide $2x^2$ into $6x^4$. Write the result, $3x^2$, above $6x^4$. Then multiply $3x^2$ times $2x^2 - x + 1$, and write the result, $6x^4 - 3x^3 + 3x^2$, below the dividend.

$$
\begin{array}{rll}
 & 3x^2 & \\
2x^2 - x + 1 & \overline{)\,6x^4 + 5x^3 - 3x^2 + x + 3} & \\
 & \underline{6x^4 - 3x^3 + 3x^2} & = 3x^2(2x^2 - x + 1)
\end{array}
$$

Now subtract, as in long division of integers.

$$
\begin{array}{rl}
 & 3x^2 \\
2x^2 - x + 1 & \overline{)\,6x^4 + 5x^3 - 3x^2 + x + 3} \\
\text{subtract} & \underline{6x^4 - 3x^3 + 3x^2} \\
 & \phantom{6x^4 -{}} \boxed{8x^3} - 6x^2
\end{array}
$$

Next, divide $2x^2$ into $8x^3$ and proceed as before, bringing down the term x from the dividend for convenience.

$$
\begin{array}{rl}
 & 3x^2 + 4x \\
2x^2 - x + 1 & \overline{)\,6x^4 + 5x^3 - 3x^2 + x + 3} \\
 & \underline{6x^4 - 3x^3 + 3x^2} \quad \downarrow \\
 & \qquad 8x^3 - 6x^2 + x \\
\text{subtract} & \qquad \underline{8x^3 - 4x^2 + 4x} \quad = 4x(2x^2 - x + 1)
\end{array}
$$

Subtract again, and repeat the process until the result of the subtraction is either 0 or of degree less than the degree of the divisor.

$$
\begin{array}{rl}
 & 3x^2 + 4x - 1 \qquad \longleftarrow \text{quotient} \\
2x^2 - x + 1 & \overline{)\,6x^4 + 5x^3 - 3x^2 + x + 3} \\
 & \underline{6x^4 - 3x^3 + 3x^2} \\
 & \qquad 8x^3 - 6x^2 + x \\
 & \qquad \underline{8x^3 - 4x^2 + 4x} \\
 & \qquad\quad -2x^2 - 3x + 3 \\
 & \qquad\quad \underline{-2x^2 + x - 1} \\
 & \qquad\qquad\quad - 4x + 4 \longleftarrow \text{remainder}
\end{array}
$$

Thus

$$\frac{6x^4 + 5x^3 - 3x^2 + x + 3}{2x^2 - x + 1} = 3x^2 + 4x - 1 + \frac{-4x + 4}{2x^2 - x + 1}$$

and

$$6x^4 + 5x^3 - 3x^2 + x + 3 = (2x^2 - x + 1)(3x^2 + 4x - 1) + (-4x + 4).$$

For the division process the terms of the divisor and dividend should be arranged so that the powers of x decrease from left to right. Also, in writing the dividend space should be left for any missing powers of the variable; in the following example space has been left for an x^3 term.

EXAMPLE 27.2 Determine the quotient and remainder when $4 + 5x - 6x^2 + x^4$ is divided by $x - 2$.

$$\begin{array}{r} 3x^2 + 4x - 1 \\ 2x^2 - x + 1 \overline{)\, 6x^4 + 5x^3 - 3x^2 + x + 3} \\ 6x^4 - 3x^3 + 3x^2 \\ \hline 8x^3 - 6x^2 + x \\ 8x^3 - 4x^2 + 4x \\ \hline 2x^2 - 3x + 3 \\ 2x^2 + x - 1 \\ \hline -4x + 4 \quad R \end{array}$$

$3x^2(2x^2 - x + 1) = 6x^4 - 3x^3 + 3x^2$

$4x(2x^2 - x + 1) = 8x^3 + 4x^2 - 4x$

$1(2x^2 - x + 1) = 2x^2 + x - 1$

$$\frac{6x^4 + 5x^3 - 3x^2 + x + 3}{2x^2 - x + 1} = 3x^2 + 4x - 1 + \frac{-4x + 4}{2x^2 - x + 1}$$

$$6x^4 + 5x^3 - 3x^2 + x + 3 = (2x^2 - x + 1)(3x^2 + 4x - 1) + (-4x + 4$$

$$x - 2 \overline{)\, 4 + 5x - 6x^2 + x^4}$$

$$\begin{array}{r} x^3 \\ x - 2 \overline{)\, x^4 \quad - 6x^2 + 5x + 4} \\ x^4 \end{array}$$

$x^3(x - 2) = x^4 - 2x^3$

$$x - 2 \overline{\smash{\big)}\ \begin{aligned} x^3 + 2x^2 - 2x + 1 \quad &\longleftarrow \text{ quotient}\\[2pt] x^4 \qquad\ \ - 6x^2 + 5x + 4\ \ \end{aligned}}$$

$$\begin{aligned} x^4 - 2x^3 &\\ \hline 2x^3 - 6x^2&\\ 2x^3 - 4x^2&\\ \hline -2x^2 + 5x&\\ -2x^2 + 4x&\\ \hline x + 4&\\ x - 2&\\ \hline 6 \quad &\longleftarrow \text{ remainder} \end{aligned}$$

B. Synthetic Division

Synthetic division is a process that shortens the work in any division problem in which the divisor has the form $x - c$, as in Example 27.2. Example 27.3 will summarize the rules that describe synthetic division. But first we run through Example 27.2 again, showing how the process of synthetic division is derived.

Synthetic division depends on two facts about division: First, there is no need to write the powers of the variable as long as the numbers in the various columns are kept carefully aligned. Second, many coefficients occur more than once—we can save time by not writing the duplications.

Here is Example 27.2 with the powers of x deleted and with the duplications in each column circled or connected by an arrow.

$$\begin{array}{ccccc} & \textcircled{1} & \textcircled{2} & \ominus{2} & \textcircled{1} \\ 1 \quad -2 \ \overline{\smash{\big)}\ } & \textcircled{1} & 0 & -6 & 5 \quad 4 \\ & \textcircled{1} & -2 & & \\ \cline{2-3} & \textcircled{2} & -6 & & \\ & \textcircled{2} & -4 & & \\ \cline{2-3} & & \ominus{2} & 5 & \\ & & \ominus{2} & 4 & \\ \cline{3-4} & & & \textcircled{1} & 4 \\ & & & \textcircled{1} & -2 \\ \cline{4-5} & & & & 6 \end{array}$$

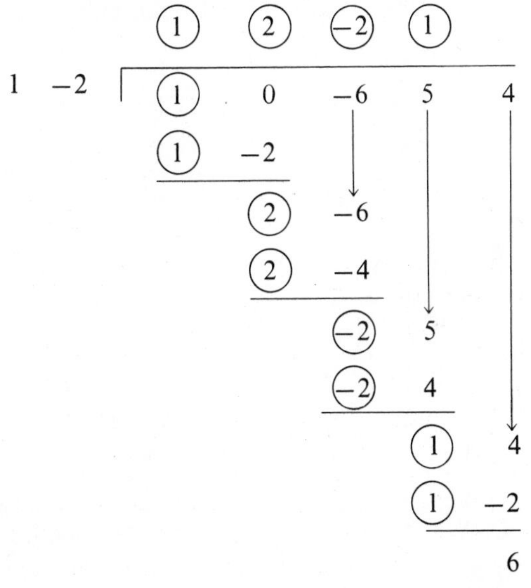

We can rewrite this keeping just one of each of the various duplicated numbers.

$$
\begin{array}{rr|rrrrr}
1 & -2 & 1 & 0 & -6 & 5 & 4 \\
 & & & -2 & & & \\
\cline{3-4}
 & & & 2 & & & \\
 & & & & -4 & & \\
\cline{4-5}
 & & & & -2 & & \\
 & & & & & 4 & \\
\cline{5-6}
 & & & & & 1 & \\
 & & & & & & -2 \\
\cline{6-7}
 & & & & & & 6
\end{array}
$$

Now we can compress what remains. Also, we can delete the 1 at the left in the first line, since it accounts only for the x in $x - c$ and we use this process only when the divisor has the form $x - c$. For convenience in what comes later, we repeat in the third row the first number under the division symbol, as indicated by the arrow.

$$
\begin{array}{r|rrrrrl}
-2 & 1 & 0 & -6 & 5 & 4 & \\
 & \downarrow & -2 & -4 & 4 & -2 & \text{subtract} \\
\hline
 & 1 & 2 & -2 & 1 & 6 &
\end{array}
$$

Each number in the second row (such as -4) arose originally from multiplying the -2 in $x - 2$ by the number in the third row and one column to the left of the specified number in the second row (thus $-4 = -2 \cdot 2$). Because addition is easier than subtraction, we now change the sign on the leading -2: this will change the sign on each number in the second row, and so we compensate by replacing subtraction by addition, as desired. When we remove the bar on top, we get this abbreviated form.

$$
\begin{array}{r|rrrrrl}
2 & 1 & 0 & -6 & 5 & 4 & \\
 & & 2 & 4 & -4 & 2 & \text{add} \\
\hline
 & 1 & 2 & -2 & 1 & 6 &
\end{array}
$$

The last number in the third row, 6, is the remainder (compare Example 27.2). The other numbers in the third row are the coefficients in the quotient, $x^3 + 2x^2 - 2x + 1$, which has degree one less than the degree of the dividend.

EXAMPLE 27.3 Use synthetic division to determine the quotient and remainder when $2x^5 - 10x^3 + 20x^2 - 9x + 5$ is divided by $x + 3$.

Solution For synthetic division the divisor has the form $x - c$. In this case, $x + 3 = x - (-3)$, so that $c = -3$. Thus -3 will be the leading number in the first row. The other numbers in the first row are the coefficients of the dividend when the terms are written with exponents decreasing from left to right, and when zeros are supplied for missing powers of the variable.

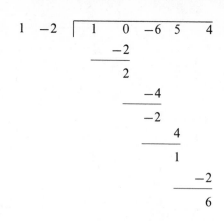

$$\begin{array}{r|rrrrrr} -3 & 2 & 0 & -10 & 20 & -9 & 5 \\ & & & & & & \\ \hline \end{array}$$

Bring down 2, multiply it by -3, and write the result, -6, as shown.

$$\begin{array}{r|rrrrrrl} -3 & 2 & 0 & -10 & 20 & -9 & 5 & \\ & \downarrow & -6 & & & & & \text{add} \\ \hline & 2 & & & & & & \end{array}$$

Add 0 and -6, multiply the result by -3, and write the result, 18, as shown.

$$\begin{array}{r|rrrrrrl} -3 & 2 & 0 & -10 & 20 & -9 & 5 & \\ & \downarrow & -6 & 18 & & & & \text{add} \\ \hline & 2 & -6 & & & & & \end{array}$$

Now repeat; alternately adding and multiplying until the end.

$$\begin{array}{r|rrrrrrl} -3 & 2 & 0 & -10 & 20 & -9 & 5 & \\ & \downarrow & -6 & 18 & -24 & 12 & -9 & \text{add} \\ \hline & 2 & -6 & 8 & -4 & 3 & -4 & \end{array}$$

The last number is the remainder. The other numbers in the third row are the coefficients of the quotient.

The quotient is $2x^4 - 6x^3 + 8x^2 - 4x + 3$.

The remainder is -4.

EXAMPLE 27.4 Determine the quotient and remainder when $x^4 + x^2 + 1$ is divided by $x - \frac{1}{2}$.

Solution

$$\begin{array}{r|rrrrr} \frac{1}{2} & 1 & 0 & 1 & 0 & 1 \\ & & \frac{1}{2} & \frac{1}{4} & \frac{5}{8} & \frac{5}{16} \\ \hline & 1 & \frac{1}{2} & \frac{5}{4} & \frac{5}{8} & \frac{21}{16} \end{array}$$

The quotient is $x^3 + \frac{1}{2}x^2 + \frac{5}{4}x + \frac{5}{8}$.

The remainder is $\frac{21}{16}$.

C. Remainder Theorem

The following theorem provides an indirect but convenient way to compute $f(c)$ for each polynomial $f(x)$ and each real number c.

Remainder Theorem

If c is a real number and a polynomial $f(x)$ is divided by $x - c$, then the remainder is $f(c)$.

Proof. By the Division Algorithm, when $f(x)$ is divided by $x - c$ (which has degree one), the remainder is either 0 or a polynomial of degree 0 (that is, a constant). Thus, in any case, the remainder is a real number, which we can denote by r. If $q(x)$ denotes the quotient, then

$$f(x) = (x - c)q(x) + r.$$

Therefore,

$$\begin{aligned} f(c) &= (c - c)q(c) + r \\ &= 0 \cdot q(c) + r \\ &= r. \end{aligned}$$ □

EXAMPLE 27.5 Use the Remainder Theorem and synthetic division to compute $f(3)$ if $f(x) = x^4 - 2x^3 + x - 1$.

Solution We want the remainder when $f(x)$ is divided by $x - 3$. Here is the calculation using synthetic division.

To compute $f(3)$, use 3 here. → 3

$$\begin{array}{r|rrrrr} 3 & 1 & -2 & 0 & 1 & -1 \\ & & 3 & 3 & 9 & 30 \\ \hline & 1 & 1 & 3 & 10 & 29 = f(3) \end{array}$$

The remainder is 29, so $f(3) = 29$, as indicated.

Here is the calculation by direct substitution:

$$f(3) = 3^4 - 2 \cdot 3^3 + 3 - 1 = 81 - 54 + 3 - 1 = 29.$$

Generally, the total number of operations (additions, subtractions, and multiplications) needed to compute $f(c)$ by synthetic division will be less than the number needed by direct substitution.

EXAMPLE 27.6 Use the Remainder Theorem and synthetic division to show that $-\frac{2}{3}$ is a solution of $3x^3 + 2x^2 + 3x + 2 = 0$.

Solution Remember that c is a solution of $f(x) = 0$ iff $f(c) = 0$. Therefore, we let $f(x) = 3x^3 + 2x^2 + 3x + 2$ and then use synthetic division to show that $f(-\frac{2}{3}) = 0$.

$$\begin{array}{r|rrrr} -\frac{2}{3} & 3 & 2 & 3 & 2 \\ & & -2 & 0 & -2 \\ \hline & 3 & 0 & 3 & 0 = f(-\frac{2}{3}) \end{array}$$

Determine the quotient and remainder when $f(x)$ is divided by $g(x)$.

1. $f(x) = 3x^3 - 2x^2 + 5x + 7,\ g(x) = 3x + 1$
2. $f(x) = 4x^3 + x - 3,\ g(x) = 2x - 1$
3. $f(x) = 2x^3 - 14x - 8,\ g(x) = 2x + 4$
4. $f(x) = 3x^4 - 10x^2 - 2x + 2,\ g(x) = x^2 + 2x + 1$
5. $f(x) = 3x^3 - 8x - 4,\ g(x) = x^2 + 2x + 1$
6. $f(x) = x^5 + 2x^3 - 1,\ g(x) = x^3 - 1$
7. $f(x) = 2x^3 + 3x^2 + x,\ g(x) = x^3 + 1$
8. $f(x) = x^3 - x^2 - 2,\ g(x) = 2x^2 - 1$
9. $f(x) = x^4 + x^3 + x + 1,\ g(x) = 5x^2 - x$
10. $f(x) = x^5 - x^3 + x + 1,\ g(x) = 4x^3 + 2$
11. $f(x) = x^3 + x^2 - 4,\ g(x) = x^4 + 1$
12. $f(x) = x^3 + x^2 + x + 1,\ g(x) = x^3 + 1$

Use synthetic division to determine the quotient and remainder when $f(x)$ is divided by $g(x)$.

13. $f(x) = 2x^3 - x^2 + 5x + 1,\ g(x) = x - 1$
14. $f(x) = x^3 - x + 4,\ g(x) = x - 3$
15. $f(x) = x^3 + 2x^2 - 1,\ g(x) = x - 2$
16. $f(x) = x^4 + x^2 + 1,\ g(x) = x + 2$
17. $f(x) = x^3 - x,\ g(x) = x + 5$
18. $f(x) = 5x^3 - 5x + 2,\ g(x) = x + 5$
19. $f(x) = x^4 + x^3,\ g(x) = x + \frac{1}{2}$
20. $f(x) = x^2 + x + 1,\ g(x) = x + 0.1$
21. $f(x) = x^3 + x,\ g(x) = x + 0.5$
22. $f(x) = x^4 - 1,\ g(x) = x + 0.2$
23. $f(x) = x^5 + x^4 + x^3 + x^2 + x + 1,\ g(x) = x - \frac{1}{2}$
24. $f(x) = 3x^4 - x^2 + 1,\ g(x) = x - \frac{1}{3}$

Use the Remainder Theorem and synthetic division to compute the indicated values $f(c)$.

25. $f(x) = x^3 - x + 1;\ f(2), f(-5)$
26. $f(x) = 2x^4 - x^2 + x;\ f(3), f(-4)$
27. $f(x) = x^5 + 2x^3 - x + 1;\ f(1), f(-2)$
28. $f(t) = t^3 + t^2 + t + 1;\ f(0.2), f(-\frac{1}{2})$
29. $f(t) = 5t^4 - 2t^2 - 1;\ f(\frac{1}{3}), f(-0.1)$
30. $f(t) = 2t^3 + t^2 - t + 3;\ f(\frac{1}{2}), f(-0.1)$

Use the Remainder Theorem and synthetic division to show that the given number is a solution of the given equation.

31. $2;\ x^3 - 3x^2 + 3x - 2 = 0$
32. $-1;\ x^4 + x^3 + 2x + 2 = 0$
33. $-\frac{1}{2};\ 2x^3 - x^2 + x + 1 = 0$
34. $0.1;\ 10x^3 - x^2 + 10x - 1 = 0$
35. $-0.4;\ 5x^3 + 2x^2 + 5x + 2 = 0$
36. $-0.3;\ 10x^4 + 3x^3 + 20x + 6 = 0$

SECTION

28 Factors and Real Zeros of Polynomials

A. Factor Theorem

Recall that a polynomial $g(x)$ is a *factor* of a polynomial $f(x)$ if $f(x) = q(x)g(x)$ for some polynomial $q(x)$ (Section 4D). The following theorem provides a quick test to determine whether a polynomial of the form $x - c$ is a factor of another polynomial.

Factor Theorem
If $f(x)$ is a polynomial and c is a real number, then

$$x - c \text{ is a factor of } f(x) \text{ iff } f(c) = 0.$$

Proof. Divide $f(x)$ by $x - c$. By the Division Algorithm and the Remainder Theorem (Section 27),

$$f(x) = q(x) \cdot (x - c) + f(c).$$

This equation shows that if $f(c) = 0$, then $f(x) = q(x)(x - c)$ so that $x - c$ is a factor of $f(x)$. Conversely, since $f(c)$ is the *unique* remainder when $f(x)$ is divided by $x - c$, the equation shows that if $x - c$ is a factor of $f(x)$, then $f(c) = 0$. □

We can compute $f(c)$ by synthetic division, and by the Factor Theorem $f(c)$ will tell us whether $x - c$ is a factor of $f(x)$. But synthetic division does more: it furnishes $q(x)$ such that $f(x) = q(x)(x - c)$ whenever $x - c$ is a factor of $f(x)$. Here is an illustration.

EXAMPLE 28.1 Use synthetic division and the Factor Theorem to show that $x - 4$ is a factor of

$$f(x) = 6x^3 - 23x^2 - 6x + 8.$$

Also determine $q(x)$ such that $f(x) = (x - 4)q(x)$.

Solution

$$\begin{array}{r|rrrr} 4 & 6 & -23 & -6 & 8 \\ & & 24 & 4 & -8 \\ \hline & 6 & 1 & -2 & 0 = f(4) \end{array}$$

This shows that $f(4) = 0$, and thus $x - 4$ is a factor of $f(x)$ by the Factor Theorem. Moreover, the numbers 6 1 −2 in the third row show that the quotient is $q(x) = 6x^2 + x - 2$. Thus

$$6x^3 - 23x^2 - 6x + 8 = (x - 4)(6x^2 + x - 2).$$

EXAMPLE 28.2 Is $x + 0.1$ a factor of

$$f(x) = x^3 + 0.1x^2 + 0.1x - 0.01?$$

Solution Because $x + 0.1 = x - (-0.1)$, we use -0.1 (not 0.1) in testing by synthetic division.

$$\begin{array}{r|rrrr} -0.1 & 1 & 0.1 & 0.1 & -0.01 \\ & & -0.1 & 0 & -0.01 \\ \hline & 1 & 0 & 0.1 & -0.02 \end{array}$$

The remainder, $f(-0.1) = -0.02$, is not zero. Thus $x + 0.1$ is not a factor of $f(x)$.

EXAMPLE 28.3 Use the Factor Theorem to prove that $x - a$ is a factor of $x^n - a^n$ for each real number a and each positive integer n.

Solution Let $f(x) = x^n - a^n$. Then $f(a) = a^n - a^n = 0$. Therefore, by the Factor Theorem, $x - a$ is a factor of $f(x)$.

B. Factors and Zeros

A number c is called a **zero** of a polynomial $f(x)$ if $f(c) = 0$. Thus c is a zero of a polynomial $f(x)$ iff c is a solution of the equation $f(x) = 0$. For example, 2 is a zero of $f(x) = x^2 - 4$ because $f(2) = 2^2 - 4 = 0$.

The conclusion of the Factor Theorem can be restated like this:

> $x - c$ is a factor of $f(x)$
> iff
> c is a zero of $f(x)$.

This means that every factor of the form $x - c$ determines a zero, and, conversely, every zero determines a factor $x - c$.

EXAMPLE 28.4 Find the three zeros of $f(x) = x^3 - x^2 - 2x$ by factoring.

Solution Because

$$f(x) = x(x^2 - x - 2) = x(x - 2)(x + 1),$$

the Factor Theorem tells us that 0, 2, and -1 are zeros.

EXAMPLE 28.5 Form a polynomial having -3, 2, and 5 as zeros.

Solution By the Factor Theorem, it suffices to form a polynomial having $x + 3$, $x - 2$, and $x - 5$ as factors. Thus

$$\begin{aligned} f(x) &= (x+3)(x-2)(x-5) \\ &= (x+3)(x^2-7x+10) \\ &= x^3-4x^2-11x+30 \end{aligned}$$

has the required zeros.

C. Complete Factorization

Assume that c is a zero of the polynomial $f(x)$. Then $x - c$ is a factor of $f(x)$, and it may even be true that $(x-c)^2$ or $(x-c)^3$ or some higher power of $x - c$ is a factor of $f(x)$. If $(x-c)^m$ is the highest power of $x - c$ that is a factor of $f(x)$, then c is said to be a zero of **multiplicity** m.

EXAMPLE 28.6 Determine the multiplicity of each zero of $x^3(x+4)(x-5)^2$.

Solution

0 is a zero of multiplicity 3	[from $x^3 = (x-0)^3$]
-4 is a zero of multiplicity 1	[from $x + 4 = x - (-4)$]
5 is a zero of multiplicity 2	[from $(x-5)^2$].

EXAMPLE 28.7 Form a polynomial having 0, -4, and 1 as zeros of multiplicities 1, 2, and 3, respectively.

Solution The answer must have x^1, $(x+4)^2$, and $(x-1)^3$ as factors. The simplest such polynomial is

$$x(x+4)^2(x-1)^3.$$

It will be instructive to expand the answer to the last example:

$$\begin{aligned} x(x+4)^2(x-1)^3 &= x(x^2+8x+16)(x^3-3x^2+3x-1) \\ &= (x^3+8x^2+16x)(x^3-3x^2+3x-1) \\ &= x^6+5x^5-5x^4-25x^3+40x^2-16x. \end{aligned}$$

The degree of this polynomial is 6, which is the sum of the multiplicities of the zeros $(6 = 1 + 2 + 3)$. We can see that, in general, a zero of multiplicity m will contribute m to the degree of a polynomial, because the degree of $(x-c)^m$ is m. This leads to the following conclusion.

> The degree of a polynomial must be at least as great as the sum of the multiplicities of its real zeros. (28.1)

The next example shows that we cannot replace "at least as great as" by "equal to" in Statement (28.1). We can make such a replacement in the next chapter, however, after we extend the number system beyond the real numbers to the complex numbers and replace "real zeros" by "complex zeros." All of the polynomials in this chapter have real numbers as coefficients.

EXAMPLE 28.8 Determine the zeros of $x^3 + x$, along with their multiplicities.

Solution

$$x^3 + x = 0 \text{ iff } x(x^2 + 1) = 0$$
$$\text{iff } x = 0 \text{ or } x^2 + 1 = 0.$$

The factor $x^2 + 1$ is never 0 for x a real number (because $x^2 \geq 0$ for all x). Thus the only real zero of $x^3 + x$ is 0, and its multiplicity is 1.

The degree of $x^3 + x$ is 3, and that is greater than the sum of the degrees of the multiplicities of the real zeros (which is 1).

If a zero of multiplicity m is counted as a zero m times, then we can rephrase Statement (28.1) as follows.

> Each polynomial of degree n has at most n real zeros. (28.2)

Very often we are interested in finding *all* of the zeros of a polynomial. The higher the degree of the polynomial the more difficult this problem tends to be. The next example illustrates that once we have a zero we can use synthetic division and factorization to reduce the degree of our problem by one: in this example we are given a zero of a polynomial of degree three and we use it to reduce the problem to that of a polynomial of degree two.

EXAMPLE 28.9 Given that -1 is a zero of

$$f(x) = x^3 + 4x^2 - 7x - 10,$$

determine all of the zeros along with their multiplicities.

Solution If -1 is a zero, then $x + 1$ must be a factor. We can use synthetic division to find $q(x)$ such that $f(x) = (x + 1)q(x)$.

$$\begin{array}{r|rrrr} -1 & 1 & 4 & -7 & -10 \\ & & -1 & -3 & 10 \\ \hline & 1 & 3 & -10 & 0 \end{array}$$

The coefficients 1 3 -10 from the last row show that

$$x^3 + 4x^2 - 7x - 10 = (x + 1)(x^2 + 3x - 10).$$

The factor $x^2 + 3x - 10$ contributes two zeros:

$$x^2 + 3x - 10 = 0$$
$$(x + 5)(x - 2) = 0$$
$$x = -5, 2.$$

The zeros of $f(x)$ are, therefore, -1, -5, and 2, each of multiplicity 1. That is,

$$f(x) = (x + 1)(x + 5)(x - 2).$$

Recall from Section 14D that a polynomial has been *factored completely* when it has been written as a product of irreducible polynomials, where a polynomial is irreducible if it cannot be written as a product of two other polynomials that are both of positive degree. Since each linear polynomial is irreducible, the factorization

$$x^3 + 4x^2 - 7x - 10 = (x + 1)(x + 5)(x - 2)$$

in Example 28.9 is complete. As stated in Section 14D, some second-degree polynomials are irreducible and some are not; a second-degree polynomial is reducible iff it has a real zero. It can be shown that each polynomial of degree greater than two is reducible.

EXAMPLE 28.10 Given that 3 is a zero of

$$f(x) = x^3 - 5x^2 + 8x - 6,$$

determine all of the real zeros along with their multiplicities. Also, factor $f(x)$ completely.

Solution

$$\begin{array}{r|rrrr} 3 & 1 & -5 & 8 & -6 \\ & & 3 & -6 & 6 \\ \hline & 1 & -2 & 2 & 0 \end{array}$$

This shows that $f(x) = (x - 3)(x^2 - 2x + 2)$. The quadratic factor $x^2 - 2x + 2$ has no real zero, because its discriminant is negative: $b^2 - 4ac = (-2)^2 - 4(1)(2) = -4 < 0$. Therefore, 3 is the only real zero of $f(x)$; its multiplicity is 1. The quadratic factor is irreducible since it has no real zero. The complete factorization is

$$x^3 - 5x^2 + 8x - 6 = (x - 3)(x^2 - 2x + 2).$$

EXERCISES FOR SECTION 28

Use synthetic division and the Factor Theorem to determine whether $x - c$ is a factor of $f(x)$. If $x - c$ is a factor, determine $q(x)$ such that $f(x) = (x - c)q(x)$.

1. $x - 2$; $f(x) = 6x^3 - 11x^2 - 4x + 4$
2. $x - 3$; $f(x) = 5x^3 - 8x^2 - 13x - 20$
3. $x - 1$; $f(x) = x^4 - 3x^3 + 2x^2 + x - 1$
4. $x + 1$; $f(x) = 2x^3 - x^2 + 10$

5. $x + 5$; $f(x) = x^4 + 5x^3 - x^2 - x + 20$
6. $x + 4$; $f(x) = x^3 + 4x^2 + 5x + 20$
7. $x - 0.1$; $f(x) = 10x^4 - x^3 - 2x$
8. $x - 0.2$; $f(x) = 5x^6 - x^5 + 10x^2 - 2x$
9. $x + 0.4$; $f(x) = 5x^3 + x + 1$

Determine the multiplicity of each zero of each polynomial.

10. $(x - 1)^3(x + 2)^5$
11. $x(x + 1)(x - 5)^2$
12. $(x - 10)^3(x + 2)^2$
13. $x^3(x - 10)^2(x + 0.1)^5$
14. $(x - 0.5)^2(x + 3)^5$
15. $x(x + 2)(x + 0.4)^2$

Form a polynomial having the specified zeros with the specified multiplicities.

16. 0, 1, and -3 of multiplicities 2, 2, and 1, respectively.
17. 0, $\frac{1}{2}$, and -5 of multiplicities 1, 2, and 3, respectively.
18. $\sqrt{2}$, -1, and 3 of multiplicities 2, 1, and 2, respectively.
19. 0, -0.2, and 2 of multiplicities 4, 2, and 1, respectively.
20. 4, $-\sqrt{3}$, and 3 of multiplicities 1, 1, and 4, respectively.
21. $-\frac{1}{4}$, 1.2, and 5 of multiplicities 3, 2, and 1, respectively.

Determine the real zeros of each polynomial, along with their multiplicities; one zero is given in each case. Also, factor each polynomial completely.

22. $x^3 + x^2 - 5x - 5$; -1 is a zero.
23. $x^3 - x^2 - 4x + 4$; 1 is a zero.
24. $x^3 - 5x^2 - 4x + 20$; 5 is a zero.
25. $2x^3 - 5x^2 + 4x - 1$; $\frac{1}{2}$ is a zero.
26. $5x^3 - x^2 - 10x + 2$; 0.2 is a zero.
27. $5x^3 + 12x^2 + 9x + 2$; -0.4 is a zero.
28. $x^3 + 3x^2 - 2x - 6$; -3 is a zero.
29. $x^3 - 5x^2 - 5x + 25$; 5 is a zero.
30. $2x^3 - 9x^2 + 11x - 2$; 2 is a zero.
31. $x^3 + 2x^2 + x + 2$; -2 is a zero.
32. $2x^3 - x^2 + 8x - 4$; $\frac{1}{2}$ is a zero.
33. $3x^3 - 8x^2 + 6x + 3$; $-\frac{1}{3}$ is a zero.

34. By Example 28.3, $x - 2$ is a factor of $x^6 - 64$. Use synthetic division to find $q(x)$ such that $x^6 - 64 = (x - 2)q(x)$.
35. Use the Factor Theorem to show that $x - 2$ is not a factor of $x^6 + 64$. (Compare Exercise 34.)
36. (a) Use the Factor Theorem and direct substitution (as in Example 28.3) to prove that $x + 2$ is a factor of $x^5 + 32$.
(b) Use synthetic division to find $q(x)$ such that $x^5 + 32 = (x + 2)q(x)$.
37. For which positive integers n (if any) is $x - a$ a factor of $x^n + a^n$? Assume $a \neq 0$.
38. For which positive integers n (if any) is $x + a$ a factor of $x^n - a^n$? Assume $a \neq 0$.
39. For which positive integers n (if any) is $x + a$ a factor of $x^n + a^n$? Assume $a \neq 0$.

SECTION

29 Finding Real Zeros of Polynomials

A. Rational Zeros

The general form for a polynomial $f(x)$ of degree n is

$$f(x) = a_n x^n + a_{n-1}x^{n-1} + \cdots + a_1 x + a_0, \tag{29.1}$$

where $a_n, a_{n-1}, \ldots, a_1$, and a_0 denote the coefficients and $a_n \neq 0$. In the case of $f(x) = 3x^4 - \sqrt{2}x^2 + \frac{1}{2}x - 5$, for example,

$$n = 4, \quad a_4 = 3, \quad a_3 = 0, \quad a_2 = -\sqrt{2}, \quad a_1 = \tfrac{1}{2}, \quad \text{and} \quad a_0 = -5.$$

We know how to find the zeros of polynomials of degrees one and two; the ideas in this section will help with polynomials of higher degree. The following theorem puts a narrow restriction on the possible *rational zeros* of polynomials with *integral coefficients*.

Rational Zero Theorem

Assume that the coefficients of $f(x)$ in Equation (29.1) are integers. If a rational number r/s (reduced to lowest terms) is a zero of $f(x)$, then r is a factor of a_0 (the constant term) and s is a factor of a_n (the leading coefficient).

Proof. If r/s is a zero of $f(x)$, then

$$a_n\left(\frac{r}{s}\right)^n + a_{n-1}\left(\frac{r}{s}\right)^{n-1} + \cdots + a_1\left(\frac{r}{s}\right) + a_0 = 0.$$

Multiply both sides of this equation by s^n, and then subtract $a_0 s^n$ from both sides. The result is

$$a_n r^n + a_{n-1}r^{n-1}s + \cdots + a_1 r s^{n-1} = -a_0 s^n.$$

Since r is a factor of all of the terms on the left side of this equation, r must also be a factor of $-a_0 s^n$. Therefore, since r and s have no common factor (recall that r/s is reduced to lowest terms), r must be a factor of a_0.* In the same way, from

$$a_{n-1}r^{n-1}s + \cdots + a_1 r s^{n-1} + a_0 s^n = -a_n r^n$$

*Here we have used the following fact about integers: If b, c, and d denote nonzero integers such that d is a factor of bc, and b and d have no common factor except ± 1, then d must be a factor of c. Elementary number theory books give a proof.

we can deduce that s must be a factor of a_n. □

EXAMPLE 29.1 Find the rational zeros of

$$x^3 - 3x^2 + 2x - 6.$$

Solution

Possible numerators (factors of -6): $\pm 1, \pm 2, \pm 3, \pm 6$

Possible denominators (factors of 1): ± 1

Possible rational zeros: $\pm 1, \pm 2, \pm 3, \pm 6$

We can test these possible zeros by synthetic division. If we choose the order $1, -1, 2, -2, \ldots$, then the first zero we find will be 3.

$$\begin{array}{r|rrrr} 3 & 1 & -3 & 2 & -6 \\ & & 3 & 0 & 6 \\ \hline & 1 & 0 & 2 & 0 \end{array}$$

We could continue through the list of possible rational zeros, trying -3, 6, and -6, but it is more efficient to use the Factor Theorem and reduce the degree by one. Thus $x^3 - 3x^2 + 2x - 6 = (x - 3)(x^2 + 2)$ (recall that the factor $x^2 + 2$ comes from the numbers 1 0 2 in the third row of the synthetic division). The factor $x^2 + 2$ has no real zero (because its discriminant is negative, or because $x^2 + 2 > 0$ for every real number x). Therefore, the only real zero (rational or irrational) of $x^3 - 3x^2 + 2x - 6$ is 3.

The preceding example, in which the only possible denominators for zeros were ± 1, leads to the following observation.

> If the coefficients in Equation (29.1) are integers and $a_n = 1$, then any rational zero of $f(x)$ must be an integer.

EXAMPLE 29.2 Find the rational solutions of

$$12x^3 - 8x^2 = 3x - 2.$$

Solution Remember that the zeros of $f(x)$ are the same as the solutions of $f(x) = 0$. Thus we can rewrite the given equation as $12x^3 - 8x^2 - 3x + 2 = 0$ and apply the Rational Zero Theorem.

Possible numerators (factors of 2): $\pm 1, \pm 2$

Possible denominators (factors of 12): $\pm 1, \pm 2, \pm 3, \pm 4, \pm 6, \pm 12$

Possible rational zeros: $\pm 1, \pm 2, \pm \frac{1}{2}, \pm \frac{1}{3}, \pm \frac{2}{3}, \pm \frac{1}{4}, \pm \frac{1}{6}, \pm \frac{1}{12}$

If we choose the order $1, -1, 2, -2, \ldots$, then the first zero we find will be $\frac{1}{2}$.

$$\begin{array}{r|rrrr} \frac{1}{2} & 12 & -8 & -3 & 2 \\ & & 6 & -1 & -2 \\ \hline & 12 & -2 & -4 & 0 \end{array}$$

Thus $12x^3 - 8x^2 - 3x + 2 = (x - \frac{1}{2})(12x^2 - 2x - 4)$. The factor $12x^2 - 2x - 4$ gives

$$\begin{aligned} 12x^2 - 2x - 4 &= 0 \\ 6x^2 - x - 2 &= 0 \\ (2x + 1)(3x - 2) &= 0 \\ x = -\tfrac{1}{2} \quad \text{or} \quad x &= \tfrac{2}{3}. \end{aligned}$$

Therefore, the rational zeros are $\frac{1}{2}$, $-\frac{1}{2}$, and $\frac{2}{3}$.

Sometimes we can use the Rational Zero Theorem to find rational zeros and, at the same time, reduce the polynomial by factoring so that we also get irrational zeros.

EXAMPLE 29.3 Find the real zeros of

$$f(x) = 3x^4 + 5x^3 - 8x^2 - 10x + 4.$$

Solution

Possible numerators: $\pm 1, \pm 2, \pm 4$

Possible denominators: $\pm 1, \pm 3$

Possible rational zeros: $\pm 1, \pm 2, \pm 4, \pm\frac{1}{3}, \pm\frac{2}{3}, \pm\frac{4}{3}$

If we choose the order $1, -1, 2, -2, \ldots$, then the first zero we find will be -2.

$$\begin{array}{r|rrrrr} -2 & 3 & 5 & -8 & -10 & 4 \\ & & -6 & 2 & 12 & -4 \\ \hline & 3 & -1 & -6 & 2 & 0 \end{array}$$

Thus $f(x) = (x + 2)(3x^3 - x^2 - 6x + 2)$. Now we can restrict our attention to possibe rational zeros of

$$g(x) = 3x^3 - x^2 - 6x + 2.$$

These are ± 1, ± 2, $\pm\frac{1}{3}$, and $\pm\frac{2}{3}$. Since 1, -1, and 2 are not zeros of $f(x)$, they will not be zeros of $g(x)$. Therefore, we need to try only -2, $\pm\frac{1}{3}$, and $\pm\frac{2}{3}$.

$$\begin{array}{r|rrrr} \frac{1}{3} & 3 & -1 & -6 & 2 \\ & & 1 & 0 & -2 \\ \hline & 3 & 0 & -6 & 0 \end{array}$$

Thus

$$f(x) = (x + 2)g(x)$$
$$= (x + 2)(x - \tfrac{1}{3})(3x^2 - 6).$$

The factor $3x^2 - 6$ gives

$$3x^2 - 6 = 0$$
$$x^2 = 2$$
$$x = \sqrt{2} \quad \text{or} \quad x = -\sqrt{2}.$$

Therefore, the zeros of $f(x)$ are -2, $\frac{1}{3}$, $\sqrt{2}$, and $-\sqrt{2}$.

B. Bounds on Real Zeros

A number c is an **upper bound** for the zeros of a polynomial if no zero exceeds c. A number c is a **lower bound** for the zeros of a polynomial if no zero is less than c. The following theorem can be used to find such bounds.

Theorem on Bounds

Let $f(x)$ be a polynomial whose leading coefficient is positive, and divide $f(x)$ synthetically by $x - c$.

(a) If $c > 0$ and the numbers in the third row of the synthetic division are all nonnegative, then c is an upper bound for the zeros of $f(x)$.
(b) If $c < 0$ and the numbers in the third row of the synthetic division alternate in sign (with 0 thought of as positive or negative, whichever will produce a variation in sign), then c is a lower bound for the zeros of $f(x)$.

The proof of the Theorem on Bounds will be omitted, but here is the idea. Suppose $c > 0$ and the numbers in the third row of the synthetic division are all nonnegative. If c is replaced by any larger number d, then the numbers in the third row will all increase, so that the last number in the third row cannot be zero. Thus d cannot be a zero of $f(x)$. (A specific example will make the idea clear. Try $c = 4$ and then $c = 5$ with the polynomial in Example 29.1, for example.) Similar remarks apply if $c < 0$.

EXAMPLE 29.4 Apply the Theorem on Bounds to find integral bounds for the zeros of

$$f(x) = 2x^4 - 3x^3 - 8x^2 + 9x + 6.$$

Solution We try $1, 2, 3, \ldots$ to search for an upper bound, and then $-1, -2, -3, \ldots$ to search for a lower bound.

The first row of Table 29.1 lists the coefficients of $f(x)$. Each remaining row gives the third row from the synthetic division using the number at the left end of that row; thus if c is at the left end of a row, the number at the right end is $f(c)$. The entries in such a table usually can be calculated mentally.

TABLE 29.1

	2	−3	−8	9	6	*Information about zeros*
1	2	−1	−9	0	6	No information.
2	2	1	−6	−3	0	2 is a zero.
3	2	3	1	12	42	{3 is an upper bound because these numbers are nonnegative.
−1	2	−5	−3	12	−6	No information.
−2	2	−7	6	−3	12	{−2 is a lower bound because the signs alternate.

The information from the table shows that -2 is a lower bound for the zeros and 3 is an upper bound. (The zeros of $f(x)$ are, in fact, $-\sqrt{3} \approx -1.7$, -0.5, $\sqrt{3} \approx 1.7$, and 2.)

EXAMPLE 29.5 The following computations show that -1 is a lower bound for the zeros of $x^3 + x^2 + 3x - 1$.

$$\begin{array}{r|rrrr} -1 & 1 & 1 & 3 & -1 \\ & & -1 & 0 & -3 \\ \hline & 1 & 0 & 3 & -4 \end{array} \leftarrow \begin{cases} \text{Think of 0 as negative to} \\ \text{give alternating signs.} \end{cases}$$

C. Isolating Zeros

We can use the next theorem to isolate the real zeros of a polynomial even if we cannot find the zeros exactly.

Intermediate-Value Theorem

If $f(x)$ is a polynomial and a and b are real numbers such that $f(a)$ and $f(b)$ have opposite signs, then $f(x)$ has at least one zero between $x = a$ and $x = b$.

The Intermediate-Value Theorem is proved in books on calculus. The theorem is usually stated in a more general form than that used here.

Intuitively, the theorem is true because the graph of a polynomial is an unbroken curve, so that we can trace the graph between any two points without lifting our pencil from the paper. Thus, if we move from one point where the graph of $f(x)$ is above the x-axis to another point where the graph is below the x-axis, then we must cross the axis at least once. Any crossing-point (such as at c in Figure 29.1) gives a zero of $f(x)$.

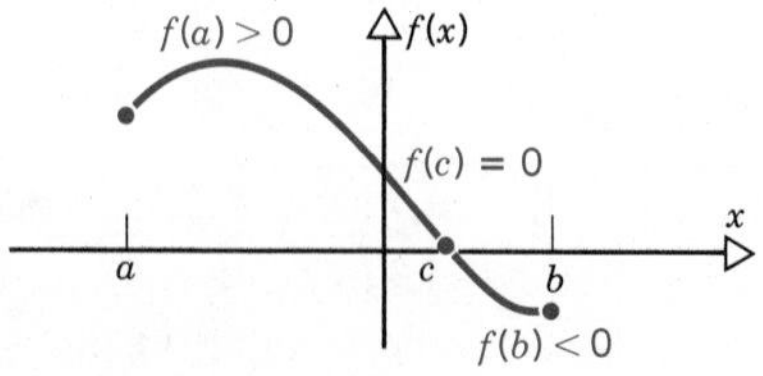

Figure 29.1

EXAMPLE 29.6 The polynomial $2x^3 + 7x^2 - 10x - 35$ has three real zeros, none of which is an integer. Isolate the zeros between consecutive integers.

Solution Either synthetic division or substitution will give values of $f(x)$ for consecutive integral values of x.

x	-4	-3	-2	-1	0	1	2	3
$f(x)$	-11	4	-3	-20	-35	-36	-11	52

The sign changes in $f(x)$ are indicated. By the Intermediate-Value Theorem there is at least one zero in each of the intervals $(-4, -3)$, $(-3, -2)$, and $(2, 3)$. In fact, since a polynomial of degree three has at most three zeros, there is exactly one zero in each interval. (You can verify that the zeros are -3.5, $-\sqrt{5} \approx -2.24$, and $\sqrt{5} \approx 2.24$.)

D. Approximating Zeros

The Intermediate-Value Theorem can be used to approximate the real zeros of a polynomial to any desired degree of accuracy. Suppose, for example, that $f(x)$ is a polynomial and that $f(2) > 0$ and $f(3) < 0$. Then there is at least one zero between 2 and 3. Compute $f(2.5)$. If $f(2.5) = 0$, then 2.5 is a zero. Otherwise, either $f(2.5) < 0$ or $f(2.5) > 0$. If $f(2.5) < 0$, then there is at least one zero between 2 and 2.5; if $f(2.5) > 0$, then there is at least one zero between 2.5 and 3. If there is a zero between 2 and 2.5, try 2.2 (or any other number between 2 and 2.5); if there is a zero between 2.5 and 3, try 2.7, for example. Continuing in this way we will either locate a zero exactly or generate a sequence of intervals of decreasing lengths, each of which must contain a zero. This method is especially practical when used with a computer or calculator.

EXAMPLE 29.7 The polynomial

$$f(x) = x^4 - 4x^3 - x^2 + 10x + 4$$

has one irrational zero between 2 and 3. Find it to the nearest 0.5.

Solution We use the Intermediate-Value Theorem and synthetic division, writing only the third row of the synthetic division in each case.

	1	-4	-1	10	4	
2	1	-2	-5	0	4	$= f(2) > 0$
3	1	-1	-4	-2	-2	$= f(3) < 0$
2.5	1	-1.5	-4.75	-1.875	-0.6875	$= f(2.5) < 0$
2.25	1	-1.75	-4.9375	-1.109375	1.503963	$= f(2.25) > 0$

Because $f(2.25) > 0$ and $f(2.5) < 0$, the zero is between 2.25 and 2.5. Therefore, the zero is closer to 2.5 than to 2, and to the nearest 0.5 the zero is 2.5. (The zeros of $f(x)$ are, in fact, $1 \pm \sqrt{2}$ and $1 \pm \sqrt{5}$. The zero between 2 and 3 is $1 + \sqrt{2} \approx 2.414$.)

E. Summary and Example

Here is an outline for searching for the real zeros of a polynomial $f(x)$ of degree greater than two. The computations are to be done by synthetic division. An example follows the outline.

- Factor the highest possible power of x from the polynomial. If this highest power is x^k $(k \geq 1)$, then 0 will be a zero of multiplicity k. Now work with the factor that remains.
- Test nonnegative integers in increasing order beginning with 0. (See Remark 1 that follows this outline.)
- If any positive integer is a zero, use the factor (quotient) produced by synthetic division to reduce the degree of the problem by one. Otherwise, continue testing positive integers until an upper bound for the zeros has been reached.
- Test negative integers in decreasing order beginning with -1. (See Remark 1 that follows this outline.)
- Continue as with the positive integers. Stop when a lower bound for the zeros has been reached.
- Use the Rational Zero Theorem and the Intermediate-Value Theorem to help determine which rational possibilities to test next (all of the possible integers between the bounds will have been tested already).
- Always factor the polynomial when a zero is found. If, at any step, only a second-degree factor remains, apply the methods for treating quadratics (Section 11).
- If the previous techniques do not yield all of the real zeros, test any remaining rational possibilities or use the Intermediate-Value Theorem to compute approximations.

Remark 1 Instead of testing every positive integer until an upper bound for the zeros has been reached, you may want to test only those that are possible zeros as dictated by the Rational Zero Theorem. The same comment applies to negative integers. The idea of not omitting integers along the way will be especially appropriate in the next section, where we will consider graphs.

Remark 2 There may be a zero between a and b even if $f(a)$ and $f(b)$ have the same sign. For example, if there are two unequal zeros between a and b, then the graph of $f(x)$ will intersect the x-axis twice in that interval. In many cases locating the zeros is not easy.

Remark 3 Remember that a zero may have multiplicity greater than 1. Therefore, if you find that c is a zero of $f(x)$ and that $f(x) = (x - c)g(x)$, then you should also test c as a possible zero of $g(x)$.

EXAMPLE 29.8 Find the real zeros of

$$f(x) = 2x^5 - 3x^4 - 11x^3 + 9x^2 + 15x.$$

Solution We work through the steps in the preceding outline. Tables 29.2 and 29.3 summarize the computations.

The number 0 is a zero of multiplicity one. Write

$$f(x) = x(2x^4 - 3x^3 - 11x^2 + 9x + 15);$$

call the second factor $g(x)$ and work only with it hereafter.

$g(0) = 15 > 0$	
$g(1) = 12 > 0$	
$g(2) = -3 < 0$	There is a zero between 1 and 2.
$g(3) = 24 > 0$	There is a zero between 2 and 3.
$g(4) = 195 > 0$	The signs are nonnegative in this row of Table 29.2, so 4 is an upper bound for the zeros of $g(x)$ and $f(x)$. Now test negative integers.
$g(-1) = 0$	Thus -1 is a zero. Also, $g(x) = (x + 1)h(x)$, where $h(x) = 2x^3 - 5x^2 - 6x + 15$. Consider $h(x)$ (Table 29.3).
$h(-2) = -9 < 0$	Because the signs alternate in this row of Table 29.3, -2 is a lower bound for the zeros of $h(x)$, $g(x)$, and $f(x)$. Now test $\frac{3}{2}$, the rational possibility between 1 and 2.
$h(\frac{3}{2}) = \frac{3}{2} > 0$	Not a zero. Test $\frac{5}{2}$, the rational possibility between 2 and 3.
$h(\frac{5}{2}) = 0$	Thus $\frac{5}{2}$ is a zero. Also, $h(x) = (x - \frac{5}{2})(2x^2 - 6)$.

Solve $2x^2 - 6 = 0$: $x^2 = 3$ and $x = \pm\sqrt{3}$. Thus the polynomial $f(x)$ has five real zeros, each of multiplicity one. In increasing order they are $-\sqrt{3}$, -1, 0, $\sqrt{3}$ and $\frac{5}{2}$.

TABLE 29.2 Synthetic Division for $g(x)$

	2	−3	−11	9	15
0	2	−3	−11	9	15
1	2	−1	−12	−3	12
2	2	1	−9	−9	−3
3	2	3	−2	3	24
4	2	5	9	45	195
−1	2	−5	−6	15	0

TABLE 29.3 Synthetic division for $h(x)$.

	2	−5	−6	15
−2	2	−9	12	−9
$\frac{3}{2}$	2	−2	−9	$\frac{3}{2}$
$\frac{5}{2}$	2	0	−6	0

EXERCISES FOR SECTION 29

Use the Rational Zero Theorem to construct a list of possible rational zeros for each polynomial.

1. $5x^3 - x^2 + 30x - 6$ **2.** $4x^3 - 2x^2 + 20x - 5$ **3.** $14x^3 + x^2 - 28x - 2$

Use the Theorem on Bounds to find integral bounds for the real zeros of each polynomial.

4. $3x^3 + x^2 - 21x - 7$ **5.** $2x^3 - 5x^2 - 10x + 25$ **6.** $x^3 + 5x^2 - 6x - 30$

Each polynomial in Exercises 7–9 has three real zeros, none of which is an integer. Use the Intermediate-Value Theorem to isolate the zeros between consecutive integers.

7. $2x^3 - x^2 - 14x + 7$ **8.** $2x^3 - 11x^2 + 14x - 3$ **9.** $3x^3 - x^2 - 9x - 3$

Find the real zeros of each polynomial.

10. $x^3 - x^2 - 5x + 5$
11. $x^3 + x^2 - 3x - 3$
12. $x^3 + x^2 - 7x - 7$
13. $4x^3 - 7x - 3$
14. $2 - 11x + 17x^2 - 6x^3$
15. $2x^3 - 7x^2 + 4x + 4$
16. $-2 + 3x - 2x^2 + 3x^3$
17. $2x^3 - x^2 + 8x - 4$
18. $-1 + 5x - x^2 + 5x^3$
19. $x^3 - 2x^2 + 2x$
20. $x^3 - 2x^2 - 3x$
21. $x^3 - 10x^2 + 22x$
22. $4x^4 - 3x^3 + 3x^2 - 3x - 1$
23. $4x^4 - 4x^3 - 15x^2 + 16x - 4$
24. $2x^4 - 7x^3 - 7x^2 + 35x - 15$
25. $x^4 - 5x^3 + 4x^2 - 20x$
26. $3x^4 + 2x^3 - 6x^2 - 4x$
27. $3x^4 + 4x^3 - 9x^2 - 12x$

28. The polynomial $f(x) = x^4 - 6x^3 + 8x^2 - 6x + 7$ has one irrational zero between 1 and 2. Use the Intermediate-Value Theorem to find it to the nearest 0.5. [C]

29. The polynomial $f(x) = x^4 - 2x^3 - 5x^2 - 2x - 6$ has one irrational zero between 3 and 4. Use the Intermediate-Value Theorem to find it to the nearest 0.5. [C]

30. The polynomial $f(x) = x^4 - 2x^3 - 2x - 1$ has one irrational zero between -1 and 0. Use the Intermediate-Value Theorem to find it to the nearest 0.5. [C]

31. The fourth power of an integer is 75 more than the integer plus its square. What is the integer?

32. The square of a rational number plus 6 times the reciprocal of the number is 2 more than 3 times the number. What is the number?

33. The cube of an integer plus 3 times the square of the integer is 12 more than 20 times the integer. What is the integer?

34. The shortest edge of a box is half as long as another edge and one inch shorter than the third edge. If the volume of the box is 504 cubic inches, how long is each edge?

35. The height of a pyramid with a square base is one meter longer than each edge of the base. If the volume of the pyramid is $1\frac{7}{8}$ cubic meters, what is the height? (Use $V = \frac{1}{3}hb^2$, where h is the height and b is the length of each edge of the base.)

36. A box with volume 72 cubic centimeters and an open top has been formed from a square piece of cardboard 10 centimeters on a side by cutting equal squares from the corners and turning up the sides (Figure 29.2).
(a) Show that if x denotes the length of each edge of the removed squares, then

$$x^3 - 10x^2 + 25x - 18 = 0.$$

(b) Solve the equation in part (a), and use the answer to determine the dimensions of the box.

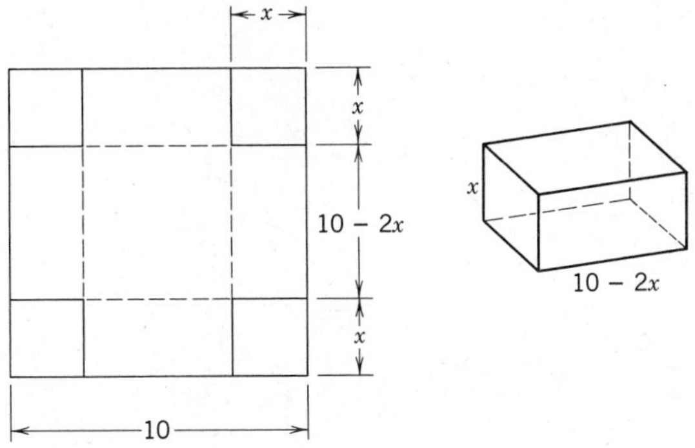

Figure 29.2

SECTION 30 Graphs of Polynomial Functions

A. Power Functions

We know that the graph of every linear (first-degree) function is a straight line and the graph of every quadratic (second-degree) function is a parabola. In this section we study the graphs of polynomial functions of degree greater than two.

A function f is called a **power function** if it has the form

$$f(x) = kx^n, \tag{30.1}$$

where k and n are nonzero real numbers.* Thus a power function of x is simply a constant times a power of x. If n is a nonnegative integer, then kx^n is a polynomial. These are the polynomials with the simplest graphs and so they'll be considered first.

* For the present n will be an integer. Other cases will be considered in Section 35.

Figures 30.1–30.4 show examples. Some coordinate pairs used to draw the graphs of $y = x^3$ and $y = x^4$ are given in Tables 30.1 and 30.2.

TABLE 30.1

x	0	0.5	-0.5	1	-1	1.5	-1.5
x^3	0	0.125	-0.125	1	-1	3.375	-3.375

TABLE 30.2

x	0	0.5	-0.5	1	-1	1.5	-1.5
x^4	0	0.0625	0.0625	1	1	5.0625	5.0625

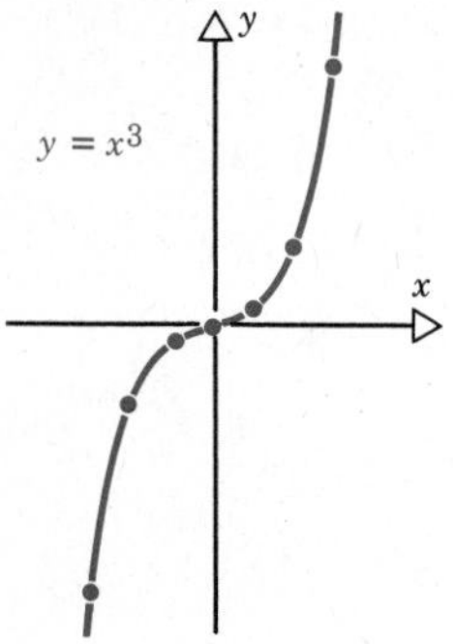

Figure 30.1

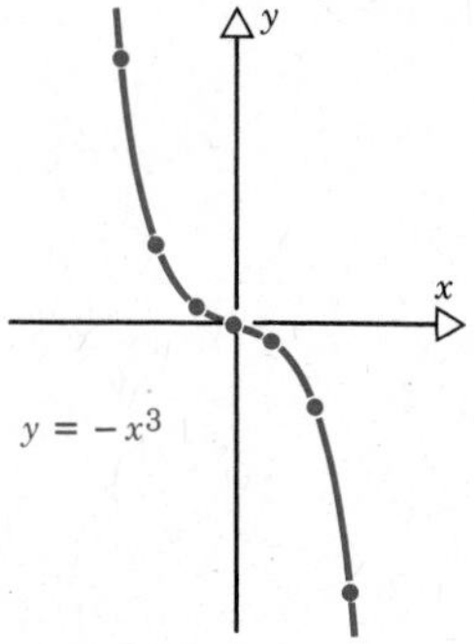

Figure 30.2

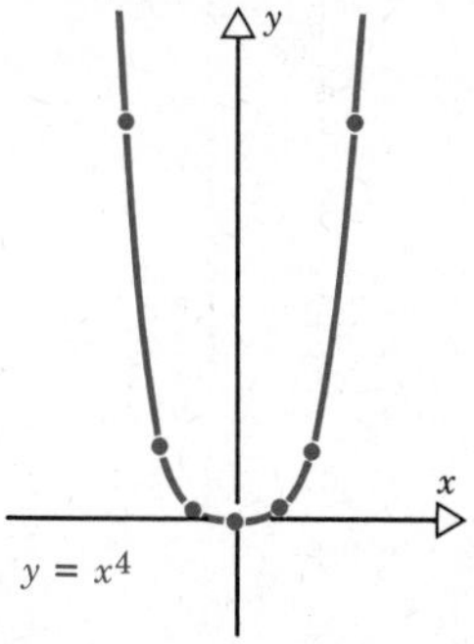

Figure 30.3

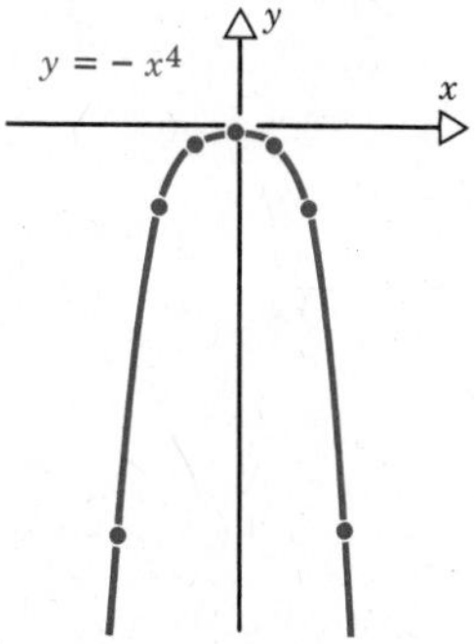

Figure 30.4

The graphs of $y = x^5, y = x^7, \ldots$ (odd powers) all have the same general shape as that of $y = x^3$; the graphs of $y = x^4, y = x^6, \ldots$ (even powers) all have the same general shape as that of $y = x^2$. The differences are that as n increases, the graph of $y = x^n$ becomes flatter near the origin and steeper away from the origin. (See Exercises 1–3.)

Changing the sign of k in kx^n will reflect the graph through the x-axis—that is, turn it upside down. Compare the graph of $y = x^3$ with the graph of $y = -x^3$ (Figures 30.1 and 30.2). Also, compare the graph of $y = x^4$ with the graph of $y = -x^4$ (Figures 30.3 and 30.4).

Changing k but not its sign in kx^n will change the details of the graph but not its general characteristics. (See Exercises 4–6.)

B. Symmetry

A graph is *symmetric about a line* if for each point on the graph there is a corresponding point on the graph an equal distance from the line but on the opposite side (Figure 30.5). Figure 30.3 shows that the graph of $f(x) = x^4$ is symmetric about the y-axis. If the graph of a function f is symmetric about the y-axis and a is in the domain of f, then the point corresponding to $(a, f(a))$ must be $(-a, f(-a))$; thus necessarily $f(-a) = f(a)$ (Figure 30.5 again). In the case of $f(x) = x^4$, the symmetry is a result of the even exponent, 4, which makes a negative sign inconsequential: for example, $(-2)^4 = 16 = 2^4$. These observations about symmetry and even powers lead to the following definition and remark.

A function f is **even** if $f(-x) = f(x)$ for all x.
The graph of f is symmetric about the y-axis
iff
f is even.

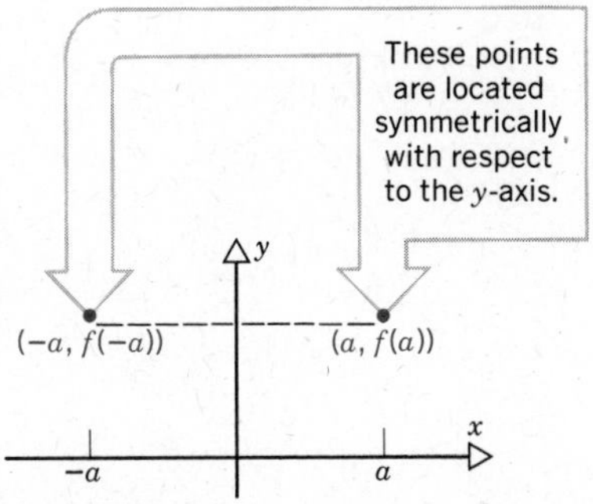

Figure 30.5

A graph is *symmetric about a point P* if for each point on the graph there is a corresponding point on the graph the same distance from P but in the opposite direction (Figure 30.6). Figure 30.1 shows that the graph of $f(x) = x^3$ is symmetric about the origin. If the graph of a function f is symmetric about the origin and a is in the domain of f, then the point corresponding to $(a, f(a))$ must be $(-a, f(-a))$, so that necessarily $f(-a) = -f(a)$ (Figure 30.6 again). In the case of $f(x) = x^3$, the symmetry is a result of the odd exponent, 3. This leads to the next definition and remark.

A function f is **odd** if $f(-x) = -f(x)$ for all x.
The graph of f is symmetric about the origin
iff
f is odd.

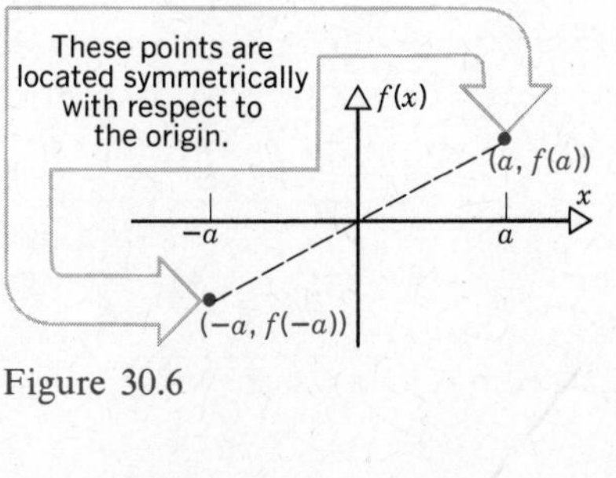

Figure 30.6

The next statements are illustrated by the examples already given and by the examples that follow.

> A polynomial function is even
> iff
> it contains only even powers of its variable.†

> A polynomial function is odd
> iff
> it contains only odd powers of its variable.

EXAMPLE 30.1 The function $f(x) = -x^4$ is even and its graph is symmetric about the y-axis (Figure 30.4). The function $f(x) = -x^3$ is odd and its graph is symmetric about the origin (Figure 30.2).

EXAMPLE 30.2 The function $f(x) = x^4 - 6x^2 + 3$ is even because it contains only even powers of x. Here is a direct verification that $f(-x) = f(x)$:

$$f(-x) = (-x)^4 - 6(-x)^2 + 3 = x^4 - 6x^2 + 3 = f(x).$$

EXAMPLE 30.3 The function $g(x) = x^3 - 2x$ is odd because it contains only odd powers of x. Here is a direct verification that $g(-x) = -g(x)$:

$$g(-x) = (-x)^3 - 2(-x) = -x^3 + 2x = -(x^3 - 2x) = -g(x).$$

EXAMPLE 30.4 The function $h(x) = 2x^3 + x^2 + 2$ is neither even nor odd, because it contains both odd and even powers of x.

There are also even and odd functions that are not polynomials; we'll meet some of those in the next section. Knowing that a function is even or odd essentially cuts in half the work of graphing the function—once we determine what happens to the right of the y-axis we automatically know what happens to the left.

C. A General Plan

Steps I–V that follow furnish a plan for graphing any polynomial function. The steps assume a knowledge of the preceding parts of this chapter.

Step I. *Check for symmetry* by determining if the function is either even or odd. If the function is either even or odd, we can concentrate on nonnegative values of x in most of the remaining steps; what happens for negative x will follow automatically.

Step II. *Compute and plot points.*

†Nonzero constant terms have degree zero, which is even.

(a) Use synthetic division to compute $f(x)$ for $x = 1, 2, 3, \ldots$. Continue at least until an upper bound for the zeros of $f(x)$ has been reached (when the numbers in the third row of the synthetic division are all nonnegative; see Section 29B).

(b) Continue with $x = -1, -2, -3, \ldots$, at least until a lower bound for the zeros of $f(x)$ has been reached (when the numbers in the third row of the synthetic division alternate in sign).

Step III. *Locate intercepts.*

(a) The y-intercept is at $(0, f(0))$.

(b) The x-intercepts are the zeros of $f(x)$. Use the methods of Section 29 to find these.

Step IV. *Plot more points.*

(a) Each computation that was done by synthetic division in the previous steps will have produced a value $f(c)$. Plot all of the corresponding points.

(b) Plot other points as needed. Besides intercepts, the *turning points* (where the function changes from increasing to decreasing or from decreasing to increasing) are especially important. With calculus these turning points can be located exactly; with only algebra we usually have to settle for estimates.

Step V. *Draw the graph* as carefully as possible, remembering that it should be smooth and unbroken.

D. Examples

EXAMPLE 30.5 Draw the graph of

$$f(x) = x^3 - 3x^2 + 2.$$

Solution

I. The function is neither even nor odd.

II. The first row of Table 30.3 lists the coefficients of $f(x)$. Each remaining row gives the third row from synthetic division using the number at the left end of that row; if c is at the left end of a row, the number at the right end is $f(c)$.

TABLE 30.3

	1	−3	0	2	
1	1	−2	−2	0	
2	1	−1	−2	−2	
3	1	0	0	2	3 is an upper bound for x-intercepts because these numbers are nonnegative.
−1	1	−4	4	−2	−1 is a lower bound for x-intercepts because these signs alternate.

III. (a) The y-intercept is at $f(0) = 2$.

(b) Table 30.3 shows that $f(1) = 0$ and that

$$x^3 - 3x^2 + 2 = (x - 1)(x^2 - 2x - 2).$$

From $x^2 - 2x - 2 = 0$ the quadratic formula gives

$$x = 1 \pm \sqrt{3}.$$

Thus the x-intercepts are at 1, $1 + \sqrt{3} \approx 2.7$, and $1 - \sqrt{3} \approx -0.7$.

IV. (a) The points from the synthetic division in Step II and the x-intercepts from Step III have been plotted in Figure 30.7. These include the turning points (0, 2) and (2, −2); the fact that these *are* the turning points is beyond the scope of our techniques, but can be proved using calculus.

(b) Figure 30.7 also shows the points computed by synthetic division in Table 30.4.

V. Figure 30.7 shows the graph.

Remark Notice that the graph shows that

$$x^3 - 3x^2 + 2 > 0 \text{ if } 1 - \sqrt{3} < x < 1 \text{ or } x > 1 + \sqrt{3},$$

and

$$x^3 - 3x^2 + 2 < 0 \text{ if } x < 1 - \sqrt{3} \text{ or } 1 < x < 1 + \sqrt{3}.$$

TABLE 30.4

	1	−3	0	2
0.5	1	−2.5	−1.25	1.375
−0.5	1	−3.5	1.75	1.125
1.5	1	−1.5	−2.25	−1.375
2.5	1	−0.5	−1.25	−1.125

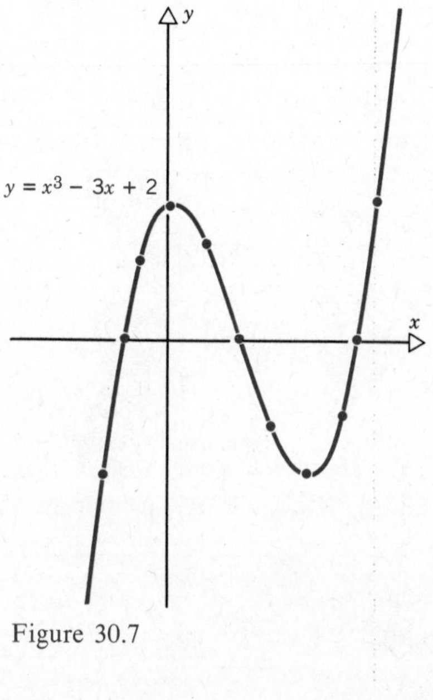

Figure 30.7

EXAMPLE 30.6 Draw the graph of $f(x) = x^4 - 3x^2 - 4$.

Solution

I. The function is even, so the graph is symmetric with respect to the y-axis.

II and III. Table 30.5 shows that $f(2) = 0$. It also shows that

$$f(x) = (x - 2)(x^3 + 2x + x + 2).$$

Call the second factor $g(x)$. Because f is even, $f(2) = 0$ implies $f(-2) = 0$. Therefore, $g(-2) = 0$. If we divide $g(x)$ by $x + 2$, we'll be able to factor $g(x)$. From

$$\begin{array}{r|rrrr} -2 & 1 & 2 & 1 & 2 \\ & & -2 & 0 & -2 \\ \hline & 1 & 0 & 1 & 0 \end{array}$$

we conclude that

$$g(x) = (x + 2)(x^2 + 1).$$

The factor $x^2 + 1$ has no real zeros. Thus

$$f(x) = (x - 2)(x + 2)(x^2 + 1),$$

and the x-intercepts (real zeros) of $f(x)$ are at 2 and -2. The y-intercept is at $f(0) = -4$.

TABLE 30.5 Synthetic Division of $f(x) = x^4 - 3x^2 - 4$

	1	0	−3	0	−4
1	1	1	−2	−2	−6
2	1	2	1	2	0

IV. Table 30.6 gives the calculation of the coordinates of more points to go with the y-intercept and the points from Table 30.5. (It can be shown with calculus that the turning points for the graph, besides the y-intercept, are at $x = \pm\sqrt{\frac{3}{2}} \approx \pm 1.22$ and $y = -6.25$.)

TABLE 30.6

	1	0	−3	0	−4
0.5	1	0.5	−2.75	−1.375	−4.6875
1.5	1	1.5	−0.75	−1.125	−5.6875
2.5	1	2.5	3.25	8.125	16.3125
3	1	3	6	18	50

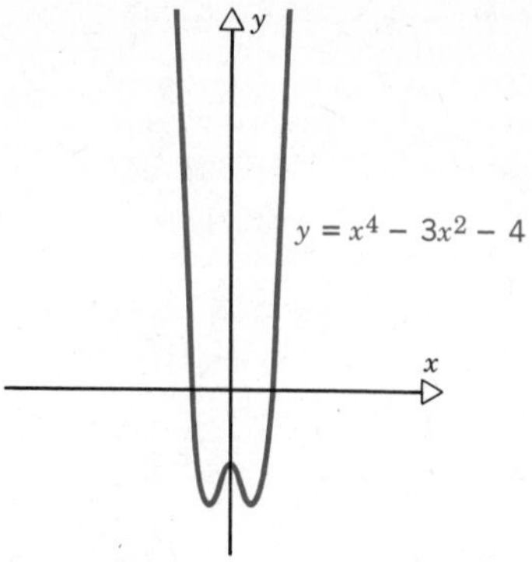

Figure 30.8

V. Figure 30.8 shows the graph.

It can be shown that the graph of any cubic (third-degree) function will have one of the two general forms in Figure 30.9, depending on the sign of a, the leading coefficient. (Figures 30.1 and 30.2 represent special cases that are either never decreasing or never increasing.) In particular, if $a > 0$, then the function is increasing if we move sufficiently far to the left or right, and if $a < 0$, then the function is decreasing if we move sufficiently far to the left or right. Compare Figures 30.1 and 30.7 (leading coefficients positive), and Figure 30.2 (leading coefficient negative).

The graph of any quartic (fourth-degree) function will have one of the two general forms in Figure 30.10, depending again on the sign of a, the leading coefficient. (Figures 30.3 and 30.4 represent special cases.) If $a > 0$, then the function is decreasing if we move sufficiently far to the left and increasing if we move sufficiently far to the right; for $a < 0$ these patterns are reversed. Compare Figures 30.3 and 30.8 (leading coefficients positive) and Figure 30.4 (leading coefficient negative).

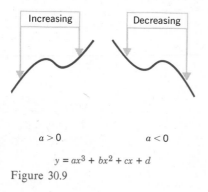

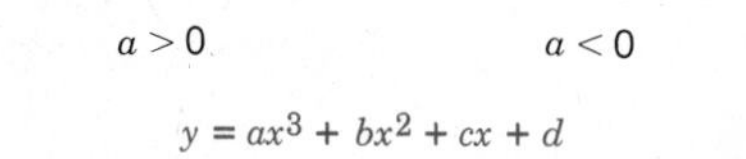

Figure 30.9

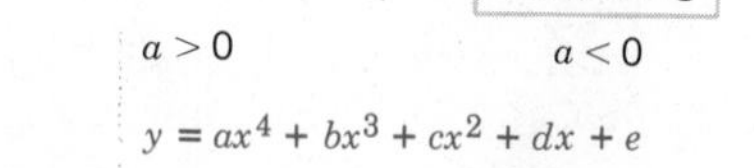

Figure 30.10

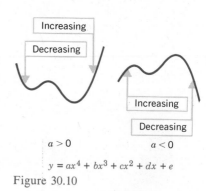

EXERCISES FOR SECTION 30

1. On a single Cartesian plane draw the graphs of $y = x^2$, $y = x^3$, and $y = x^4$, for $-1 \leq x \leq 1$. Make the unit of length as long as possible.
2. On a single Cartesian plane draw the graphs of $y = x^2$ and $y = x^4$, for $-2 \leq x \leq 2$.

3. On a single Cartesian plane draw the graphs of $y = x^3$ and $y = x^5$, for $-1.5 \le x \le 1.5$.
4. On a single Cartesian plane draw the graphs of $y = x^3$, $y = 2x^3$, and $y = \frac{1}{2}x^3$. How does changing k affect the graph of $y = kx^3$ for $k > 0$?
5. On a single Cartesian plane draw the graphs of $y = x^4$, $y = 2x^4$, and $y = \frac{1}{2}x^4$. How does changing k affect the graph of $y = kx^4$ for $k > 0$?
6. On a single Cartesian plane draw the graphs of $y = -x^3$, $y = -2x^3$, and $y = -\frac{1}{2}x^3$. How does changing k affect the graph of $y = kx^3$ for $k < 0$?

In each exercise, state whether the function is even, odd, or neither. Also state what your answer indicates about the symmetry of the graph. In Exercises 13–18 you'll need to compute $f(-x)$.

7. $f(x) = 2x^3 - 2x$
8. $f(x) = x^3 + x^2 + x + 1$
9. $f(x) = x^4 + 1$
10. $f(x) = x^3 - 1$
11. $f(x) = 4x^5 + 2x$
12. $f(x) = 3x^4 + 5x$
13. $f(x) = |x|$
14. $f(x) = |2x| + 1$
15. $f(x) = |x + 1|$
16. $f(x) = 1/x$
17. $f(x) = x + (1/x^2)$
18. $f(x) = \sqrt{x^2 + 1}$

Draw the graph of each function.

19. $f(x) = x^3 - 3x^2$
20. $f(x) = -x^3 + x$
21. $f(x) = 2x^3 + 5x^2 - 2x - 5$
22. $f(x) = x^4 - x$
23. $f(x) = x^3 + x^2 - 7x - 7$
24. $f(x) = -x^3 - x^2 + 9x + 9$
25. $f(x) = -x^3 + 6x^2 - 9x$
26. $f(x) = -x^3 + 2x^2 + 7x + 4$
27. $f(x) = x^4 - 5x^2 + 4$
28. $f(x) = -4x^4 + 17x^2 - 4$
29. $f(x) = x^4 - 4x^3 + 6x^2 - 4x - 1$
30. $f(x) = -4x^4 + 4x^3 + 15x^2 - 16x + 4$

Exercises 31–33 are based on the fact that for any real numbers k, a, b, and c ($k \neq 0$), the x-intercepts of the graph of $f(x) = k(x - a)(x - b)(x - c)$ are a, b, and c (Factor Theorem).

31. Find a cubic function f with leading coefficient 1 whose graph has x-intercepts -2, 3, and 4.
32. Find a cubic function f whose graph has x-intercepts -1, 2, and 3 and y-intercept 12. [The last condition will determine k in $k(x - a)(x - b)(x - c)$.]
33. Find a cubic function f whose graph has x-intercepts -1, 0, and 2, and also passes through $(1, -6)$.

SECTION

31 Graphs of Rational Functions

A. Power Functions

DEFINITION. A function of a variable x is called a **rational function** if it can be written as a fraction in which both the numerator and denominator are polynomials in x.

The simplest examples of rational functions are the power functions that have the form

$$f(x) = kx^{-n} = \frac{k}{x^n}, \tag{31.1}$$

where n is a positive integer. Figures 31.1 and 31.2 show the graphs of the power functions $f(x) = 1/x$ and $f(x) = 1/x^2$. These cases will be analyzed in detail because they illustrate the two significant ways in which graphs of rational functions can differ from graphs of polynomials: First, a polynomial function of x is defined for every value of x, but a rational function will be undefined for any value of x that is a zero of the denominator; special care is required in graphing the function near such a value of x. Second, the graph of a polynomial function will either increase or decrease beyond all bounds as we move to the left or right along the graph, but the graph of a rational function can approach a horizontal line as we move to the left or right.

EXAMPLE 31.1 Draw the graph of $f(x) = \dfrac{1}{x}$.

Solution First, notice that $f(x)$ is an odd function, because

$$f(-x) = \frac{1}{-x} = -\frac{1}{x} = -f(x).$$

Therefore, the graph is symmetric about the origin, so we can concentrate on positive values of x and then use symmetry to take care of negative values of x. Table 31.1 gives some coordinate pairs for points on the graph, and these points have been plotted in Figure 31.1.

TABLE 31.1

x	1	2	3	4	$\frac{1}{2}$	$\frac{1}{3}$	$\frac{1}{4}$
$1/x$	1	$\frac{1}{2}$	$\frac{1}{3}$	$\frac{1}{4}$	2	3	4

As we compute $f(1), f(2), \ldots$, we quickly realize that $f(x)$ approaches 0 as x gets larger. We abbreviate this by

$$f(x) \longrightarrow 0 \text{ as } x \longrightarrow \infty,$$

which we read

"$f(x)$ approaches 0 as x approaches infinity."

Notice that we do *not* write $x = \infty$. Infinity is not a real number; it merely provides a shorthand way to indicate what happens as a variable increases without bound—that is, beyond every real number, no matter how large.

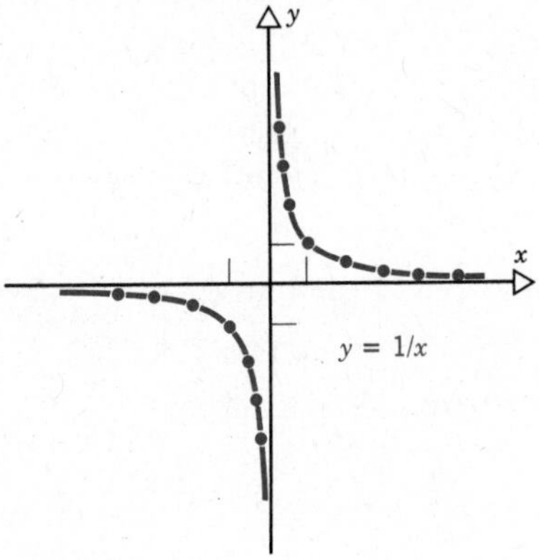

Figure 31.1

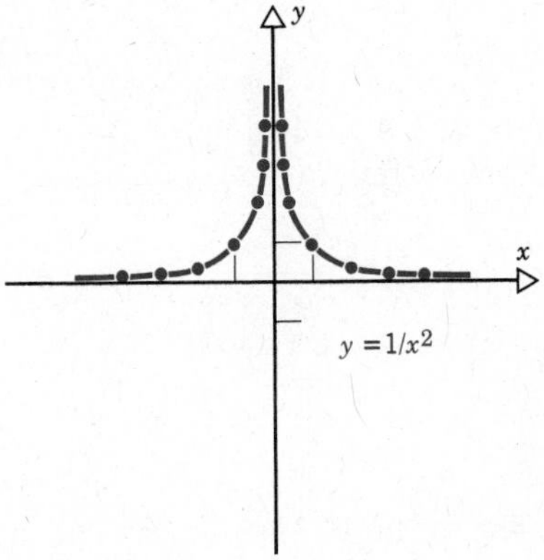

Figure 31.2

When a graph approaches a horizontal line, as in the case of $1/x \to 0$ as $x \to \infty$, we say that the line is a **horizontal asymptote** for the graph. This means that the distance between the line (asymptote) and the points on the graph approaches 0 as $x \to \infty$ or as $x \to -\infty$.

Now consider $f(\frac{1}{2}), f(\frac{1}{3}), \ldots$. Here we see that $f(x) \to \infty$ as $x \to 0^+$. The notation $x \to 0^+$ is used to indicate that x is approaching 0 from the right. (We use $x \to 0^-$ when we approach 0 from the left.) As we move along the x-axis from the right toward the origin, the corresponding points on the graph rise without bound.

When a graph approaches a vertical line, as in the case of $1/x \to \infty$ as $x \to 0^+$, we say that the line is a **vertical asymptote** for the graph. Thus $x = 0$ is a vertical asymptote. In general, a line $x = c$ is a vertical asymptote for a graph if the distance between the line and the points on the graph approaches 0 as $x \to c^+$ or as $x \to c^-$.

Making use of the symmetry about the y-axis, we obtain the portion of Figure 31.1 that lies in the third quadrant. Again, $y = 0$ is a horizontal asymptote:

$$\frac{1}{x} \longrightarrow 0 \text{ as } x \longrightarrow -\infty.$$

And $x = 0$ is a vertical asymptote:

$$\frac{1}{x} \longrightarrow -\infty \text{ as } x \longrightarrow 0^-.$$

EXAMPLE 31.2 Draw the graph of $f(x) = \dfrac{1}{x^2}$.

Solution This function is even, because

$$f(-x) = \frac{1}{(-x)^2} = \frac{1}{x^2} = f(x).$$

Thus the graph is symmetric about the y-axis.

TABLE 31.2

x	1	2	3	4	$\frac{1}{2}$	$\frac{1}{3}$	$\frac{1}{4}$
$1/x^2$	1	$\frac{1}{4}$	$\frac{1}{9}$	$\frac{1}{16}$	4	9	16

Table 31.2 gives some coordinate pairs for points on the graph. These points have been plotted in Figure 31.2. Also, symmetry has been used to plot the corresponding points in the second quadrant. Observe the following similarities with Figure 31.1:

$f(x) \to 0$ as $x \to \infty$, so that $y = 0$ is a horizontal asymptote.

$f(x) \to \infty$ as $x \to 0^+$, so that $x = 0$ is a vertical asymptote.

$f(x) \to 0$ as $x \to -\infty$, so that $y = 0$ is a horizontal asymptote.

$f(x) \to \infty$ as $x \to 0^-$, so that $x = 0$ is a vertical asymptote.

The graph of any function of the form

$$f(x) = \frac{k}{(x-a)^n} \quad (n > 0) \tag{31.2}$$

will be similar to the graph of $f(x) = 1/x$ if n is odd, and similar to the graph of $f(x) = 1/x^2$ if n is even. Figures 31.3 and 31.4 give two typical examples. The effect of replacing x by $x - a$ is to shift the graph to the right (if $a > 0$) or left (if $a < 0$) so that $x = a$ becomes an asymptote.

> The graph of
>
> $$f(x) = \frac{k}{(x-a)^n} \quad (n > 0)$$
>
> has horizontal asymptote $y = 0$ and vertical asymptote $x = a$.

Changing k in $k/(x - a)^n$ will change the details of the graph but not its asymptotes. In particular, changing the sign of k will reflect the graph through the x-axis—that is, turn it upside down.

EXAMPLE 31.3 Draw the graph of $f(x) = \dfrac{2}{x+3}$.

Solution The graph is shown in Figure 31.3. Here are some of its features:

Horizontal asymptote: $y = 0$.
Vertical asymptote: $x = -3$ (shown as a dashed line).
Symmetry about the point $(-3, 0)$.

x	0	1	2	-1	-2	$-\frac{5}{2}$	$-\frac{7}{2}$	-4	-5	-6	-7	-8
$2/(x+3)$	$\frac{2}{3}$	$\frac{1}{2}$	$\frac{2}{5}$	1	2	4	-4	-2	-1	$-\frac{2}{3}$	$-\frac{1}{2}$	$-\frac{2}{5}$

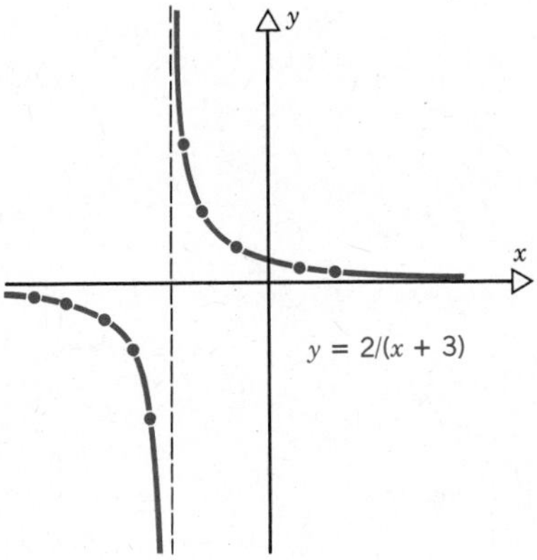

Figure 31.3

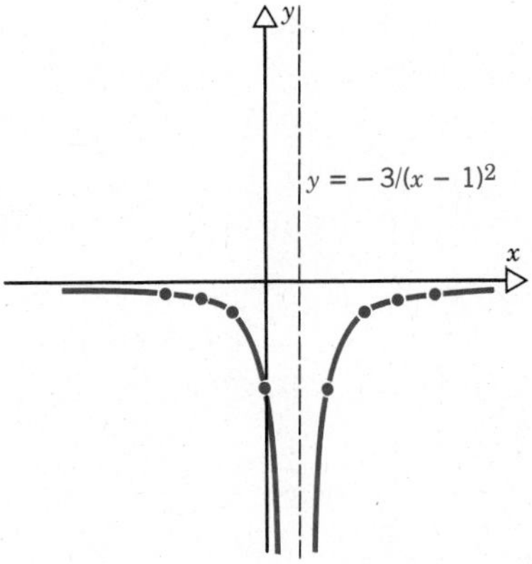

Figure 31.4

EXAMPLE 31.4 Draw the graph of $f(x) = \dfrac{-3}{(x-1)^2}$.

Solution The graph is shown in Figure 31.4. Here are some of its features:

Horizontal asymptote: $y = 0$.
Vertical asymptote: $x = 1$ (shown as a dashed line).
Symmetry about the line $x = 1$.

x	$\frac{3}{2}$	2	3	4	5	$\frac{1}{2}$	0	-1	-2	-3
$-3/(x-1)^2$	-12	-3	$-\frac{3}{4}$	$-\frac{1}{3}$	$-\frac{3}{16}$	-12	-3	$-\frac{3}{4}$	$-\frac{1}{3}$	$-\frac{3}{16}$

B. A General Plan

The following plan for graphing rational functions is based on what we learned from the preceding examples. It is modified slightly to account for a wider range of functions. The examples in Subsection C will follow the plan step-by-step. *If you have trouble with Steps I–V, move ahead to Step VI and plot some points; this may help you see through I–V.*

Step I. *Check for symmetry* by determining if the function is either even or odd.

Step II. *Check for horizontal asymptotes.* Assume that $a(x)$ and $b(x)$ are polynomials, so that $f(x) = a(x)/b(x)$ is a rational function.

(a) If degree $a(x) <$ degree $b(x)$, then $y = 0$ is a horizontal asymptote. (See the examples in Subsection A.)

(b) If degree $a(x) =$ degree $b(x)$ and a_n and b_n are the leading coefficients of $a(x)$ and $b(x)$, respectively, then $y = a_n/b_n$ is a horizontal asymptote. [For instance, in Example 31.5 we'll see that $f(x) = 3x/(x + 1)$ has $y = 3$ as a horizontal asymptote; in this case $a_n = 3$ and $b_n = 1$.]

Step III. *Locate the intercepts.*

(a) The y-intercept is at $(0, f(0))$, provided $f(0)$ is defined.

(b) The x-intercepts occur at the zeros of $f(x)$; these are the zeros of the numerator that are not also zeros of the denominator. (For what happens when a number is a zero of both the numerator and the denominator, see Exercises 13–18.)

Step IV. *Check for vertical asymptotes.* These will occur at any value $x = c$ where the denominator is zero but the numerator is not zero.

Step V. *Determine the behavior of the function near the vertical asymptotes.* Consider $x \to c^+$ (x approaching c from the right) and $x \to c^-$ (x approaching c from the left). For example, determine whether $f(x) \to \infty$ or $f(x) \to -\infty$ as $x \to c^+$.

Step VI. *Plot points.* The number and location of required points will depend on the particular function. We will at least need points near and on both sides of any vertical asymptote.

Step VII. *Draw the graph* as carefully as possible. Breaks can occur only at x-values for which the denominator is zero.

C. More Examples

EXAMPLE 31.5 Draw the graph of $f(x) = \dfrac{3x}{x + 1}$.

Solution Follow the steps in Subsection B.

I. $f(-x) = 3(-x)/[(-x) + 1] = -3x/(-x + 1)$. Thus $f(-x) \neq f(x)$ and $f(-x) \neq -f(x)$ so that $f(x)$ is neither even nor odd.

II. Because the degrees of the numerator and denominator are equal, there is a horizontal asymptote. To see why this is at $y = 3$, divide both the numerator and denominator by the highest power of x in each [this amounts to multiplying by $(1/x)/(1/x)$ with $x \neq 0$, and thus leaves $f(x)$ unchanged]:

$$f(x) = \frac{3x}{x + 1} \cdot \frac{1/x}{1/x} = \frac{3}{1 + (1/x)}.$$

Because $1/x \to 0$ as $x \to \infty$, it is clear that

$$f(x) \longrightarrow 3 \text{ as } x \longrightarrow \infty.$$

Similarly, $f(x) \to 3$ as $x \to -\infty$. The asymptote $y = 3$ is shown as a dashed line in Figure 31.5.

III. (a) There is a y-intercept at $(0, 0)$ because $f(0) = 0$. (b) Because $3x = 0$ iff $x = 0$, the (only) x-intercept is also at $(0, 0)$.

IV. $x + 1 = 0$ iff $x = -1$, and so $x = -1$ is the (only) vertical asymptote.

V. For $x > -1$ but x close to -1, $3x < 0$ but $x + 1 > 0$. Thus $f(x) = 3x/(x + 1) < 0$ and $f(x) \to -\infty$ as $x \to -1^+$. For $x < -1$ but x close to -1, $3x < 0$ and $x + 1 < 0$. Thus $f(x) = 3x/(x + 1) > 0$ and $f(x) \to \infty$ as $x \to -1^-$.

VI. Here is a table of coordinate pairs.

x	0	1	2	4	5	6	7	$-\frac{1}{2}$	$-\frac{3}{4}$	$-\frac{5}{4}$	$-\frac{3}{2}$	-2	-3	-4	-5
$3x/(x + 1)$	0	$\frac{3}{2}$	2	$\frac{12}{5}$	$\frac{5}{2}$	$\frac{18}{7}$	$\frac{21}{8}$	-3	-9	15	9	6	$\frac{9}{2}$	4	$\frac{15}{4}$

VII. The graph is shown in Figure 31.5.

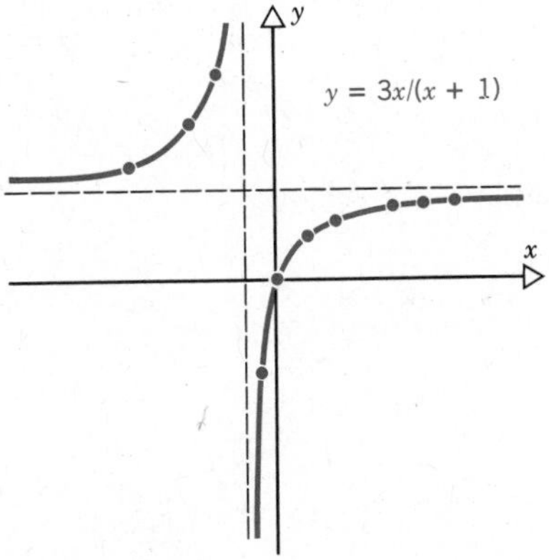

Figure 31.5

EXAMPLE 31.6 Draw the graph of $f(x) = \dfrac{x}{x^2 - 1}$.

Solution

I. $f(-x) = \dfrac{-x}{(-x)^2 - 1} = -\dfrac{x}{x^2 - 1} = -f(x)$. Therefore, f is odd and the graph is symmetric about the origin.

II. The degree of the numerator is less than the degree of the denominator; thus $x = 0$ is a horizontal asymptote.

III. (a) The y-intercept is at $(0, 0)$ because $f(0) = 0$. (b) The x-intercept is also at $(0, 0)$.

IV. Since $f(x) = \dfrac{x}{(x + 1)(x - 1)}$, there are vertical asymptotes at $x = -1$ and $x = 1$.

V. For $x > 1$ but x close to 1, $f(x) > 0$, and $f(x) \to \infty$ as $x \to 1^+$. For $x < 1$ but x close to 1, $f(x) < 0$, and $f(x) \to -\infty$ as $x \to 1^-$. (The behavior near $x = -1$ will follow by symmetry.)

VI. Here is a table of coordinate pairs.

x	0	$\frac{1}{2}$	$\frac{3}{4}$	$\frac{5}{4}$	$\frac{3}{2}$	2	3
$x/(x^2-1)$	0	$-\frac{2}{3}$	$-\frac{12}{7}$	$\frac{20}{9}$	$\frac{6}{5}$	$\frac{2}{3}$	$\frac{3}{8}$

VII. The graph is shown in Figure 31.6. The points in Step VI have been plotted and symmetry has been used.

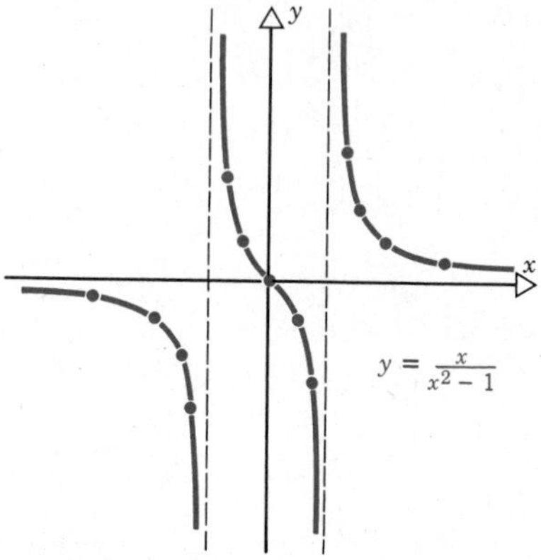

Figure 31.6

EXERCISES FOR SECTION 31

Draw the graph of each function.

1. $f(x) = \dfrac{-1}{x}$ **2.** $f(x) = \dfrac{-1}{x^2}$ **3.** $f(x) = \dfrac{-1}{x^3}$

4. $f(x) = \dfrac{1}{(x-1)^3}$ **5.** $f(x) = \dfrac{3}{x-3}$ **6.** $f(x) = \dfrac{2}{(x+2)^2}$

7. $f(x) = \dfrac{2x}{x-2}$ **8.** $f(x) = \dfrac{-x}{x+1}$ **9.** $f(x) = \dfrac{-x}{x-1}$

10. $f(x) = \dfrac{x}{(x-1)(x+2)}$ **11.** $f(x) = \dfrac{-x}{x^2-4}$ **12.** $f(x) = \dfrac{-x}{(x+1)(x-3)}$

Suppose that $f(x) = a(x)/b(x)$ and that c is a zero of both $a(x)$ and $b(x)$. Then $f(x)$ is undefined at $x = c$, but the line $x = c$ will not be a vertical asymptote as it is when c is zero of $b(x)$ but not $a(x)$. Exercises 13–18 provide illustrations. In Exercise 13, for example, $f(0)$ is undefined, but $f(x) = x + 1$ for $x \neq 0$; thus the graph is a straight line with one point removed. Draw the graph of each function.

13. $f(x) = \dfrac{x^2+x}{x}$ **14.** $f(x) = \dfrac{x^2-1}{x+1}$ **15.** $f(x) = \dfrac{2x^2-3x}{x}$

16. $f(x) = \dfrac{x-1}{x^2-x}$ **17.** $f(x) = \dfrac{x}{x^2-2x}$ **18.** $f(x) = \dfrac{x^4-x^2}{x^2-1}$

Suppose that $f(x) = a(x)/b(x)$, that $a(x)$ and $b(x)$ have no common zero, and that the degree of $a(x)$ is one more than the degree of $b(x)$. Then division will yield $f(x) = cx + d + [r(x)/b(x)]$ where c and d are real numbers, $c \neq 0$, and

$\deg r(x) < \deg b(x)$. In this case $r(x)/b(x) \to 0$ as $x \to \infty$ or as $x \to -\infty$, so that $f(x) \to cx + d$. The graph of $y = cx + d$, which is a straight line, is called an **oblique asymptote** for the graph of $f(x)$. Exercises 19–24 provide illustrations. In Exercise 19, for example, $f(x) = (x^2 + 1)/x = x + (1/x)$, so that $y = x$ gives an oblique asymptote. Draw the graph of each function.

19. $f(x) = \dfrac{x^2 + 1}{x}$ **20.** $f(x) = \dfrac{-x^2 + 1}{x}$ **21.** $f(x) = \dfrac{x^2 - 1}{2x}$

22. $f(x) = \dfrac{x^3}{x^2 + 1}$ **23.** $f(x) = \dfrac{x^3}{x^2 - 1}$ **24.** $f(x) = \dfrac{2x^2 + 1}{x + 1}$

25. Boyle's law states that if the temperature is constant, then the pressure P of a confined gas varies inversely as the volume of the gas: $P = k/V$ (Example 9.12). Graph P as a function of V in the special case $k = 4000$, with V measured in cubic inches and P in pounds per square inch. (You will need to choose the unit distance on each axis so that the graph is manageable.)

26. The intensity I of a sound is inversely proportional to the square of the distance r from the source: $I = k/r^2$ (Example 9.13). Graph I as a function of r in the special case $k = 45$, with r measured in meters and I in watts per square meter.

27. The time t required to fill a tank from a water hose is inversely proportional to the square of the diameter of the hose: $t = k/d^2$. Graph t as a function of d in the special case $k = 2$, with d measured in inches and t measured in minutes.

SECTION 32 Partial Fractions

A. Introduction

To add rational functions we convert to common denominators, add, and then simplify. For example,

$$\frac{3}{x+1} - \frac{x}{x^2 - x + 2} = \frac{3(x^2 - x + 2) - x(x+1)}{(x+1)(x^2 - x + 2)}$$

$$= \frac{2x^2 - 4x + 6}{x^3 + x + 2}. \qquad (32.1)$$

In this section we consider how to reverse that process—how to begin with an expression like that on the right in (32.1) and end with an expression like that on the left. This reversed process is needed in calculus, for example.

The method we use rests on the following fact, whose proof is omitted: With real coefficients, every polynomial can be written as a product of linear and irreducible quadratic factors. That fact, together with the division algorithm for polynomials, can be used to justify the following statement.

> Every rational function can be written uniquely as a sum of a polynomial (which may be identically zero) and fractions of the form
>
> $$\frac{A}{(ax + b)^m} \quad \text{and} \quad \frac{Bx + C}{(ax^2 + bx + c)^n},$$
>
> where $ax^2 + bx + c$ is irreducible.

The resulting form is called the **partial fraction decomposition** of the original rational function. The linear $(ax + b)$ and quadratic $(ax^2 + bx + c)$ factors in the decomposition are the irreducible factors of the original denominator. The first step is always to factor the original denominator into powers of irreducible factors. (Remember that a quadratic factor is irreducible iff its discriminant is negative.)

B. Distinct Linear Factors

We call an irreducible factor of the denominator *distinct* (as opposed to *repeated*) if it occurs only with exponent one when we factor the denominator.

> Each distinct linear factor $ax + b$ contributes a term of the form
>
> $$\frac{A}{ax + b}$$
>
> to the decomposition.

EXAMPLE 32.1 Decompose $\dfrac{x + 12}{(x - 2)(x + 5)}$ into partial fractions.

Solution The factors $x - 2$ and $x + 5$ will contribute terms

$$\frac{A}{x - 2} \quad \text{and} \quad \frac{B}{x + 5}$$

respectively. If

$$\frac{x + 12}{(x - 2)(x + 5)} = \frac{A}{x - 2} + \frac{B}{x + 5} \tag{32.2}$$

then on multiplying by $(x - 2)(x + 5)$ to clear the denominators, we obtain

$$x + 12 = A(x + 5) + B(x - 2). \tag{32.3}$$

The last equation must be true for all values of x, so in particular it must be true for $x = -5$ and $x = 2$, the zeros of the two linear factors. By substituting these in turn we can determine A and B.

Use $x = -5$ in (32.3):

$$-5 + 12 = A(0) + B(-7)$$
$$B = -1.$$

Use $x = 2$ in (32.3):

$$2 + 12 = A(7) + B(0)$$
$$A = 2.$$

With $A = 2$ and $B = -1$, Equation (32.2) gives the answer:

$$\frac{x + 12}{(x - 2)(x + 5)} = \frac{2}{x - 2} - \frac{1}{x + 5}.$$

To check the answer perform the subtraction indicated on the right and show that the result simplifies to the expression on the left.

C. Repeated Linear Factors

Each repeated linear factor $(ax + b)^m$ contributes a sum of the form

$$\frac{A_1}{ax + b} + \frac{A_2}{(ax + b)^2} + \cdots + \frac{A_m}{(ax + b)^m}$$

to the decomposition.

EXAMPLE 32.2 Decompose $\dfrac{x^2 - 5x - 2}{x^3 + 2x^2 + x}$ into partial fractions.

Solution First, factor the denominator completely:

$$x^3 + 2x^2 + x = x(x^2 + 2x + 1) = x(x + 1)^2.$$

The distinct linear factor x contributes

$$\frac{A}{x}.$$

The repeated linear factor $(x + 1)^2$ contributes

$$\frac{B}{x + 1} + \frac{C}{(x + 1)^2}.$$

If

$$\frac{x^2 - 5x - 2}{x(x + 1)^2} = \frac{A}{x} + \frac{B}{x + 1} + \frac{C}{(x + 1)^2} \tag{32.4}$$

then

$$x^2 - 5x - 2 = A(x + 1)^2 + Bx(x + 1) + Cx. \tag{32.5}$$

With $x = 0$ in (32.5) we will get $A = -2$. With $x = -1$ in (32.5) we will get $C = -4$. Now we use any other value of x in (32.5), along with $A = -2$ and $C = -4$, to get B. With $x = 1$ we will get $B = 3$.

If we use $A = -2$, $B = 3$, and $C = -4$ in Equation (32.4), and revert to the original form of the denominator on the left, we get

$$\frac{x^2 - 5x - 2}{x^3 + 2x^2 + x} = -\frac{2}{x} + \frac{3}{x+1} - \frac{4}{(x+1)^2}.$$

D. Distinct Quadratic Factors

Each distinct quadratic factor $ax^2 + bx + c$ contributes a term of the form

$$\frac{Ax + B}{ax^2 + bx + c}$$

to the decomposition.

EXAMPLE 32.3 Decompose $\dfrac{3x^2 + 2x}{(x+1)(x^2 + x + 1)}$ into partial fractions.

Solution The factor $x^2 + x + 1$ is irreducible. The factors $x + 1$ and $x^2 + x + 1$ contribute terms

$$\frac{A}{x+1} \quad \text{and} \quad \frac{Bx + C}{x^2 + x + 1}$$

respectively. If

$$\frac{3x^2 + 2x}{(x+1)(x^2 + x + 1)} = \frac{A}{x+1} + \frac{Bx + C}{x^2 + x + 1} \tag{32.6}$$

then

$$3x^2 + 2x = A(x^2 + x + 1) + (Bx + C)(x + 1). \tag{32.7}$$

With $x = -1$ in (32.7) we will get $A = 1$. With $x = 0$ and $A = 1$ in (32.7) we will get $C = -1$. With $x = 1$, $A = 1$, and $C = -1$ in (32.7) we will get $B = 2$. Thus

$$\frac{3x^2 + 2x}{(x+1)(x^2 + x + 1)} = \frac{1}{x+1} + \frac{2x - 1}{x^2 + x + 1}.$$

E. Repeated Quadratic Factors

Each repeated quadratic factor $(ax^2 + bx + c)^n$ contributes a sum of the form

$$\frac{A_1x + B_1}{ax^2 + bx + c} + \frac{A_2x + B_2}{(ax^2 + bx + c)^2} + \cdots + \frac{A_nx + B_n}{(ax^2 + bx + c)^n}$$

to the decomposition.

In the previous examples the coefficients in the partial fraction decompositions have been determined by substituting appropriate numbers for x. The next example uses a method based on the fact that a polynomial in x is identically zero (that is, zero for every value of x) iff each of its coefficients is zero. (This will be proved in Section 34.)

EXAMPLE 32.4 Decompose $\dfrac{1}{x(x^2+2)^2}$ into partial fractions.

Solution The factors x and $(x^2 + 2)^2$ contribute

$$\frac{A}{x} \quad \text{and} \quad \frac{Bx + C}{x^2 + 2} + \frac{Dx + E}{(x^2 + 2)^2}$$

respectively. If

$$\frac{1}{x(x^2 + 2)^2} = \frac{A}{x} + \frac{Bx + C}{x^2 + 2} + \frac{Dx + E}{(x^2 + 2)^2} \tag{32.8}$$

then

$$1 = A(x^2 + 2)^2 + (Bx + C)x(x^2 + 2) + (Dx + E)x$$
$$1 = A(x^4 + 4x^2 + 4) + B(x^4 + 2x^2) + C(x^3 + 2x) + Dx^2 + Ex$$

$$(A + B)x^4 + Cx^3 + (4A + 2B + D)x^2 + (2C + E)x + 4A - 1 = 0. \tag{32.9}$$

Since Equation (32.9) is to be satisfied for every value of x, each of the coefficients must be zero. (See the remark preceding this example.)

$$\begin{aligned} x^4&:\ A + B = 0 & x^3&:\ C = 0 \\ x^2&:\ 4A + 2B + D = 0 & x&:\ 2C + E = 0 \\ x^0&:\ 4A - 1 = 0 \end{aligned}$$

The equation for x^0 yields $A = \frac{1}{4}$. The equation for x^4 then yields $B = -\frac{1}{4}$. The equation for x^3 yields $C = 0$. The equation for x then yields $E = 0$. With $A = \frac{1}{4}$ and $B = -\frac{1}{4}$, the equation for x^2 yields $D = -\frac{1}{2}$. With these values for A, B, C, D, and E, Equation (32.8) becomes

$$\frac{1}{x(x^2 + 2)^2} = \frac{1}{4x} - \frac{x}{4(x^2 + 2)} - \frac{x}{2(x^2 + 2)^2}.$$

F. Improper Fractions

A rational function is called a **proper fraction** if the degree of the numerator is less than the degree of the denominator; otherwise it is called an **improper fraction.** By division, an improper fraction can be written as a sum of a polynomial and a proper fraction. The preceding methods can then be applied to decompose the proper fraction into partial fractions.

EXAMPLE 32.5 Decompose

$$f(x) = \frac{x^4 + 3x^3 - 11x^2 - 2x + 22}{x^2 + 3x - 10}$$

into partial fractions.

Solution Division yields

$$f(x) = x^2 - 1 + \frac{x + 12}{x^2 + 3x - 10}.$$

The proper fraction on the right is the same as the fraction in Example 32.1. Therefore, using the solution of that example, we have

$$f(x) = x^2 - 1 + \frac{2}{x - 2} - \frac{1}{x + 5}.$$

EXERCISES FOR SECTION 32

Decompose into partial fractions.

1. $\dfrac{x - 8}{(x + 1)(x - 2)}$
2. $\dfrac{3x + 3}{x(x - 3)}$
3. $\dfrac{6x - 18}{(x + 2)(x - 4)}$
4. $\dfrac{2}{3x^2 + 4x}$
5. $\dfrac{-1}{2x^2 + x - 1}$
6. $\dfrac{8x + 3}{6x^2 + 2x}$
7. $\dfrac{3x - 4}{(x - 1)^2}$
8. $\dfrac{4x + 5}{(2x + 1)^2}$
9. $\dfrac{-4x + 11}{(2x - 3)^2}$
10. $\dfrac{-x^2 + 2x + 4}{x^3 + x^2}$
11. $\dfrac{2x^2 - 2x + 3}{x^3 - 2x^2 + x}$
12. $\dfrac{7x^2 - 9x}{(x + 1)(x - 1)^2}$
13. $\dfrac{4x^2 + x + 4}{x^3 + x^2 + x}$
14. $\dfrac{2x^2 + 5x + 2}{x^3 + x}$
15. $\dfrac{-3x^2 + 4x - 6}{x^3 + 2x}$
16. $\dfrac{7x^2 + 13}{(x^2 + 1)(x^2 + 2)}$
17. $\dfrac{x^2 - 2}{x^4 + x^2}$
18. $\dfrac{-x^2}{(x^2 + 1)(x^2 + 3)}$
19. $\dfrac{x^3 - x + 1}{(x^2 + 1)^2}$
20. $\dfrac{3x^2 + 4x + 5}{(x^2 + 2)^2}$
21. $\dfrac{2x^3 + x^2 + 4x - 1}{(x^2 + x + 1)^2}$
22. $\dfrac{-2x^2 + 2x + 3}{x^4 + x^3}$
23. $\dfrac{4x^2 - 11x + 8}{(x - 1)^3}$
24. $\dfrac{-2x^3 + x^2 + 2x + 4}{(x^2 + 1)^3}$

25. $\dfrac{3x^3 + 2x^2 - 2x + 1}{x^2 + x}$

26. $\dfrac{x^4 - 2x^3 - x^2 + 6x - 5}{(x - 1)^2}$

27. $\dfrac{2x^5 - x^4 + 5x^3 - 3x^2 + x - 2}{(x^2 + 1)^2}$

REVIEW EXERCISES FOR CHAPTER VII

Determine the quotient and remainder when $f(x)$ is divided by $g(x)$.

1. $f(x) = 2x^3 - 6x^2 + 5, \quad g(x) = 2x^2 - 1$
2. $f(x) = x^4 - x^3 + 3x^2 - x + 2, \quad g(x) = x^2 + 2$

Use synthetic division to determine the quotient and remainder when $f(x)$ is divided by $g(x)$.

3. $f(x) = x^4 - 3x^2 + x + 4, \quad g(x) = x - 2$
4. $f(x) = 3x^3 - 5x^2 + 2, \quad g(x) = x + 1$

Use the Remainder Theorem and synthetic division to compute the indicated values of $f(c)$.

5. $f(x) = 2x^3 + x^2 - 5; f(3), f(-1)$
6. $f(t) = t^4 + 3t^2 - t; f(2), f(-2)$

Use the Remainder Theorem and synthetic division to show that the given value of c is a zero of the given polynomial. Also determine $q(x)$ such that $f(x) = (x - c)q(x)$.

7. $c = 4; f(x) = x^3 - 3x^2 - 2x - 8$
8. $c = -\frac{1}{2}; f(x) = 4x^4 + 2x^3 - 2x - 1$

Determine the real zeros of each polynomial along with their multiplicities. Also, factor each polynomial completely.

9. $x^3 - 4x^2 + x + 6$
10. $x^3 - 2x^2 - 3x + 6$
11. $2x^4 - 3x^3 - 3x^2 + 6x - 2$
12. $3x^4 + x^3 - 11x^2 - 3x + 6$

Each polynomial in Exercises 13 and 14 has three real zeros, none of which is an integer. Use the Intermediate-Value Theorem to isolate the zeros between consecutive integers.

13. $2x^3 + 7x^2 - 10x - 35$
14. $3x^3 + 5x^2 - 30x - 50$

In each exercise, state whether the function is even, odd, or neither. Also state what your answer indicates about the symmetry of the graph of the function.

15. $f(x) = x^3 - x + 1$
16. $f(x) = 3x^6 + x^4 - 2x^2 - 1$
17. $f(x) = \dfrac{2x}{x^3 - x}$
18. $f(x) = \dfrac{x^2 + 1}{x^3 - 5x}$

Draw the graph of each function.

19. $f(x) = x^3 - 4x$
20. $f(x) = -2x^3 + 3x^2$
21. $f(x) = -x^4 + x^3 + x^2 - x$
22. $f(x) = 2x^4 + x^3 - 6x^2$
23. $f(x) = \dfrac{3}{(x - 1)^2}$
24. $f(x) = \dfrac{-2x}{x + 2}$

Decompose into partial fractions.

25. $\dfrac{3x + 2}{(x + 2)^2}$

26. $\dfrac{4x^2 - 3x + 4}{x^3 + x}$

27. $\dfrac{x^3 + x^2 + 2x + 5}{x^2 + x - 2}$

28. $\dfrac{x^5 - x^4 + 3x^3 - 4x^2 + 7x - 9}{(x - 1)(x^2 + 2)}$

29. If 3 times the cube of a rational number minus 9 times the number is 6 less than 2 times the square of the number, what is the number?

30. The shortest edge of a box is one-third as long as another edge and two centimeters shorter than the third edge. If the volume of the box is 1323 cubic centimeters, how long is each edge?

CHAPTER VIII
COMPLEX NUMBERS

Because $x^2 \geq 0$ for every real number x, the quadratic equation $x^2 = -1$ has no solution among the real numbers. The quadratic formula shows that, more generally, no quadratic equation has a solution among the real numbers if its discriminant is negative (Section 11A). The real numbers also fail to provide a solution for many polynomial equations of degree greater than two. Mathematicians have overcome this problem by extending the system of real numbers to a larger system—the system of *complex numbers*—in which every polynomial equation does have a solution. Section 33 will introduce the complex numbers and Section 34 will show how the complex numbers help to complete the study of polynomial equations that was begun in Chapter VII.

SECTION

33 Complex Numbers

A. Introduction

The system of complex numbers, to be introduced in this section, has the following fundamental properties:

The system contains the system of real numbers.

The system also contains other numbers, including a number i such that

$$i^2 = -1.$$

The system has operations $(+, -, \times, \div)$ that satisfy all of the properties listed in Section 1C; more precisely, the system satisfies the field axioms (see page 8) and all of the properties that can be proved from the field axioms.

If the system is to have the properties listed, then for each real number b the product bi must also be a complex number (because b and i are both complex numbers). For each real number a, then, the sum $a + bi$ must be a complex number (because a and bi are both complex numbers). It turns out that every complex number can be written in this form $a + bi$, where a and b are real numbers. In fact, this form $a + bi$ will be our starting point in studying the complex numbers.

B. Definitions and Operations

DEFINITION. The set of **complex numbers** is the set of all expressions of the form $a + bi$, where a and b are real numbers and $i^2 = -1$.

Statement (33.1), which follows, summarizes the way in which complex numbers are combined to form a system with the properties listed in Subsection A. Equations (33.2)–(33.5), which give explicit formulas for the fundamental operations $(+, -, \times, \div)$ between complex numbers, are consequences of this statement.

The number i is combined with real numbers to form complex numbers in the same way that a variable x is combined with real numbers to form polynomials, except that i^2 can always be replaced by -1.	(33.1)

EXAMPLE 33.1

(a) Each real number is a complex number because $a = a + 0i$ (just as $a = a + 0x$ for polynomials).

(b) i is a complex number because $i = 0 + 1i$ (just as $x = 0 + 1x$ for polynomials).
(c) $a - bi = a + (-b)i$.
(d) $0 = 0 + 0i$ and $1 = 1 + 0i$.

When we refer to the form $a + bi$ we always assume that a and b are real numbers. We call a the **real part** of $a + bi$ and b the **imaginary part.** If $b \neq 0$, then $a + bi$ is said to be **imaginary.** Both $\sqrt{2} + 4i$ and $-2i$ are imaginary. If $b \neq 0$ and $a = 0$, then $a + bi$ is said to be a **pure imaginary number.** Thus $-2i$ is pure imaginary but $\sqrt{2} + 4i$ is not.

Two complex numbers are **equal** iff their real parts are equal and their imaginary parts are equal. That is:

$$a + bi = c + di \quad \text{iff} \quad a = c \quad \text{and} \quad b = d.$$

Compare the following equations for addition and subtraction with the corresponding equations for polynomials: for example, $(a + bx) + (c + dx) = (a + c) + (b + d)x$.

Addition	$(a + bi) + (c + di) = (a + c) + (b + d)i$	(33.2)
Subtraction	$(a + bi) - (c + di) = (a - c) + (b - d)i$	(33.3)

The next computation is a direct consequence of Statement (33.1) and will give the equation for multiplication:

$$\begin{aligned}(a + bi)(c + di) &= a(c + di) + bi(c + di)\\ &= ac + adi + bci + bdi^2\\ &= (ac - bd) + (ad + bc)i.\end{aligned}$$

Multiplication $(a + bi)(c + di) = (ac - bd) + (ad + bc)i$ (33.4)

EXAMPLE 33.2 Write each expression in the form $a + bi$ and simplify.

(a) $(2 + i) + (3 + 6i) = (2 + 3) + (1 + 6)i = 5 + 7i$
(b) $(2 + i) - (8 - 3i) = (2 - 8) + (1 + 3)i = -6 + 4i$
(c) $(3 + 2i)(1 + 4i) = (3 - 8) + (12 + 2)i = -5 + 14i$
(d) $(7 - i)(2 - 5i) = (14 - 5) + (-35 - 2)i = 9 - 37i$

The equations $i^2 = -1$ and $(-i)^2 = (-1)^2(i)^2 = -1$ show that both i and $-i$ are square roots of -1. We choose i as the principal square root. That is,

$$\sqrt{-1} = i.$$

The square root of any negative real number can now be simplified as follows.

$$\sqrt{-a} = \sqrt{a}\sqrt{-1} = \sqrt{a}\,i \text{ for each positive real number } a.$$

EXAMPLE 33.3

(a) $\sqrt{-9} = \sqrt{9}\sqrt{-1} = 3i$
(b) $(\sqrt{2} + \sqrt{-4}) - (1 + 5i) = (\sqrt{2} + \sqrt{4}\sqrt{-1}) - (1 + 5i)$
$= (\sqrt{2} + 2i) - (1 + 5i) = (\sqrt{2} - 1) - 3i$

In simplifying a product of imaginary numbers, it is important to reduce each factor to the form $a + bi$ before multiplying. For example,

$$\sqrt{(-4)(-4)} = \sqrt{16} = 4 \text{ but } \sqrt{-4}\sqrt{-4} = 2i \cdot 2i = 4i^2 = -4,$$

so that

$$\sqrt{(-4)(-4)} \neq \sqrt{-4}\sqrt{-4}.$$

Thus the law $\sqrt{ab} = \sqrt{a}\sqrt{b}$, which holds when both a and b are positive and when one is positive and the other is negative, does not hold when both a and b are negative.

The powers in the sequence $i, i^2, i^3, \ldots$ can be simplified easily by using

$$i^2 = -1,$$

$$i^3 = i^2 i = (-1)i = -i,$$

and

$$i^4 = i^3 i = (-i)i = -i^2 = -(-1) = 1.$$

Specifically, any power i^n can be reduced to either 1, i, -1, or $-i$ by separating the highest multiple of 4 in the exponent, as in the following example.

EXAMPLE 33.4

(a) $i^5 = i^4 i = 1i = i$
(b) $i^{20} = (i^4)^5 = 1^5 = 1$
(c) $i^{1935} = i^{4(483)+3} = (i^4)^{483} i^3 = 1(-i) = -i$
(d) $(2 + i)^3 = 2^3 + 3(2)^2(i) + 3(2)(i)^2 + i^3$ [by Equation (14.6)]
$= 8 + 12i - 6 - i$
$= 2 + 11i$

To see how to divide complex numbers we need the following idea.

The **conjugate** of a complex number $a + bi$ is $a - bi$.

Thus the conjugate of $2 + 3i$ is $2 - 3i$, the conjugate of $4 - i$ is $4 + i$, the conjugate of i is $-i$, and the conjugate of 6 is 6.

The product of a complex number and its conjugate is always a real number:

$$\begin{aligned}(c + di)(c - di) &= c^2 - (di)^2 \\ &= c^2 - d^2i^2 \\ &= c^2 - d^2(-1) \\ &= c^2 + d^2.\end{aligned}$$

We can use the equation $(c + di)(c - di) = c^2 + d^2$ to determine an equation for division.

$$\begin{aligned}\frac{a + bi}{c + di} &= \frac{a + bi}{c + di} \cdot \frac{c - di}{c - di} \\ &= \frac{a(c - di) + bi(c - di)}{c^2 + d^2} \\ &= \frac{ac - adi + bci - bdi^2}{c^2 + d^2} \\ &= \frac{ac + bd}{c^2 + d^2} + \frac{bc - ad}{c^2 + d^2}i.\end{aligned}$$

Division $\dfrac{a + bi}{c + di} = \dfrac{ac + bd}{c^2 + d^2} + \dfrac{bc - ad}{c^2 + d^2}i$	(33.5)

To simplify a fraction we use the idea behind Equation (33.5) rather than the equation itself. That is, if the denominator is $c + di$, multiply by $(c - di)/(c - di)$ and simplify.

EXAMPLE 33.5 Write each number in the form $a + bi$ and simplify.

(a) $\dfrac{1}{1 + i} = \dfrac{1}{1 + i} \cdot \dfrac{1 - i}{1 - i} = \dfrac{1 - i}{1 + 1} = \dfrac{1}{2} - \dfrac{1}{2}i$

(b) $\dfrac{2}{i} = \dfrac{2}{i} \cdot \dfrac{-i}{-i} = \dfrac{-2i}{-i^2} = \dfrac{-2i}{1} = -2i$

(c) $\dfrac{2 + i}{3 - 2i} = \dfrac{2 + i}{3 - 2i} \cdot \dfrac{3 + 2i}{3 + 2i} = \dfrac{6 + 7i - 2}{9 + 4} = \dfrac{4 + 7i}{13} = \dfrac{4}{13} + \dfrac{7}{13}i$

C. Solutions of Quadratic Equations

The quadratic formula (Section 11A) shows that if a quadratic equation has real coefficients and its discriminant is negative, then the solutions of the equation are complex conjugates.

EXAMPLE 33.6 The discriminant of

$$x^2 + 2x + 5 = 0$$

is

$$2^2 - 4(1)(5) = 4 - 20 = -16 < 0.$$

The solutions of the equation are

$$x = \frac{-2 \pm \sqrt{-16}}{2} = \frac{-2 \pm \sqrt{16}\,\sqrt{-1}}{2} = \frac{-2 \pm 4i}{2} = -1 \pm 2i.$$

Table 33.1 gives a summary concerning the solutions of a quadratic equation $ax^2 + bx + c = 0$ with real coefficients a, b, and c. (Compare Table 11.1.) Quadratic equations with imaginary coefficients can lead to square roots of imaginary numbers, which are beyond the scope of this book.

TABLE 33.1

Discriminant	*Character of the solutions*
$b^2 - 4ac < 0$	Two conjugate imaginary solutions
$b^2 - 4ac = 0$	One real solution
$b^2 - 4ac > 0$	Two unequal real solutions

D. Complex Planes

Real numbers are represented geometrically by using a real line (Section 1). To represent complex numbers geometrically we use a plane: Choose a Cartesian plane and assign the complex number $a + bi$ to the point with coordinates (a, b). This establishes a one-to-one correspondence between the set of complex numbers and the set of points in the plane. A plane together with such a correspondence is called a **complex plane.** Figure 33.1 shows examples of points with the corresponding complex numbers.

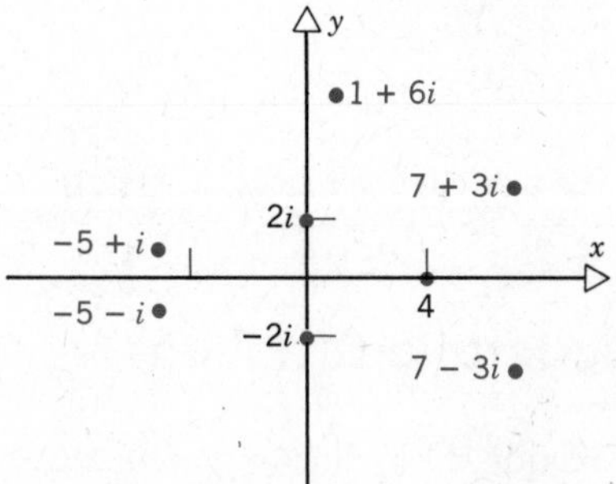

Figure 33.1

Notice that any complex number and its conjugate are located symmetrically with respect to the horizontal axis: compare the conjugate pairs $-5 \pm i$, $\pm 2i$, and $7 \pm 3i$ in Figure 33.1.

The geometrical representation of complex numbers is not essential for our purposes, but it is very important for more extensive work with complex numbers.

EXERCISES FOR SECTION 33

In each of Exercises 1–3 make three lists, one of the *real numbers,* one of the *imaginary numbers,* and one of the *pure imaginary numbers.* A number may be in more than one list.

1. $4 - i,\ \sqrt{3},\ \pi i,\ 2 + \sqrt{-4},\ 0$
2. $5i,\ 1,\ \sqrt{-25},\ 5 + i,\ -\sqrt{2}$
3. $3 + \sqrt{-15},\ \pi,\ \sqrt{2} + i,\ -1,\ 4i$

Write each expression in the form $a + bi$ and simplify.

4. $(9 + i) + (3 - 2i)$
5. $(7 - i) - (6 - i)$
6. $(14 + 5i) + (-8 + 2i)$
7. $5i - (4 - i)$
8. $\sqrt{-1} + \sqrt{25}$
9. $-2 + (7 - \sqrt{-2})$
10. $(1 + \sqrt{-16}) - (\pi + 4i)$
11. $10 - (1 - \sqrt{-9})$
12. $(\sqrt{2} + i) - (1 - 2\sqrt{-4})$
13. $i(2 + 3i)$
14. $2i(5 + i)$
15. $-3i(8 + i)$
16. $(2 + i)(1 - 2i)$
17. $(-4 + i)(2 + i)$
18. $(6 - i)(-1 - i)$
19. $(1 + i)(1 - i)$
20. $(5 - i)^2$
21. $(-1 - 3i)(1 - 3i)$
22. $(3 + i)^2$
23. $(-2 + i)(2 + i)$
24. $(-1 - 2i)^2$
25. i^{15}
26. i^{18}
27. i^{32}
28. $\dfrac{2}{2 + i}$
29. $\dfrac{1}{3 - i}$
30. $\dfrac{1}{1 + 2i}$
31. $\dfrac{i}{1 - i}$
32. $\dfrac{-2i}{1 + i}$
33. $\dfrac{-5i}{3 - 4i}$
34. $\dfrac{1 - i}{2 + i} + \dfrac{2i}{3 - i}$
35. $\dfrac{4}{i} - \dfrac{3 + i}{2 - i}$
36. $\dfrac{i}{(1 - i)^2} - \dfrac{1}{2i}$
37. $(1 + i)i^{-5}$
38. $(-i)^{-10}(2 - i)$
39. $(-i)^{-7}(-1 + i)$
40. $(1 - i)^3$
41. $(1 + 2i)^3$
42. $(3 - 2i)^3$

Solve for x.

43. $x^2 - x + 2 = 0$
44. $x^2 + 2x + 2 = 0$
45. $x^2 + x + 3 = 0$
46. $2x^2 + x = -1$
47. $3x^2 = 5x - 3$
48. $4x^2 + 1 = 0$
49. $2x^2 = -18$
50. $x^2 + 10 = 0$
51. $2x^2 - 3x = -2$

Solve for x and y, assuming that both are real numbers. (Remember: For a, b, c, and d real, $a + bi = c + di$ iff $a = c$ and $b = d$).

52. $2x + i = 5 + yi$
53. $x + 4 + i = 2 - yi$
54. $(x - y) + (x + 2y)i = 1 + 10i$

55. Is i a solution of $x^4 + x^3 + x^2 + x + 1 = 0$? Justify your answer.

56. Is $-i$ a solution of $x^3 + x^2 + x + 1 = 0$? Justify your answer.

57. Is i a solution of $x^6 + x^4 + x^2 + 1 = 0$? Justify your answer.

58. Verify that both of the numbers $-\frac{1}{2} \pm \frac{1}{2}\sqrt{3}i$ are cube roots of 1, that is, that both are solutions of $x^3 = 1$.

59. Verify that both of the numbers $\frac{1}{2} \pm \frac{1}{2}\sqrt{3}i$ are cube roots of -1, that is, that both are solutions of $x^3 = -1$.

60. Verify that $-i$, $\frac{1}{2}\sqrt{3} + \frac{1}{2}i$, and $-\frac{1}{2}\sqrt{3} + \frac{1}{2}i$ are cube roots of i, that is, that each number is a solution of $x^3 = i$.

61. (a) Show that if a and b are both real, then the sum of $a + bi$ and its conjugate is a real number.

(b) Show that if a and b are both real and $b \neq 0$, then $a + bi$ minus its conjugate is a pure imaginary number.

62. If the sum of a complex number and its reciprocal is 1, what is the number?

63. If the sum of a complex number and its reciprocal is 0, what is the number?

Let $\bar{z}$ denote the conjugate of a complex number z. That is, if $z = a + bi$, then $\bar{z} = a - bi$ (where a and b are real). Assume that $z = a + bi$ and $w = c + di$, and then prove each statement in Exercises 64–72.

64. $\bar{z} = z$ iff z is real (that is, iff $b = 0$) **65.** $\bar{\bar{z}} = z$ [where $\bar{\bar{z}}$ means $\overline{(\bar{z})}$]

66. $\bar{z} + z = 0$ iff the real part of z is 0 (that is, iff $a = 0$)

67. $\overline{z + w} = \bar{z} + \bar{w}$ **68.** $\overline{z - w} = \bar{z} - \bar{w}$ **69.** $\overline{\left(\frac{z}{w}\right)} = \frac{\bar{z}}{\bar{w}}$ **70.** $\overline{zw} = \bar{z}\bar{w}$ **71.** $\overline{z^2} = (\bar{z})^2$

72. $\overline{z^n} = (\bar{z})^n$ (Use Exercise 71.)

In each of Exercises 73–75, draw a complex plane and label the points with the given numbers as coordinates.

73. $3 + 2i,\ -2,\ 4i,\ -1 - 2i,\ -4 + i$ **74.** $1 - i,\ 6 + i,\ -2i,\ 5,\ -3 - i$

75. $-i,\ 2 - i,\ -4,\ -5 + i,\ 4 + 3i$

The absolute value of a complex number $z = a + bi$ is denoted $|z|$ and is defined by

$$|z| = \sqrt{a^2 + b^2}.$$

Prove each statement in Exercises 76–81. [Notice that $|z|$ is the distance from the origin to the point that represents $a + bi$ in a complex plane. See the remarks preceding Exercise 64 for the meaning of $\bar{z}$.]

76. $|z| = |-z|$ **77.** $|z| = |\bar{z}|$ **78.** $|z| = 0$ iff $z = 0$

79. $|z| = \sqrt{z\bar{z}}$ **80.** $|zw| = |z| \cdot |w|$ **81.** $\left|\frac{z}{w}\right| = \frac{|z|}{|w|}$

SECTION

34 Complex Zeros of Polynomials

A. Complex Polynomials

Previously we have considered only **real polynomials,** that is, polynomials whose coefficients are real numbers. In this section we consider the more general class of **complex polynomials**—polynomials whose coefficients are complex numbers. (Every real polynomial is also complex, of course, since every real number is a complex number.)

All of the results about divisibility, factors, and zeros from Sections 27 and 28 hold for complex polynomials as well as for real polynomials. This subsection contains examples to illustrate the highlights, without proofs. For proofs it suffices to replace real numbers by complex numbers in the proofs in Sections 27 and 28.

Remember that to divide by $x - c$ using synthetic division the leading number in the first row must be c.

EXAMPLE 34.1 Use synthetic division to determine the quotient and remainder when $z^4 - 3iz^2 + 6z + i$ is divided by $z + 2i$.

Solution

$$\begin{array}{r|rrrrr} -2i & 1 & 0 & -3i & 6 & i \\ & & -2i & -4 & -6+8i & 16 \\ \hline & 1 & -2i & -4-3i & 8i & 16+i \end{array}$$

The quotient is $z^3 - 2iz^2 + (-4 - 3i)z + 8i$. The remainder is $16 + i$.

The Remainder Theorem states that if $f(z)$ is divided by $z - c$, then the remainder is $f(c)$.

EXAMPLE 34.2 Use the Remainder Theorem and synthetic division to compute $f(1 + i)$ if $f(z) = iz^3 + z^2 - iz$.

Solution

$$\begin{array}{r|rrrr} 1+i & i & 1 & -i & 0 \\ & & -1+i & -1+i & -1-i \\ \hline & i & i & -1 & -1-i \end{array}$$

The remainder is $-1 - i$, so $f(1 + i) = -1 - i$.

The Factor Theorem states that $z - c$ is a factor of $f(z)$ iff $f(c) = 0$.

EXAMPLE 34.3 Use synthetic division and the Factor Theorem to show that $z - 2i$ is a factor of $f(z) = z^4 + 6z^2 + 8$.

Solution

$2i$	1	0	6	0	8
		$2i$	-4	$4i$	-8
	1	$2i$	2	$4i$	0

Thus $f(2i) = 0$, so $z - 2i$ is a factor of $f(z)$ by the Factor Theorem. The numbers in the third row of the synthetic division show that, in fact,

$$z^4 + 6z^2 + 8 = (z - 2i)(z^3 + 2iz^2 + 2z + 4i).$$

If c is a zero of $f(z)$ and $(z - c)^m$ is the highest power of $z - c$ that divides $f(z)$, then c is said to be a zero of *multiplicity m*.

EXAMPLE 34.4 Form a complex polynomial of lowest degree having 0, i, and $-2i$ as zeros of multiplicities 3, 2, and 1, respectively.

Solution The answer must have z^3, $(z - i)^2$, and $z + 2i$ as factors. Thus an acceptable answer is $z^3(z - i)^2(z + 2i)$. To expand this, write

$$\begin{aligned} z^3(z - i)^2(z + 2i) &= z^3(z^2 - 2iz - 1)(z + 2i) \\ &= z^3(z^3 + 3z - 2i) \\ &= z^6 + 3z^4 - 2iz^3. \end{aligned}$$

B. The Fundamental Theorem of Algebra

From Section 33 we know that every second-degree real polynomial has a zero among the complex numbers. The following important theorem extends that result from *real* polynomials of degree *two* to *complex* polynomials of *every* positive degree.

The Fundamental Theorem of Algebra
Every complex polynomial whose degree is at least one has a zero among the complex numbers.

Again, notice the implications of this theorem. To have zeros for all real polynomials we must go beyond the real numbers to the complex numbers. The Fundamental Theorem of Algebra asserts that to find zeros for complex polynomials there will be no need to go further: the complex numbers contain a zero not only for every real polynomial, but also for every complex polynomial.

Proofs of the Fundamental Theorem of Algebra lie outside the scope of this book. However, we can give a proof of a useful corollary based on the theorem. First, recall that in Section 28 we showed that each *real* polynomial of degree n has at most n *real* zeros. The argument used there can also be used to show that each *complex* polynomial of degree n has at most n *complex* zeros. If, as before, each zero of multiplicity m is counted m times, we can prove the following more precise result.

Corollary
Each polynomial of degree $n \geq 1$ has exactly n complex zeros.

Proof. Assume that

$$f(z) = a_n z^n + a_{n-1} z^{n-1} + \cdots + a_1 z + a_0 \text{ with } a_n \neq 0.$$

If $n = 1$, then $f(z) = a_1 z + a_0$. By inspection there is one zero, $-a_0/a_1$, as required.

Assume that $n > 1$. By the Fundamental Theorem of Algebra $f(z)$ has at least one zero. If c_1 is a zero, then $z - c_1$ is a factor of $f(z)$, and thus $f(z) = (z - c_1)f_1(z)$ for some complex polynomial $f_1(z)$ of degree $n - 1$. If $n = 2$, then $f_1(z)$ has degree one; therefore, as required, $f(z)$ has two zeros, c_1 together with the zero of $f_1(z)$. If $n > 2$, then $f_1(z)$ has degree greater than one. Thus $f_1(z)$ has a zero, say c_2, and $f(z) = (z - c_1)(z - c_2)f_2(z)$ for some polynomial $f_2(z)$ of degree $n - 2$. We can continue in this way until we arrive at

$$f(z) = a_n(z - c_1)(z - c_2) \cdots (z - c_n). \tag{34.1}$$

[The factor a_n appears because a_n is the coefficient of z^n in $f(z)$, and the coefficients of $f(z)$ on the two sides of the equation must be equal.] Equation (34.1) shows that $f(z)$ has at least the n complex numbers $c_1, c_2, \ldots, c_n$ as zeros. On the other hand, Equation (34.1) also shows that if $c \neq c_k$ so that $c - c_k \neq 0$ for $1 \leq k \leq n$, then $f(c) \neq 0$; thus $f(z)$ has no other zeros. □

The following fact was used in Section 32E on partial fractions.

Corollary

A complex polynomial in z is identically zero (that is, zero for every value of z) iff each of its coefficients is zero.

Proof. It is obvious that if a polynomial $f(z)$ has only zero coefficients, then $f(c) = 0$ for every c. Suppose, on the other hand, that $f(z)$ is identically zero but has one or more nonzero coefficients. Then $f(z)$ has degree n for some $n \geq 1$, since a nonzero constant polynomial is not identically zero. But if $f(z)$ has degree n, then (by the preceding corollary) $f(z)$ has at most n zeros, which contradicts the fact that $f(z)$ is identically zero. This proves that if $f(z)$ is identically zero, then each of its coefficients must be zero. □

In Section 14 we defined a polynomial to be *reducible* if it can be written as a product of two other polynomials that are both of positive degree; otherwise it is *irreducible*. A polynomial has been *factored completely* when it has been written as a product of irreducible factors. The reducibility or irreducibility of a polynomial depends on whether the factors are required to be real polynomials or whether they can be complex polynomials. By the Factor Theorem and the Fundamental Theorem of Algebra, a complex polynomial is irreducible relative to complex factors iff the polynomial is linear, that is, of degree one.

EXAMPLE 34.5

(a) The polynomial $x^2 + 1$ is reducible with complex factors:

$$x^2 + 1 = (x + i)(x - i).$$

(b) The polynomial $x^2 + 1$ is irreducible if the factors are required to be real polynomials.

EXAMPLE 34.6 Given that -3 is a zero of

$$f(z) = z^3 + 3z^2 + 4z + 12,$$

factor $f(z)$ completely using complex factors.

Solution To determine $q(z)$ such that $f(z) = (z + 3)q(z)$, use synthetic division.

$$\begin{array}{r|rrrr} -3 & 1 & 3 & 4 & 12 \\ & & -3 & 0 & -12 \\ \hline & 1 & 0 & 4 & 0 \end{array}$$

Thus

$$f(z) = (z + 3)(z^2 + 4).$$

The zeros of $z^2 + 4$ are the solutions of $z^2 + 4 = 0$, which are $\pm 2i$. Thus

$$z^2 + 4 = (z - 2i)(z + 2i)$$

and

$$f(z) = (z + 3)(z - 2i)(z + 2i).$$

EXAMPLE 34.7 A complex number z is a cube root of -1 iff $z^3 = -1$. One cube root of -1 is -1. Find the others.

Solution The answer will be the zeros of $g(z)$, where $z^3 + 1 = (z + 1)g(z)$. The computation

$$\begin{array}{r|rrrr} -1 & 1 & 0 & 0 & 1 \\ & & -1 & 1 & -1 \\ \hline & 1 & -1 & 1 & 0 \end{array}$$

shows that $g(z) = z^2 - z + 1$. By the quadratic formula $g(z)$ has zeros

$$z = \frac{1 \pm \sqrt{1 - 4}}{2} = \frac{1 \pm \sqrt{3}i}{2}.$$

Thus the cube roots of -1 are -1 and $(1 \pm \sqrt{3}i)/2$. (You can check each one by cubing.)

C. Complex Zeros of Real Polynomials

The zeros of a real quadratic polynomial are either real, or imaginary conjugates (Table 33.1). Therefore, if $a + bi$ is a zero of such a polynomial, then $a - bi$ is also a zero. This is a special case of the following theorem, which will not be proved.

Conjugate Zero Theorem

If $a + bi$ is a zero of a real polynomial $f(x)$, then its conjugate $a - bi$ is also a zero of $f(x)$.

To paraphrase the Conjugate Zero Theorem: *Imaginary zeros of real polynomi-*

als occur in conjugate pairs. A proof of the Conjugate Zero Theorem is outlined in Exercise 27.

EXAMPLE 34.8 Given that $1 + i$ is a zero of

$$f(x) = x^4 - 2x^3 + 5x^2 - 6x + 6,$$

factor $f(x)$ completely using complex factors.

Solution If $1 + i$ is a zero, then $x - (1 + i)$ must be a factor. To find $g(x)$ such that $f(x) = [x - (1 + i)]g(x)$, we use synthetic division.

$$\begin{array}{r|rrrrr} 1+i & 1 & -2 & 5 & -6 & 6 \\ & & 1+i & -2 & 3+3i & -6 \\ \hline & 1 & -1+i & 3 & -3+3i & 0 \end{array}$$

$$\underbrace{1 \quad -1+i \quad 3 \quad -3+3i}_{\text{coefficients of } g(x)}$$

By the Conjugate Zero Theorem, since $1 + i$ is a zero of the real polynomial $f(x)$, its conjugate $1 - i$ must also be a zero of $f(x)$. Therefore, $1 - i$ is a zero of $g(x)$. With synthetic division we can determine $h(x)$ such that $g(x) = [x - (1 - i)]h(x)$.

$$\begin{array}{r|rrrr} 1-i & 1 & -1+i & 3 & -3+3i \\ & & 1-i & 0 & 3-3i \\ \hline & 1 & 0 & 3 & 0 \end{array}$$

$$\underbrace{1 \quad 0 \quad 3}_{\text{coefficients of } h(x)}$$

Since $h(x) = x^2 + 3$, the zeros of $h(x)$ are $\pm\sqrt{3}i$, so that $h(x) = (x - \sqrt{3}i)(x + \sqrt{3}i)$. Thus the complete factorization of $f(x)$ is

$$f(x) = (x - 1 - i)(x - 1 + i)(x - \sqrt{3}i)(x + \sqrt{3}i).$$

EXAMPLE 34.9 Form a *real* polynomial of lowest degree having $3i$ and -1 as zeros of multiplicities 1 and 2, respectively.

Solution The Conjugate Zero Theorem implies that if $3i$ is a zero, then its conjugate $-3i$ must also be a zero. Thus the answer must have $x - 3i$, $x + 3i$, and $(x + 1)^2$ as factors. An acceptable polynomial, in factored form, is

$$(x - 3i)(x + 3i)(x + 1)^2.$$

To make it obvious that this is a real polynomial, we can expand it to get

$$x^4 + 2x^3 + 10x^2 + 18x + 9.$$

We have seen that the only complex polynomials that are irreducible relative to complex factors are the linear polynomials. The next theorem gives the corresponding fact for real factors of real polynomials. The proof of this theorem will be omitted.

Irreducible Real Polynomials

A real polynomial is irreducible relative to real factors iff it is either linear or quadratic with a negative discriminant.

EXAMPLE 34.10 Factor $f(x) = x^4 - 2x^3 + 5x^2 - 6x + 6$ completely using real factors.

Solution The complete factorization using complex factors was given in Example 34.8. To obtain a factorization with real polynomials, multiply the linear factors arising from conjugate zeros (this will always yield real factors).

$$[x - (1 + i)][x - (1 - i)] = x^2 - 2x + 2$$
$$[x - \sqrt{3}i][x + \sqrt{3}i] = x^2 + 3$$

This gives the complete factorization

$$f(x) = (x^2 - 2x + 2)(x^2 + 3).$$

EXERCISES FOR SECTION 34

Use synthetic division to find the quotient and remainder when $f(z)$ is divided by $g(z)$.

1. $f(z) = z^3 - 2iz^2 + z + i, \quad g(z) = z - i$
2. $f(z) = iz^4 + z^2 - 2iz, \quad g(z) = z + i$
3. $f(z) = 2z^4 - z^3 + iz^2 - z + 1 + i, \quad g(z) = z - 2i$

Use the Remainder Theorem and synthetic division to compute the indicated values $f(c)$.

4. $f(z) = (1 + i)z^4 - z^3 + z - 2i$; $f(2), f(1 + i)$
5. $f(z) = 3z^3 - iz^2 + z - 1 - i$; $f(i), f(1 + i)$
6. $f(z) = iz^3 - iz^2 - z + 3i$; $f(-i), f(1 - i)$

Use synthetic division and the Factor Theorem to show that $z - c$ is a factor of $f(z)$. Also determine $q(z)$ such that $f(z) = (z - c)q(z)$.

7. $z - 2i$; $f(z) = z^4 + z^2 - 12$
8. $z - 3i$; $f(z) = z^3 - 3iz^2 + z - 3i$
9. $z + \sqrt{5}i$; $f(z) = z^4 + 4z^2 - 5$

Form a *complex* polynomial of lowest degree having the specified zeros with the specified multiplicities.

10. 0, $-i$, and $1 + i$ of multiplicities 2, 2, and 1, respectively.
11. 1, i, and $-2i$ of multiplicities 1, 1, and 2, respectively.
12. $1 - i$ and $2 + i$ of multiplicities 1 and 2, respectively.

Form a *real* polynomial of lowest degree having the specified zeros with the specified multiplicities. Write the answer so that it is obvious that it has only real coefficients.

13. $-2i$ and 2 of multiplicities 1 and 2, respectively.

14. $1 + i$ and 0 of multiplicities 1 and 3, respectively.
15. $1 - i$ and $3i$ of multiplicities 1 and 2, respectively.

For each polynomial in Exercises 16–21:

(a) Determine the complex zeros, along with their multiplicities; one zero is given in each case.
(b) Factor the polynomial completely using complex factors.
(c) Factor the polynomial completely using real factors.

16. $z^3 + 2z^2 + 5z + 10$; -2 is a zero.
17. $z^3 - 4z^2 - 4z - 5$; 5 is a zero.
18. $z^3 + z + 2$; -1 is a zero.
19. $z^4 + 5z^2 + 4$; i is a zero.
20. $z^4 + z^2 + 1$; $(-1 + \sqrt{3}i)/2$ is a zero.
21. $z^3 + z^2 - 4z + 6$; $1 + i$ is a zero.

A complex number z is an nth root of c iff $z^n = c$.

22. One cube (third) root of 1 is 1. Find the others.
23. One fourth root of 1 is i. Find the others.
24. One fourth root of -1 is $(\sqrt{2} + \sqrt{2}i)/2$. Find the others.
25. Prove that every real polynomial of odd degree has at least one real zero. (Use the Conjugate Zero Theorem.)
26. Give an example of a complex polynomial of odd degree that does not have a real zero. (Compare Exercise 25.)
27. Prove the Conjugate Zero Theorem. [Suggestion: Assume that z is a zero of

$$f(x) = a_n x^n + \cdots + a_1 x + a_0.$$

Then $f(z) = 0$. Use results from Exercises 64–72 of Section 33 to deduce that $f(\overline{z}) = 0$. For $n = 2$, the proof would look like this:

$$\begin{aligned} f(\overline{z}) &= a_2\overline{z}^2 + a_1\overline{z} + a_0 = \overline{a}_2\overline{z^2} + \overline{a}_1\overline{z} + \overline{a}_0 \\ &= \overline{a_2 z^2 + a_1 z + a_0} = \overline{f(z)} = \overline{0} = 0. \end{aligned}$$

Justify each step.]

REVIEW EXERCISES FOR CHAPTER VIII

Write each expression in the form $a + bi$ and simplify.

1. $(5 + i) - 2(-1 + i)$
2. $(\sqrt{4} - \sqrt{-4}) - (\sqrt{9} + \sqrt{-9})$
3. $i(4 + i) - 4(-1 + i)$
4. $2i[(1 - 6i) - 2i(3 + i)]$
5. $(2 + i)(-3 + 2i)$
6. $(-1 + \sqrt{-9})(4 + 2i)$
7. $\dfrac{5i}{2 + i}$
8. $\dfrac{-3}{5 - i}$
9. $i^{10} - 3i^7 + i^3 + 5$
10. $i^5(2 - i)^3$

Solve for x.

11. $2x^2 - 2x + 1 = 0$ **12.** $x^2 + 3x = -3$

Form a *complex* polynomial of lowest degree having the specified zeros with the specified multiplicities.

13. 0, $2i$, and $-i$ of multiplicities 2, 1, and 1, respectively.
14. 3, $-i$, and i of multiplicities 1, 2, and 1, respectively.

Form a *real* polynomial of lowest degree having the specified zeros with the specified multiplicities.

15. $2 + 2i$ and 0 of multiplicities 1 and 2, respectively.
16. $-i$ and $3i$ of multiplicities 2 and 1, respectively.

For each polynomial in Exercises 17–18:

(a) Determine the complex zeros, along with their multiplicities; one zero is given in each case.
(b) Factor the polynomial completely using complex factors.
(c) Factor the polynomial completely using real factors.

17. $2x^3 - 3x^2 + 8x - 12$; $2i$ is a zero.
18. $4x^4 + 9x^2 + 2$; $-\sqrt{2}i$ is a zero.

CHAPTER IX
EXPONENTIAL AND LOGARITHMIC FUNCTIONS

A *power function* of x has the form x^b where the exponent b is constant (Section 30). In this chapter we interchange the roles of x and b to obtain *exponential functions,* which have the form b^x where b is constant. We also consider the inverses of exponential functions, which are called *logarithmic functions.* We study the general properties of exponential and logarithmic functions and also look at some typical applications. This chapter will draw freely on ideas from Section 17 (Rational Exponents) and Section 26 (Inverse Functions).

SECTION

35 Exponential and Logarithmic Functions

A. Example

Let's begin with the specific function $f(x) = 2^x$. To make the domain the set of all real numbers, we must give a meaning to 2^x for *every* real number x. Recall that 2^x already has a meaning if x is rational. For example,

$$2^3 = 2 \cdot 2 \cdot 2 = 8,$$
$$2^{-4} = 1/2^4 = 1/16,$$
$$2^{5/3} = (\sqrt[3]{2})^5,$$

and, in general,

$$2^{m/n} = (\sqrt[n]{2})^m \qquad (35.1)$$

(Section 17). By looking at these powers geometrically, we'll see how to go further.

Figure 35.1 indicates the pattern of the points

$$(x, 2^x) \text{ for } x = 0, \pm 1, \pm 2, \ldots.$$

Some coordinate pairs are shown in Table 35.1.

TABLE 35.1

x	...	-3	-2	-1	0	1	2	3	...
2^x	...	$\frac{1}{8}$	$\frac{1}{4}$	$\frac{1}{2}$	1	2	4	8	...

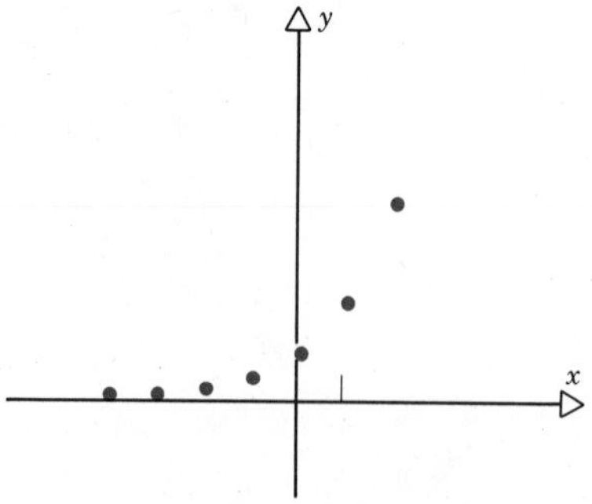

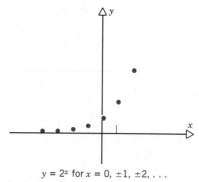

$y = 2^x$ for $x = 0, \pm 1, \pm 2, \ldots$

Figure 35.1

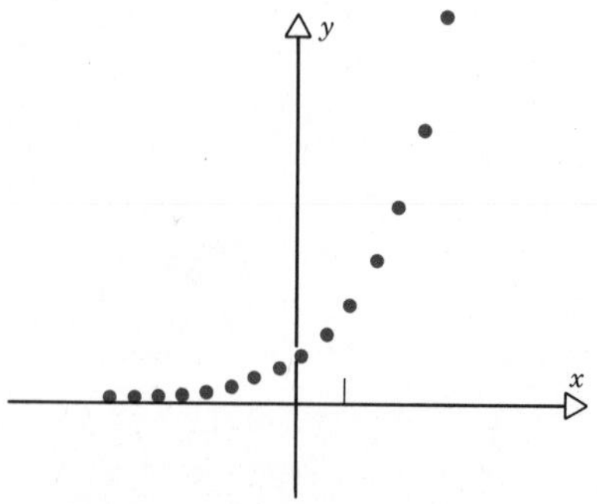

$y = 2^x$ for $x = 0, \pm 0.5, \pm 1, \pm 1.5, \pm 2, \ldots$

Figure 35.2

We can add to Figure 35.1 any point $(x, 2^x)$ for x a rational number by using Equation (35.1). Table 35.2 shows some specific calculations, and the points determined from Tables 35.1 and 35.2 are all shown in Figure 35.2.

TABLE 35.2

x	2^x
0.5	$2^{1/2} = \sqrt{2} \approx 1.414$
1.5	$2^{3/2} = \sqrt{2^3} = \sqrt{2^2}\sqrt{2} = 2\sqrt{2} \approx 2.828$
2.5	$2^{5/2} = \sqrt{2^5} = \sqrt{2^4}\sqrt{2} = 4\sqrt{2} \approx 5.656$
⋮	⋮
−0.5	$2^{-1/2} = 1/\sqrt{2} = \sqrt{2}/2 \approx 0.707$
−1.5	$2^{-3/2} = 1/2\sqrt{2} = \sqrt{2}/4 \approx 0.354$
−2.5	$2^{-5/2} = 1/4\sqrt{2} = \sqrt{2}/8 \approx 0.177$
⋮	

If we continue in this way with other pairs $(x, 2^x)$ for x rational, the points will continue to fall into the pattern suggested by Figures 35.1 and 35.2. This leads to a geometric interpretation of 2^x for every real number x. First, draw a smooth curve through the points $(x, 2^x)$ for rational values of x. This curve is shown in Figure 35.3. Then, given any real number x (rational or irrational), use the corresponding y-value on the curve as an approximation of 2^x.

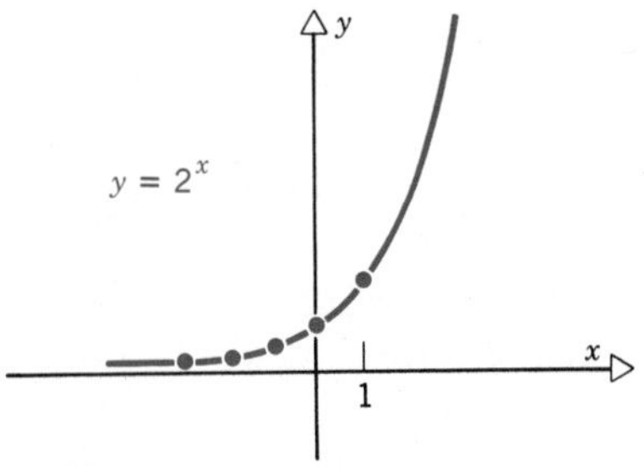

Figure 35.3

Here is the idea in nongeometric terms for the special case 2^π. The following sequence of rational numbers gives successively more accurate approximations of π:

3
3.1
3.14
3.141
3.1415
3.14159
⋮

The equation $2^{m/n} = (\sqrt[n]{2})^m$ permits us to interpret 2^x for each number x in this sequence. For instance,

$$2^{3.14} = 2^{314/100} = (\sqrt[100]{2})^{314}.$$

TABLE 35.3

x	2^x
3	8
3.1	8.57418 . . .
3.14	8.81524 . . .
3.141	8.82135 . . .
3.1415	8.82441 . . .
3.14159	8.82496 . . .
. . .	. . .

This idea leads to the sequence of pairs in Table 35.3. (The values in Table 35.3 can be verified with a calculator having a $\boxed{y^x}$ key.) It can be proved that the sequence of numbers on the right in Table 35.3 approaches a unique real number, and this unique number is taken as the definition of 2^π. To seven significant figures $2^\pi = 8.824978$. If we were to plot the pairs from Table 35.3 on a graph, they would approach the point $(\pi, 2^\pi)$ as x approaches π.

Because any real number has a decimal representation, we can apply the idea just used for 2^π to define 2^x for any real number x. This yields the function $f(x) = 2^x$ whose domain is the set of all real numbers. The graph of this function is the smooth curve in Figure 35.3.

B. Exponential Functions

The idea in Subsection A can be carried out with any other positive real number in place of 2. This leads to the class of functions in the following definition.

DEFINITION. If b is any positive real number except 1, then the function f defined by

$$f(x) = b^x \text{ for every real number } x \tag{35.2}$$

is called the **exponential function** with **base** b.

[The reason for excluding 1 in this definition is that later we'll want to consider inverses of such functions, and $b = 1$ would lead to the constant function $f(x) = 1$, which does not have an inverse.]

EXAMPLE 35.1 The graph of $f(x) = 4^x$ is shown in Figure 35.4. Some coordinate pairs for points on the graph are $(-2, \frac{1}{16})$, $(-1, \frac{1}{4})$, $(0, 1)$, $(1, 4)$ and $(2, 16)$.

EXAMPLE 35.2 The graph of $f(x) = (\frac{1}{4})^x = 4^{-x}$ is shown in Figure 35.5. Some coordinate pairs for points on the graph are $(-2, 16)$, $(-1, 4)$, $(0, 1)$, $(1, \frac{1}{4})$, and $(2, \frac{1}{16})$.

EXAMPLE 35.3 Figure 35.6 shows the graphs of the four exponential functions

$$f(x) = 3^x, \quad f(x) = (\tfrac{1}{3})^x, \quad f(x) = (\tfrac{3}{2})^x, \quad \text{and} \quad f(x) = (\tfrac{2}{3})^x.$$

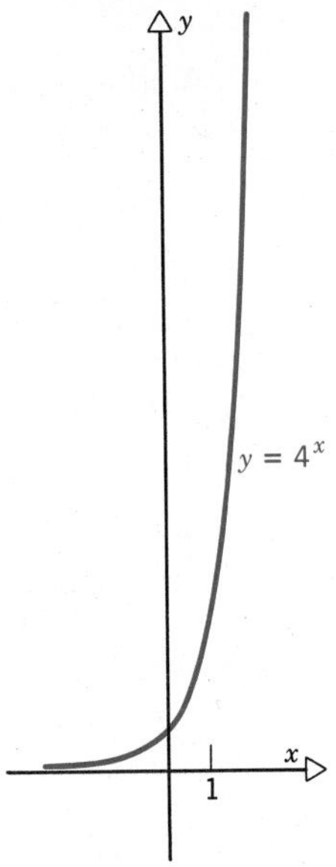

Figure 35.4

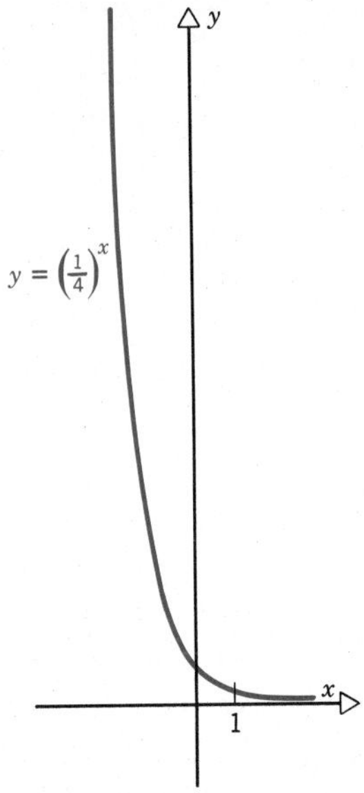

Figure 35.5

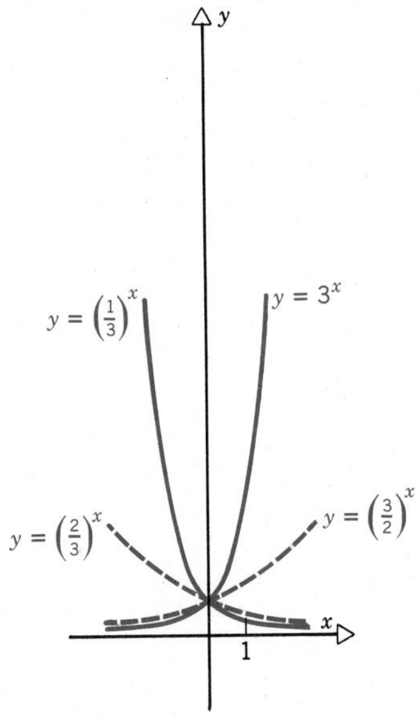

Figure 35.6

Compare the graphs of the two functions in each of the following pairs:

$$f(x) = 4^x \quad \text{and} \quad f(x) = (\tfrac{1}{4})^x = 4^{-x}$$
$$f(x) = 3^x \quad \text{and} \quad f(x) = (\tfrac{1}{3})^x = 3^{-x}$$
$$f(x) = (\tfrac{3}{2})^x \quad \text{and} \quad f(x) = (\tfrac{2}{3})^x = (\tfrac{3}{2})^{-x}.$$

These comparisons illustrate that the graph of $y = (1/b)^x = b^{-x}$ is the reflection through the y-axis of the graph of $y = b^x$. To prove that this relationship between graphs is true, let $f(x) = b^x$ and $g(x) = b^{-x}$. Then (x, y) is on the graph of $y = f(x)$ iff $(-x, y)$ is on the graph of $y = g(x)$, because $y = f(x)$ iff $y = b^x$ iff $y = b^{-(-x)}$ iff $y = g(-x)$.

Recall (from Section 24) that a function f is

increasing if $x_1 < x_2$ implies $f(x_1) < f(x_2)$,

and

decreasing if $x_1 < x_2$ implies $f(x_1) > f(x_2)$.

The increasing and decreasing properties of exponential functions are included in the following list. The properties in this list are suggested and illustrated by the examples in Figures 35.3–35.6. There is no need to memorize these properties; instead, just remember the representative graphs in Figure 35.7.

Let $f(x) = b^x$ be an exponential function. The *domain* of f is the set of all real numbers. The *range* of f is the set of all positive real numbers. If $b > 1$, then

f is increasing,
$f(x) \to \infty$ as $x \to \infty$, and
$f(x) \to 0$ as $x \to -\infty$.

If $0 < b < 1$, then

f is decreasing,
$f(x) \to 0$ as $x \to \infty$, and
$f(x) \to \infty$ as $x \to -\infty$.

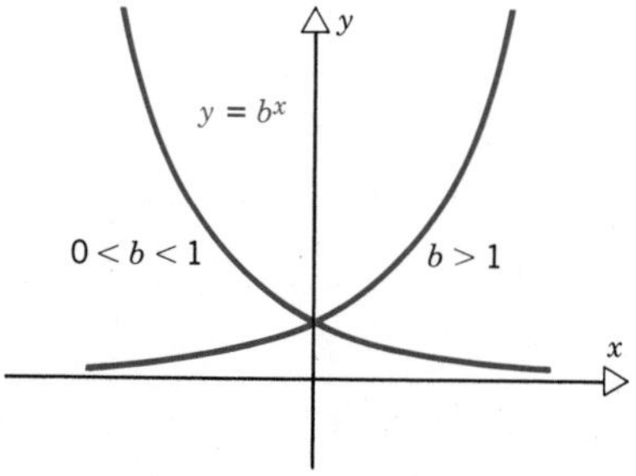

Figure 35.7

It can be proved that the laws for rational exponents (Section 17) carry over to any real exponents. Here is a summary.

If a, b, r, and s denote real numbers with $a > 0$ and $b > 0$, then

$$a^r a^s = a^{r+s} \tag{35.3}$$

$$(a^r)^s = a^{rs} \tag{35.4}$$

$$(ab)^r = a^r b^r \tag{35.5}$$

$$\frac{a^r}{b^r} = \left(\frac{a}{b}\right)^r \tag{35.6}$$

$$\frac{a^r}{a^s} = a^{r-s} \tag{35.7}$$

Applications of exponential functions will come later in the chapter.

C. Logarithmic Functions

Let b denote any positive number except 1. We know from Subsection B that the graph of the exponential function $f(x) = b^x$ is either increasing (if $b > 1$) or decreasing (if $b < 1$). Therefore, in either case the function f has an *inverse function* (Section 26).

DEFINITION. The inverse of the exponential function $f(x) = b^x$ ($b \neq 1$) is called the **logarithmic function** with **base** b. If we denote this inverse function by g, then

$$y = f(x) \quad \text{iff} \quad g(y) = x. \tag{35.8}$$

The output of this inverse function g for any input element y is denoted by $\log_b y$ (which is read "log to the base b of y"). With this notation, Relation (35.8) becomes

$$\boxed{y = b^x \quad \text{iff} \quad \log_b y = x.} \tag{35.9}$$

EXAMPLE 35.4

(a) $\log_2 8 = 3$ because $2^3 = 8$.
(b) $\log_5 25 = 2$ because $5^2 = 25$
(c) $\log_{10} 10000 = 4$ because $10^4 = 10000$.
(d) $\log_5 1 = 0$ because $5^0 = 1$.
(e) $\log_3(\frac{1}{9}) = -2$ because $3^{-2} = \frac{1}{9}$.
(f) $\log_4 8 = \frac{3}{2}$ because $4^{3/2} = \sqrt{4^3} = 8$.
(g) $\log_{0.2} 25 = -2$ because $(0.2)^{-2} = (\frac{1}{5})^{-2} = 5^2 = 25$.

Relation (35.9) allows us to write many equations in either an exponential form or a logarithmic form. Here are some illustrations.

EXAMPLE 35.5 Solve each equation for x by using the logarithmic form.

(a) $6^x = 3$ **(b)** $b = a^{2x}$ **(c)** $\sqrt{2} = 3 \cdot 5^x$

Solution
(a) The logarithmic form gives the solution:

$$x = \log_6 3.$$

(b) The logarithmic form of $b = a^{2x}$ is $\log_a b = 2x$. Therefore, $x = \frac{1}{2}\log_a b$.
(c) To use Relation (35.9) we first write the given equation as $\sqrt{2}/3 = 5^x$. This gives $x = \log_5(\sqrt{2}/3)$.

EXAMPLE 35.6 Solve each equation for x by using the exponential form.

(a) $\log_2 x = \sqrt{2}$ **(b)** $u = \log_b(x/2)$ **(c)** $4 \log_3 x = 5$

Solution
(a) The exponential form gives the solution:

$$x = 2^{\sqrt{2}}.$$

(b) The exponential form of the equation is $x/2 = b^u$. Therefore, $x = 2b^u$.
(c) To use Relation (35.9) we first rewrite the given equation as $\log_3 x = \frac{5}{4}$. This gives

$$x = 3^{5/4} = 3^{1.25}.$$

We can describe the graph of any logarithmic function by using the information about the graphs of exponential functions from Subsection B along with what we know about the graphs of inverse functions from Section 26. Specifically, we know that if g is the inverse of a function f, then the graph of g can be obtained from the graph of f by reflection through the line $y = x$. Figure 35.8 shows the result for a typical case with $b > 1$.

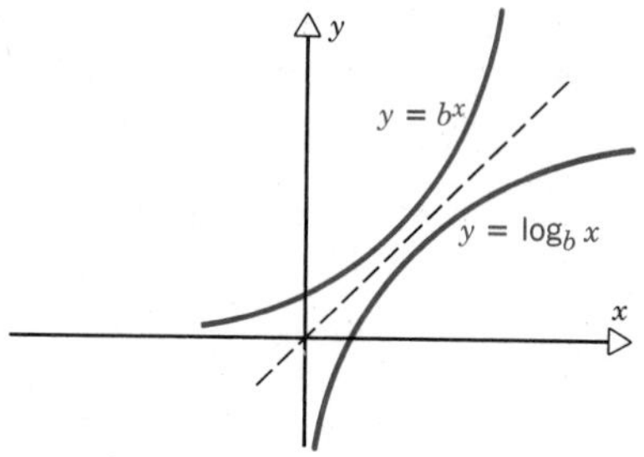

Figure 35.8

EXAMPLE 35.7 Draw the graphs of $y = 4^x$ and $y = \log_4 x$ on the same Cartesian plane.

Solution Figure 35.9 shows the solution. The graph of $y = 4^x$ has been taken from Figure 35.4. The graph of $y = \log_4 x$ results from the graph of $y = 4^x$ by reflection through the line $y = x$.

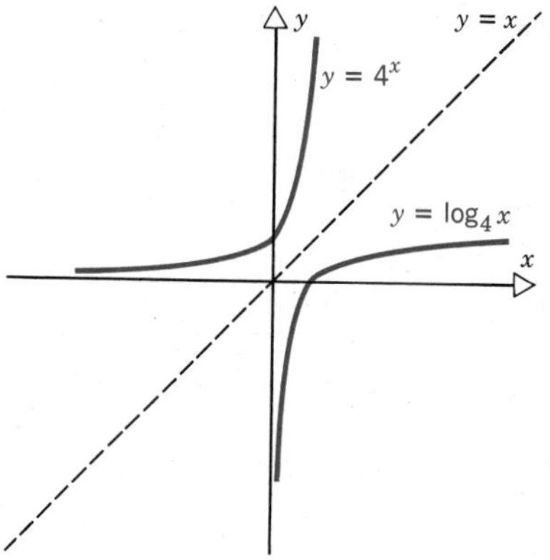

Figure 35.9

Other properties of logarithmic functions will be considered in Section 36.

EXERCISES FOR SECTION 35

For each function in Exercises 1–6, compute decimal values or approximations of $f(x)$ for $x = 0, \pm\frac{1}{2}, \pm 1, \pm\frac{3}{2}, \pm 2, \pm\frac{5}{2}$, and ± 3. (Compare Tables 35.1 and 35.2.) Then draw the graph of each function.

1. $f(x) = 5^x$ (Use $\sqrt{5} \approx 2.24$.)

2. $f(x) = 2.5^x$ (Use $\sqrt{2.5} \approx 1.58$.)
3. $f(x) = 3.5^x$ (Use $\sqrt{3.5} \approx 1.87$.)
4. $f(x) = (\frac{1}{5})^x$ (Use $\sqrt{1/5} \approx 0.45$.)
5. $f(x) = 0.4^x$ (Use $\sqrt{0.4} \approx 0.63$.)
6. $f(x) = (\frac{2}{7})^x$ (Use $\sqrt{2/7} \approx 0.53$.)

Determine each of the following logarithms.

7. $\log_2 2$ **8.** $\log_3 9$ **9.** $\log_4 4$ **10.** $\log_2 16$ **11.** $\log_3 3$
12. $\log_5 125$ **13.** $\log_5(\frac{1}{25})$ **14.** $\log_7(\frac{1}{49})$ **15.** $\log_8 2$ **16.** $\log_{100} 10$
17. $\log_6 1$ **18.** $\log_3(\frac{1}{27})$ **19.** $\log_{10} 1$ **20.** $\log_{0.1} 100$ **21.** $\log_{0.5} 16$
22. $\log_9 27$ **23.** $\log_{25} 125$ **24.** $\log_4 1$ **25.** $\log_{0.2} 125$ **26.** $\log_{25} 5$
27. $\log_{16} 8$

Solve each equation for x by using the logarithmic form.

28. $2^x = 3$ **29.** $u^x = v$ **30.** $5^x = 10$ **31.** $4^{3x} = 2$ **32.** $3 = \pi^{2x}$
33. $y^{4x} = z$ **34.** $a = b \cdot c^x$ **35.** $2 \cdot 5^x = 4$ **36.** $6 = 2 \cdot \pi^x$

Solve each equation for x by using the exponential form.

37. $\log_3 x = \pi$ **38.** $\log_5 x = -2$ **39.** $\log_{10} x = \sqrt{2}$ **40.** $y = \log_4 3x$
41. $2y = \log_3 2x$ **42.** $v = \log_b cx$ **43.** $5 \log_a x = 20$ **44.** $4 = 3 \log_b x$
45. $y = 3 \log_2 x$

46. Draw the graphs of $y = 2^x$ and $y = \log_2 x$ on the same Cartesian plane.
47. Draw the graphs of $y = (\frac{1}{3})^x$ and $y = \log_{1/3} x$ on the same Cartesian plane.
48. Draw the graphs of $y = 5^x$ and $y = \log_5 x$ on the same Cartesian plane.
49. Draw the graph of $y = 2^{|x|}$.
50. Draw the graph of $y = \log_4 |x|$, for x both positive and negative.
51. Construct a coordinate system with each unit on the y-axis one-fifth as long as each unit on the x-axis. Use this coordinate system to draw the graphs of both $y = 2^x$ and $y = x^2$ for $0 \leq x \leq 6$. For which values of x in this interval is $2^x = x^2$?
52. Show that, for all x and for all $b > 0$,

$$(b^x + b^{-x})^2 - (b^x - b^{-x})^2 = 4.$$

53. Prove that if both a and b are positive and different from 1, then $\log_a(1/b) = \log_{1/a}(b)$.
54. If $f(x) = a^x$, then Equation (35.3) can be written as $f(r)f(s) = f(r + s)$. Rewrite Equations (35.4) and (35.7) using the same idea.

SECTION

36 Properties of Logarithmic Functions

A. Basic Properties

Laws of logarithms. Each of the following equations is valid for all real numbers b, x, y, and r for which both sides are defined. Thus the only restrictions are $b > 0$, $b \neq 1$, $x > 0$, and $y > 0$.

$$\log_b b = 1 \tag{36.1}$$

$$\log_b 1 = 0 \tag{36.2}$$

$$\log_b xy = \log_b x + \log_b y \tag{36.3}$$

$$\log_b \left(\frac{x}{y}\right) = \log_b x - \log_b y \tag{36.4}$$

$$\log_b x^r = r \cdot \log_b x \tag{36.5}$$

Proof. Each law can be proved by using an appropriate law of exponents together with the defining condition

$$\log_b x = y \quad \text{iff} \quad b^y = x. \tag{36.6}$$

Following are proofs of Laws (36.1), (36.2), and (36.3). The proofs of (36.4) and (36.5) will be left as exercises. In (36.5) $\log_b x^r$ means $\log_b (x^r)$, *not* $(\log_b x)^r$.

Law (36.1) is simply the logarithmic form of $b^1 = b$. Law (36.2) is the logarithmic form of $b^0 = 1$.

To prove Law (36.3), let

$$u = \log_b x \quad \text{and} \quad v = \log_b y.$$

Then

$$b^u = x \quad \text{and} \quad b^v = y.$$

Therefore, by a law of exponents [Equation (35.3)],

$$xy = b^u b^v = b^{u+v}.$$

In logarithmic form, this is

$$\log_b xy = u + v = \log_b x + \log_b y,$$

as required. □

The next two laws merely restate that the logarithmic function with base b is the inverse of the exponential function with base b. (See the proof that follows the laws.)

$$\log_b b^x = x \qquad (36.7)$$
$$b^{\log_b x} = x \qquad (36.8)$$

Proof. Let $f(x) = b^x$ and $g(x) = \log_b x$. Then

$$g(f(x)) = x,$$

so that

$$g(b^x) = x \quad \text{and} \quad \log_b b^x = x,$$

which is Equation (36.7). Also,

$$f(g(x)) = x,$$

so that

$$f(\log_b x) = x \quad \text{and} \quad b^{\log_b x} = x,$$

which is Equation (36.8) □

B. Examples

EXAMPLE 36.1 Express each logarithm in terms of $\log_b x$, $\log_b y$, and $\log_b z$.

(a) $\log_b(1/x)$ **(b)** $\log_b \sqrt{x}$
(c) $\log_b(x^2y/z)$ **(d)** $\log_b \sqrt[3]{yz^4/x}$

Solution

(a) $\log_b(1/x) = \log_b x^{-1}$
$= -\log_b x$ by (36.5) with $r = -1$

(b) $\log_b \sqrt{x} = \log_b x^{1/2}$
$= \frac{1}{2}\log_b x$ by (36.5) with $r = \frac{1}{2}$

(c) $\log_b(x^2y/z) = \log_b x^2y - \log_b z$ by (36.4)
$= \log_b x^2 + \log_b y - \log_b z$ by (36.3)
$= 2 \cdot \log_b x + \log_b y - \log_b z$ by (36.5)

(d) $\log_b \sqrt[3]{yz^4/x} = \log_b(yz^4/x)^{1/3}$
$= \frac{1}{3}\log_b(yz^4/x)$ by (36.5) with $r = \frac{1}{3}$
$= \frac{1}{3}(\log_b y + 4 \cdot \log_b z - \log_b x)$ by (36.3), (36.4), and (36.5)

EXAMPLE 36.2 Rewrite each expression as a single logarithm.

(a) $\log_b u - \log_b v$ **(b)** $\log_b u + 3 \cdot \log_b v$
(c) $\frac{1}{2}(\log_b u^2 + 6 \cdot \log_b v - 4 \cdot \log_b u)$ **(d)** $\log_b 4 + \log_b 3 - \log_b 6$

Solution

(a) $\log_b u - \log_b v = \log_b(u/v)$ by (36.4)

(b) $\log_b u + 3 \cdot \log_b v = \log_b u + \log_b v^3$ by (36.5)
$= \log_b uv^3$ by (36.3)

(c) $\frac{1}{2}(\log_b u^2 + 6 \cdot \log_b v - 4 \cdot \log_b u)$
$= \frac{1}{2}(\log_b u^2 + \log_b v^6 - \log_b u^4)$ by (36.5)

$= \frac{1}{2} \log_b(u^2v^6/u^4)$ by (36.3) and (36.4)
$= \frac{1}{2} \log_b(v^6/u^2)$ by a law of exponents
$= \log_b(v^6/u^2)^{1/2}$ by (36.5)
$= \log_b(v^3/u)$ by laws of exponents

(d) $\log_b 4 + \log_b 3 - \log_b 6 = \log_b(4 \cdot \frac{3}{6})$ by (36.3) and (36.4)
$= \log_b 2$

EXAMPLE 36.3 Determine each logarithm that follows, given that $\log_7 3 = 0.5646$ and $\log_7 5 = 0.8271$.

(a) $\log_7 15$ **(b)** $\log_7 25$ **(c)** $\log_7 \sqrt{3}$ **(d)** $\log_7 21$

Solution
(a) $\log_7 15 = \log_7 3 \cdot 5 = \log_7 3 + \log_7 5 = 0.5646 + 0.8271 = 1.3917$
(b) $\log_7 25 = \log_7 5^2 = 2 \log_7 5 = 2(0.8271) = 1.6542$
(c) $\log_7 \sqrt{3} = \log_7 3^{1/2} = \frac{1}{2} \log_7 3 = \frac{1}{2}(0.5646) = 0.2823$
(d) $\log_7 21 = \log_7 3 \cdot 7 = \log_7 3 + \log_7 7 = 0.5646 + 1 = 1.5646$

EXAMPLE 36.4 Simplify.

(a) $\log_2 2^5 = 5$ by (36.7)
(b) $\log_{10} 10^{-\pi} = -\pi$ by (36.7)
(c) $3^{\log_3 7} = 7$ by (36.8)
(d) $10^{\log_{10} \sqrt{2}} = \sqrt{2}$ by (36.8)

C. Change of Base and Other Properties

The next two equations allow us to change from one logarithmic base to another.

$$\log_a c = (\log_a b)(\log_b c) \tag{36.9}$$

$$\log_a b = \frac{1}{\log_b a} \tag{36.10}$$

To verify Equation (36.9), let

$$u = \log_a b \quad \text{and} \quad v = \log_b c.$$

Then

$$a^u = b \quad \text{and} \quad b^v = c.$$

Therefore,

$$c = b^v = (a^u)^v = a^{uv},$$

so that

$$\log_a c = uv = (\log_a b)(\log_b c).$$

Equation (36.10) is a special case of (36.9); the proof is left as an exercise.

Equation (36.9) shows that there is a close relationship between any two logarithmic functions. If

$$f(x) = \log_a x \quad \text{and} \quad g(x) = \log_b x,$$

for example, then by Equation (36.9) (with x in place of c)

$$f(x) = (\log_a b)g(x) \quad \text{for all } x$$

or

$$\boxed{\log_a x = (\log_a b) \log_b x \quad \text{for all } x.} \tag{36.11}$$

The number $\log_a b$ is a constant that depends only on a and b.

EXAMPLE 36.5 Determine a constant k such that

$$\log_2 x = k \cdot \log_8 x \quad \text{for all } x.$$

Solution By Equation (36.11) the proper value of k is $\log_2 8 = 3$. That is,

$$\log_2 x = 3 \cdot \log_8 x \quad \text{for all } x.$$

The graphs of the logarithmic functions for different values of b will furnish convincing evidence of the following properties. For the case $b > 1$, see Figure 35.8.

> Let $f(x) = \log_b x$ be a logarithmic function. The *domain* of f is the set of all positive real numbers. The *range* of f is the set of all real numbers. If $b > 1$, then
>
> f is increasing,
> $f(x) \to \infty$ as $x \to \infty$, and
> $f(x) \to -\infty$ as $x \to 0$.
>
> If $0 < b < 1$, then
>
> f is decreasing,
> $f(x) \to -\infty$ as $x \to \infty$, and
> $f(x) \to \infty$ as $x \to 0$.

EXERCISES FOR SECTION 36

Express each logarithm in terms of $\log_b x$, $\log_b y$, and $\log_b z$.

1. $\log_b(xy/z)$
2. $\log_b(x/yz)$
3. $\log_b(1/xy)$
4. $\log_b \sqrt[3]{xy}$
5. $\log_b x\sqrt{z}$
6. $\log_b \sqrt{xyz}$
7. $\log_b (\sqrt{x/yz})^3$
8. $\log_b \sqrt{xy}/z$
9. $\log_b \sqrt[3]{x}\,\sqrt[4]{y}\,\sqrt[5]{z}$ mult rule ③
10. $\log_b \sqrt[4]{x^2y^4}/\sqrt{x}$
11. $\log_b \sqrt{x/(yz)^3}$
12. $\log_b x\sqrt[3]{1/(x^6y)}$

Rewrite each expression as a single logarithm and simplify.

13. $\log_b u^2 + \log_b v^2$
14. $\log_b v - 2 \cdot \log_b u$
15. $2 \cdot \log_b u - \log_b u^2 v$
16. $\frac{1}{2}(\log_b u^4 - \log_b v^2)$
17. $\frac{1}{3}(9 \cdot \log_b v + 3 \cdot \log_b u)$
18. $\frac{1}{2}(\log_b u + 2 \cdot \log_b v + 3 \cdot \log_b u)$
19. $\log_b 5 + \log_b 12 - \log_b 15$
20. $\log_b 1 + \log_b 6 - \log_b 12$
21. $\log_b 0.2 + \log_b 5$
22. $2 \cdot \log_b b - \log_b b^3 - \log_b(1/b)$
23. $\log_b \sqrt{b} + 3 \cdot \log_b \sqrt{b}$
24. $\log_b \sqrt[3]{b} + \frac{1}{3}\log_b b^2$

Determine the logarithms in Exercises 25–33 given that $\log_7 2 = 0.3562$, $\log_7 3 = 0.5646$, and $\log_7 5 = 0.8271$.

25. $\log_7 6$
26. $\log_7 10$
27. $\log_7 9$
28. $\log_7 \sqrt[3]{2}$
29. $\log_7 \sqrt{2}$
30. $\log_7 \sqrt[3]{5}$
31. $\log_7 \frac{1}{8}$
32. $\log_7 \frac{1}{50}$
33. $\log_7 \frac{5}{6}$

Simplify.

34. $\log_5 5^{-2}$
35. $2^{\log_2 3}$
36. $\log_{10} 10^{12}$
37. $7^{\log_7 \pi}$
38. $\log_3 3^{-0.5}$
39. $10^{\log_{10} 1.73}$
40. $a^{-\log_a \pi}$
41. $\log_a a^{-x}$
42. $a^{2 \log_a x}$
43. $\log_a x + \log_a(1/x)$
44. $b^{(\log_b a)(\log_a x)}$
45. $\log_a(x + \sqrt{x^2 - 1}) + \log_a(x - \sqrt{x^2 - 1})$

46. Determine a constant k such that $\log_{100} x = k \cdot \log_{10} x$ for all x.
47. Determine a constant k such that $\log_{0.1} x = k \cdot \log_{10} x$ for all x.
48. Determine a constant k such that $\log_8 x = k \cdot \log_{16} x$ for all x.
49. Prove Law (36.4).
50. Prove Law (36.5).
51. Use Equation (36.9) to verify that Equation (36.10) is true.
52. Assume that $0 < b < 1$ and that $f(x) = b^x$ has a graph as shown in Figure 35.7. Draw the graph of $y = \log_b x$.
53. Assume that $0 < b < 1$.
(a) For which values of x is $\log_b x > 0$?
(b) For which values of x is $\log_b x < 0$?
54. Assume that $b > 1$.
(a) For which values of x is $\log_b x > 0$?
(b) For which values of x is $\log_b x < 0$?

SECTION

37 Common Logarithms. Exponential Equations

A. Definition and Examples

Although any positive number except 1 can be used as the base for a logarithmic function, two particular bases are of paramount importance. One is the number 10, which will be used in this section. The other is the number e, which will be defined and used in the section that follows. Base 10 logarithms are important because our number system is based on 10. The importance of base e logarithms lies deeper; we'll see, for example, that the number e has the same kind of intrinsic significance in mathematics as the number π.

A logarithm with base 10 is called a **common logarithm.** These logarithms will be written with the base omitted. That is:

> In this book $\log x$ will mean $\log_{10} x$.

(This is a widely used convention. On the other hand, in higher mathematics the notation $\log x$ is often reserved for $\log_e x$.) From the definition of logarithm, it follows that

$$\log x = y \quad \text{iff} \quad 10^y = x.$$

Therefore, $\log x$ is the power to which 10 must be raised to produce x.

EXAMPLE 37.1

(a) $\log 10 = 1$ because $10^1 = 10$.
(b) $\log 100{,}000 = 5$ because $10^5 = 100{,}000$.
(c) $\log 0.001 = -3$ because $10^{-3} = 0.001$.
(d) $\log 10^n = n$ because $10^n = 10^n$.

If x is not a readily recognizable power of 10, then more advanced ideas are needed to compute $\log x$. Fortunately, these ideas have been programmed into calculators and used to construct logarithm tables. Table 37.1 is a condensed version of Table II at the back of the book. The values in these tables are not exact, but have been rounded off to four significant figures. Although we should write $\log 2.5 \approx 0.3979$, for example, we follow the more customary practice of writing $\log 2.5 = 0.3979$. Just remember that most of our answers based on these tables will be approximate.

TABLE 37.1 Common Logarithms to Four Significant Figures (Also see Table II)

x	$\log x$	x	$\log x$
1.0	0.0000	5.5	0.7404
1.5	0.1761	6.0	0.7782
2.0	0.3010	6.5	0.8129
2.5	0.3979	7.0	0.8451
3.0	0.4771	7.5	0.8751
3.5	0.5441	8.0	0.9031
4.0	0.6021	8.5	0.9294
4.5	0.6532	9.0	0.9542
5.0	0.6990	9.5	0.9777

With the equation $\log 10^n = n$ [Example 37.1(d)] and the property $\log xy = \log x + \log y$, we can determine the common logarithm of any positive number if we know the common logarithms of the numbers between 1 and 10. Here's how: First write x in scientific notation (Section 3C),

$$x = a \times 10^n, \quad \text{with} \quad 1 \leq a < 10.$$

Then Law (36.3) implies that

$$\log x = \log a + \log 10^n,$$

so that

$$\log x = n + \log a.$$

That is:

If $x = a \times 10^n$, with $1 \leq a < 10$, then

$$\log x = n + \log a. \tag{37.1}$$

The number n in Equation (37.1) is called the **characteristic** of $\log x$, and $\log a$ is called the **mantissa.** By the definition of scientific notation, n is the number of places the decimal point in x must be moved to bring it just to the right of the leftmost nonzero digit, counting movement to the left as positive and to the right as negative. For example, $4735 = 4.735 \times 10^3$, so the characteristic of $\log 4735$ is 3.

EXAMPLE 37.2 Use Table 37.1 to determine each logarithm.

(a) $\log 450$ **(b)** $\log 0.0007$

Solution

(a) Because $450 = 4.5 \times 10^2$, the characteristic is 2. From Table 37.1, the mantissa is 0.6532. Therefore,

$$\log 450 = 2 + \log 4.5 = 2 + 0.6532 = 2.6532.$$

(b) The characteristic of log 0.0007 is -4 and (from Table 37.1) the mantissa is 0.8451. Therefore,

$$\log 0.0007 = -4 + \log 7 = -4 + 0.8451 = -3.1549.$$

Table II in the back of the book gives a more extensive list of common logarithms than Table 37.1. In Section 39 we'll see what to do about mantissas that are not included in Table II; until then Table II will suffice.

EXAMPLE 37.3 Use Table II to determine each logarithm.

(a) log 57100 **(b)** log 0.0369

Solution

(a) The characteristic is 4 and (from Table II) the mantissa is 0.7566. Therefore,

$$\log 57100 = 4.7566.$$

(b) The characteristic is -2 and (from Table II) the mantissa is 0.5670. Therefore,

$$\log 0.0369 = -2 + 0.5670 = -1.4330.$$

To determine a logarithm on a calculator with a [log] key, the characteristic and mantissa do not have to be treated separately. Simply enter the given number (such as 0.0369), press the [log] key, and then read the logarithm in the display (for log 0.0369 a calculator with eight-digit display will show -1.4329736). [C]

B. Exponential Equations

An **exponential equation** is an equation with an unknown variable in an exponent, such as $2^x = 5$. Many exponential equations can be solved using logarithms and the property

$$\log b^y = y \log b. \tag{37.2}$$

[This is Law (36.5) with a change of letters.]

EXAMPLE 37.4 Solve the equation $2^x = 5$.

Solution If $2^x = 5$, then $\log 2^x = \log 5$. Therefore, since $\log 2^x = x \log 2$,

$$x \log 2 = \log 5$$
$$x = \frac{\log 5}{\log 2}.$$

For an approximate solution we can use the values for log 5 and log 2 from Table 37.1:

$$x = \frac{\log 5}{\log 2} = \frac{0.6990}{0.3010} \approx 2.32.$$ [C]

The main idea in the next example is the same as in Example 37.4: take the logarithm of both sides of the equation, and then use Law (37.2) to remove x as an exponent and obtain a linear equation in x. Other properties of logarithms are then used to reduce the equation to the form $ax = b$.

EXAMPLE 37.5 Solve the equation

$$5^{2x-1} = 3^{1-x}.$$

Solution If $5^{2x-1} = 3^{1-x}$, then

$$\begin{aligned}
\log 5^{2x-1} &= \log 3^{1-x} \\
(2x-1)\log 5 &= (1-x)\log 3 \qquad \text{by (37.2)} \\
2x \cdot \log 5 - \log 5 &= \log 3 - x \cdot \log 3 \\
(2 \cdot \log 5 + \log 3)x &= \log 3 + \log 5 \\
(\log 5^2 + \log 3)x &= \log (3 \cdot 5) \qquad \text{by (37.2) and (36.3)} \\
x \cdot \log (5^2 \cdot 3) &= \log 15 \qquad \text{by (36.3)} \\
x &= \frac{\log 15}{\log 75}.
\end{aligned}$$

The solution (log 15)/(log 75) is exact. For an approximate solution use Table 37.1 or Table II:

$$x = \frac{\log 15}{\log 75} = \frac{1.1761}{1.8751} \approx 0.627.$$ C

C. Application: Compound Interest

In Section 18C we derived the following equation for the compound amount S from a principal amount P invested at an annual rate of $r\%$ compounded n times per year for t years:

$$S = P\left(1 + \frac{r}{n}\right)^{nt}. \tag{37.3}$$

EXAMPLE 37.6 Solve Equation (37.3) for t.

Solution

$$\begin{aligned}
\frac{S}{P} &= \left(1 + \frac{r}{n}\right)^{nt} \\
\log\left(\frac{S}{P}\right) &= \log\left(1 + \frac{r}{n}\right)^{nt} \\
\log\left(\frac{S}{P}\right) &= nt \log\left(1 + \frac{r}{n}\right)
\end{aligned}$$

$$t = \frac{\log\left(\dfrac{S}{P}\right)}{n \log\left(1 + \dfrac{r}{n}\right)} \tag{37.4}$$

EXAMPLE 37.7 Determine the minimum number of whole years that \$1000 must be invested at 6% compounded annually so that the compound amount will be \$2000.

Solution Use $S = \$2000$, $P = \$1000$, $r = 0.06$, and $n = 1$ in Equation (37.4):

$$t = \frac{\log(2000/1000)}{\log 1.06} = \frac{\log 2}{\log 1.06} = \frac{0.3010}{0.0253} \approx 11.9. \qquad \boxed{C}$$

Therefore, the minimum number of whole years is 12.

Notice that in Equation (37.4) only the *ratio* of S to P matters, not the specific values of S and P. For instance, the answer to Example 37.7 would be the same for any amounts S and P such that $S/P = 2$. In other words, Example 37.7 shows that it takes money approximately 12 years to double at 6% compounded annually. This is a special case of the following rule of thumb.

> **Rule of 72.** If a quantity is increasing at a constant rate of p% per year, then the quantity will double in approximately $72/p$ years.

[Here p has been used in place of r to denote rate in order to stress that p is not to be changed to decimal form, in contrast to r in Equations (37.3) and (37.4).]

Although the value $72/p$ is only approximate, it is reasonably close for the rates that arise in most interest rate problems. For example, if $5 \leq p \leq 12$, the Rule of 72 will give the doubling time with at most a 2% error. Thus money invested at 12% compounded annually will double in approximately $72/12 = 6$ years, and 6 years is off by at most 2%, that is, by at most $0.02 \times 6 = 0.12$ years.

For the rates of increase that occur in most population problems (which we'll consider shortly) a closer approximation is given by using 70 in place of 72; the result is called the **Rule of 70.** The Rule of 70 will give the doubling time with at most a 2% error if $0.1 \leq p \leq 5$.

In Example 37.6 we proved that Equation (37.4) is correct. In contrast, the Rule of 72 will not be proved; however, in the next example and the exercises the reliability of the rule will be verified in special cases.

EXAMPLE 37.8 Suppose that money is invested in an account that pays 9% interest compounded annually.

(a) Compute the time required for doubling, using Equation (37.4).
(b) Compute the time required for doubling, using the Rule of 72.
(c) Compute the percentage error from using the Rule of 72.

Solution

(a) Use Equation (37.4) with $S/P = 2$, $r = 0.09$, and $n = 1$:

$$t = \frac{\log 2}{\log 1.09} = \frac{0.3010}{0.0374} \approx 8.04 \text{ years}. \qquad \boxed{C}$$

(b) The Rule of 72 gives doubling in $72/9 = 8$ years.

(c) The *percentage error* is $100 \cdot \left|\frac{\text{error}}{\text{correct value}}\right|$. In this case the correct value is 8.04 [from part (a)], and the error is $8.04 - 8 = 0.04$. Thus the percentage error is

$$100 \cdot \left|\frac{0.04}{8.04}\right| \approx 0.50\%, \qquad \boxed{\text{C}}$$

or approximately one-half of one percent.

D. Application: Population Growth

Compound interest on money is just one example of **exponential growth** or **decay,** which occurs whenever a quantity is increasing or decreasing at a constant percentage rate. The basic ideas and formulas are similar in all examples of exponential growth or decay—the notation and the specific numbers and units are all that change from one example to another. We illustrate this now with population growth.

Suppose that P represents the population of a country at a certain time, and that the population is increasing at a constant rate of $r\%$ per period of time (such as a year). Then the population at the end of the first period, which we denote by P_1, will be

$$P_1 = P + Pr = P(1 + r).$$

The population at the end of the second period will be

$$\begin{aligned} P_2 &= P_1 + P_1 r = P(1 + r) + P(1 + r)r \\ &= P(1 + r)(1 + r) = P(1 + r)^2. \end{aligned}$$

If you compare this with the derivation of Equation (18.9) [which is the same as Equation (37.3)], you'll see that after t periods the population will be

$$\boxed{P_t = P(1 + r)^t.} \tag{37.5}$$

EXAMPLE 37.9 The 1970 population of America (both continents and the adjacent areas) was 509 million. Between 1970 and 1976, the population increased at an average rate of 2.0% per year. *If* this trend remained constant, what would the population be in the year 2000?

Solution Use Equation (37.5) with $P = 509$, $r = 0.02$, and $t = 30$:

$$P_{30} = 509(1.02)^{30} \approx 922. \qquad \boxed{\text{C}}$$

That is, in the year 2000 the population would be approximately 922 million.

EXAMPLE 37.10 The population of the U.S. was 76.0 million in 1900 and 92.0 million in 1910. If the population had continued to grow at the same percentage rate for each ten-year period, what would the population have been in 1970?

Solution The increase from 1900 to 1910 was $92.0 - 76.0 = 16.0$ million. Since $16.0/76.0 \approx 0.21$, this means that the increase between 1900 and 1910 was approximately 21%.

Now use Equation (37.5) with $P = 76.0$, $r = 0.21$, and $t = 7$ (each time period is ten years). The answer will be in millions:

$$P_7 = 76.0(1.21)^7 \approx 289 \text{ million.}$$ [C]

In contrast, the actual population in 1970 was only 203 million. (See Figure 37.1, where the projections are based on the percentage increase from 1900 to 1910 that was used in this example.)

The U.S. population increased by at least 25% over each decade from 1800 to 1890, but the increase has been less than 20% for each decade since 1910. This brings out the extremely important point that a past rate of growth will not necessarily continue into the future.

Problems in which t must be determined from Equation (37.5) can be solved just like Examples 37.6 and 37.7.

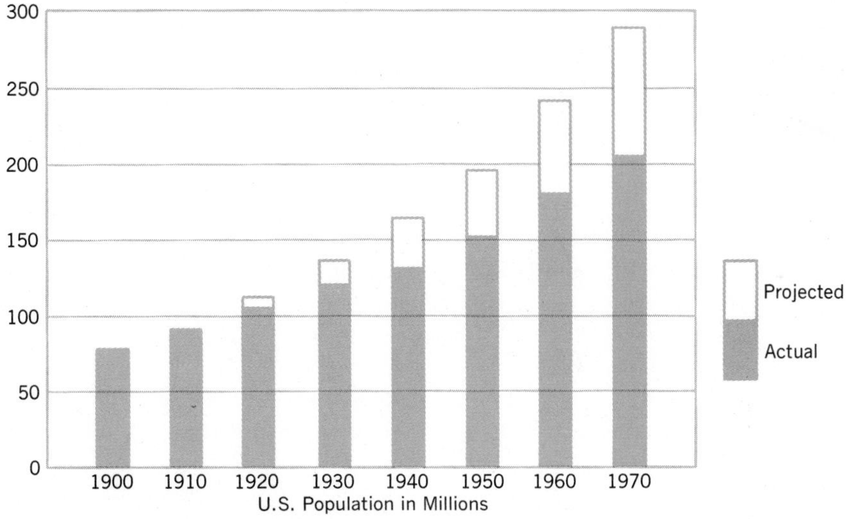

Figure 37.1

EXERCISES FOR SECTION 37

Use Table II to determine the common logarithm of each number.

1. (a) 8.51 (b) 851 (c) 0.0851
2. (a) 6.7 (b) 0.00067 (c) 67
3. (a) 3.4 (b) 3400 (c) 0.34
4. (a) 570 (b) 5700 (c) 0.57
5. (a) 16200 (b) 1.620 (c) 0.0162
6. (a) 0.000222 (b) 222 (c) 22200

Solve each equation for x. Give both an exact answer (involving logarithms) and an approximate decimal answer. (Table 37.1 will provide the logarithms you need.)

7. $3^x = 2$ **8.** $5^x = 7$ **9.** $2^x = 10$
10. $4^{3x-1} = 5^x$ **11.** $9^{x-1} = 2^{x+1}$ **12.** $7^{2x} = 5^{x-3}$
13. $5 = 2 \cdot 3^{2x}$ **14.** $11 = 2 \cdot 5^{-x}$ **15.** $3 = 2 \cdot 4^{5x}$

Determine the minimum number of whole years that the given principal P must be invested at the given rate r compounded annually to produce the given compound amount S.

16. $P = \$1000$, $r = 6\%$, $S = \$1500$ **17.** $P = \$1000$, $r = 8\%$, $S = \$1800$
18. $P = \$2000$, $r = 10\%$, $S = \$4500$ **19.** $P = \$500$, $r = 6.5\%$, $S = \$950$
20. $P = \$1500$, $r = 7\%$, $S = \$2500$ **21.** $P = \$1000$, $r = 5\%$, $S = \$3000$

In each of Exercises 22–27, assume that money is invested in an account that pays the given rate of interest compounded annually. Then do each of the following.

(a) Compute the time required for doubling by using Equation (37.4).
(b) Compute the time required for doubling by using the Rule of 72.
(c) Compute the percentage error from using the Rule of 72.

22. 6% **23.** 8% **24.** 12% **25.** 1% **26.** 24% **27.** 18% C

Exercises 28–48* are based on Table 37.2, which gives population figures for the six major areas of the world used in U.N. statistical summaries. Notice that the rates of increase are *annual* rates. In each computation, take 1970 as year 0 and *assume that the annual rate of increase shown in the table will continue throughout the period being considered.* (In most cases this will probably give a population estimate that is slightly high.) C

TABLE 37.2 Estimates of Mid-year Population in Millions for Major Areas of the World

Area	*1970*	*1976*	*Annual rate of increase 1970–1976*
Africa	352	412	2.7%
America	509	572	2.0%
Asia	2027	2304	2.2%
Europe	459	476	0.6%
Oceania	19.3	21.7	2.0%
USSR	243	258	1.0%
World[a]	3610	4044	1.9%

[a] *The slight discrepancy in the totals is due to rounding.*
SOURCE: *1977 Statistical Yearbook of the United Nations.*

28. Estimate the population of Africa in the year 1990.
29. Estimate the population of America in the year 1990.
30. Estimate the population of Asia in the year 1990.
31. Estimate the population of Europe in the year 2000.

*The powers needed for Exercises 28–33 are given in a separate list at the end of the exercises.

32. Estimate the population of Oceania in the year 2000.
33. Estimate the population of the USSR in the year 2000.
34. When will the population of Africa reach 500 million?
35. When will the population of America reach 700 million?
36. When will the population of Asia reach 3500 million?
37. When will the population of Europe reach 600 million?
38. When will the poulation of Oceania reach 25 million?
39. When will the population of the USSR reach 300 million?
40. By the Rule of 70, in how many years will the population of Africa double?
41. By the Rule of 70, in how many years will the population of America double?
42. By the Rule of 70, in how many years will the population of Asia double?
43. By the Rule of 70, in how many years will the population of Europe double?
44. By the Rule of 70, in how many years will the population of Oceania double?
45. By the Rule of 70, in how many years will the population of the USSR double?
46. Estimate the world population in the year 2000.
47. When will the world population reach 5 billion?
48. (a) In how many years will the world population double according to Equation (37.4)?
(b) In how many years will the world population double according to the Rule of 70?
(c) What do you get if you use the Rule of 72 rather than the Rule of 70 in part (b)?

Here are four-place approximations for the powers needed in Exercises 28–33. (The order is random.)

$$1.022^{20} = 1.5453 \qquad 1.020^{20} = 1.4859$$
$$1.020^{30} = 1.8114 \qquad 1.010^{30} = 1.3478$$
$$1.027^{20} = 1.7038 \qquad 1.006^{30} = 1.1966$$

SECTION

38 Applications Involving e

A. The Number e

The small letter e is used to denote a special irrational number that is approximately equal to 2.71828. This number, like π, arises naturally throughout mathematics and its applications. The following example contains a more precise definition of e.

EXAMPLE 38.1 Suppose that \$1 is invested at 100% interest compounded n times per year for one year. The amount accumulated is given by either Equation (18.9) or Equation (37.3): the interest rate (in decimal form) per conversion period is $1/n$, so the compound amount, in dollars, is

$$S = 1\left(1 + \frac{1}{n}\right)^n = \left(1 + \frac{1}{n}\right)^n.$$

Table 38.1 gives some specific examples. The table shows, for instance, that quarterly compounding would yield \$2.44. Compounding either hourly or each minute or each second would yield \$2.72 (rounded to the nearest cent). (A calculator with a $\boxed{y^x}$ key can be used to verify the entries in the table.)

TABLE 38.1

Compounding interval	n	$[1 + (1/n)]^n$ *(to five decimal places)*
annually	1	2
semiannually	2	2.25
quarterly	4	2.44141
monthly	12	2.61304
daily	365	2.71457
hourly	8,760	2.71813
each minute	525,600	2.71828
each second	31,536,000	2.71828

It can be proved that the values of $[1 + (1/n)]^n$ approach an irrational number as n increases without bound. This number is called the *limit of* $[1 + (1/n)]^n$ *as n approaches infinity,* and it is the number e. Concisely,

$$e = \lim_{n\to\infty}\left(1 + \frac{1}{n}\right)^n. \tag{38.1}$$

To 20 decimal places,

$$e = 2.71828\ 18284\ 59045\ 23536.$$

Besides giving a definition of e, this example illustrates a general point about compound interest: For a fixed principal and rate of interest, an increase in the frequency of compounding will increase the compound amount; but beyond a certain point this increase in compound amount will be negligible. In **continuous compounding** the compound amount is the limit of the compound amounts as $n \to \infty$, where n is the number of compounding periods. Continuous compounding might make an investor feel better, but it won't make him or her significantly richer than daily compounding. The next subsection will give a more general formula for continuous compounding.

Any precise definition of e must involve some kind of limit, as in Equation (38.1). Such limits are studied in more detail in calculus.

B. More about Continuous Compounding

Example 38.1 shows that if \$1 is invested at 100% interest compounded continuously for one year, then the compound amount will be \$$e$ (rounded off). We now generalize this example to derive a formula for the compound amount from continuous compounding for any amount invested at any annual rate for any number of years.

First, from either Equation (18.9) or Equation (37.3), we know that if a principal P is invested at an interest rate r compounded n times per year for t years, then the compound amount is

$$S = P\left(1 + \frac{r}{n}\right)^{nt}. \tag{38.2}$$

We can analyze the last equation more easily if we let $x = n/r$. Then $r/n = 1/x$. Also, $n = xr$ so that $nt = xrt$. Thus Equation (38.2) can be rewritten as

$$S = P\left(1 + \frac{1}{x}\right)^{xrt}$$

$$S = P\left[\left(1 + \frac{1}{x}\right)^{x}\right]^{rt}. \tag{38.3}$$

For continuous compounding we let the number n of conversion periods per year increase without bound. This will not change P, r, or t. However, $x \to \infty$ as $n \to \infty$, and by Equation (38.1) (with x in place of n)

$$e = \lim_{x\to\infty} \left(1 + \frac{1}{x}\right)^{x}.$$

Thus the right-hand side of Equation (38.3) approaches Pe^{rt} as $x \to \infty$. This leads to the following conclusion.

If a principal amount P is invested for t years at an annual rate r compounded continuously, then the compound amount will be

$$S = Pe^{rt}. \tag{38.4}$$

As usual, the rate r must be in decimal form in this equation. The special case of Equation (38.4) with $P = 1$, $t = 1$, and $r = 1$ is Example 38.1.

Powers of e needed to apply Equation (38.4) can be obtained from Table III or from a calculator having an $\boxed{e^x}$ key or a $\boxed{y^x}$ key.

EXAMPLE 38.2 What is the compound amount if \$500 is invested for two years at 12% compounded continuously?

Solution Use Equation (38.4) with $P = \$500$, $r = 0.12$, and $t = 2$. Table III gives $e^{0.24} = 1.2712$. Using this,

$$S = 500 \times 1.2712$$
$$S = \$635.60.$$

In Example 18.6 we solved this problem with quarterly compounding in place of continuous compounding. The compound amount under quarterly compounding was \$633.39.

C. Natural Logarithms

A logarithm with base e is called a **natural logarithm.** We use the widely adopted notation $\ln x$ for these logarithms. That is:

In this book $\ln x$ will mean $\log_e x$.

The next subsection will give an application of natural logarithms, but first let's see how they can be computed.

One way to compute $\ln x$ is to use a table of common logarithms and the equation for changing from one logarithmic base to another. Specifically, use $a = 10$ and $b = e$ in

$$\log_b x = \frac{\log_a x}{\log_a b}$$

[from Equation (36.11)]. The result is

$$\log_e x = \frac{\log_{10} x}{\log_{10} e}$$

or

$$\ln x = \frac{\log x}{\log e}.$$

It can be shown that $\log e = 0.4343$ and $1/\log e = 2.3026$ (to four decimal places). Therefore:

$$\ln x = 2.3026 \log x. \tag{38.5}$$

EXAMPLE 38.3 Use Table II and Equation (38.5) to compute ln 5130.

Solution Using Table II, we find that $\log 5130 = 3.7101$. Therefore, by Equation (38.5),

$$\begin{aligned}\ln 5130 &= 2.3026 \log 5130\\ &= 2.3026 \times 3.7101\\ &= 8.5429.\end{aligned}$$

[C]

A second way to compute $\ln x$ is to use Table IV, which gives the natural logarithms of numbers between 1 and 10. First, write x in scientific notation,

$$x = a \times 10^n, \quad 1 \leq a < 10.$$

Then

$$\begin{aligned}\ln x &= \ln a + \ln 10^n\\ \ln x &= n \ln 10 + \ln a.\end{aligned}$$

Since

$$\log_e 10 = \frac{\log_{10} 10}{\log_{10} e} = \frac{1}{\log_{10} e} = 2.3026,$$

we have the following conclusion.

If $x = a \times 10^n$, with $1 \leq a < 10$, then

$$\ln x = 2.3026n + \ln a. \tag{38.6}$$

Notice that this is similar to Equation (37.1), which states that $\log x = n + \log a$, except that in the present case $2.3026n$ appears in place of n.

EXAMPLE 38.4 Use Table IV and Equation (38.6) to compute ln 0.064.

Solution First, write $0.064 = 6.4 \times 10^{-2}$. From Table IV, $\ln 6.4 = 1.8563$. Therefore,

$$\begin{aligned}\ln 0.064 &= 2.3026(-2) + 1.8563\\ \ln 0.064 &= -2.7489.\end{aligned}$$

To determine a natural logarithm on a calculator with an [ln] key, simply enter the given number (such as 0.064), press the [ln] key, and then read the logarithm in the display. For ln 0.064 a calculator with eight-digit display will show -2.7488722.

[C]

D. Continuous Growth and Decay

In Section 37 we used population growth to illustrate that the basic ideas and formulas for compound interest carry over to any example of exponential growth or decay, that is, to any example where a quantity is increasing or decreasing at a constant percentage rate. In the same way, the ideas and formulas for *continuous* compounding carry over to other examples; this produces what is called **continuous growth** or **decay.** The basic formulas for continuous growth and decay are

$$Q = Q_0e^{kt} \quad \text{(growth)} \tag{38.7}$$

and

$$Q = Q_0e^{-kt} \quad \text{(decay).} \tag{38.8}$$

In these equations Q represents a quantity that changes with time, k is a positive constant that depends on the particular problem, t denotes time expressed in the appropriate units, and Q_0 is the **initial value** of Q, that is, the value of Q when $t = 0$.

EXAMPLE 38.5 If a principal amount P of money is invested for t years at an annual rate r compounded continuously, then the compound amount will be

$$S = Pe^{rt}.$$

This equation is just Equation (38.4). It is the special case of Equation (38.7) with S in place of Q, r in place of k, and P in place of Q_0.

The decay of any radioactive substance is described by Equation (38.8) with the constant k depending on the particular substance. An especially useful number connected with a radioactive substance is the **half-life** of the substance, which is the time required for an amount of the substance to be reduced by half. The following example will show how the half-life of a substance is related to the constant k in Equation (38.8).

EXAMPLE 38.6 Suppose that the amount Q of a radioactive substance remaining after t units of time is determined by

$$Q = Q_0e^{-kt}.$$

Let T denote the half-life of the substance. Then $Q = \frac{1}{2}Q_0$ when $t = T$. Therefore,

$$\begin{aligned} \tfrac{1}{2}Q_0 &= Q_0e^{-kT} \\ \tfrac{1}{2} &= e^{-kT}. \end{aligned}$$

The logarithmic form of the last equation is

$$\ln \tfrac{1}{2} = -kT.$$

Since $\ln \frac{1}{2} = \ln 2^{-1} = -\ln 2$, we see that $-\ln 2 = -kT$, so that

$$T = \frac{\ln 2}{k}. \tag{38.9}$$

Also, $k = (\ln 2)/T$, so that the equation describing radioactive decay can be written as

$$Q = Q_0 e^{-[(\ln 2)/T]t}, \tag{38.10}$$

where T denotes the half-life of the substance.

EXAMPLE 38.7 The half-life of carbon-14 is 5730 years. Answer each question based on an initial amount of 1 gram of the substance.

(a) How much will remain after two half-lives?
(b) How much will remain after three half-lives?
(c) How much will remain after 1000 years?
(d) When will 0.9 grams remain?

Solution

(a) After one half-life $\frac{1}{2}$ gram will remain. After another half-life $\frac{1}{2}$ of $\frac{1}{2}$, or $\frac{1}{4}$ gram, will remain.
(b) Continuing as in part (a), we see that after three half-lives $\frac{1}{2}$ of $\frac{1}{4}$, or $\frac{1}{8}$ gram, will remain. (Notice that three half-lives is more than 17 thousand years.)
(c) The amount Q in this example is given by Equation (38.10) with $T = 5730$ years and $Q_0 = 1$ gram:

$$Q = e^{-[(\ln 2)/5730]t}. \tag{38.11}$$

From Table IV, $\ln 2 = 0.6931$, and we are asked to find Q when $t = 1000$. The result is

$$\begin{aligned} Q &= e^{-(0.6931/5730)1000} \\ &\approx e^{-0.12} \\ &\approx 0.887 \text{ grams} \qquad \text{(from Table III).} \end{aligned}$$

(d) To determine when 0.9 grams will remain, let $Q = 0.9$ in Equation (38.11):

$$0.9 = e^{-[(\ln 2)/5730]t}.$$

We must solve this equation for t. The logarithmic form of the equation is

$$-[(\ln 2)/5730]t = \ln 0.9.$$

Thus

$$t = -5730(\ln 0.9)/(\ln 2).$$

With $\ln 2 = 0.6931$ and $\ln 0.9 = -0.1054$, you can verify that this gives

$$t \approx 871 \text{ years.}$$ C

(The decaying property of carbon-14 is critical in carbon-dating, which is used to determine the age of once-living objects found at archaeological sites.)

In Section 37 we used the equation

$$P_t = P(1 + r)^t$$

to study population growth [Equation (37.5)]. This equation was essentially the same as the formula for interest compounded over fixed time intervals. We can also study population growth by using Equation (38.7),

$$Q = Q_0 e^{kt},$$

the equation for continuous growth. Here is an illustration.

EXAMPLE 38.8 The world population was estimated to be 2.5 billion in 1950 and 3.6 billion in 1970. Assume that after 1950 world population has grown and will continue to grow according to Equation (38.7). Compute an estimate for the world population in the year 2000.

Solution Use the equation

$$Q = Q_0 e^{kt}$$

with $t = 0$ in 1950, and with Q measured in billions and t in years. Then $Q_0 = 2.5$, and $Q = 3.6$ when $t = 20$ (that is, in 1970). Therefore,

$$3.6 = 2.5e^{20k}$$
$$3.6/2.5 = e^{20k}.$$

The logarithmic form of this equation is

$$20k = \ln(3.6/2.5),$$

so that

$$k = \tfrac{1}{20} \ln(3.6/2.5).$$

This gives

$$k \approx 0.0182.$$

C

Therefore, our equation for estimating population becomes

$$Q = 2.5e^{0.0182t}, \tag{38.12}$$

where t denotes years measured from 1950. For the year 2000 we use $t = 50$. The result is

$$Q = 2.5e^{50(0.0182)}$$
$$Q \approx 6.2 \text{ billion} \qquad \text{(using Table III).}$$

Remember that this estimate is based on the assumptions that the world population will continue to increase between 1970 and 2000 according to Equation (38.7) and at the same rate that it increased between 1950 and 1970. Such assumptions are safe only for short-range projections. The difficulty of making projections about population growth is illustrated by the fact that in 1977 the U.N. offered three estimates for the world population in the year 2000: one low (5.4 billion), one medium (6.1 billion), and one high (7.0 billion). Which estimate turns out to be best will depend on such unpredictable factors as birth control and the supply and distribution of food.

EXERCISES FOR SECTION 38

For Exercises 1–3 use Example 38.1, including Table 38.1.

1. What is the compound amount if \$10 is invested for one year at 100% compounded quarterly? monthly? daily?

2. What is the compound amount if \$100 is invested for one year at 100% compounded semiannually? hourly? each second?

3. What is the compound amount if \$1000 is invested for one year at 100% compounded monthly? daily? each minute?

Determine the compound amount for the given principal invested for the given time period at the given annual rate compounded *continuously*.

4. \$1000 for 3 years at 6%.

5. \$5000 for 5 years at 10%.

6. \$4000 for 2 years at 8%.

7. \$200 for 6 months at 10%.

8. \$3000 for 18 months at 7%.

9. \$5000 for 30 months at 12%.

Compute each logarithm in Exercises 10–18 by two methods:

(a) using Table II and Equation (38.5), and
(b) using Table IV and Equation (38.6).

(The two answers may differ slightly in each case because of rounding errors.)

10. $\ln 470$ **11.** $\ln 92$ **12.** $\ln 4500$

13. $\ln 0.04$ **14.** $\ln 0.61$ **15.** $\ln 0.00035$

16. $\ln(34 \times 10^5)$ **17.** $\ln(0.07 \times 10^{-2})$ **18.** $\ln(550 \times 10^{-3})$

19. The half-life of krypton-85 is 11 years. Write the equation that gives the amount of the substance after t years if the initial amount is 0.1 grams.

20. The half-life of radon-222 is 3.8 days. Write the equation that gives the amount of the substance after t days if the initial amount is 10^{-5} grams.

21. The half-life of plutonium-239 is 2.4×10^4 years. Write the equation that gives the amount of the substance after t years if the initial amount is 0.001 grams.

22. The half-life of radium-226 is 1600 years. Answer each question based on an initial amount of 2 grams.
(a) How much will remain after three half-lives?
(b) How much will remain after 500 years?
(c) When will 1.5 grams remain?

23. The half-life of strontium-90 is 29 years. Answer each question based on an initial amount of 1 milligram.
(a) How much will remain after four half-lives.

(b) How much will remain after 12 years?
(c) When will 0.7 milligrams remain?

24. The half-life of polonium-210 is 139 days. Answer each question based on an initial amount of 1 gram.
(a) How much will remain after two half-lives?
(b) How much will remain after 200 days?
(c) When will 0.2 grams remain?

Use the growth equation (38.12) in Example 38.8 to estimate the world population in each of the following years.

25. 1990 **26.** 2005 **27.** 2010

28. Estimate the population in 20 years for a country whose population is currently 10 million and growing 1.5% annually. (Use a continuous growth equation.)

29. Estimate the population in 15 years for a city whose population is currently 500,000 and growing 3% annually. (Use a continuous growth equation.)

30. At its present rate of growth the population of a city will double in the next 20 years. Estimate the population 8 years from now if the population is currently 200,000. [Begin by determining the growth equation (38.7) based on the data given.]

31. Prove that Equation (38.10) can be rewritten as $Q = Q_0 2^{-t/T}$.

32. Prove that if the growth of a quantity is given by Equation (38.7) and Q has the values Q_1 and Q_2 at times t_1 and t_2, respectively, then

$$k = \frac{1}{t_1 - t_2} \ln\left(\frac{Q_1}{Q_2}\right).$$

33. Prove that

$$\text{if } I = \frac{E}{R}(1 - e^{-Rt/L}), \text{ then } t = -\frac{L}{R}\ln\left(1 - \frac{IR}{E}\right).$$

(The first equation expresses the electric current I in a series circuit in terms of the voltage V, the resistance R, the inductance L, and the time t, when all are expressed in standard units.)

SECTION

39 Computing with Logarithms

A. Antilogs

Logarithms were first developed in the seventeenth century to make computing less tedious. Until recently, computing continued to be a primary use of logarithms. Now, however, computations that were formerly done with logarithms can be carried out much more easily with a computer or a hand held calculator. This section is devoted to computing with logarithms in spite of their declining use for this purpose; this is primarily so you will have seen the idea if you encounter it elsewhere. *Throughout this section* logarithm *will mean* common logarithm.

The **antilog** of a number x is the number whose common logarithm is x. Thus:

$$y = \text{antilog}\, x \ \text{ iff } \ x = \log y.$$

In fact, then, antilog $x = 10^x$, because $x = \log y$ iff $y = 10^x$.

EXAMPLE 39.1 Determine antilog 3.

Solution antilog $3 = 10^3 = 1000$.

If x is not an integer, then we need either a table or a calculator to determine antilog x. The next four examples show how to determine antilogs using a table.

EXAMPLE 39.2 Determine antilog 0.8513.

Solution Viewed as a common logarithm, the number 0.8513 has characteristic 0 and mantissa 0.8513. Therefore, we look for 0.8513 among the logarithms in the body of Table II. We find $\log 7.1 = 0.8513$, so that antilog $0.8513 = 7.1$.

If $x > 1$ and x is not an integer, write x as a characteristic plus a mantissa and then apply the idea in Example 39.2 to the mantissa.

EXAMPLE 39.3 Determine antilog 2.5527.

Solution If $\log x = 2.5527$, then 2 is the characteristic and 0.5527 is the mantissa. Search through the body of Table II until you find 0.5527. You'll see that $\log 3.57 = 0.5527$. To get from 3.57 to a number whose logarithm has characteristic 2 (without changing the mantissa), we move the decimal point two places to the right. Thus antilog $2.5527 = 357$.

If we want antilog x and x is negative, then we again write x as a characteristic plus a mantissa. Remember, however, that *a mantissa is never negative;* every number in Table II is positive. To get around this problem we consider $-n + n + x$ for the smallest integer n such that $n + x$ is positive, as in the following example.

EXAMPLE 39.4 Determine antilog -1.7545.

Solution We subtract and add 2, since 2 is the smallest integral value of n such that $n - 1.7545$ is positive. This gives $-1.7545 = -2 + 2 - 1.7545 = -2 + 0.2455$, so that

$$\text{antilog } -1.7545 = \text{antilog}(-2 + 0.2455).$$

To determine antilog$(-2 + 0.2455)$, think of -2 as the characteristic and 0.2455 as the mantissa. From Table II, antilog 0.2455 = 1.76. Now, because the characteristic is -2, we move the decimal point two places to the left. Thus antilog $-1.7455 = 0.0176$.

EXAMPLE 39.5 Determine antilog$(9.7782 - 10)$.

Solution To get a characteristic plus a mantissa, write

$$9.7782 - 10 = 9 + 0.7782 - 10 = -1 + 0.7782.$$

For the mantissa 0.7782, Table II gives antilog 0.7782 = 6. Since the characteristic is -1, we see that

$$\text{antilog}(9.7782 - 10) = \text{antilog}(-1 + 0.7782) = 0.6.$$

We can also determine antilogs with a calculator if it has the appropriate key or keys.

C

Method I. *For a calculator with a* [y^x] *key.* Since antilog $x = 10^x$, to find antilog x we simply *enter* 10, *press* [y^x], *enter* x, and read antilog x in the display.

Method II. *For a calculator with an* [inv] *(inverse function) key and a* [log] *(common logarithm) key.* Notice that, as a function, *antilog* is the inverse of *log*, because $y = \text{antilog } x$ iff $x = \log y$. Therefore, the *antilog* of a number is the same as the *inverse log* of the number. This means that to find antilog x we can *enter* x, *press* [inv], *press* [log], and read antilog x in the display.

For practice with either calculator method you can redo Examples 39.1–39.5. If x is negative, as in Example 39.4, be sure to enter a negative sign as part of x. To do Example 39.5, let the calculator compute $9.7782 - 10$ as the first step; the characteristic and mantissa do not have to be treated separately as they do when we use a table.

B. Interpolation

EXAMPLE 39.6 Suppose that we need log 7.135. Table II does not give this directly, but it does give log 7.13 and log 7.14.

$$\begin{aligned} \log 7.13 &= 0.8531 \\ \log 7.135 &= \quad ? \\ \log 7.14 &= 0.8537 \end{aligned}$$

Since 7.135 is halfway between 7.13 and 7.14, it is reasonable to assume that log 7.135 is approximately halfway between log 7.13 and log 7.14. That is, approximately,

$$\log 7.135 = 0.8534.$$

Example 39.6 is a special case of **linear interpolation,** by which we obtain an approximate value for the output of a function based on the assumption that as we move from one input value to another input value, the change in the output is proportional to the change in the input. Geometrically, this amounts to assuming that the graph of the function between two known points is approximately a straight line. (Section 24D gives a general explanation of linear interpolation; here we concentrate on examples.)

EXAMPLE 39.7 Use Table II and interpolation to determine log 6.853.

Solution Since Table II does not give log 6.853, we work with the two choices that are closest to 6.853, which are 6.85 and 6.86. Table II gives the values for log 6.85 and log 6.86 that are shown below. To determine log 6.853 we first determine the

6.853 is 0.003/0.010 of the way from 6.85 to 6.86

$$0.010\left[\begin{array}{l} 0.003\left[\begin{array}{ll} \log 6.85 & = 0.8357 \\ \log 6.853 = & ? \end{array}\right]d \\ \qquad \log 6.86 \quad = 0.8363 \end{array}\right]0.0006$$

log 6.853 is $d/0.0006$ of the way from 0.8357 to 0.8363

increment d satisfying the proportion

$$\frac{d}{0.0006} = \frac{0.003}{0.010} \quad \text{or} \quad \frac{d}{0.0006} = \frac{3}{10}.$$

We find

$$d = \frac{0.0006 \times 3}{10} = \frac{0.0018}{10} \approx 0.0002.$$

Since $0.8357 + 0.0002 = 0.8359$, we get $\log 6.853 = 0.8359$.

EXAMPLE 39.8 Use Table II and interpolation to determine log 1234.

Solution The characteristic is 3. To determine the mantissa use interpolation, with log 1.23 and log 1.24 taken from Table II.

$$0.010\left[\begin{array}{l} 0.004\left[\begin{array}{ll} \log 1.23 & = 0.0899 \\ \log 1.234 = & ? \end{array}\right]d \\ \qquad \log 1.24 \quad = 0.0934 \end{array}\right]0.0035$$

$$\frac{d}{0.0035} = \frac{0.004}{0.010} = \frac{4}{10} \quad \text{so} \quad d = \frac{0.0035 \times 4}{10} \approx 0.0014.$$

Therefore, since $0.0899 + 0.0014 = 0.0913$, we find that $\log 1.234 = 0.0913$. Thus $\log 1234 = 3.0913$.

Because Table II is accurate to only four decimal places, we do not write the answers from interpolation to more than four decimal places. Also, we interpolate for only one significant figure more than provided directly by the table: Table II gives logarithms of numbers with *three* significant figures, and thus by interpolation we can reasonably compute approximate values for logarithms of numbers with *four* significant figures.

Interpolation can also be applied to antilogs.

EXAMPLE 39.9 Use Table II and interpolation to determine antilog 0.1738.

Solution Although 0.1738 does not appear in the body of Table II, we do find log 1.49 = 0.1732 and log 1.50 = 0.1761.

$$0.0029\left[0.0006\left[\begin{array}{l}\text{antilog}\,0.1732 = 1.49\\ \text{antilog}\,0.1738 = \ ?\end{array}\right]d\quad\begin{array}{l}\\ \\ \text{antilog}\,0.1761 = 1.50\end{array}\right]0.01$$

$$\frac{d}{0.01} = \frac{0.0006}{0.0029} \quad \text{so} \quad d = \frac{0.01 \times 6}{29} \approx 0.002.$$

Thus antilog 0.1738 = 1.49 + 0.002 = 1.492.

EXAMPLE 39.10 Determine $10^{-2.6449}$.

Solution The logarithmic form of $x = 10^{-2.6449}$ is $\log x = -2.6449$ or $x = \text{antilog}\ -2.6449$. Thus we are really being asked for an antilog. As in Example 39.4, we must subtract and add a positive integer to determine the mantissa:

$$\begin{aligned}\text{antilog}\ -2.6449 &= \text{antilog}(-3 + 3 - 2.6449)\\ &= \text{antilog}(-3 + 0.3551).\end{aligned}$$

The characteristic is −3 and the mantissa is 0.3551. We now determine antilog 0.3551 by interpolation.

$$0.0019\left[0.0010\left[\begin{array}{l}\text{antilog}\,0.3541 = 2.26\\ \text{antilog}\,0.3551 = \ ?\end{array}\right]d\quad\begin{array}{l}\\ \\ \text{antilog}\,0.3560 = 2.27\end{array}\right]0.01$$

$$\frac{d}{0.01} = \frac{0.0010}{0.0019} \quad \text{so} \quad d = \frac{0.01 \times 10}{19} \approx 0.005.$$

Thus antilog 0.3551 = 2.26 + 0.005 = 2.265. Therefore, antilog(−3 + 0.3551) = 0.002265 and $10^{-2.6449} = 0.002265$.

C. Computation

To compute with logarithms we exploit the following three properties from Section 36.

$$\log xy = \log x + \log y \tag{39.1}$$

$$\log \frac{x}{y} = \log x - \log y \tag{39.2}$$

$$\log x^r = r \log x \tag{39.3}$$

With the first property we can replace a multiplication problem by an addition problem; with the second we can replace a division problem by a subtraction problem; and with the third we can replace an exponentiation problem by a multiplication problem. For products or quotients of numbers with only a few significant digits, this method offers no advantage, especially since it requires logarithms and antilogs. But for more complicated cases the idea becomes very useful. And for exponentiation with large or fractional powers it generally is the best method available unless you have a calculator with a y^x key.

EXAMPLE 39.11 Use logarithms to compute 41.8×0.00359.

Solution By Equation (39.1),

$$\log(41.8 \times 0.00359) = \log 41.8 + \log 0.00359.$$

$$\begin{aligned} \log 41.8 &= 1.6212 \\ \log 0.00359 &= \underline{0.5551 - 3} \quad \text{add} \\ & 2.1763 - 3 \end{aligned}$$

Therefore,

$$41.8 \times 0.00359 = \text{antilog}(2.1763 - 3) = 0.150.$$

(The answer is accurate to the three significant figures shown.)

EXAMPLE 39.12 On the average, the sun is 92,900,000 miles from the earth. The speed of light is 186,000 miles per second. How long does it take the light of the sun to reach the earth?

Solution Use the equation $D = RT$ (*distance* equals *rate* times *time*) with $D = 92{,}900{,}000$ miles and $R = 186{,}000$ miles per second. The answer will be

$$\frac{D}{R} = \frac{92{,}900{,}000}{186{,}000} \text{ seconds.}$$

By Equation (39.2), $\log \dfrac{92{,}900{,}000}{186{,}000} = \log 92{,}900{,}000 - \log 186{,}000.$

$$\begin{aligned} \log 92{,}900{,}000 &= 7.9680 \\ \log 186{,}000 &= \underline{5.2695} \quad \text{subtract} \\ & 2.6985 \end{aligned}$$

Therefore,

$$\frac{D}{R} = \text{antilog } 2.6985 = 499.$$

Thus the answer is 499 seconds, or approximately $8\frac{1}{3}$ minutes.

EXAMPLE 39.13 Use logarithms to compute $\sqrt[12]{2}$.

Solution From $\sqrt[12]{2} = 2^{1/12}$ and Equation (39.3) we have

$$\log \sqrt[12]{2} = \log 2^{1/12} = \tfrac{1}{12} \log 2.$$

$$\log 2 = 0.3010$$

$$\tfrac{1}{12} \log 2 = 0.0251$$

Therefore, with interpolation,

$$\sqrt[12]{2} = \text{antilog}\, 0.0251 = 1.060.$$

[To six significant figures, $\sqrt[12]{2} = 1.05946$. The number $\sqrt[12]{2}$ is basic to the well-tempered scale in music, which consists of 12 equal intervals or semitones within one octave. The interval between two semitones is determined by the ratio of their frequencies. If the frequency of C is taken as 1 and the frequency of the adjacent C$^\#$ is r, then the frequency of D must be r^2, that of D$^\#$ must be r^3, and so on, so that the frequency of C′ (the next higher C) is r^{12}. But the ratio for the octave is 2, so that $r^{12} = 2$. Thus $r = \sqrt[12]{2}$.]

EXAMPLE 39.14 Use logarithms to compute $\sqrt[3]{0.05}$.

Solution $\log \sqrt[3]{0.05} = \log 0.05^{1/3} = \frac{1}{3} \log 0.05$. The characteristic of log 0.05 is -2 and the mantissa is 0.6990, so that

$$\log 0.05 = -2 + 0.6990.$$

If we divide $-2 + 0.6990$ by 3, we'll be left without an integer to use as characteristic. To remedy this we rewrite -2 as $-3 + 1$. This gives

$$\log 0.05 = -3 + 1.6990$$

$$\tfrac{1}{3} \log 0.05 = \tfrac{1}{3}(-3 + 1.6990) = -1 + 0.5663.$$

Thus

$$\sqrt[3]{0.05} = \text{antilog}(-1 + 0.5663) = 0.368.$$

EXERCISES FOR SECTION 39

Determine the antilog of each number. (No interpolation is needed.)

1. 4	**2.** -2	**3.** 0	**4.** 0.7825	**5.** 0.0645
6. 0.9542	**7.** 1.5224	**8.** 6.3820	**9.** 4.9773	**10.** -1.6925
11. -0.2596	**12.** -2.3979	**13.** $9.3010 - 10$	**14.** $2.7701 - 3$	**15.** $4.7007 - 5$

Use Table II and interpolation to determine each log or antilog.

16. log 3.535	**17.** log 2.825	**18.** log 8.335
19. log 29.54	**20.** log 8426	**21.** log 0.04477
22. log 0.1111	**23.** log 75.92	**24.** log 1997

25. antilog 0.6583 26. antilog 0.2146 27. antilog 0.9316
28. antilog 2.6431 29. antilog 1.9570 30. antilog 6.3308
31. antilog -2.1907 32. antilog -3.4605 33. antilog -0.1570

Perform each computation using logarithms.

34. 629×118 **35.** 0.129×461 **36.** 0.0861×914 **37.** $384/0.708$

38. $0.19/369$ **39.** $6.65/37.8$ **40.** $\dfrac{401 \times 3.7}{0.7555}$ **41.** $\dfrac{46 \times 0.85}{5.021}$

42. $\dfrac{5642}{387 \times 0.014}$ **43.** $\sqrt[3]{50}$ **44.** $\sqrt[4]{10}$ **45.** $\sqrt{1980}$

46. $\sqrt[5]{0.01}$ **47.** $\sqrt{0.2}$ **48.** $\sqrt[3]{0.25}$

49. The area of a triangle is $A = \sqrt{s(s-a)(s-b)(s-c)}$, where a, b, and c are the lengths of the sides and $s = (a + b + c)/2$. Use this formula and logarithms to compute the area of a triangle whose edges have lengths 10, 11, and 12 centimeters, respectively.

50. The circumference of a basketball is 30 inches. There are approximately 61.0 cubic inches in a liter. Use logarithms to compute the volume of a basketball in liters. ($C = 2\pi r$ and $V = \frac{4}{3}\pi r^3$, where r, C, and V denote radius, circumference, and volume, respectively.)

51. Use logarithms to determine the length in inches of each edge of a cube whose volume is one liter. (See Exercise 50.)

REVIEW EXERCISES FOR CHAPTER IX

Determine each of the following logarithms without using a calculator or tables.

1. $\log_3 27$ **2.** $\log_6 \frac{1}{36}$ **3.** $\log_{100} 1000$ **4.** $\log_{0.4} 2.5$

Solve for x.

5. $\log_7 x = 2$ **6.** $\log_{0.2} x = -1$ **7.** $\log_{15} 3x = 2$ **8.** $a \log_b cx = d$

Express in terms of $\log_b x$, $\log_b y$, and $\log_b z$.

9. $\log_b(xz/y)$ **10.** $\log_b \sqrt[5]{yz}$

Rewrite as a single logarithm and simplify.

11. $2 \log_b uv - \log_b(u^3/v)$ **12.** $\log_b 25 - \log_b \frac{125}{3} + \frac{1}{2} \log_b \frac{125}{3}$

Solve each equation for x. Give both an exact answer (involving logarithms) and an approximate decimal answer.

13. $5^x = 4$ **14.** $3^{2x} = 0.5$ **15.** $5^x = 3^{2+x}$ **16.** $4 \cdot 7^{-x} = 5$

Determine the antilog of each number.

17. 3.6785 **18.** -3.0665

Perform each computation using logarithms.

19. 0.221×42 **20.** $5391/10.18$ **21.** 2^{100} **22.** $\sqrt[5]{20}$

Compute each logarithm by two methods:

(a) using Table II and Equation (38.5), and
(b) using Table IV and Equation (38.6).

23. ln 31 **24.** ln 365

25. The world population in 1970 was approximately 3.61 billion. Write an expression that will give the approximate world population in 1995 if the population were to increase 1.9% per year from 1970 through 1995. If you have a calculator, compute a decimal approximation. Use a continuous growth equation.

26. Assume that money is invested in an account that pays 10% interest compounded annually.
(a) Compute the time required for doubling by using Equation (37.4).
(b) Compute the time required for doubling by using the Rule of 72.
(c) Compute the percentage error from using the Rule of 72.

27. Determine the compound amount if $1000 is invested for 4 years at an annual rate of 6% compounded continuously.

28. Show that if

$$x = \frac{e^y - e^{-y}}{2},$$

then

$$y = \ln(x + \sqrt{x^2 + 1}).$$

[Begin by multiplying both sides of the first equation by $2e^y$. The result can be viewed as a quadratic equation with e^y as the variable (see Section 12C); solve for e^y and then convert to logarithmic form to get y.]

CHAPTER X
SYSTEMS OF EQUATIONS AND INEQUALITIES

The solution of many problems involves one equation with one unknown variable. Other problems involve more than one unknown variable; the solution of such problems generally requires that we write more than one equation or inequality involving the unknown variables. This creates the need for a systematic analysis of *systems* (sets) of equations and inequalities; such an analysis is the purpose of this chapter.

SECTION

40 Systems of Linear Equations with Two Variables

A. Introduction

A **solution** of a system of two linear equations such as

$$\begin{aligned} a_1x + b_1y &= c_1 \\ a_2x + b_2y &= c_2 \end{aligned} \tag{40.1}$$

is an ordered pair of numbers (a, b) such that both equations are satisfied when $x = a$ and $y = b$.

EXAMPLE 40.1 The pair $(2, -3)$ is a solution of the system

$$\begin{aligned} 4x + y &= 5 \\ x + 2y &= -4 \end{aligned}$$

because

$$4(2) + (-3) = 5$$

and

$$(2) + 2(-3) = -4.$$

A system of two linear equations with two variables may have one solution, infinitely many solutions, or no solution. By thinking geometrically it will be easy for us to see why. First, recall that the *graph* of an equation in x and y is the set of all points whose coordinates satisfy the equation. Therefore, a system of two such equations will have a solution iff the graphs of the equations have at least one point in common; any point they have in common will provide a solution. For instance, the graphs for Example 40.1 are shown in Figure 40.1. The coordinates of the point of intersection, $(2, -3)$, give the solution of the system.

Since the graph of each linear equation in x and y is a straight line (Section 20), there are three possibilities for a system of two such equations:

I. The lines intersect in a single point. In this case there is exactly one solution.
II. The lines coincide. In this case there are infinitely many solutions.
III. The lines are distinct and parallel. In this case there is no solution.

If there is no solution the system is said to be **inconsistent** (possibility III). If there is at least one solution the system is said to be **consistent** (possibilities I and II). If there are infinitely many solutions the system is also said to be **dependent** (possibility II).

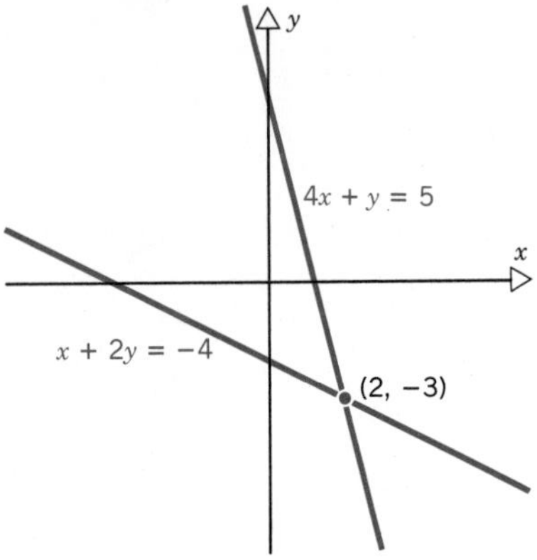

Figure 40.1

EXAMPLE 40.2 The line in Figure 40.2 is the graph of both the equations in the system

$$\begin{aligned} 2x - 3y &= -7 \\ -4x + 6y &= 14. \end{aligned}$$

Therefore, the system has infinitely many solutions and is dependent. The coordinates of any point on the graph provide a solution. Some examples are $(-\frac{7}{2}, 0)$, $(0, \frac{7}{3})$, and $(1, 3)$. Notice that the second equation is the result of multiplying both sides of the first equation by -2. That is typical of a dependent system of two linear equations; either one will result from multiplying both sides of the other by an appropriate nonzero constant.

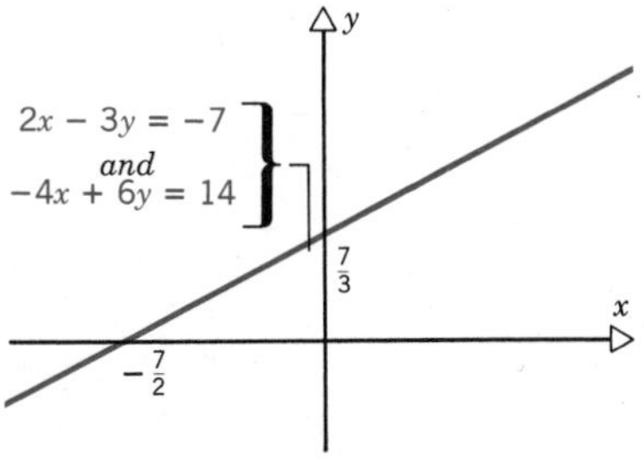

Figure 40.2

EXAMPLE 40.3 The lines in Figure 40.3 are the graphs of the equations in the system

$$\begin{aligned} x + 2y &= 2 \\ 2x + 4y &= -3. \end{aligned}$$

The lines are parallel, because they both have slope $-\frac{1}{2}$. (Remember from Section 20 that the slope is the coefficient of x after the equation has been solved for y.) The system has no solution; it is inconsistent.

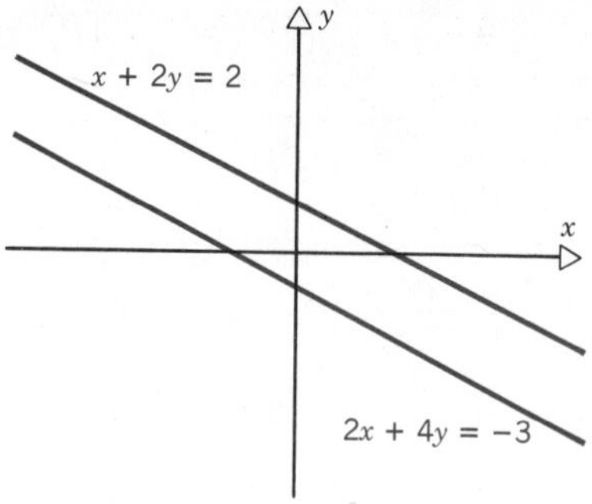

Figure 40.3

B. Solution by Substitution

One way to solve a system like (40.1) is by **substitution:** Solve one of the equations for either variable, x or y. Substitute the answer in place of that variable in the other equation; this will give a linear equation with only one variable. Solve for that variable, and substitute the result in the answer from the first step to determine the other variable.

EXAMPLE 40.4 Solve the system

$$x - 2y = 6 \qquad \text{(A)}$$

$$2x + y = 7 \qquad \text{(B)}$$

by substitution.

Solution Solve Equation (A) for x:

$$x = 2y + 6. \qquad \text{(C)}$$

Use (C) to replace x in (B):

$$2(2y + 6) + y = 7.$$

Solve for y:

$$\begin{aligned} 4y + 12 + y &= 7 \\ 5y &= -5 \\ y &= -1. \end{aligned}$$

Now use $y = -1$ in (C) to get x:

$$x = 2(-1) + 6 = 4.$$

The system has one solution, $(4, -1)$.

Check Substitute $x = 4$ and $y = -1$ in both (A) and (B):

$$\text{(A)}\quad 4 - 2(-1) \stackrel{?}{=} 6 \qquad\qquad \text{(B)}\quad 2(4) + (-1) \stackrel{?}{=} 7$$

$$6 = 6 \qquad\qquad 7 = 7$$

Remark. The only reason to check the solutions obtained from a linear system is to catch possible mistakes in computation; our methods will not lead to extraneous solutions. Hereafter, the step of checking a solution will not be shown.

For a dependent system the method of substitution will lead to the equation $0 = 0$ during the solution process, as in the next example.

EXAMPLE 40.5 Solve the system

$$2x - 3y = -7 \qquad \text{(A)}$$

$$-4x + 6y = 14 \qquad \text{(B)}$$

by substitution. (This is the system in Example 40.2.)

Solution Solve Equation (A) for y:

$$3y = 2x + 7$$

$$y = \tfrac{1}{3}(2x + 7). \qquad \text{(C)}$$

Use (C) to replace y in (B):

$$-4x + 6[\tfrac{1}{3}(2x + 7)] = 14$$

$$-4x + 4x + 14 = 14$$

$$0 = 0.$$

The last equation is true for every value of x. To get a solution of the original system, assign x arbitrarily and determine the corresponding value of y from (C); any pair obtained in this way will be a solution. For example, if $x = 4$, then (C) gives

$$y = \tfrac{1}{3}(2 \cdot 4 + 7) = 5.$$

Thus (4, 5) is one of the infinitely many solutions. In set-builder notation the set of all solutions is

$$\{(x, \tfrac{1}{3}(2x + 7)): \quad x \text{ is a real number}\}.$$

For an inconsistent system the method of substitution will lead to an impossible equation, as in the following example.

EXAMPLE 40.6 Solve the system

$$x + 2y = 2 \qquad \text{(A)}$$

$$2x + 4y = -3 \qquad \text{(B)}$$

by substitution. (This is the system in Example 40.3.)

Solution Solve Equation (A) for x:

$$x = -2y + 2. \qquad \text{(C)}$$

Use (C) to replace x in (B).

$$
\begin{aligned}
2(-2y + 2) + 4y &= -3 \\
-4y + 4 + 4y &= -3 \\
4 &= -3. \qquad \text{(D)}
\end{aligned}
$$

Equation (D) is impossible, so the given system has no solution.

C. Solution by Elimination

One system of equations is **equivalent** to another if the two systems have the same solutions. In the method of **elimination** we solve a system by replacing it by a succession of equivalent systems until we arrive at a system for which the solutions are obvious. This method depends on the fact that we arrive at an equivalent system if we perform any combination of the following operations on a given system:

I. Interchange the equations.
II. Multiply an equation by a nonzero constant.
III. Add a constant multiple of one of the equations to the other equation.

In II, to multiply an equation by a constant means to multiply both sides of the equation by the constant. Operations I–III can be used to try to arrive at equations with only one variable each; that is, equations from which all but one variable has been *eliminated*. Before we look at an example, let's see why operations I–III produce equivalent systems.

It is obvious that operation I will not change the solutions of a system. Operation II will not change the solutions because if $k \neq 0$, then

$$ax + by = c \text{ iff } k(ax + by) = kc.$$

Finally, operation III will not change the solutions because if

$$a_1x + b_1y = c_1 \quad \text{and} \quad a_2x + b_2y = c_2 \qquad (40.2)$$

and k is a real number, then

$$a_1x + b_1y = c_1 \quad \text{and} \quad k(a_1x + b_1y) + a_2x + b_2y = kc_1 + c_2; \qquad (40.3)$$

and, conversely, if (40.3) is true, then (40.2) is true [add $-k$ times the first equation in (40.3) to the second equation in (40.3) to produce the second equation in (40.2)].

EXAMPLE 40.7 Solve the system

$$
\begin{aligned}
3x - 2y &= 5 \\
6x + 5y &= 7
\end{aligned}
$$

by elimination.

Solution Add -2 times the first equation to the second equation. This will eliminate x from the second equation.

$$\begin{aligned} 3x - 2y &= 5 \\ 9y &= -3 \end{aligned}$$

Multiply the second equation by $\frac{1}{9}$ to get y.

$$\begin{aligned} 3x - 2y &= 5 \\ y &= -\tfrac{1}{3} \end{aligned}$$

Add 2 times the second equation to the first equation. This will eliminate y from the first equation.

$$\begin{aligned} 3x \quad &= \tfrac{13}{3} \\ y &= -\tfrac{1}{3} \end{aligned}$$

Multiply the first equation by $\frac{1}{3}$ to get x.

$$\begin{aligned} x \quad &= \tfrac{13}{9} \\ y &= -\tfrac{1}{3} \end{aligned}$$

The solution is now obvious: $(\frac{13}{9}, -\frac{1}{3})$.

EXAMPLE 40.8 Solve the system

$$\begin{aligned} 4x + y &= 5 \\ x + 2y &= -4 \end{aligned}$$

by elimination. (This is the system in Example 40.1.)

Solution First we replace the system by an equivalent system in which the coefficient of the first variable in the first equation is 1. This is not an essential step, but it fits a systematic pattern that will arise later when we consider systems with more equations and more variables. Interchange the two equations.

$$\begin{aligned} x + 2y &= -4 \\ 4x + y &= 5 \end{aligned}$$

Add -4 times the first equation to the second equation. This will eliminate x from the second equation.

$$\begin{aligned} x + 2y &= -4 \\ -7y &= 21 \end{aligned}$$

Multiply the second equation by $-\frac{1}{7}$, so that the coefficient on y will be 1.

$$\begin{aligned} x + 2y &= -4 \\ y &= -3 \end{aligned}$$

Add -2 times the second equation to the first equation. This will eliminate y from the first equation.

$$\begin{aligned} x \quad &= 2 \\ y &= -3 \end{aligned}$$

Thus the solution is $(2, -3)$.

The method of elimination will uncover dependent and inconsistent systems in the same way as the method of substitution: a dependent system will lead to the equation $0 = 0$, and an inconsistent system will lead to an impossible equation.

D. Applications

EXAMPLE 40.9 A total principal of \$1500 is invested in two savings accounts, part in an account that pays 6% simple annual interest and the remainder in an account that pays 8% simple annual interest. If the accounts earn a total of \$108 interest per year, how much is invested in each account?

Solution From what is given we can write two linear equations, one involving the principal and the other involving the interest. Let

$$\begin{aligned} x &= \text{amount invested at } 6\% \\ y &= \text{amount invested at } 8\%. \end{aligned}$$

Then

$$\begin{aligned} x + y &= 1500 \quad \text{(principal)} && \text{(A)} \\ 0.06x + 0.08y &= 108 \quad \text{(interest).} && \text{(B)} \end{aligned}$$

Solve (A) for x and substitute the result in (B):

$$\begin{aligned} 0.06(1500 - y) + 0.08y &= 108 \\ 0.02y &= 18 \\ y &= 900. \end{aligned}$$

With $y = 900$ in (A) we find that $x = 1500 - 900 = 600$. That is, \$600 is invested at 6% and \$900 is invested at 8%.

We handled this same type of problem in Chapter II by using only one variable. Either way, the main idea is to write two equations, one involving whole amounts (in the example, the principal), and the other involving fractions or percentages (in the example, the interest earned on the principal).

EXAMPLE 40.10 A plane flies 560 miles with the wind in 2 hours and 20 minutes. On its return trip, which is against the same wind but with the same constant air speed, the trip takes 2 hours and 40 minutes. Determine the air speed of the plane and the speed of the wind. (The air speed is the speed the plane would be traveling relative to the ground if there were no wind.)

Solution We can use $D = RT$ (distance equals rate times time) and write two equations, one for the trip *with* the wind and the other for the trip *against* the wind. Let p denote the air speed of the plane and w the speed of the wind. Then the rate of the plane is $p + w$ when it flies with the wind, and $p - w$ when it flies against the wind. With the wind the time is $\frac{7}{3}$ hours; against the wind the time is $\frac{8}{3}$ hours. Therefore,

$$560 = (p + w)\tfrac{7}{3} \quad \text{(with the wind)}$$
$$560 = (p - w)\tfrac{8}{3} \quad \text{(against the wind).}$$

The solution of this system is $p = 225$ and $w = 15$, as can be determined by either substitution or elimination. Thus the air speed is 225 mph and the wind speed is 15 mph.

EXERCISES FOR SECTION 40

Each system in Exercises 1–6 has exactly one solution. Find it by *substitution.* Also, draw the graphs of the two equations and use the point of intersection to check your answer.

1. $x - 3y = -1$, $3x + 2y = 8$

2. $x + 2y = 3$, $-2x - y = 6$

3. $x + 5y = 15$, $4x - 2y = -6$

4. $3x - 5y = -1$, $4x + 2y = 3$

5. $5x - 2y = -4$, $3x - 2y = 0$

6. $2x - y = -10$, $7x + 5y = -1$

Each system in Exercises 7–12 has exactly one solution. Find it by *elimination.*

7. $x + 3y = -2$, $7x - 2y = 9$

8. $x - 3y = -5$, $-x + 4y = 9$

9. $x - 4y = 8$, $5x + 3y = -6$

10. $15x + 4y - 5 = 0$, $5x - 2y + 5 = 0$

11. $3x - 2y - 12 = 0$, $6x + 5y - 24 = 0$

12. $5x - 7y + 3 = 0$, $6x + 3y - 4 = 0$

Each system in Exercises 13–15 has infinitely many solutions. Find the solutions by *substitution* and express each answer in set-builder notation. (See the answer to Example 40.5.)

13. $3x - y = 1$, $-6x + 2y = -2$

14. $4x - 2y = 4$, $10x - 5y = 10$

15. $x + 2y + 3 = 0$, $3x + 6y + 9 = 0$

Verify by *elimination* that each system in Exercises 16–18 is inconsistent.

16. $x - 3y = 2$, $2x - 6y = 2$

17. $4x + 2y - 5 = 0$, $6x + 3y + 4 = 0$

18. $x - 2y = 5$, $-2x + 4y = -8$

Use either substitution or elimination to determine all of the solutions of each system.

19. $5x - 2y = 1$, $-x + 3y = 5$

20. $2x - 4y + 1 = 0$, $3x - 6y + 2 = 0$

21. $6x + 3y = 3$, $8x + 4y = 4$

22. $10u - 5v + 5 = 0$, $-4u + 2v - 2 = 0$

23. $4w - 7z = -1$, $-2w + 5z = 1$

24. $-0.25p + 0.5q + 0.2 = 0$
$p - 2q + 1 = 0$

25. $0.5x - 1.5y = 1.0$
$-0.4x + 1.2y = -0.6$

26. $-\frac{1}{3}x + \frac{1}{2}y + \frac{1}{6} = 0$
$\frac{1}{2}x - \frac{3}{4}y - \frac{1}{4} = 0$

27. $0.25x + 0.5y = 0$
$\frac{1}{6}x - \frac{1}{3}y = -4$

28. Part of \$8000 is invested in an account that pays 10% simple annual interest and the remainder is invested in an account that pays 8% simple annual interest. The total interest earned from the two accounts in one year is \$669. How much is invested in each account?

29. Tickets to a theater cost \$2.00 for children and \$3.50 for adults. The receipts for one evening total \$1075 from 335 total tickets sold. How many of the tickets were for children and how many were for adults?

30. A retiree wants to earn \$5000 per year from investing \$50,000. She can earn 7% annually from an insured savings account and 12% from slightly risky bonds. How much should she invest in each so as to earn the \$5000 but minimize the risk?

31. A plane flies 350 miles *against* the wind in 1 hour and 15 minutes. On its return trip, which is *with* the same wind, the trip takes 1 hour and 10 minutes. Determine the air speed of the plane and the speed of the wind.

32. A riverboat travels 6 miles downstream in 20 minutes. The return trip, upstream, takes 30 minutes. Determine the speed of the river current and the speed of the boat in still water.

33. When a walker and a runner start together on a quarter-mile track and go in the same direction, the runner passes the walker after three minutes. When they start together and go in opposite directions they meet after one minute. Determine the rate of each in miles per hour.

34. Wilt Chamberlain once scored 100 points in a single National Basketball Association game, from a total of 64 field goals and free throws. How many field goals (worth two points each) and how many free throws (worth one point each) did he make? (Use linear equations; not trial and error.)

35. A jeweler wants to mix 14-karat gold wih 24-karat gold to obtain 20 grams of 18-karat gold. How many grams each of 14- and 24-karat gold should the jeweler use? (*Karats* are discussed in Review Exercise 23 at the end of Chapter II.)

36. Container A holds a solution that is 20% alcohol and container B holds a solution that is 60% alcohol. How much solution from each should we use to produce 2 liters of solution that is 50% alcohol?

37. Three times the tens digit of a two-digit integer is five more than the units digit. If the digits are reversed, the resulting integer exceeds by five the sum of the original integer added to twice the sum of the two digits. Find the original integer. (If t is the tens digit and u is the units digit, then the original integer is $10t + u$.)

38. If both the numerator and denominator of a fraction are increased by 1, the resulting fraction equals $\frac{2}{3}$. If both the numerator and denominator of the original fraction are decreasd by 13, the resulting fraction equals $\frac{1}{2}$. Find the fraction.

39. In the 75 years from 1905 through 1979, American League teams won the World Series 13 years more than National League teams. How many years did each league have the winning team?

Exercises 40–42 concern the system

$$\begin{aligned} a_1x + b_1y &= c_1 \\ a_2x + b_2y &= c_2 \end{aligned} \tag{40.4}$$

where a_1, b_1, c_1, a_2, b_2, and c_2 are all assumed to be nonzero. To answer each exercise, argue geometrically after writing the equations in slope-intercept form [Equation (20.5)].

40. Explain why the system is inconsistent iff

$$\frac{a_1}{a_2} = \frac{b_1}{b_2} \neq \frac{c_1}{c_2}.$$

41. Explain why the system is dependent iff

$$\frac{a_1}{a_2} = \frac{b_1}{b_2} = \frac{c_1}{c_2}.$$

42. Explain why the system has exactly one solution iff

$$\frac{a_1}{a_2} \neq \frac{b_1}{b_2}.$$

SECTION 41 Systems of Linear Equations with More Than Two Variables

A. Introduction

The idea of a linear equation in two variables extends in a natural way to more than two variables. For example, a **linear equation** in the variables x, y, and z is an equation that can be written in the form

$$ax + by + cz = d$$

for some constant real numbers a, b, c, and d, with a, b, and c not all 0. In this section we consider systems of such linear equations.

A **solution** of a system such as

$$\begin{aligned} a_1x + b_1y + c_1z &= d_1 \\ a_2x + b_2y + c_2z &= d_2 \\ a_3x + b_3y + c_3z &= d_3 \end{aligned} \tag{41.1}$$

is an ordered triple (a, b, c) of real numbers that satisfies each of the equations in the system. If necessary, we can be more explicit and write a solution in the form $x = a$, $y = b$, and $z = c$. The definition of a *solution* of a system with more variables or different variables, or with nonlinear equations, is defined in the same way.

EXAMPLE 41.1 The triple $(4, 0, -3)$ is a solution of the system

$$\begin{aligned} x + 3y - z &= 7 \\ 2x - y + 3z &= -1 \\ -x + 5y - 2z &= 2. \end{aligned}$$

If we substitute $x = 4$, $y = 0$, and $z = -3$ in the second equation, for instance, the result is

$$\begin{aligned} 2(4) - (0) + 3(-3) &\stackrel{?}{=} -1 \\ -1 &= -1. \end{aligned}$$

You can check the first and third equations in the same way.

We concentrate primarily on three linear equations with three variables, but the method we use will carry over to systems with more linear equations and variables. As with systems of two linear equations with two variables, any system of linear equations will have either one solution, infinitely many solutions, or no solution. If a system has at least one solution, then the system is said to be **consistent.** If there is no solution, then the system is said to be **inconsistent.**

For linear systems with three variables, the solutions have a geometrical interpretation similar to that in the case of two variables. It can be shown that relative to a three-dimensional coordinate system, each linear equation in x, y, and z is represented by a plane. A solution for a system corresponds to a point of intersection of the planes that represent the system. If there is a unique point of intersection, then the system has a unique solution. If the planes intersect in a line or a plane, then there are infinitely many solutions. If the planes have no point in common, then there is no solution.

B. Gaussian Elimination

In Section 40 we solved linear systems by two methods, substitution and elimination. The method we use in this section is a combination of the two; we begin with elimination and then complete the process by substitution. This method is called **Gaussian elimination** [after Karl Friedrich Gauss (1777–1855)].

As with two variables, for more than two variables the method of elimination depends on the fact that we arrive at an equivalent system if we perform any combination of the following operations on a given system.

I. Interchange two equations.
II. Multiply an equation by a nonzero constant.
III. Add a constant multiple of one equation to another equation.

Operations I–III are used for all but the last two steps in the following outline; the last two steps use substitution.

Outline of Gaussian Elimination

*The goal of the step on the left is to produce the form on the right, where each * denotes an unspecified constant.*

Obtain an equivalent system with 1 as the coefficient of x in the first equation. This may require no change or it may require the use of I and II.

$$\begin{aligned} x + {*y} + {*z} &= * \\ {*x} + {*y} + {*z} &= * \\ {*x} + {*y} + {*z} &= * \end{aligned}$$

Eliminate x from the second and third equations by adding appropriate multiples of the first equation to the second and third equations.

$$\begin{aligned} x + {*y} + {*z} &= * \\ {*y} + {*z} &= * \\ {*y} + {*z} &= * \end{aligned}$$

Obtain an equivalent system with 1 as the coefficient of y in the second equation, without changing the first equation.

$$\begin{aligned} x + {*y} + {*z} &= * \\ y + {*z} &= * \\ {*y} + {*z} &= * \end{aligned}$$

Eliminate y from the third equation by adding an appropriate multiple of the second equation to the third equation.

$$\begin{aligned} x + {*y} + {*z} &= * \\ y + {*z} &= * \\ {*z} &= * \end{aligned}$$

Obtain an equivalent system with 1 as the coefficient of z in the third equation, without changing the first two equations. The value of z will now be obvious.

$$\begin{aligned} x + {*y} + {*z} &= * \\ y + {*z} &= * \\ z &= * \end{aligned}$$

Substitute the value of z into the second equation and solve for y.

$$\begin{aligned} x + {*y} + {*z} &= * \\ y \phantom{+ {*z}} &= * \\ z &= * \end{aligned}$$

Substitute the values of y and z into the first equation and solve for x.

$$\begin{aligned} x \phantom{+ {*y} + {*z}} &= * \\ y \phantom{+ {*z}} &= * \\ z &= * \end{aligned}$$

Gaussian elimination can also be used for more than three variables and equations. For four linear equations in w, x, y, and z, for example, work for the form

$$\begin{aligned} w + {*x} + {*y} + {*z} &= * \\ x + {*y} + {*z} &= * \\ y + {*z} &= * \\ z &= * \end{aligned} \quad \text{and then} \quad \begin{aligned} w \phantom{+ {*x} + {*y} + {*z}} &= * \\ x \phantom{+ {*y} + {*z}} &= * \\ y \phantom{+ {*z}} &= * \\ z &= * \end{aligned}$$

The steps listed should be thought of as an outline that can be varied whenever it is convenient or necessary. For example, if a system is inconsistent, an impossible equation will arise somewhere along the way, and then we simply stop. Or, for a system with infinitely many solutions the equation $0 = 0$ will arise somewhere along the way; Example 41.4 will show how to handle that.

EXAMPLE 41.2 Solve the system

$$\begin{aligned} 3x - y - 2z &= 8 \\ x + 2y - z &= -1 \\ -2x - 5y + z &= -1. \end{aligned}$$

Solution We follow the plan outlined above, with a variation where noted. Interchange the first two equations.

$$\begin{aligned} x + 2y - z &= -1 \\ 3x - y - 2z &= 8 \\ -2x - 5y + z &= -1 \end{aligned}$$

Add -3 times the first equation to the second equation.
Add 2 times the first equation to the third equation.

$$\begin{aligned} x + 2y - z &= -1 \\ -7y + z &= 11 \\ -y - z &= -3 \end{aligned}$$

We could now multiply the second equation by $-\frac{1}{7}$ so that the coefficient on y would be 1. But that would introduce fractions, which can be avoided in this problem as follows. (Fractions cannot always be avoided, however.) Multiply the third equation by -1 to make 1 the coefficient of y; then interchange the second and third equations.

$$\begin{aligned} x + 2y - z &= -1 \\ y + z &= 3 \\ -7y + z &= 11 \end{aligned}$$

Add 7 times the second equation to the third equation.

$$\begin{aligned} x + 2y - z &= -1 \\ y + z &= 3 \\ 8z &= 32 \end{aligned}$$

Multiply the third equation by $\frac{1}{8}$.

$$\begin{aligned} x + 2y - z &= -1 \\ y + z &= 3 \\ z &= 4 \end{aligned}$$

Thus $z = 4$. With $z = 4$ the second equation becomes $y + 4 = 3$, so that $y = -1$. Finally, with $y = -1$ and $z = 4$ the first equation becomes $x + 2(-1) - 4 = -1$, so that $x = 5$. Therefore, the solution of the system is $(5, -1, 4)$.

EXAMPLE 41.3 Solve the system

$$\begin{aligned} x - 2y + 4z &= 9 \\ 2x + y + 3z &= 3 \\ -x - 4y - 2z &= 4. \end{aligned}$$

Solution Again we follow the plan preceding Example 41.2. The first step there requires no change in this problem.

Add -2 times the first equation to the second equation.
Add the first equation to the third equation.

$$\begin{aligned} x - 2y + 4z &= 9 \\ 5y - 5z &= -15 \\ -6y + 2z &= 13 \end{aligned}$$

Multiply the second equation by $\frac{1}{5}$.

$$\begin{aligned} x - 2y + 4z &= 9 \\ y - z &= -3 \\ -6y + 6z &= 13 \end{aligned}$$

Add 6 times the second equation to the third equation.

$$\begin{aligned} x - 2y + 4z &= 9 \\ y - z &= -3 \\ 0 &= -5 \end{aligned}$$

The equation $0 = -5$ is impossible, so the system is inconsistent.

EXAMPLE 41.4 Solve the system

$$\begin{aligned} x - y + 3z &= 0 \\ 2x - y + 4z &= 1 \\ 5x - 3y + 11z &= 2. \end{aligned}$$

Solution Add -2 times the first equation to the second equation. Add -5 times the first equation to the third equation.

$$\begin{aligned} x - y + 3z &= 0 \\ y - 2z &= 1 \\ 2y - 4z &= 2 \end{aligned}$$

Add -2 times the second equation to the third equation.

$$\begin{aligned} x - y + 3z &= 0 \\ y - 2z &= 1 \\ 0 &= 0. \end{aligned}$$

The last set of equations shows that if t is any real number then we get a solution with

$$\begin{aligned} z &= t \\ y &= 2t + 1 \qquad \text{(from the second equation)} \\ x &= y - 3z \\ &= (2t + 1) - 3t \\ &= -t + 1 \qquad \text{(from the first equation).} \end{aligned}$$

For example, $t = 0$ gives the solution $(1, 1, 0)$. And $t = -2$ gives the solution $(3, -3, -2)$. In set-builder notation, the set of all solutions is

$$\{(-t + 1, 2t + 1, t)\colon \quad t \text{ is a real number}\}.$$

C. Applications

EXAMPLE 41.5 A farmer needs 1000 pounds of fertilizer that is 50% nitrogen, 15% phosphorous, and 35% potassium. He can draw from mixtures A, B, and C, whose compositions are shown in Table 41.1. How many pounds of each mixture should he use?

TABLE 41.1

	Nitrogen	*Phosphorus*	*Potassium*
A	40%	20%	40%
B	60%		40%
C	50%	20%	30%

Solution Let x, y, and z denote the required number of pounds of mixtures A, B, and C, respectively. We can write one equation for each nutrient. For example, the total amount of nitrogen, which is 50% of 1000 pounds, or 500 pounds, must equal the sum of the amounts from mixture A (40% of x, or $0.4x$), mixture B (60% of y, or $0.6y$), and mixture C (50% of z, or $0.5z$). The equations that arise in this way are

$$\begin{aligned} 0.4x + 0.6y + 0.5z &= 500 \qquad \text{nitrogen} \\ 0.2x \qquad\quad + 0.2z &= 150 \qquad \text{phosphorous} \\ 0.4x + 0.4y + 0.3z &= 350 \qquad \text{potassium.} \end{aligned}$$

The solution of this system is $x = 250$, $y = 250$, and $z = 500$, so the farmer should use 250 pounds each of mixtures A and B, and 500 pounds of mixture C.

If two points are in a Cartesian plane but are not on the same vertical line, they determine a unique *linear* function—the function whose graph passes through the two points (Section 24). In the same way, if three noncollinear points are in a Cartesian plane but no two are on the same vertical line, they determine a unique *quadratic* function—the function whose graph passes through the three points. The following example shows how to find the quadratic function from the coordinates of the three points.

EXAMPLE 41.6 There is a unique quadratic function

$$f(x) = ax^2 + bx + c$$

such that $f(-1) = 0$, $f(1) = 2$, and $f(2) = 9$. What is it? [Geometrically, the problem asks for the equation of the parabola through $(-1, 0)$, $(1, 2)$, and $(2, 9)$. See Figure 41.1.]

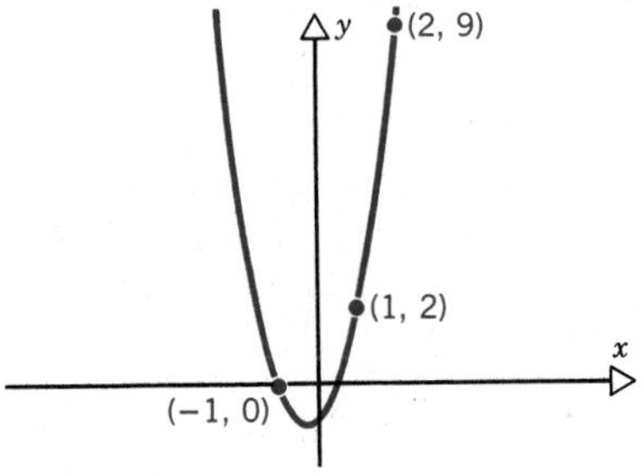

Figure 41.1

Solution If $f(-1) = 0$, then

$$0 = a(-1)^2 + b(-1) + c \text{ so that } a - b + c = 0.$$

Similarly, $f(1) = 2$ implies $a + b + c = 2$, and $f(2) = 9$ implies $4a + 2b + c = 9$. Therefore, a, b, and c must satisfy each of the following equations.

$$\begin{aligned} a - b + c &= 0 \\ a + b + c &= 2 \\ 4a + 2b + c &= 9 \end{aligned}$$

The solution of this system is $a = 2$, $b = 1$, and $c = -1$, so the function is

$$f(x) = 2x^2 + x - 1.$$

Example 41.6 is related to the following general fact: If $b_1, b_2, \ldots, b_{n+1}$ are any real numbers and $a_1, a_2, \ldots, a_{n+1}$ are any distinct real numbers, then there is a unique polynomial function of degree not exceeding n such that $f(a_i) = b_i$ for $1 \le i \le n + 1$. Exercise 31 gives an illustration with $n = 3$.

EXERCISES FOR SECTION 41

Solve each system by Gaussian elimination. If there are infinitely many solutions, give the answer in set-builder notation. If a system is inconsistent, so state.

1. $$\begin{aligned} x + 2y + z &= 1 \\ y + 3z &= -7 \\ 4x + 8y + 3z &= 6 \end{aligned}$$

2. $$\begin{aligned} x + 5y - 2z &= -5 \\ -3x + 2y - 4z &= 2 \\ y + z &= 4 \end{aligned}$$

3. $\begin{aligned} x + 2y - 2z &= -2 \\ x - 4y + z &= 4 \\ 2x + 4y - z &= 5 \end{aligned}$

4. $\begin{aligned} x - 3y + 2z &= 1 \\ 2x - 5y + 6z &= 3 \\ 4x - 11y + 10z &= 5 \end{aligned}$

5. $\begin{aligned} 2u + 7v + 4w &= 2 \\ u + 4v + 3w &= 1 \\ u + 5v + 5w &= 1 \end{aligned}$

6. $\begin{aligned} 2r \quad\quad + 6t &= -7 \\ s + 2t &= 2 \\ r + s + 5t &= 1 \end{aligned}$

7. $\begin{aligned} 3p + q + 14r &= 10 \\ q - 4r &= -12 \\ p + q + 2r &= 1 \end{aligned}$

8. $\begin{aligned} -x + 3y - 3z &= 2 \\ y - z &= 1 \\ 2x - y + 2z &= 7 \end{aligned}$

9. $\begin{aligned} 5x + 2y - z &= 0 \\ y + 2z &= -10 \\ x - y - 3z &= 14 \end{aligned}$

10. $\begin{aligned} 2a + 3b - c \quad\quad &= 0 \\ a + 2b + c - 7 &= 0 \\ 5a + 8b - c - 7 &= 0 \end{aligned}$

11. $\begin{aligned} 2a \quad\quad + 36c &= 7 \\ 4a + 6b + 3c &= -1 \\ a + 2b - 5c &= 2 \end{aligned}$

12. $\begin{aligned} p + 9q + 3r - 5 &= 0 \\ p \quad\quad + 6r - 6 &= 0 \\ 3p + 9q + 15r - 17 &= 0 \end{aligned}$

13. $\begin{aligned} x + 4y - z &= 2 \\ -2x + 3y + z &= 0 \\ 4x + 5y - 3z &= 4 \end{aligned}$

14. $\begin{aligned} x + y - 2z - 2 &= 0 \\ 4x + 3y - 4z - 7 &= 0 \\ 6x - y + 4z - 2 &= 0 \end{aligned}$

15. $\begin{aligned} -x + 4y + 2z &= -1 \\ 3x + 2y - 2z &= 2 \\ 9x - 8y - 10z &= 6 \end{aligned}$

16. $\begin{aligned} 9x - y + 5z &= 2 \\ 4x + 2y + 3z &= 2 \\ -2x + 3y + z &= 1 \end{aligned}$

17. $\begin{aligned} 2x + 2y - 2z &= 3 \\ -3x + y + 2z &= 1 \\ 7x + 3y - 6z &= 5 \end{aligned}$

18. $\begin{aligned} 2x - y + 2z &= 3 \\ -3x + 2y - z &= -3 \\ 4x + 2y - z &= -1 \end{aligned}$

19. $\begin{aligned} x + y + 2z - w &= 1 \\ y - z + w &= 8 \\ 2x \quad\quad\quad + 2w &= 10 \\ x - y \quad\quad + w &= 1 \end{aligned}$

20. $\begin{aligned} w - 2x + 3y \quad\quad &= 4 \\ w - 3x + 4y - z &= 0 \\ 4x - 3y \quad\quad &= 2 \\ 5w - 4x + 11y - 2z &= 16 \end{aligned}$

21. $\begin{aligned} y - 2z + w &= 3 \\ x - y - 3z - w &= 1 \\ 2x + 4y \quad\quad + 2w &= 5 \\ 4x + 3y - 8z + w &= 9 \end{aligned}$

22. Suppose the farmer in Example 41.5 needs 1000 pounds of fertilizer that is 53% nitrogen, 10% phosphorous, and 37% potassium. How many pounds of each mixture (A, B, and C) should he use?

23. Suppose the farmer in Example 41.5 has the same requirements as in the example, except that he must draw from mixtures A and B (as in the example) and a mixture D that is 40% nitrogen, 30% phosphorous, and 30% potassium. How many pounds of each mixture (A, B, and D) should he use?

24. A total of \$1000 is divided in three accounts, which pay 6%, 7%, and 9% simple annual interest respectively. The total interest earned in one year is \$75. The sum of the amounts in the 6% and 9% accounts is \$100 more than the amount in the 7% account. How much is invested in each account?

25. In the 50 years from 1926 through 1975, National League teams won the World Series one more time than the Yankees (an American League team), and the

Yankees won eight more times than the other American League teams combined. Determine how many times the Series was won by National League. teams, by the Yankees, and by American League teams other than the Yankees. (It's irrelevant to the solution of this problem, but during 1926–1975 the Yankees lost the series seven times.)

26. Of the 50 states in the U.S., the number entering the union from 1800 through 1850 was one less than the number entering before 1800, one more than the number entering from 1851 through 1900, and three times the number entering after 1900. How many entered in each of the four periods: before 1800, 1800–1850, 1851–1900, and after 1900?

27. The normal annual precipitation amounts in Cleveland, Phoenix, and Mobile satisfy these relationships: the amount in Cleveland is five times the amount in Phoenix, the amount in Mobile is just three inches less than ten times the amount in Phoenix, and the total amount in the three cities is 109 inches. Find the amount in each city.

28. There is a unique quadratic function f such that $f(-1) = 0$, $f(1) = 1$, and $f(4) = 0$. What is it?

29. Find the equation of the parabola through the points $(-2, 6)$, $(1, 3)$, and $(2, 10)$.

30. Find the equation of the parabola that passes through the points $(0, 4)$ and $(1, \frac{3}{2})$ and is symmetric with respect to the y-axis.

31. There is a unique polynomial function f of degree three such that $f(-1) = -2$, $f(1) = 2$, $f(2) = 4$, and $f(3) = 14$. What is it?

32. Prove that there is no quadratic function f such that $f(0) = -3$, $f(2) = 1$, and $f(3) = 3$. (Suggestion: The method of Example 41.6 will lead to $a = 0$.)

33. A small rocket is to be fired upward near the earth's surface with initial velocity v_0 feet per second and initial height s_0 feet, in such a way that its maximum altitude will be 225 feet and it will reach the ground 4.5 seconds after it is fired. Determine what v_0 and s_0 should be. (See Example 25.5. Use $g = 32$.)

SECTION

42 Linear Systems and Row-Equivalent Matrices

A. Augmented Matrices

The essential information from the system

$$\begin{aligned} a_1x + b_1y + c_1z &= d_1 \\ a_2x + b_2y + c_2z &= d_2 \\ a_3x + b_3y + c_3z &= d_3 \end{aligned}$$

is contained in the rectangular array

$$\text{Row 1} \rightarrow \begin{bmatrix} a_1 & b_1 & c_1 & d_1 \\ a_2 & b_2 & c_2 & d_2 \\ a_3 & b_3 & c_3 & \underset{\substack{\uparrow \\ \text{Column 4}}}{d_3} \end{bmatrix}, \tag{42.1}$$

which is called the **augmented matrix** of the system. In general, the term *matrix* is used to describe any rectangular array of numbers. (Very little about matrices will be needed in this section; they will be treated more fully in Chapter XI.) Any matrix consists of (horizontal) **rows** and (vertical) **columns,** numbered respectively from top to bottom and from left to right as suggested in (42.1).

EXAMPLE 42.1 The augmented matrix of the system

$$\begin{aligned} 2x + y - 8z &= 7 \\ y - 2z &= 2 \\ x - 2y + 6z &= 1 \end{aligned}$$

is

$$\begin{bmatrix} 2 & 1 & -8 & 7 \\ 0 & 1 & -2 & 2 \\ 1 & -2 & 6 & 1 \end{bmatrix}.$$

The 0 in the second row and first column is the coefficient of the missing x in the second equation.

Augmented matrices for linear systems with more variables or different variables are defined in the same way. Just remember that before writing the augmented matrix the variables must be in the same order in all of the equations. Also, the constant terms are to be on the right of the equality signs.

Linear Systems and Row-Equivalent Matrices

Rather than writing out all of the equations at each step in solving a system, we can simply write the corresponding augmented matrix. Each of the three types of operations used to solve a system by elimination produces a corresponding operation on the augmented matrix. If we interchange the first and third equations, for example, then we correspondingly interchange the first and third rows of the augmented matrix. The operations on matrices that can arise in this way are called **elementary row operations.** They are of three types (compare I, II, and III in Section 41B).

I. Interchange two rows.
II. Multiply a row by a nonzero constant.
III. Add a constant multiple of one row to another row.

To solve systems in this section we use I–III on the augmented matrix until the solution is obvious. This amounts to modifying Gaussian elimination so that substitution is not used. If possible, transform the augmented matrix as follows (for the case of three equations with three variables; other cases are similar).

$$\begin{bmatrix} * & * & * & * \\ * & * & * & * \\ * & * & * & * \end{bmatrix} \longrightarrow \begin{bmatrix} 1 & * & * & * \\ * & * & * & * \\ * & * & * & * \end{bmatrix} \longrightarrow \begin{bmatrix} 1 & * & * & * \\ 0 & * & * & * \\ 0 & * & * & * \end{bmatrix} \longrightarrow \begin{bmatrix} 1 & * & * & * \\ 0 & 1 & * & * \\ 0 & * & * & * \end{bmatrix}$$

$$\longrightarrow \begin{bmatrix} 1 & 0 & * & * \\ 0 & 1 & * & * \\ 0 & 0 & * & * \end{bmatrix} \longrightarrow \begin{bmatrix} 1 & 0 & * & * \\ 0 & 1 & * & * \\ 0 & 0 & 1 & * \end{bmatrix} \longrightarrow \begin{bmatrix} 1 & 0 & 0 & * \\ 0 & 1 & 0 & * \\ 0 & 0 & 1 & * \end{bmatrix}$$

In some cases the final form will necessarily be different, as Examples 42.3 and 42.4 will show. To keep track of the operations we use abbreviations like these:

(Row 1) $\leftrightarrow$ (Row 2) will mean to interchange Rows 1 and 2.

5(Row 1) will mean to multiply each element of Row 1 by 5.

(Row 3) − 6(Row 1) will mean to add −6 times Row 1 to Row 3.

EXAMPLE 42.2 Use augmented matrices to solve the system in Example 42.1.

Solution

$$\begin{bmatrix} 2 & 1 & -8 & 7 \\ 0 & 1 & -2 & 2 \\ 1 & -2 & 6 & 1 \end{bmatrix} \xrightarrow{\text{(Row 1)} \leftrightarrow \text{(Row 3)}} \begin{bmatrix} 1 & -2 & 6 & 1 \\ 0 & 1 & -2 & 2 \\ 2 & 1 & -8 & 7 \end{bmatrix}$$

$$\xrightarrow{\text{(Row 3)} - 2\text{(Row 1)}} \begin{bmatrix} 1 & -2 & 6 & 1 \\ 0 & 1 & -2 & 2 \\ 0 & 5 & -20 & 5 \end{bmatrix}$$

$$\xrightarrow[\text{(Row 3)} - 5\text{(Row 2)}]{\text{(Row 1)} + 2\text{(Row 2)}} \begin{bmatrix} 1 & 0 & 2 & 5 \\ 0 & 1 & -2 & 2 \\ 0 & 0 & -10 & -5 \end{bmatrix}$$

$$\xrightarrow{-\frac{1}{10}(\text{Row } 3)} \left[\begin{array}{ccc|c} 1 & 0 & 2 & 5 \\ 0 & 1 & -2 & 2 \\ 0 & 0 & 1 & \frac{1}{2} \end{array}\right]$$

$$\xrightarrow[(\text{Row } 2) + 2(\text{Row } 3)]{(\text{Row } 1) - 2(\text{Row } 3)} \left[\begin{array}{ccc|c} 1 & 0 & 0 & 4 \\ 0 & 1 & 0 & 3 \\ 0 & 0 & 1 & \frac{1}{2} \end{array}\right]$$

The last matrix is the augmented matrix of

$$\begin{aligned} x \qquad &= 4 \\ y \quad &= 3 \\ z &= \tfrac{1}{2}. \end{aligned}$$

Thus the solution is $(4, 3, \frac{1}{2})$.

EXAMPLE 42.3 Use augmented matrices to solve the system

$$\begin{aligned} x - 3y - 5z &= 2 \\ 2x - 5y + z &= 3 \\ x - 2y + 6z &= 2. \end{aligned}$$

Solution

$$\left[\begin{array}{ccc|c} 1 & -3 & -5 & 2 \\ 2 & -5 & 1 & 3 \\ 1 & -2 & 6 & 2 \end{array}\right] \xrightarrow[(\text{Row } 3) - (\text{Row } 1)]{(\text{Row } 2) - 2(\text{Row } 1)} \left[\begin{array}{ccc|c} 1 & -3 & -5 & 2 \\ 0 & 1 & 11 & -1 \\ 0 & 1 & 11 & 0 \end{array}\right]$$

$$\xrightarrow[(\text{Row } 3) - (\text{Row } 2)]{(\text{Row } 1) + 3(\text{Row } 2)} \left[\begin{array}{ccc|c} 1 & 0 & 28 & -1 \\ 0 & 1 & 11 & -1 \\ 0 & 0 & 0 & 1 \end{array}\right]$$

The third row of the last matrix corresponds to the equaion $0x + 0y + 0z = 1$, which implies $0 = 1$, an impossibility. Therefore, the system is inconsistent.

EXAMPLE 42.4 Use augmented matrices to solve the system

$$\begin{aligned} x - 2y + z &= 4 \\ 2x - 4y + 3z &= 14 \\ -x + 2y \qquad &= 2. \end{aligned}$$

Solution

$$\left[\begin{array}{ccc|c} 1 & -2 & 1 & 4 \\ 2 & -4 & 3 & 14 \\ -1 & 2 & 0 & 2 \end{array}\right] \xrightarrow[(\text{Row } 3) + (\text{Row } 1)]{(\text{Row } 2) - 2(\text{Row } 1)} \left[\begin{array}{ccc|c} 1 & -2 & 1 & 4 \\ 0 & 0 & 1 & 6 \\ 0 & 0 & 1 & 6 \end{array}\right]$$

No row operation will transform the last matrix to the form

$$\begin{bmatrix} 1 & * & * & * \\ 0 & 1 & * & * \\ 0 & * & * & * \end{bmatrix}.$$

But we can get the form

$$\begin{bmatrix} 1 & * & * & * \\ 0 & 0 & 1 & * \\ 0 & 0 & 0 & 0 \end{bmatrix}.$$

$$\begin{bmatrix} 1 & -2 & 1 & 4 \\ 0 & 0 & 1 & 6 \\ 0 & 0 & 1 & 6 \end{bmatrix} \xrightarrow[\text{(Row 3) − (Row 2)}]{\text{(Row 1) − (Row 2)}} \begin{bmatrix} 1 & -2 & 0 & -2 \\ 0 & 0 & 1 & 6 \\ 0 & 0 & 0 & 0 \end{bmatrix}$$

The last matrix translates to

$$\begin{aligned} x - 2y \qquad &= -2 \\ z &= 6. \end{aligned}$$

The first equation will give a value of x for each value of y. The second equation dictates $z = 6$. Therefore, the complete solution is

$$\{(2t - 2, t, 6): \quad t \text{ is a real number}\}.$$

The ideas in the preceding examples can be summarized as follows. If one matrix can be obtained from another by a finite sequence of elementary row operations, then the two matrices are said to be **row equivalent.** It follows that two systems of linear equations are equivalent iff their augmented matrices are row equivalent. Therefore, we can solve a system by transforming its augmented matrix by a sequence of elementary row operations until the solutions become obvious. An appropriate form is given in the following statement, which is presented here without proof.

Each matrix is row equivalent to a unique matrix in **reduced echelon form,** which is, by definition, a matrix such that

1. the first nonzero entry (the **leading entry**) of each row is 1;
2. the other entries in any column containing such a leading entry are 0;
3. the leading entry in each row is to the right of the leading entry in each preceding row; and
4. rows containing only 0's are below rows with nonzero entries.

EXAMPLE 42.5

(a) For the reasons indicated, these matrices are not in reduced echelon form.

$$\begin{bmatrix} 1 & 0 & 4 & 3 \\ 0 & 2 & -4 & 0 \\ 0 & 0 & 0 & 0 \end{bmatrix} \begin{bmatrix} 1 & 0 & 5 \\ 1 & 0 & 10 \end{bmatrix} \begin{bmatrix} 1 & 0 & 0 & 8 \\ 0 & 0 & 1 & 4 \\ 0 & 1 & 0 & -7 \end{bmatrix} \begin{bmatrix} 1 & 0 & 0 & 0 & -9 \\ 0 & 0 & 0 & 0 & 0 \\ 0 & 0 & 1 & 0 & 6 \\ 0 & 0 & 0 & 0 & 0 \end{bmatrix}$$

Violates 1 Violates 2 Violates 3 Violates 4

(b) Each of these matrices is in reduced echelon form and is equivalent to the corresponding matrix in part (a).

$$\begin{bmatrix} 1 & 0 & 4 & 3 \\ 0 & 1 & -2 & 0 \\ 0 & 0 & 0 & 0 \end{bmatrix} \begin{bmatrix} 1 & 0 & 5 \\ 0 & 0 & 1 \end{bmatrix} \begin{bmatrix} 1 & 0 & 0 & 8 \\ 0 & 1 & 0 & -7 \\ 0 & 0 & 1 & 4 \end{bmatrix} \begin{bmatrix} 1 & 0 & 0 & 0 & -9 \\ 0 & 0 & 1 & 0 & 6 \\ 0 & 0 & 0 & 0 & 0 \\ 0 & 0 & 0 & 0 & 0 \end{bmatrix}$$

C. More Examples

Linear systems with two variables or with more than three variables can also be solved by the method of this section. The method applies as well when the number of equations is different from the number of variables, as illustrated by the following two examples.

EXAMPLE 42.6 Solve

$$\begin{aligned} 2x \phantom{{}-3y} + 5z &= 6 \\ 3y \phantom{{}+z} &= -1 \\ 4x - 3y + z &= 4 \\ 2x + 3y + z &= 1. \end{aligned}$$

Solution The augmented matrix

$$\begin{bmatrix} 2 & 0 & 5 & 6 \\ 0 & 3 & 0 & -1 \\ 4 & -3 & 1 & 4 \\ 2 & 3 & 1 & 1 \end{bmatrix}$$

is row equivalent to the reduced echelon matrix

$$\begin{bmatrix} 1 & 0 & 0 & \frac{1}{2} \\ 0 & 1 & 0 & -\frac{1}{3} \\ 0 & 0 & 1 & 1 \\ 0 & 0 & 0 & 0 \end{bmatrix}$$

Therefore, the solution is $(\frac{1}{2}, -\frac{1}{3}, 1)$.

EXAMPLE 42.7 Solve

$$\begin{aligned} w + 2x + 7y \qquad\quad - 1 &= 0 \\ x + 2y \qquad\quad - 1 &= 0 \\ -w + x - y + 5z - 6 &= 0. \end{aligned}$$

Solution The augmented matrix

$$\begin{bmatrix} 1 & 2 & 7 & 0 & 1 \\ 0 & 1 & 2 & 0 & 1 \\ -1 & 1 & -1 & 5 & 6 \end{bmatrix}$$

is row equivalent to the reduced echelon matrix

$$\begin{bmatrix} 1 & 0 & 3 & 0 & -1 \\ 0 & 1 & 2 & 0 & 1 \\ 0 & 0 & 0 & 1 & \frac{4}{5} \end{bmatrix}$$

The last row gives $z = \frac{4}{5}$. No leading coefficient occurs in the third column, so y can be assigned arbitrarily. If $y = t$, then the first and second rows give

$$w + 3t = -1 \quad \text{and} \quad x + 2t = 1,$$

Therefore,

$$w = -3t - 1 \quad \text{and} \quad x = -2t + 1.$$

so that

$$w = -3t - 1, \quad x = -2t + 1, \quad y = t, \quad \text{and} \quad z = \tfrac{4}{5}$$

is a solution for each real number t. Or, the set of all solutions is

$$\{(-3t - 1, -2t + 1, t, \tfrac{4}{5}): \quad t \text{ is a real number}\}.$$

The following statements can be proved about a system of m linear equations with n unknown variables. If $m > n$, then the system may have one solution, infinitely many solutions, or no solution, depending on the system. If $m < n$, then the system may have infinitely many solutions or no solution, depending on the system; there will never be just one solution in this case.

EXERCISES FOR SECTION 42

Each matrix in Exercises 1–9 is the reduced echelon matrix for a system with variables w, x, y, and z (in that order). Write the solution or solutions, or indicate that the system is inconsistent.

1. $\begin{bmatrix} 1 & 0 & 0 & 0 & -5 \\ 0 & 1 & 0 & 0 & 0 \\ 0 & 0 & 1 & 0 & 6 \\ 0 & 0 & 0 & 0 & 1 \end{bmatrix}$

2. $\begin{bmatrix} 1 & 0 & 0 & 0 & 0 \\ 0 & 1 & 0 & 0 & 8 \\ 0 & 0 & 1 & 0 & -3 \\ 0 & 0 & 0 & 1 & 12 \end{bmatrix}$

3. $\begin{bmatrix} 1 & 0 & 0 & 0 & 4 \\ 0 & 1 & 0 & 3 & -1 \\ 0 & 0 & 1 & 2 & 7 \\ 0 & 0 & 0 & 0 & 0 \end{bmatrix}$

4. $\begin{bmatrix} 1 & 0 & 0 & 0 & -1 \\ 0 & 1 & 8 & 0 & -2 \\ 0 & 0 & 0 & 1 & 23 \\ 0 & 0 & 0 & 0 & 0 \end{bmatrix}$

5. $\begin{bmatrix} 1 & 0 & 0 & 0 & 13 \\ 0 & 1 & 0 & 0 & 4 \\ 0 & 0 & 0 & 0 & 1 \\ 0 & 0 & 0 & 0 & 0 \end{bmatrix}$

6. $\begin{bmatrix} 1 & 0 & 0 & 0 & 8 \\ 0 & 1 & 0 & 0 & -9 \\ 0 & 0 & 1 & 0 & 7 \\ 0 & 0 & 0 & 1 & 23 \end{bmatrix}$

7. $\begin{bmatrix} 1 & 0 & 0 & 0 & -3 \\ 0 & 1 & 0 & 0 & 8 \\ 0 & 0 & 1 & 0 & 6 \\ 0 & 0 & 0 & 1 & 6 \end{bmatrix}$

8. $\begin{bmatrix} 1 & 4 & 0 & 0 & 7 \\ 0 & 0 & 1 & 0 & 4 \\ 0 & 0 & 0 & 1 & -3 \\ 0 & 0 & 0 & 0 & 0 \end{bmatrix}$

9. $\begin{bmatrix} 1 & 2 & 0 & 0 & 11 \\ 0 & 0 & 1 & 0 & -2 \\ 0 & 0 & 0 & 0 & 1 \\ 0 & 0 & 0 & 0 & 0 \end{bmatrix}$

In Exercises 10–30 solve each system by reducing the augmented matrix to reduced echelon form.

10. $\begin{aligned} x - 3y &= -6 \\ 3x + 2y &= 26 \end{aligned}$

11. $\begin{aligned} 2x + y &= -2 \\ 3x + 2y &= 0 \end{aligned}$

12. $\begin{aligned} x + 5y &= -2 \\ 4x + 2y &= -1 \end{aligned}$

13. $\begin{aligned} x - 7y + 2z &= -6 \\ 5x - 5y + 3z &= 2 \\ -2x + y - z &= -2 \end{aligned}$

14. $\begin{aligned} x + y + z &= 3 \\ x - y - z &= -1 \\ 2x + 3y \phantom{{}-z} &= 4 \end{aligned}$

15. $\begin{aligned} x + 5y + 10z &= 2 \\ -3x \phantom{{}+5y} + 15z &= 4 \\ -x + 10y + 35z &= 8 \end{aligned}$

16. $\begin{aligned} -a - 2b \phantom{{}-c} &= 3 \\ 2a + 5b - c &= -4 \\ -3a - 5b - c &= 12 \end{aligned}$

17. $\begin{aligned} p - 4q + r &= 0 \\ 3p - 11q + 4r &= 1 \\ 2p - 7q + 3r &= -2 \end{aligned}$

18. $\begin{aligned} -t + 4u - v &= 2 \\ 3t - u + 2v &= 1 \\ -4t + u - v &= -1 \end{aligned}$

19. $\begin{aligned} 2x + y - z - 2 &= 0 \\ -5x + y + 3z \phantom{{}-2} &= 0 \\ 3x + 5y - z - 8 &= 0 \end{aligned}$

20. $\begin{aligned} 3x - 3y + z - 6 &= 0 \\ 2x \phantom{{}-3y} + 2z - 5 &= 0 \\ -4x + 6y \phantom{{}+z} + 7 &= 0 \end{aligned}$

21. $\begin{aligned} 5x - 2y + z - 2 &= 0 \\ 2x + y - z - 2 &= 0 \\ 3x + 6y - 5z + 6 &= 0 \end{aligned}$

22.
$$\begin{aligned} w - 2x + 3y + 7z &= 8 \\ x + z &= -3 \\ w \qquad\quad y + 3z &= 2 \\ w + x + y + 4z &= -1 \end{aligned}$$

23.
$$\begin{aligned} w + 2x - y &= -6 \\ w + 3x + 3z &= -1 \\ -2x + y &= 2 \\ 2w - 2y + 5z &= -7 \end{aligned}$$

24.
$$\begin{aligned} w + 2y &= 1 \\ x + z &= -2 \\ w + 2x + 2y + 2z &= 0 \\ -w + x - 2y &= 0 \end{aligned}$$

25.
$$\begin{aligned} x + 5y &= 8 \\ 2x + 4y &= 7 \\ -x + 3y &= 4 \end{aligned}$$

26.
$$\begin{aligned} 2x - y &= 5 \\ x + 2y &= 4 \\ -x + 8y &= 3 \end{aligned}$$

27.
$$\begin{aligned} x - y &= 4 \\ -2x + 2y &= -8 \\ 5x - 10y &= 20 \end{aligned}$$

28.
$$\begin{aligned} p - q + 2r - 4 &= 0 \\ -2p + 2q - 4r + 7 &= 0 \end{aligned}$$

29.
$$\begin{aligned} t - v &= -3 \\ u + 2v &= 1 \\ w &= 2 \\ 2t + u &= -5 \end{aligned}$$

30.
$$\begin{aligned} 2a + b + 2c &= 0 \\ b + 2c &= 1 \\ 2a + b + 2c - 3d &= -1 \\ 2a + 2b + 4c &= 1 \end{aligned}$$

Solve each exercise by writing a system of equations and then solving the system with augmented matrices.

31. The perimeter of a triangle is 45 centimeters. The length of the longest side is 3 centimeters less than the sum of the lengths of the other two sides. The longest side is 3 times as long as the shortest side. Find the length of each side.

32. The sum of three numbers is 85. The sum of 10% of the first, 20% of the second, and 25% of the third is 17. The sum of 25% of the first, 40% of the second, and 90% of the third is 51. Find the three numbers.

33. The sum of three real numbers is 3, the largest of the three is 5 times the smallest of the three, and the remaining number is 3 times the smallest of the three. Find the numbers.

34. The three largest lakes in the world, by area, are the Caspian Sea, Lake Superior, and Lake Victoria. The area of the Caspian Sea is 16 thousand square miles more than 4 times the area of Lake Superior and 9 thousand square miles more than 5 times the area of Lake Victoria. The area of Lake Superior is 5 thousand square miles more than the area of Lake Victoria. Find the area of each lake. (All areas are rounded to the nearest thousand square miles. The Caspian Sea is a saltwater lake; Superior and Victoria are both freshwater lakes.)

35. In the 50 years from 1926 through 1975, the number of seven-game World Series was four more than twice the number of four-game Series, the number of seven-game Series was six more than twice the number of six-game Series, and the number of Series that lasted either four or five games was two-thirds the number that lasted either six or seven games. Find the number of Series of each length. (A World Series lasts either four, five, six, or seven games.)

36. The population of the U.S. in 1950 was 1 million less than twice the population in 1900. The population east of the Mississippi River increased by 45 million from 1900 to 1950, and the population west of the river increased by 30 million from 1900 to 1950. In 1950 the population east of the river was 2 million less than twice the population west of the river. Determine the population both east and west of the river in 1900, and both east and west of the river in 1950. Then determine the percentage increases both east and west of the river between 1900 and 1950. (All populaitons are rounded to the nearest million.)

37. There is a unique polynomial function f of degree three such that $f(-2) = 10$, $f(-1) = 4$, $f(1) = 4$, and $f(2) = -2$. What is it?

38. There is exactly one circle that passes through the three points $(1, 3)$, $(3, -1)$, and $(4, 2)$. Find its equation. [Compare Example 41.6. Begin with Equation (19.5), $x^2 + y^2 + ax + by + c = 0$, and determine a, b, and c.]

39. Prove that there is no circle that passes through the three points $(-3, 2)$, $(-2, 1)$, and $(1, -2)$. (The method suggested in Exercise 38 will lead to an inconsistent system.)

SECTION 43 Nonlinear Systems of Equations

A. Solution by Substitution

A system of equations is said to be **nonlinear** if at least one of its equations is nonlinear. A nonlinear system may have imaginary solutions (Chapter VIII) even if its coefficients are all real, but *only real solutions will be considered in this chapter.*

Many two-variable systems composed of a linear equation and one other equation can be solved by substitution.

EXAMPLE 43.1 Solve the system

$$x - 2y = 1 \qquad \text{(A)}$$

$$x^2 + y^2 = 29 \qquad \text{(B)}$$

by substitution.

Solution Solve the linear equation, (A), for x:

$$x = 2y + 1. \qquad \text{(C)}$$

Substitute $2y + 1$ for x in (B):

$$(2y + 1)^2 + y^2 = 29.$$

Solve for y:

$$4y^2 + 4y + 1 + y^2 = 29$$
$$5y^2 + 4y - 28 = 0$$
$$(5y + 14)(y - 2) = 0$$
$$y = -\tfrac{14}{5} \quad \text{or} \quad y = 2.$$

Now use $y = -\frac{14}{5}$ and then $y = 2$ in (C) to determine the corresponding values of x:

If $y = -\frac{14}{5}$, then $x = 2(-\frac{14}{5}) + 1 = -\frac{23}{5}$.
If $y = 2$, then $x = 2(2) + 1 = 5$.

The solutions of the system are $(-\frac{23}{5}, -\frac{14}{5})$ and $(5, 2)$. They can be checked by substitution in both of the original equations.

Figure 43.1 shows a geometrical interpretation of this example: the straight line is the graph of Equation (A) and the circle is the graph of Equation (B). The solutions of the system are the coordinates of the points where the line and the circle intersect.

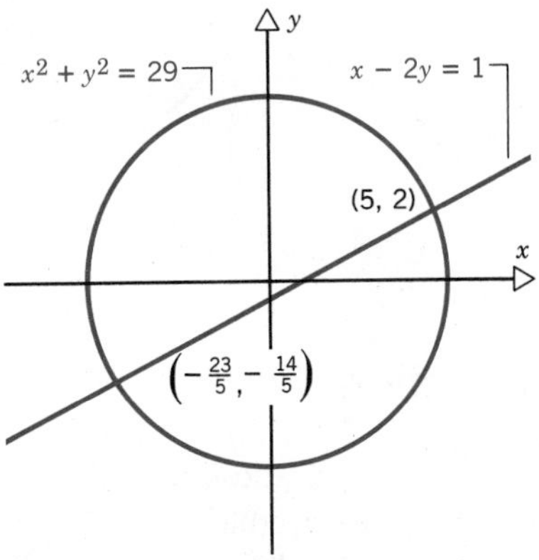

Figure 43.1

In Section 21 we saw that the graph of a quadratic equation in x and y is either a conic section, two intersecting lines, two parallel lines, one line, or the empty set. By thinking about the possibilities for graphs of quadratic equations one at a time, you'll see that the graph of a linear equation (a line) will intersect the graph of a quadratic equation in either two points, one point, or no points. It follows that a system composed of a quadratic equation and a linear equation has either two solutions, one solution, or no solution. In Example 43.1 there were two solutions; in the next example there is no solution.

EXAMPLE 43.2 Solve the system

$$2x^2 - y + 1 = 0 \qquad \text{(A)}$$
$$x + y + 1 = 0 \qquad \text{(B)}$$

by substitution.

Solution Solve (B) for y:

$$y = -x - 1.$$

Substitute $-x - 1$ for y in (A):

$$2x^2 - (-x - 1) + 1 = 0$$
$$2x^2 + x + 2 = 0.$$

The discriminant of this quadratic equation is $1^2 - 4(2)(2) = -15$, which is negative. Therefore, there is no real solution for x and the original system has no real solution. Figure 43.2 shows the graph of the system.

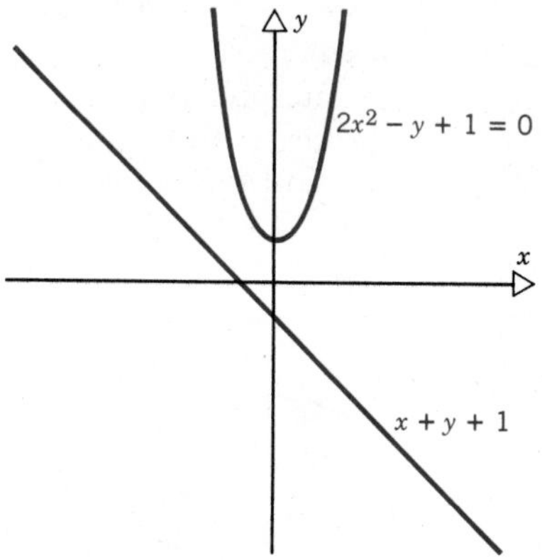

Figure 43.2

B. Solution by Elimination

Many two-variable systems of two quadratic equations can be solved by elimination. The idea here is the same as for two-variable systems of two linear equations (Section 40): obtain an equation with only one variable by adding an appropriate constant multiple of one of the equations to the other equation.

EXAMPLE 43.3 Solve the system

$$4x^2 + 9y^2 = 36 \quad \text{(A)}$$
$$2x^2 - 3y^2 = -2 \quad \text{(B)}$$

by elimination.

Solution To eliminate y, add 3 times Equation (B) to Equation (A). This gives

$$10x^2 = 30, \quad x^2 = 3, \quad x = \pm\sqrt{3}.$$

Now substitute $x = \sqrt{3}$ and then $x = -\sqrt{3}$ in Equation (A) to determine the corresponding values of y. [We could also substitute in (B) to determine y.] With either $x = \sqrt{3}$ or $x = -\sqrt{3}$ we get

$$4 \cdot 3 + 9y^2 = 36, \quad 9y^2 = 24, \quad y = \pm\sqrt{\tfrac{24}{9}} = \pm 2\sqrt{6}/3.$$

This yields the four solutions $(\sqrt{3}, 2\sqrt{6}/3)$, $(\sqrt{3}, -2\sqrt{6}/3)$, $(-\sqrt{3}, 2\sqrt{6}/3)$, and $(-\sqrt{3}, -2\sqrt{6}/3)$.

The graphs of Equations (A) and (B) are shown in Figure 43.3. The four solutions are the coordinates of the points where the graphs intersect. If you think about the possibilities for the graphs of quadratic equations in x and y (Section 21), you'll see that a two-variable system composed of two (nonequivalent) quadratic equations will have either four, three, two, one, or no solutions.

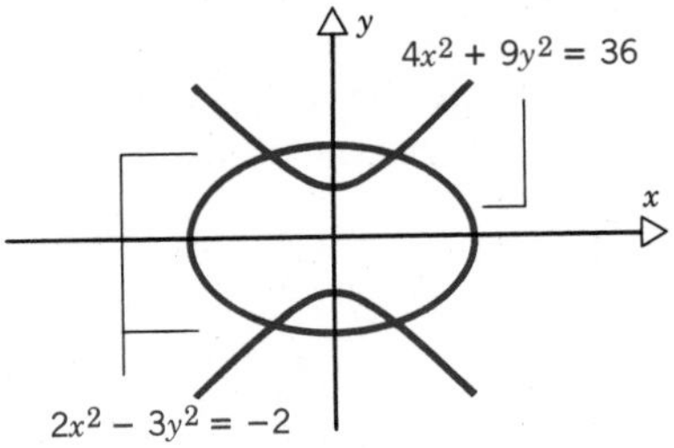

Figure 43.3

EXAMPLE 43.4 Solve the system

$$x^2 - xy + y^2 = 19 \qquad \text{(A)}$$
$$x^2 + y^2 = 13. \qquad \text{(B)}$$

Solution Subtract (A) from (B) to eliminate the x^2 and y^2 terms:

$$xy = -6$$
$$y = -6/x. \qquad \text{(C)}$$

Substitute $y = -6/x$ in (B) to get an equation that involves only one variable:

$$x^2 + (-6/x)^2 = 13$$
$$x^4 - 13x^2 + 36 = 0$$
$$(x^2 - 4)(x^2 - 9) = 0$$
$$x = \pm 2 \quad \text{or} \quad \pm 3.$$

Equation (C) gives a value of y to go with each of these values of x. Therefore, the system has four solutions, which can be checked by substitution: $(2, -3)$, $(-2, 3)$, $(3, -2)$, and $(-3, 2)$.

C. Solution by Graphing

If we cannot find exact solutions for a system, then we may still be able to find approximate solutions. The following example illustrates how to find an approximate solution graphically.

EXAMPLE 43.5 Use graphs to find an approximate solution for the system

$$y = 2^x$$
$$y = x^2$$

with $x < 0$. (A solution with $x > 0$ is $x = 2$, $y = 4$.)

Solution The idea is to draw the graphs of the equations as accurately as possible and then locate their points of intersection, if there are any. Figure 43.4*a* shows the graphs. (The graph of $y = 2^x$ is from Figure 35.3.) It is clear that there is only one point of intersection for $x < 0$. Figure 43.4*b* is a magnification of the portions of the graphs where the intersection occurs. The solution is seen to be approximately $x = -0.8$, $y = 0.6$. (A closer approximation is $x = -0.767$, $y = 0.588$, as can be shown with a calculator having a $\boxed{y^x}$ key.

D. Application

EXAMPLE 43.6 The perimeter of a rectangle is 35 centimeters and the area is 49 square centimeters. What are the dimensions?

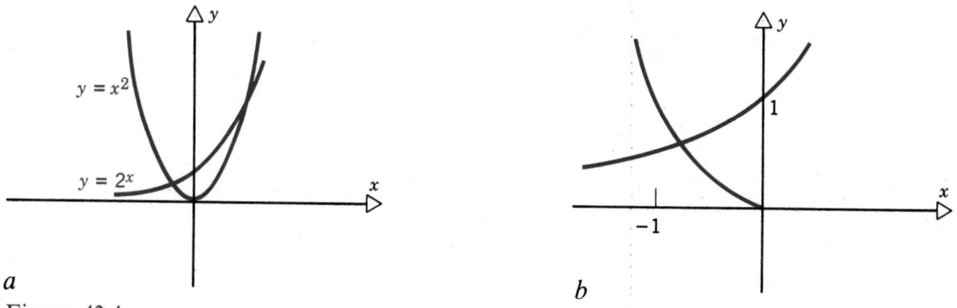

Figure 43.4

Solution Let x and y denote the dimensions. Then the area is xy and the perimeter is $2x + 2y$. This gives the system

$$xy = 49$$
$$2x + 2y = 35.$$

The method of substitution will give the two solutions (3.5, 14) and (14, 3.5). Either solution implies that the rectangle is 3.5 centimeters by 14 centimeters.

EXERCISES FOR SECTION 43

Find the real solutions of each system.

1. $y - x = 6$
   $x^2 + y^2 = 18$
2. $x + 4y = 17$
   $x^2 + y^2 = 34$
3. $x + y = 3$
   $x^2 - y^2 = 1$
4. $2x - y = -4$
   $y = 8x^2 + 1$
5. $2x + 2y = 1$
   $4x^2 - y^2 = 8$
6. $3x - y = -1$
   $y = x^2 - 2$

7. $3y = x^2 - 6$
$4x^2 + 9y^2 = 36$

8. $2x^2 + y^2 = 1$
$y = x^2 + 1$

9. $2x^2 + y^2 = 1$
$x^2 - 2y^2 = 1$

10. $4x^2 + 9y^2 = 36$
$y^2 - x^2 = 1$

11. $y = x^3$
$x = 2y^2$

12. $y = 2x^3$
$y = -x^2$

13. $x^2 - xy = 2$
$x - xy = -4$

14. $xy = 1$
$8y = x^2$

15. $xy = 1$
$y = x + 2$

16. $3x^2 + xy + 3y^2 = 52$
$x^2 + y^2 = 20$

17. $\frac{1}{x} + \frac{1}{y} = 2$
$\frac{1}{x} - \frac{1}{y} = 1$

18. $\log x + \log 2y = 1$
$\log 8x + 2 \log y = 0$

Use graphs to find an approximate solution for each system.

19. $y = 2^x$
$y = -x + 2$

20. $y = \log x$
$xy = 1$

21. $y = \log x$
$y = -x$

22. $y = 2 \log x$
$y = -x + 2$

23. $y = 2^{-x}$
$y = x$

24. $y = 10^x$
$y = x^2$

25. The sum of two real numbers is 4 and the difference of their squares is also 4. Find the numbers.

26. The perimeter of a rectangle is 17 inches and the area is 15 square inches. What are the dimensions?

27. The areas of two circles differ by 144π square centimeters and their circumferences differ by 12π centimeters. Find the radius of each circle.

28. The perimeter of a right triangle is 60 centimeters and the hypotenuse is 26 centimeters. How long are the legs? (Suggestion: $P = a + b + c$ and $a^2 + b^2 = c^2$, where a and b denote the lengths of the legs.)

29. If a plane's usual speed for the trip between two cities 420 miles apart is increased by 60 miles per hour, the trip will take 10 minutes less than usual. Determine the usual speed and flying time. [Suggestion: $420 = RT$ and $420 = (R + 60)(T - \frac{1}{6})$.]

30. Savings Accounts A and B each earn \$100 simple interest annually. The rate for Account B is 2% less than the rate for Account A, and the principal for Account B is \$250 more than the principal for Account A. Determine the principal and rate for each account. [Suggestion: If P and R denote the principal and rate for Account A, then $100 = PR$ and $100 = (P + 250)(R - 0.02)$.]

31. For which values of k does the system

$$x^2 + y^2 = 9$$
$$y = x + k$$

have two solutions? one solution? no solution? Interpret your answer geometrically.

32. For which values of k does the system

$$x^2 + (y - 3)^2 = 4$$
$$y = kx$$

have two solutions? one solution? no solution? Interpret your answers geometrically. [See Equation (19.4).]

33. Two circles are to be placed inside a larger circle so that the centers are collinear and the three circles are tangent as shown in Figure 43.5. Show that this can be done in such a way that the area of the larger circle is k times the sum of the areas of the smaller circles, provided $\frac{1}{2} \leq k \leq 1$. (Allow 0 as a radius.) Determine a and b in terms of c and k. What does the figure look like in the case $k = \frac{1}{2}$? $k = 1$?

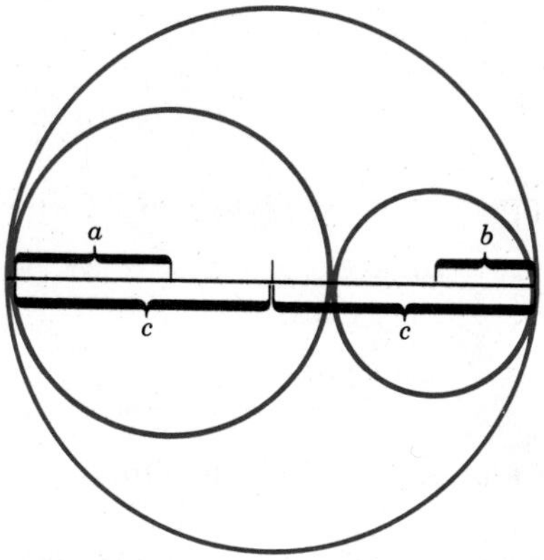

Figure 43.5

SECTION 44 Systems of Inequalities

A. Graphs of Inequalities

The next section will show that applications can involve systems of inequalities as well as systems of equations. Before looking at such applications, however, we need more facts about inequalities that contain variables.

A **solution** of an inequality in variables x and y is an ordered pair (a, b) of numbers that satisfies the inequality; that is, such that a true statement results if x is replaced by a and y is replaced by b.

EXAMPLE 44.1 Consider the inequality $2x - 3y < 6$.

(a) The pair $(0, 0)$ is a solution because $2(0) - 3(0) < 6$.
(b) The pair $(2, -1)$ is not a solution because $2(2) - 3(-1) \not< 6$.

The **graph** of an inequality in x and y is the set of all points in the Cartesian plane whose coordinates are solutions of the inequality. To graph an inequality we make use of the equation that results when the inequality sign ($<$, $>$, $\leq$, or $\geq$) is replaced by an equality sign; we call this equation the **associated equation** of the inequality. Thus the associated equation of $2x - 3y < 6$ is $2x - 3y = 6$. An inequality is said to be **linear** if its associated equation is linear; otherwise the inequality is **nonlinear.**

The first step in graphing an inequality is to graph its associated equation. If the sign in the inequality is either $<$ or $>$, then the points on the graph of the associated equation *will not* belong to the graph of the inequality; we indicate this by graphing the associated equation as a *dashed* line or curve (as in Figure 44.1). If the sign in the inequality is either $\leq$ or $\geq$, then the points on the graph of the associated equation *will* belong to the graph of the inequality; in this case we graph the associated equation as a *solid* line or curve (as in Figure 44.2).

EXAMPLE 44.2 Draw the graph of $2x - 3y < 6$.

Solution The graph of the associated equation is the dashed line in Figure 44.1. It will be easier to see just what the graph of the inequality is if we solve it for y:

$$-3y < 6 - 2x$$
$$y > (6 - 2x)/-3$$
$$y > (2x - 6)/3.$$

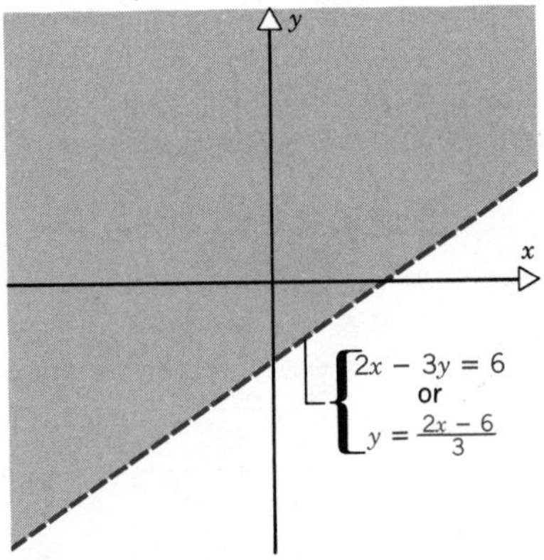

Figure 44.1

For each x, a point with coordinates (x, y) will be

on the dashed line if $y = (2x - 6)/3$

and

above the dashed line if $y > (2x - 6)/3$.

Therefore, the graph of the inequality is the shaded region consisting of all of the points above the dashed line.

Any line in a plane divides the points of the plane into three sets: the points on one side of the line, the points on the other side of the line, and the points on the dividing line itself. Each of the first two sets is called an **open half-plane.** The graph of any linear inequality with $<$ or $>$ (like Figure 44.1) will be an open half-plane. An open half-plane together with the points on the dividing line is called a **closed half-plane.** The graph of any linear inequality with $\leq$ or $\geq$ (like Figure 44.2, which follows) will be a closed half-plane. The graph of any linear inequality can be drawn quickly with the following two steps.

Step 1. Draw the graph of the associated equation, making the appropriate choice of a dashed line or a solid line.

Step 2. Choose any point (a, b) not on the graph of the associated equation. If (a, b) *satisfies* the inequality, then the graph is the half-plane *containing* (a, b) (open or closed, as appropriate). If (a, b) *does not satisfy* the inequality, then the graph is the half-plane *not containing* (a, b) (again, open or closed, as appropriate).

For Step 1, recall that the quickest way to graph most linear equations is to use the intercepts, but the important thing is to use points that are reasonably far apart. For Step 2, the easiest point to check will be $(0, 0)$; if $(0, 0)$ satisfies the associated equation, then we must use some other point to help decide which half-plane to use.

EXAMPLE 44.3 Draw the graph of $6x + 10y + 15 \geq 0$.

Solution The y-intercept of the associated equation is $(0, -1.5)$. The x-intercept is $(-2.5, 0)$. Draw the solid line through these points (Figure 44.2). The pair $(0, 0)$ satisfies the inequality, so the graph is the closed half-plane containing the origin, as shown.

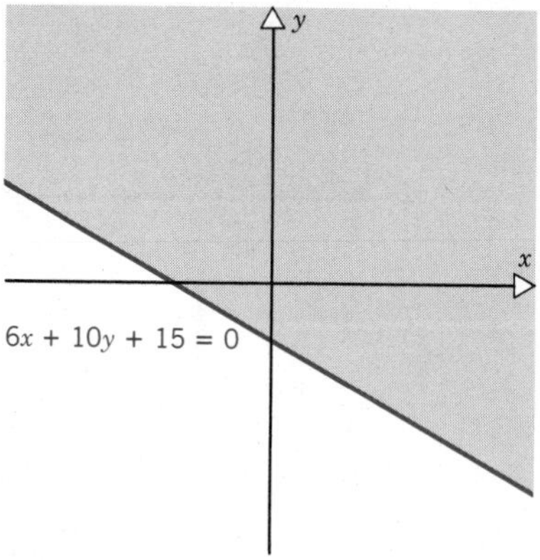

Figure 44.2

The two-step method given for graphing linear inequalities can be modified to handle many nonlinear inequalities, especially when the graph of the associated equation divides the other points of the plane into two easily identifiable parts. This includes most of the cases where the associated equation can be written as either $y = f(x)$ or $x = g(y)$ for a function f or g. Here is an example.

EXAMPLE 44.4 Draw the graph of $x^2 + y < 0$.

Solution This can be rewritten as $y < -x^2$. The graph of the associated equation is the dashed parabola in Figure 44.3, which divides the plane into two parts. It is clear that the points for which $y < -x^2$ are those below the parabola; thus they make up the graph of the inequality. We could also determine which of the two parts of the plane to use by testing a pair such as (1, 1); since this pair does not satisfy $y < -x^2$, the graph is the part of the plane not containing (1, 1).

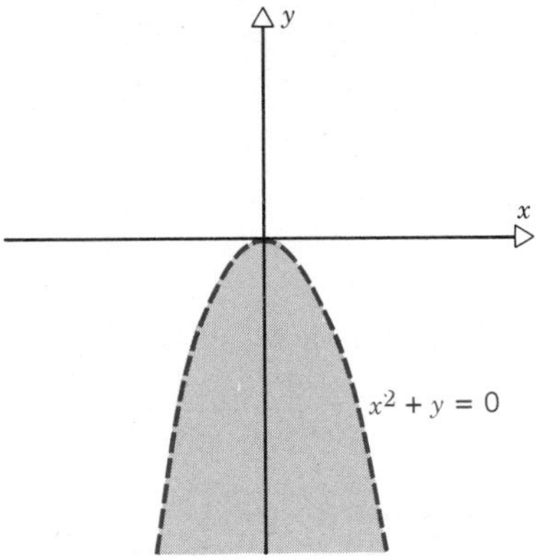

Figure 44.3

B. Systems of Inequalities

A **solution** of a system of inequalities in x and y is an ordered pair of numbers that satisfies every inequality in the system. The set of all solutions of such a system is called the **solution set** of the system.

In many cases this solution set can be described most easily with a graph. An element is in the **intersection** of a collection of sets iff it is in every set of the collection (Appendix B). It follows that the graph of the solution set of a system of inequalities is the intersection of the graphs of all of the inequalities in the system.

EXAMPLE 44.5 Graph the solution set of the system

$$x^2 + y^2 < 25$$
$$x - y^2 \geq 0.$$

Solution The graph of $x^2 + y^2 = 25$ is the dashed circle in Figure 44.4. The graph of $x^2 + y^2 < 25$ is the set of all points inside that circle.

The graph of $x = y^2$ is the parabola in Figure 44.4. The parabola divides the other points of the plane into two parts; the graph of $x - y^2 \geq 0$ includes the part to the right of the parabola, as can be seen by verifying that the pair (1, 0) satisfies the inequality.

The graph of the solution set of the system is the intersection of the graphs of the two inequalities.

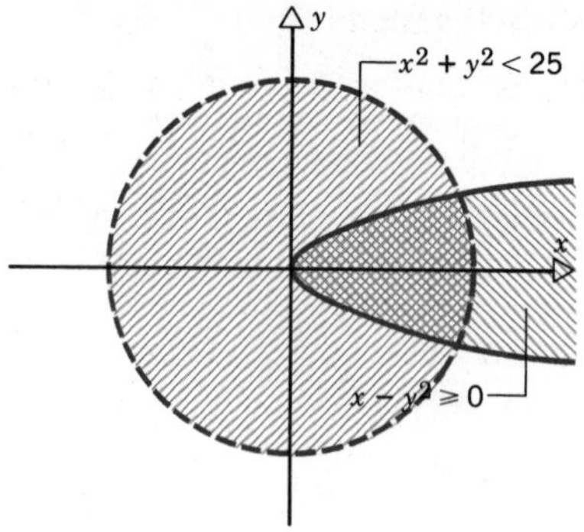

Figure 44.4

EXAMPLE 44.6 Graph the solution set of the system

$$x \geq 0, \quad y \geq 0, \quad x + 2y \leq 4.$$

Solution The graph of $x \geq 0$ is the closed half-plane of all points on or to the right of the y-axis. The graph of $y \geq 0$ is the closed half-plane of all points on or above the x-axis. The graph of $x + 2y \leq 4$ is the closed half-plane of all points on or below the line $x + 2y = 4$. The intersection of these graphs is the triangular region in Figure 44.5.

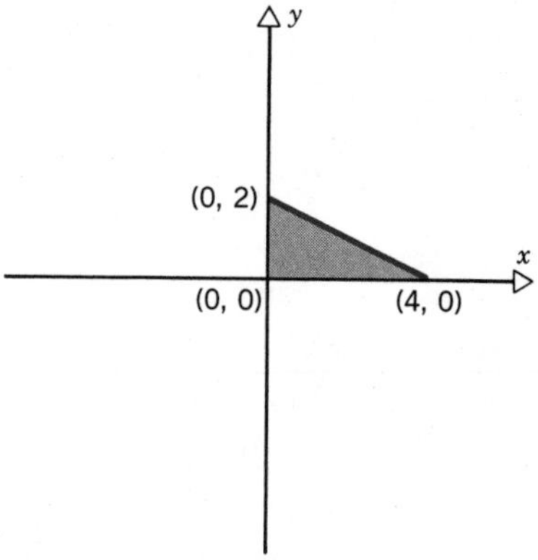

Figure 44.5

EXERCISES FOR SECTION 44

Draw the graph of each inequality.

1. $5x + 2y \geq 10$
2. $3x - 4y > 12$
3. $x - 2y \leq 4$
4. $-x + 2y < 6$
5. $2x + 2y \leq 5$
6. $4x + 3y > 8$
7. $x^2 + y^2 < 4$
8. $y - 2x^2 \geq 0$
9. $y + x^2 < 1$
10. $y \geq |x|$
11. $y < -|x|$
12. $x \geq |y|$

Graph the solution set of each system.

13. $y - x^2 \geq 0$
$y < 2$

14. $x^2 + y^2 < 9$
$x \geq 0$

15. $x^2 + y^2 \leq 25$
$y - x \geq 0$

16. $5x - 6y > -3$
$y > 0$
$x \leq 0$

17. $-x + y \geq -2$
$x > 0$
$y < 0$

18. $x + y < 3$
$y > 1$
$x \geq 0$

19. $x^2 + y^2 > 2$
$-3 \leq x \leq 3$
$-3 \leq y \leq 3$

20. $y < -x^2 + 2$
$y > x^2 - 2$

21. $y - x^2 \geq 0$
$x - y^2 \geq 0$

22. $y + 2x \geq 4$
$x \geq 0$
$y \geq 0$

23. $y - x^2 > 0$
$x - y \geq -2$

24. $x^2 + y^2 > 4$
$x^2 + y^2 < 9$
$y \geq 0$

25. $y - x < 2$
$y + x < 2$
$y - x > -2$
$y + x > -2$

26. $2y + 3x \geq 6$
$4y + x \geq 4$
$4y + 3x \leq 12$

27. $2x + 5y \geq 10$
$6x + 2y \geq 12$
$x \geq 0$
$y \geq 0$

In each of Exercises 28–33, write a system of inequalities having the given set as its solution set. (The points on the x- and y-axes are not in any quadrant.)

28. The points in the first quadrant.
29. The points in the second quadrant.
30. The points in the fourth quadrant.
31. The points in the third quadrant that are more than 10 units from the origin.
32. The points above the x-axis that are within 5 units of the origin.
33. The points that are closer to the x-axis than to the y-axis.

SECTION

45 Linear Programming

A. Linear Programming Theorem

Many applied problems involve maximizing or minimizing the output of a function subject to certain constraints on the input of the function. We now consider several examples of such problems that can be solved by a method known as *linear programming*. For motivation, before reading Example 45.1 you may want to glance at the problems in Examples 45.2 and 45.3; Example 45.1 is less interesting than the later examples, but it is free of the distractions of "word" problems.

EXAMPLE 45.1 Consider the function of x and y defined by $f(x, y) = 3x + 5y + 2$ for each pair (x, y) in the solution set of Example 44.6 (Figure 44.5). For instance,

$$f(2.5, 0) = 3(2.5) + 5(0) + 2 = 9.5$$

and

$$f(2, 1) = 3(2) + 5(1) + 2 = 13.$$

Problem For which pair or pairs (x, y), if any, does $f(x, y)$ attain a maximum value? What about a minimum value?

Solution The triangular region in Figure 44.5 contains an infinite number of points, so we certainly cannot answer the question by checking every possible pair (x, y). The theorem following this example will tell us that, in fact, we need to check only the vertices of the triangular region; the function f will attain both its maximum and minimum values when it is evaluated at appropriate vertices. In the present case, the vertices are at $(0, 0)$, $(0, 2)$, and $(4, 0)$, and

$$f(0, 0) = 3(0) + 5(0) + 2 = 2$$
$$f(0, 2) = 3(0) + 5(2) + 2 = 12$$
$$f(4, 0) = 3(4) + 5(0) + 2 = 14.$$

Thus f attains its maximum value, which is 14, at $(4, 0)$. And f attains its minimum value, which is 2, at $(0, 0)$.

Before we look at the theorem used in Example 45.1 we need several preliminary remarks. A **linear function** of x and y is a function of the form $f(x, y) = ax + by + c$, where a, b, and c are real numbers. Each linear programming problem that we consider will involve such a function and a corresponding **feasible set,** which will be the solution set of a finite system of linear inequalities in x and y. Geometrically, each feasible set will be the intersection of a finite number of closed half-planes; these half-planes are bounded by lines (the graphs of the associated equations), and the points where these lines intersect are called the **vertices** of the feasible set. Figures 44.5 and 45.1 show feasible sets with three vertices each; Figure 45.2 shows a feasible set with four vertices.

Linear Programming Theorem

Suppose that f is a linear function of x and y and that S is the feasible set of f. If f attains a maximum (minimum) value for (x, y) in S, then that maximum (minimum) value is attained at a vertex of S.

Subsection C will indicate why the theorem is true.

B. Applications

EXAMPLE 45.2 A hospital dietician must plan a day's menu using some combination of roasted chicken and canned tuna as a primary source of protein. Each person must receive at least 25 grams of protein from the combination, and the cost must not exceed \$0.45 per person. Subject to those restrictions the number of calories is to be made as small as possible. Use the data in Table 45.1 to help decide the amount of chicken and tuna that the dietician should choose.

TABLE 45.1

	Chicken (per unit)	*Tuna (per unit)*
Protein	30	20
Calories	180	140
Cost	\$0.70	\$0.30

Solution Let x and y denote respectively the number of units of chicken and tuna to be used. Then chicken will contribute $30x$ grams of protein and tuna will contribute $20y$ grams of protein. Chicken will contribute $\$0.70x$ to the cost and tuna will contribute $\$0.30y$ to the cost. Using the given restrictions on the protein and cost, and the fact that x and y cannot be negative, we arrive at the following constraints.

$$x \geq 0$$
$$y \geq 0$$
$$30x + 20y \geq 25$$
$$0.70x + 0.30y \leq 0.45$$

The feasible set is the solution set of this system, which is shown in Figure 45.1. Any point in the feasible set will give a combination that meets the requirements on protein and cost, but we want a point where the number of calories is as small as possible.

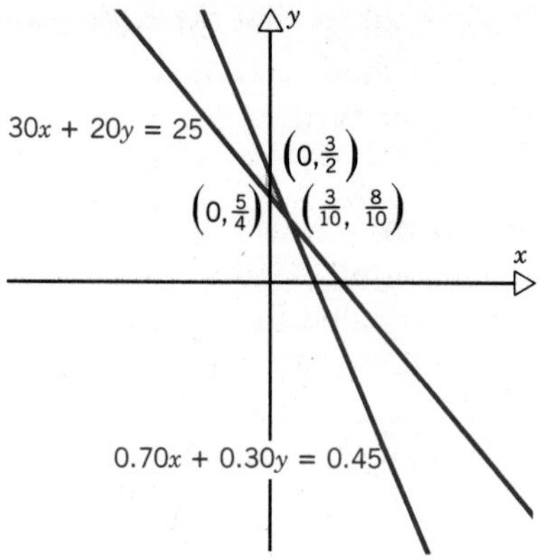

Figure 45.1

Chicken will contribute $180x$ calories and tuna will contribute $140y$ calories. To make the number of calories as small as possible we must minimize the function

$$f(x, y) = 180x + 140y.$$

By the Linear Programming Theorem the minimum will occur at a vertex of the feasible set. From

$$f(0, \tfrac{3}{2}) = 180(0) + 140(\tfrac{3}{2}) = 210$$
$$f(0, \tfrac{5}{4}) = 180(0) + 140(\tfrac{5}{4}) = 175$$
$$f(\tfrac{3}{10}, \tfrac{8}{10}) = 180(\tfrac{3}{10}) + 140(\tfrac{8}{10}) = 166$$

we see that the dietician should use $\frac{3}{10}$ units of chicken and $\frac{8}{10}$ units of tuna. This will give 166 calories.

EXAMPLE 45.3 Table 45.2 shows the nutrient contents and costs for two brands of fertilizer. A farmer wants to mix the two brands in the most economical way possible, subject to the restrictions that the mixture have at least 13 units of nitrogen, at least 8 units of phosphorus, and at least 9 units of potassium. What combination should the farmer choose?

TABLE 45.2

	A	*B*
Nitrogen	3 units per bag	2 units per bag
Phosphorus	3 units per bag	1 unit per bag
Potassium	1 unit per bag	3 units per bag
Cost	$15 per bag	$20 per bag

Solution Let x and y denote respectively the number of bags of brands A and B to be used. The restrictions give three inequalities.

$$3x + 2y \geq 13 \quad \text{(nitrogen)}$$
$$3x + y \geq 8 \quad \text{(phosphorus)}$$
$$x + 3y \geq 9 \quad \text{(potassium)}$$

Also, $x \geq 0$ and $y \geq 0$. The feasible set that these inequalities produce is shown in Figure 45.2.

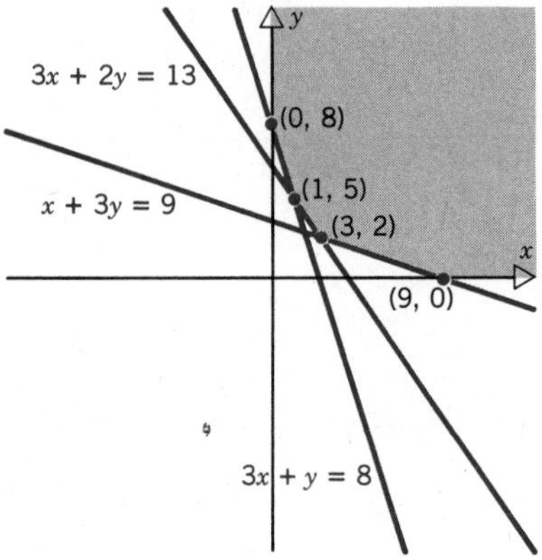

Figure 45.2

Brand A will contribute $15x$ to the total cost and brand B will contribute $20y$ to the total cost. To minimize the cost we minimize the function

$$f(x, y) = 15x + 20y.$$

By the Linear Programming Theorem, if there is a minimum it will occur at one of the four vertices. From

$$\begin{aligned} f(0, 8) &= 15(0) + 20(8) = 160 \\ f(1, 5) &= 15(1) + 20(5) = 115 \\ f(3, 2) &= 15(3) + 20(2) = 85 \\ f(9, 0) &= 15(9) + 20(0) = 135 \end{aligned}$$

we see that the farmer should use 3 bags of brand A and 2 bags of brand B. Then the total cost will be \$85.

C. Remarks

Notice that the Linear Programming Theorem states that *if* the linear function f attains a maximum (or a minimum) value, then that value is attained at a vertex. However, the function need not attain a maximum (or a minimum). A simple example is provided by $f(x, y) = x + y$ with the feasible set $x \geq 0$, $y \geq 0$; in this case there is clearly no maximum. For some feasible sets both a maximum and a minimum are guaranteed, whatever the linear function may be. Specifically, a subset of the plane is said to be **bounded** if there is a circle containing all of its points; otherwise the set is said to be **unbounded.** If the feasible set is bounded (as in Figures 44.5 and 45.1), then the linear function will attain a maximum and a minimum value. However, if the feasible set is unbounded (as in Figure 45.2), then the function may not attain a maximum (or a minimum) value.

In some cases a maximum or a minimum value is attained at two different vertices. When this happens $f(x, y)$ is constant on the segment connecting the two vertices. Exercise 5 gives an example.

The Linear Programming Theorem will not be proved here, but one way to think about the theorem is as follows. Imagine a z-axis, perpendicular to the Cartesian plane, directed out from the page, and with 0 on the z-axis at the origin of the plane. For the function f in a linear programming problem consider the equation $z = f(x, y)$. This equation will have a graph in three-dimensional space just as an equation in x and y has a graph in the Cartesian plane. It can be proved that, because f is linear, the graph of $z = f(x, y)$ is a plane. The Linear Programming Theorem is a consequence of the fact that among the points of this plane corresponding to (x, y) in a feasible set (an intersection of closed half-planes), the ones farthest from and closest to the Cartesian (x and y) plane occur for (x, y) at the vertices of the feasible set.

The theory of linear programming can handle much more general problems than those stated here. For example, many problems covered by the theory involve more than two variables.

EXERCISES FOR SECTION 45

The feasible set for Exercises 1–6 is the solution set of the system $x \geq 0$, $y \geq 0$, and $2x + y \leq 4$. In each case determine the maximum and minimum values of the function f for (x, y) in the feasible set, and also the pairs (x, y) where those maximum and minimum values are attained. Remember that if a maximum or a minimum value is attained at two different vertices, then it is also attained at all points on the segment connecting those vertices.

1. $f(x, y) = x + 3y - 1$
2. $f(x, y) = x - 5y + 5$
3. $f(x, y) = -3x - y - 2$
4. $f(x, y) = x - 7$
5. $f(x, y) = 4x + 2y$
6. $f(x, y) = -y + 1$

The feasible set for Exercises 7–12 is the solution set of the system $x \geq 0$, $y \geq 0$, $x - 5y \geq -15$, and $4x + y \leq 24$. In each case determine the maximum and minimum values of the function f for (x, y) in the feasible set, and also the pairs (x, y) where those maximum and minimum values are attained.

7. $f(x, y) = 2x - 2y + 1$
8. $f(x, y) = -x + y$
9. $f(x, y) = 5x + y - 2$
10. $f(x, y) = 5$
11. $f(x, y) = 8x + 2y$
12. $f(x, y) = -2x + 10y + 1$

13. Redo Example 45.2 assuming that the cost of chicken has decreased to \$0.50 per unit and the cost of tuna has increased to \$0.40 per unit.

14. Redo Example 45.2 assuming that the cost of chicken has increased to \$0.80 per unit and the cost of tuna has decreased to \$0.20 per unit.

15. Redo Example 45.3 assuming that the cost of both brands A and B is \$20 per bag.

16. Consider Example 45.3 and assume that the cost of brand A increases while the cost of brand B remains constant. At what point would it be to the farmer's advantage to use only brand B?

17. Refer to Table 45.3. How many cups each of orange juice and tomato juice would provide the most iron subject to the restrictions that the total calories not exceed 300 and the total amount of vitamin A be at least 2000 IU (international units)? How much iron would be provided?

18. Refer to Table 45.3. How many cups each of orange juice and tomato juice would provide the most vitamin C subject to the restrictions that the total calories not exceed 300 and the total amount of vitamin A be at least 2000 IU (international units)?

TABLE 45.3

	Orange juice (one cup)	*Tomato juice (one cup)*
Calories	120	50
Vitamin A (IU)	550	2000
Vitamin C (mg)	120	40
Iron (mg)	0.2	2.2

The feasible set for Exercises 19–21 is the solution set for the system $x \geq 0$, $y \geq 0$, $3x + y \geq 5$, and $2x + 5y \geq 12$. Notice that this set is unbounded, and remember that a linear function with an unbounded feasible set may not attain a maximum or a minimum value (Subsection C).

19. Consider the function $f(x, y) = x - y + 5$.
(a) Verify that f does not attain a maximum value for (x, y) in the given feasible set. [What happens to $f(x, 0)$ as x increases?]
(b) Verify that f does not attain a minimum value for (x, y) in the given feasible set. [What happens to $f(0, y)$ as y increases?]

20. Consider the function $f(x, y) = -x - y$.
(a) Verify that f does not attain a minimum value for (x, y) in the given feasible set. [What happens to $f(x, 0)$ as x increases?]
(b) Verify that $f(1, 2)$ is more than the output of f at any other vertex of the feasible set. [In fact, $f(1, 2)$ is the maximum value of f.]

21. Determine a and b so that $f(x, y) = ax + by$ has no maximum value but has the same value for all (x, y) on the line $3x + y = 5$. (This common value will be a minimum value for the function.)

REVIEW EXERCISES FOR CHAPTER X

Use either substitution or elimination to determine all of the solutions of each system.

1. $\begin{aligned} x - 3y &= 2 \\ -2x + 6y &= -4 \end{aligned}$

2. $\begin{aligned} 3x - y &= 5 \\ x - 2y &= -10 \end{aligned}$

3. $\begin{aligned} 3x + 4y &= -1 \\ 6x - 2y &= 3 \end{aligned}$

4. $\begin{aligned} -5x + 3y &= 2 \\ 15x - 9y &= -5 \end{aligned}$

5. $\begin{aligned} 4x - y - 1 &= 0 \\ -8x + 2y - 3 &= 0 \end{aligned}$

6. $\begin{aligned} 12x - 6y - 9 &= 0 \\ -4x + 2y + 3 &= 0 \end{aligned}$

Use Gaussian elimination to determine all of the solutions of each system.

7. $\begin{aligned} x - 2y + z &= 8 \\ -x \phantom{{}- 2y} - 2z &= -8 \\ 3x + 4y + 3z &= 4 \end{aligned}$

8. $\begin{aligned} 2x - 3y + z &= 1 \\ 2x + 2y + 2z &= 3 \\ -2x - 4y + 4z &= 1 \end{aligned}$

9. $\begin{aligned} x - y + z &= 3 \\ 2x + y - z &= 1 \\ 3y - 3z &= -5 \end{aligned}$

10. $\begin{aligned} 3x - 2y - z &= 4 \\ x + y + z &= 18 \\ -2x + y + z &= -5 \end{aligned}$

11. $\begin{aligned} 2x - 3y + 4z - 10 &= 0 \\ 2x + y - 5z - 18 &= 0 \\ -x + 4y + z &= 0 \end{aligned}$

12. $\begin{aligned} x - y - 5z - 1 &= 0 \\ 2x + 5y + 4z - 6 &= 0 \\ 7y + 14z - 3 &= 0 \end{aligned}$

Solve each system by reducing the augmented matrix to reduced echelon form.

13. $\begin{aligned} x + y + z &= 7 \\ x - 3y - z &= 5 \\ 3x \phantom{{}+ y} + z &= 17 \end{aligned}$

14. $\begin{aligned} x + y + 2z &= 8 \\ 3x - y - z &= -3 \\ 2x + 4y + z &= 3 \end{aligned}$

15. $\begin{aligned} y - 9z &= 5 \\ 3x + 2y + 3z &= 1 \\ x \phantom{{}+ 2y} - 2z &= 3 \end{aligned}$

16. $\begin{aligned} -x + 7y - 3z &= 4 \\ x + y - z &= 2 \\ 2x - 2y &= 1 \end{aligned}$

17. $\begin{aligned} 4x - 2y + 3z - 2 &= 0 \\ y - 3z - 12 &= 0 \\ x + 3y + 10z - 3 &= 0 \end{aligned}$

18. $\begin{aligned} 2x \phantom{{}+ y} - 3z - 4 &= 0 \\ -x + y + 2z - 9 &= 0 \\ x - 2y - z + 5 &= 0 \end{aligned}$

Find the real solutions of each system.

19. $\begin{aligned} 3x - y &= -7 \\ y &= 3x^2 + 1 \end{aligned}$

20. $\begin{aligned} 2x - y &= -4 \\ y &= 2x^2 \end{aligned}$

21. $\begin{aligned} x^2 + y^2 &= 25 \\ x^2 + 9y^2 &= 36 \end{aligned}$

22. $\begin{aligned} x^2 - y^2 &= 5 \\ x^2 + 4y^2 &= 25 \end{aligned}$

Graph the solution set of each system.

23. $\begin{aligned} x + 2y &\geq 10 \\ x^2 + y^2 &< 25 \end{aligned}$

24. $\begin{aligned} x &< y^2 \\ x^2 + 2y^2 &\leq 24 \end{aligned}$

25. Determine the maximum and minimum values of $f(x, y) = 3x - 9y$ for (x, y) restricted to the feasible set $x \geq 0$, $y \geq 0$, $x + y \geq 1$, $x + 3y \leq 3$.

26. Determine the maximum and minimum values of $f(x, y) = x + 2y - 3$ for (x, y) restricted to the feasible set $y \leq 2x$, $3y \geq x$, $x + 2y \leq 5$.

CHAPTER XI
MATRICES AND DETERMINANTS

This chapter continues the study of matrices that was begun in Chapter X. It includes definitions, properties, and applications of *determinants,* which are special functions of matrices.

SECTION

46 Matrix Algebra

A. Terminology

In Section 42 we used matrices merely as a convenience—to avoid writing variables, +'s, and ='s in solving systems of linear equations. In many other applications matrices are more than a mere convenience. The most significant applications depend on combining matrices through the operations of *addition* and *multiplication,* which will be studied in this section. We begin with some general terminology.

A matrix having m rows and n columns is said to be $\mathbf{m \times n}$ (read "m by n"). We also say the matrix has **size** $m \times n$. The number in the ith row and jth column of a matrix A is called its (i, j)-entry and is denoted a_{ij}. Thus, if A is an $m \times n$ matrix, then

$$A = \begin{bmatrix} a_{11} & a_{12} & \cdots & a_{1n} \\ a_{21} & a_{22} & \cdots & a_{2n} \\ \vdots & \vdots & & \vdots \\ a_{m1} & a_{m2} & \cdots & a_{mn} \end{bmatrix}$$

This is often abbreviated $A = [a_{ij}]$. In this book matrix entries will be real numbers.

Two matrices A and B are said to be **equal** if they have the same size (same number of rows and columns) and all of their corresponding entries are equal—that is, if $a_{ij} = b_{ij}$ for all i and j.

EXAMPLE 46.1 Let

$$A = \begin{bmatrix} 3 & 7 \\ -1 & 0 \end{bmatrix}, \quad B = \begin{bmatrix} 3 & 7 & -4 \\ -1 & 0 & 5 \end{bmatrix}, \quad C = \begin{bmatrix} c_{11} & c_{12} \\ -1 & 0 \end{bmatrix}.$$

Then $A \neq B$ and $B \neq C$ because the sizes are different in each case. And

$$A = C \text{ iff } c_{11} = 3 \text{ and } c_{12} = 7.$$

B. Matrix Addition

DEFINITION. The **sum** of two $m \times n$ matrices A and B, denoted $A + B$, is the $m \times n$ matrix whose (i, j)-entry is $a_{ij} + b_{ij}$.

In other words, matrices of the same size are added by adding their corresponding entries. Matrices can be added only if they have the same size.

EXAMPLE 46.2

$$\begin{bmatrix} 3 & 0 \\ 6 & 2 \\ -8 & 4 \end{bmatrix} + \begin{bmatrix} 1 & 8 \\ -1 & 3 \\ 1 & -5 \end{bmatrix} = \begin{bmatrix} 3+1 & 0+8 \\ 6+(-1) & 2+3 \\ -8+1 & 4+(-5) \end{bmatrix} = \begin{bmatrix} 4 & 8 \\ 5 & 5 \\ -7 & -1 \end{bmatrix}$$

Matrix addition satisfies many of the properties of addition of real numbers. Here are the first two such properties.

Commutative law for matrix addition. If A and B are $m \times n$ matrices, then

$$A + B = B + A.$$

Associative law for matrix addition. If A, B, and C are $m \times n$ matrices, then

$$A + (B + C) = (A + B) + C.$$

Partial Proof. The (i, j)-entry of $A + B$ is $a_{ij} + b_{ij}$. The (i, j)-entry of $B + A$ is $b_{ij} + a_{ij}$. These entries are equal because $a_{ij} + b_{ij} = b_{ij} + a_{ij}$ for all a_{ij} and b_{ij} by the commutative law for addition of real numbers (Section 1). The proof of the associative law is similar. □

For each pair (m, n), the $m \times n$ matrix whose entries are all 0's is called the $m \times n$ **zero matrix.** For example, the 2×3 zero matrix is

$$\begin{bmatrix} 0 & 0 & 0 \\ 0 & 0 & 0 \end{bmatrix}.$$

Each zero matrix will be denoted by 0; if there is a chance of confusion about which zero matrix is meant, the $m \times n$ zero matrix can be denoted $0_{m \times n}$. The following property of zero matrices is obvious.

If A is any $m \times n$ matrix and 0 is the $m \times n$ zero matrix, then

$$A + 0 = A \quad \text{and} \quad 0 + A = A.$$

The **negative** of a matrix $A = [a_{ij}]$ is the matrix whose (i, j)-entry is $-a_{ij}$ for all i and j. The negative of A is denoted $-A$.

EXAMPLE 46.3

$$\text{If } A = \begin{bmatrix} \pi & 0 \\ -1 & \sqrt{2} \end{bmatrix}, \text{ then } -A = \begin{bmatrix} -\pi & 0 \\ 1 & -\sqrt{2} \end{bmatrix}.$$

If A is any $m \times n$ matrix, then

$$A + (-A) = (-A) + A = 0_{m \times n}$$

and

$$-(-A) = A.$$

The **difference** $A - B$ of two $m \times n$ matrices is defined by

$$A - B = A + (-B).$$

We now know enough about matrix addition to be able to solve any matrix equation of the form $A + X = B$ (provided A and B have the same size; otherwise the equation is impossible). By subtracting $-A$ from both sides of $A + X = B$, we can conclude that if there is a solution, then it must be $X = (-A) + B$. Substitution shows that $(-A) + B$ is indeed a solution:

$$\begin{aligned} A + X &= B \\ A + [(-A) + B] &\stackrel{?}{=} B && \text{Substitution.} \\ [A + (-A)] + B &\stackrel{?}{=} B && \text{Associative law.} \\ 0 + B &\stackrel{?}{=} B && A + (-A) = 0. \\ B &= B && 0 + B = B. \end{aligned}$$

EXAMPLE 46.4 Solve $A + X = B$ if

$$A = \begin{bmatrix} 0.06 & 0.18 \\ -0.1 & 0.4 \end{bmatrix} \quad \text{and} \quad B = \begin{bmatrix} 2 & -1 \\ 3 & 2 \end{bmatrix}.$$

Solution The solution is

$$X = (-A) + B = \begin{bmatrix} -0.06 & -0.18 \\ 0.1 & -0.4 \end{bmatrix} + \begin{bmatrix} 2 & -1 \\ 3 & 2 \end{bmatrix} = \begin{bmatrix} 1.94 & -1.18 \\ 3.1 & 1.6 \end{bmatrix}.$$

C. Matrix Multiplication

Two matrices can be added if they have the same size. In contrast, we'll see that the *product* of two matrices is defined iff the number of columns in the first matrix is equal to the number of rows in the second matrix. When the matrix product AB is defined, its size will satisfy the following pattern.

$$\underset{m \times n}{A} \quad \underset{n \times p}{B} \quad = \quad \underset{m \times p}{C}$$

These must be equal. (n and n)

Number of rows in A. (m)

Number of columns in B. (p)

EXAMPLE 46.5

(a) If A is 3×2 and B is 2×4, then AB will be 3×4 and BA will be undefined.
(b) If A is 3×2 and B is 3×4, then neither AB nor BA will be defined.
(c) If A is 2×2 and B is 2×2, then both AB and BA will be 2×2.

In the definition of the product AB, which follows, we'll see that the (i, j)-entry of AB depends only on the *ith row* of A and the *jth column* of B.

$$\begin{array}{c} \\ \\ i\text{th row} \longrightarrow \\ \\ \\ \end{array} \begin{bmatrix} a_{11} & a_{12} & \cdots & a_{1n} \\ \vdots & \vdots & & \vdots \\ a_{i1} & a_{i2} & \cdots & a_{in} \\ \vdots & \vdots & & \vdots \\ a_{m1} & a_{m2} & \cdots & a_{mn} \end{bmatrix} \begin{bmatrix} b_{11} & \cdots & \overset{j\text{th column}}{\overset{\downarrow}{b_{1j}}} & \cdots & b_{1p} \\ b_{21} & \cdots & b_{2j} & \cdots & b_{2p} \\ \vdots & & \vdots & & \vdots \\ b_{n1} & \cdots & b_{nj} & \cdots & b_{np} \end{bmatrix}$$

DEFINITION. Assume that A is an $m \times n$ matrix and that B is an $n \times p$ matrix. The **product** of A and B, denoted AB, is the $m \times p$ matrix whose (i, j)-entry is

$$a_{i1}b_{1j} + a_{i2}b_{2j} + \cdots + a_{in}b_{nj}.$$

EXAMPLE 46.6 Assume that A is 4×3, B is 3×2, and $AB = C$. Then C will be 4×2, and, for example,

$$c_{21} = \underbrace{a_{21}b_{11}}_{1} + \underbrace{a_{22}b_{21}}_{2} + \underbrace{a_{23}b_{31}}_{3}.$$

That is, to compute the (2, 1)-entry of the product AB, use the second row of A and the first column of B: form the products of the pairs of entries that are connected in Figure 46.1, and then add the results.

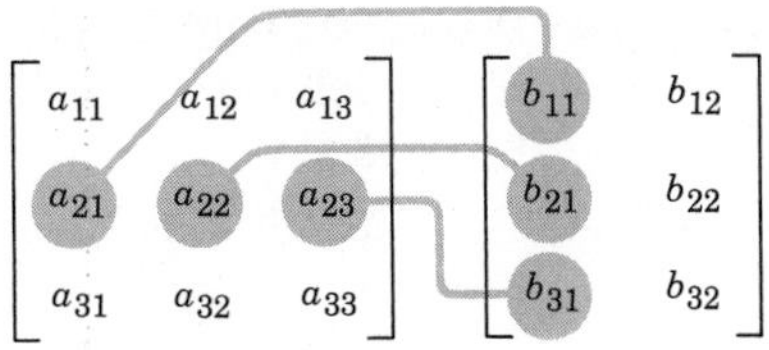

Figure 46.1

EXAMPLE 46.7 If

$$\begin{bmatrix} 2 & 0 & -5 \\ 3 & 7 & 9 \end{bmatrix} \begin{bmatrix} 4 & 8 \\ 2 & -3 \\ -1 & 5 \end{bmatrix} = \begin{bmatrix} c_{11} & c_{12} \\ c_{21} & c_{22} \end{bmatrix}$$

then

$$\begin{aligned} c_{11} &= 2 \cdot 4 + 0 \cdot 2 + (-5) \cdot (-1) = 13 \\ c_{12} &= 2 \cdot 8 + 0 \cdot (-3) + (-5) \cdot (5) = -9 \\ c_{21} &= 3 \cdot 4 + 7 \cdot 2 + 9 \cdot (-1) = 17 \\ c_{22} &= 3 \cdot 8 + 7 \cdot (-3) + 9 \cdot 5 = 48. \end{aligned}$$

Therefore,

$$\begin{bmatrix} 2 & 0 & -5 \\ 3 & 7 & 9 \end{bmatrix} \begin{bmatrix} 4 & 8 \\ 2 & -3 \\ -1 & 5 \end{bmatrix} = \begin{bmatrix} 13 & -9 \\ 17 & 48 \end{bmatrix}.$$

The commutative law for addition states that $A + B = B + A$ whenever both sums are defined (that is, whenever A and B have the same size). The corresponding law for multiplication, $AB = BA$, is *not* universally true, even if both AB and BA are defined. For one thing, both AB and BA may be defined and yet have different sizes: for example, if A is 2×3 and B is 3×2, then AB will be 2×2 but BA will be 3×3. But even if AB and BA are both defined and of the same size, they may be unequal, as in the next example.

EXAMPLE 46.8 This example shows that the commutative law is not true for matrix multiplication:

$$\begin{bmatrix} 2 & -4 \\ 1 & -2 \end{bmatrix} \begin{bmatrix} 1 & -2 \\ -3 & 6 \end{bmatrix} = \begin{bmatrix} 14 & -28 \\ 7 & -14 \end{bmatrix}$$

but

$$\begin{bmatrix} 1 & -2 \\ -3 & 6 \end{bmatrix} \begin{bmatrix} 2 & -4 \\ 1 & -2 \end{bmatrix} = \begin{bmatrix} 0 & 0 \\ 0 & 0 \end{bmatrix}.$$

The preceding equation also shows that the product of two matrices may be the zero matrix even though neither of the two matrices is the zero matrix. This contrasts with the property of real numbers that guarantees $ab = 0$ iff $a = 0$ or $b = 0$.

Here is the associative law for matrix multiplication, which is true even though the commutative law is not. Exercise 34 asks you to prove it in the special case where A, B, and C are all 2×2 matrices; the proof for all cases can be found in books on matrices or linear algebra.

> **Associative law for matrix multiplication.** If A, B, and C are matrices such that both $A(BC)$ and $(AB)C$ are defined, then
>
> $$A(BC) = (AB)C.$$

A **square matrix** is one in which the number of rows equals the number of columns. The **main diagonal** of an $n \times n$ (square) matrix consists of the elements a_{11}, $a_{22}, \ldots, a_{nn}$. For each n, the $n \times n$ matrix that contains all 1's on the main diagonal and all 0's elsewhere is called the $n \times n$ **identity matrix.** For example, the 3×3 identity matrix is

$$\begin{bmatrix} 1 & 0 & 0 \\ 0 & 1 & 0 \\ 0 & 0 & 1 \end{bmatrix}.$$

Each identity matrix will be denoted by I; if there is a chance of confusion about which identity matrix is meant, the $n \times n$ identity matrix can be denoted I_n.

The next property shows that identity matrices are to matrix multiplication what the number 1 is to multiplication of real numbers.

> If A is an $m \times n$ matrix, then
>
> $$I_m A = A \quad \text{and} \quad AI_n = A.$$

Exercise 35 asks you to prove the preceding statement for $m = n = 2$. Exercises 36 and 37 ask you to prove special cases of the following distributive laws.

> **Distributive laws.** Each of the following equations is true for all matrices A, B, and C for which the sums and products in the equation are defined.
>
> $$A(B + C) = AB + AC$$
> $$(A + B)C = AC + BC.$$

If c is a real number and $A = [a_{ij}]$ is a matrix, then cA is defined to be the matrix whose (i, j)-entry is ca_{ij}.

EXAMPLE 46.9 If $A = \begin{bmatrix} \sqrt{2} & 4 \\ -1 & 0 \end{bmatrix}$,

then

$$3A = \begin{bmatrix} 3\sqrt{2} & 12 \\ -3 & 0 \end{bmatrix}, \quad \emptyset A = \begin{bmatrix} 0 & 0 \\ 0 & 0 \end{bmatrix}, \quad \text{and} \quad (-1)A = \begin{bmatrix} -\sqrt{2} & -4 \\ 1 & 0 \end{bmatrix}.$$

The last equation illustrates the general property that $(-1)A = -A$ for every matrix A. Other properties of the operation cA are stated in the exercises.

Further properties of matrix multiplication, along with an application, will be discussed in the next section.

EXERCISES FOR SECTION 46

Determine the variables in each case so that the matrices are equal.

1. $\begin{bmatrix} 7 & -5 \\ -4 & 3 \end{bmatrix} = \begin{bmatrix} 7 & x \\ y & 3 \end{bmatrix}$

2. $\begin{bmatrix} 4 & a & -1 \\ 6 & -9 & b \end{bmatrix} = \begin{bmatrix} x & 2 & -1 \\ y & z & 8 \end{bmatrix}$

3. $\begin{bmatrix} 1 & 2 \\ c & -\frac{1}{2} \\ \frac{7}{8} & d \end{bmatrix} = \begin{bmatrix} x & 2 \\ \frac{3}{4} & y \\ z & 7 \end{bmatrix}$

Perform the indicated operations.

4. $\begin{bmatrix} 0 & -9 \\ 3 & 8 \end{bmatrix} + \begin{bmatrix} 4 & 5 \\ -3 & 4 \end{bmatrix}$

5. $\begin{bmatrix} 5 & 27 \\ 14 & -13 \end{bmatrix} + \begin{bmatrix} -6 & 1 \\ 0 & 5 \end{bmatrix}$

6. $\begin{bmatrix} 7 & \frac{4}{5} \\ \frac{1}{4} & -5 \\ 0 & 2 \end{bmatrix} + \begin{bmatrix} -7 & \frac{2}{3} \\ -\frac{1}{3} & 5 \\ \frac{1}{8} & -6 \end{bmatrix}$

7. $\begin{bmatrix} 9 & 5 & -3 \\ 2 & 0 & 17 \end{bmatrix} - \begin{bmatrix} 7 & -1 & 9 \\ 6 & 4 & 1 \end{bmatrix}$

8. $[7 \quad 12 \quad -4] - [8 \quad 9 \quad -1]$

9. $\begin{bmatrix} 6 \\ 13 \\ -7 \end{bmatrix} - \begin{bmatrix} -7 \\ 10 \\ -2 \end{bmatrix}$

Solve for X in each case.

10. $\begin{bmatrix} 9 & -6 \\ 0 & 2 \end{bmatrix} + X = \begin{bmatrix} 3 & 12 \\ -11 & 10 \end{bmatrix}$

11. $X + [4 \quad 0 \quad -7] = [14 \quad 1 \quad 8]$

12. $\begin{bmatrix} -2 & 5 & 6 \\ 9 & 7 & 11 \end{bmatrix} + X = \begin{bmatrix} 12 & 9 & -3 \\ -5 & -1 & 0 \end{bmatrix}$

13. If A is 3×5 and B is 5×1, what are the dimensions of AB?

14. If A is 3×1, B is $m \times 4$, and AB is defined, what is m and what are the dimensions of AB?

15. If A is $m \times 3$, B is $3 \times p$, and AB is 6×7, what are m and p?

Compute the matrices indicated in Exercises 16–33 with A, B, C, D, E, and F defined as follows.

$$A = \begin{bmatrix} 3 & 6 \\ 1 & -4 \end{bmatrix} \quad B = \begin{bmatrix} -8 & 1 \\ 1 & 6 \end{bmatrix} \quad C = \begin{bmatrix} 6 & 5 & -2 \\ 3 & 0 & -1 \\ 2 & 1 & 2 \end{bmatrix}$$

$$D = \begin{bmatrix} 0 & 1 & 4 \\ -7 & 1 & 2 \\ 3 & -2 & 0 \end{bmatrix} \quad E = \begin{bmatrix} 7 & 5 \\ 0 & 0 \\ -5 & 6 \end{bmatrix} \quad F = \begin{bmatrix} 2 & 9 & -5 \\ 0 & 3 & -6 \end{bmatrix}$$

16. AB **17.** BA **18.** AF **19.** BF **20.** EA **21.** EB
22. FC **23.** CD **24.** DC **25.** CE **26.** FE **27.** DE
28. $2A$ **29.** $-2E$ **30.** $3F$ **31.** $E(A + B)$ **32.** $(A + B)F$ **33.** $F(C + D)$

In Exercises 34–42, use

$$A = \begin{bmatrix} a_{11} & a_{12} \\ a_{21} & a_{22} \end{bmatrix}, \quad B = \begin{bmatrix} b_{11} & b_{12} \\ b_{21} & b_{22} \end{bmatrix}, \quad C = \begin{bmatrix} c_{11} & c_{12} \\ c_{21} & c_{22} \end{bmatrix}.$$

34. Prove that $A(BC) = (AB)C$.

35. Prove that $AI_2 = A$ and $I_2A = A$.

36. Prove that $A(B + C) = AB + AC$.

37. Prove that $(A + B)C = AC + BC$.

38. It can be proved that if A is any matrix and 0 is the appropriately sized zero matrix in each case, then $A0 = 0$ and $0A = 0$. Write a proof for the special cases where A is 2×2.

39. Prove that $(-A)B = A(-B) = -(AB)$.

40. Prove that $(-A)(-B) = AB$.

41. Prove that $A(B - C) = AB - AC$.

42. Prove that $(A - B)C = AC - BC$.

43. For real numbers, $(a + b)^2 = a^2 + 2ab + b^2$. This exercise examines the analogous relationship for matrices.
(a) Show that $(A + B)^2 \neq A^2 + 2AB + B^2$ for the matrices A and B preceding Exercise 16.
(b) Prove that if A and B are $n \times n$ matrices, then $(A + B)^2 = A^2 + AB + BA + B^2$. (Suggestion: Use the distributive laws.)

44. For real numbers, $(a + b)(a - b) = a^2 - b^2$.
(a) Show that $(A + B)(A - B) \neq A^2 - B^2$ for the matrices A and B preceding Exercise 16.
(b) Prove that if A and B are $n \times n$ matrices, then $(A + B)(A - B) = A^2 - AB + BA - B^2$.

45. For real numbers, if $ab = ac$ and $a \neq 0$, then $b = c$. Show that the analogous fact does not hold for matrices, by finding 2×2 matrices A, B, and C such that $AB = AC$ and $A \neq 0$ but $B \neq C$. (Suggestion: Let A and B be the two matrices on the left in the second equation of Example 46.8, and let C be the zero matrix.)

SECTION 47 The Inverse of a Matrix. Application to Linear Systems

A. The Inverse of a Matrix

We know that identity matrices have the following similarity with the real number 1:

$$I_m A = A = AI_n \quad \text{for every } m \times n \text{ matrix } A$$

just as

$$1a = a = a1 \quad \text{for every real number } a.$$

Real numbers have the further property that if $a \neq 0$, then a has an inverse (reciprocal), a^{-1}, such that

$$aa^{-1} = 1 = a^{-1}a.$$

Some nonzero matrices do not have an inverse in this sense, however. Those matrices that do are singled out by the following definition. Notice that the definition applies only to square matrices.

DEFINITION. An $n \times n$ matrix A is said to be **invertible** if there is an $n \times n$ matrix B such that

$$AB = I_n = BA.$$

The matrix B is called an **inverse** of A in this case.

It can be proved that if a matrix is invertible, then its inverse is unique (see Exercise 28). Therefore, we can refer to *the* inverse of an invertible matrix A; this unique inverse is denoted A^{-1}.

EXAMPLE 47.1 The matrix $\begin{bmatrix} 2 & -3 \\ 3 & -4 \end{bmatrix}$ is the inverse of $\begin{bmatrix} -4 & 3 \\ -3 & 2 \end{bmatrix}$, because

$$\begin{bmatrix} -4 & 3 \\ -3 & 2 \end{bmatrix} \begin{bmatrix} 2 & -3 \\ 3 & -4 \end{bmatrix} = \begin{bmatrix} 1 & 0 \\ 0 & 1 \end{bmatrix}$$

and

$$\begin{bmatrix} 2 & -3 \\ 3 & -4 \end{bmatrix} \begin{bmatrix} -4 & 3 \\ -3 & 2 \end{bmatrix} = \begin{bmatrix} 1 & 0 \\ 0 & 1 \end{bmatrix}.$$

EXAMPLE 47.2 Prove that the matrix $A = \begin{bmatrix} 0 & 1 \\ 0 & 0 \end{bmatrix}$ is not invertible.

Solution If $AB = I$, then

$$\begin{bmatrix} 0 & 1 \\ 0 & 0 \end{bmatrix} \begin{bmatrix} b_{11} & b_{12} \\ b_{21} & b_{22} \end{bmatrix} = \begin{bmatrix} 1 & 0 \\ 0 & 1 \end{bmatrix}$$

and thus

$$\begin{bmatrix} b_{21} & b_{22} \\ 0 & 0 \end{bmatrix} = \begin{bmatrix} 1 & 0 \\ 0 & 1 \end{bmatrix}.$$

The last equation is impossible since the (2,2)-entries, 0 and 1, are unequal. Thus there is no matrix B such that $AB = I$, and A is not invertible.

The last example points up the general fact that a matrix A is not invertible if either $AB = I$ is impossible or $BA = I$ is impossible. It can be proved that, in fact, for square matrices the two equations $AB = I$ and $BA = I$ are equivalent, in the sense that if either one is true, then the other is also true. This means that to show that B is the inverse of A we need to check only one of the two equations, $AB = I$ or $BA = I$. (Most books on matrices or linear algebra give a proof.)

B. Computing Inverses: The 2 × 2 Case by Determinants

Following is an easy-to-use criterion for determining whether a 2×2 matrix is invertible, together with a formula for the inverse if the matrix *is* invertible.

A matrix $A = \begin{bmatrix} a & b \\ c & d \end{bmatrix}$ is invertible if $ad - bc \neq 0$. If $ad - bc \neq 0$, then

$$A^{-1} = \frac{1}{ad - bc}\begin{bmatrix} d & -b \\ -c & a \end{bmatrix}. \tag{47.1}$$

The number $ad - bc$ in the denominator in (47.1) is called the **determinant** of the matrix A; notice that it is the product of the two elements on the main diagonal minus the product of the two elements on the other diagonal. The matrix shown in (47.1) results from interchanging the two entries on the main diagonal of A and changing the signs of the other two entries of A.

Proof. If $AB = I$ with $B = \begin{bmatrix} w & x \\ y & z \end{bmatrix}$, then

$$\begin{bmatrix} aw + by & ax + bz \\ cw + dy & cx + dz \end{bmatrix} = \begin{bmatrix} 1 & 0 \\ 0 & 1 \end{bmatrix}. \tag{47.2}$$

The equality of corresponding entries in the first columns in (47.2) dictates that

$$aw + by = 1 \quad \text{and} \quad cw + dy = 0. \tag{47.3}$$

The equality of the entries in the second columns in (47.2) dictates that

$$ax + bz = 0 \quad \text{and} \quad cx + dz = 1. \tag{47.4}$$

Since the matrix A is given, we think of w, x, y, and z as unknowns in the Systems (47.3) and (47.4).

To solve System (47.3) for y, compute a times the second equation minus c times the first equation; this eliminates w and leads to

$$(ad - bc)y = -c,$$

from which

$$y = \frac{-c}{ad - bc} \text{ if } ad - bc \neq 0.$$

To solve System (47.3) for w, compute d times the first equation minus b times the second equation; this eliminates y and leads to

$$(ad - bc)w = d,$$

from which

$$w = \frac{d}{ad - bc} \text{ if } ad - bc \neq 0.$$

The System (47.4) can be solved in the same way, giving

$$x = \frac{-b}{ad - bc} \text{ if } ad - bc \neq 0$$

and

$$z = \frac{a}{ad - bc} \text{ if } ad - bc \neq 0.$$

Substitution in B of these solutions for w, x, y, and z will give A^{-1} in Equation (47.1). Exercise 29 asks you to verify that $A^{-1}A = I$ and $AA^{-1} = I$ for A^{-1} as in (47.1). □

It can also be proved that if $ad - bc = 0$, then A is *not* invertible. Thus we can make the following statement.

> A 2×2 matrix is invertible
> iff
> its determinant is nonzero.

The next two sections will consider determinants of larger square matrices and will give a generalization of the preceding statement.

EXAMPLE 47.3 Determine whether $A = \begin{bmatrix} 5 & 1 \\ 6 & -3 \end{bmatrix}$ is invertible, and if it is invertible compute its inverse.

Solution The determinant of A is $(5)(-3) - (1)(6) = -21$, which is nonzero; therefore, A is invertible. To compute the inverse, apply (47.1):

$$A^{-1} = \begin{bmatrix} 5 & 1 \\ 6 & -3 \end{bmatrix}^{-1} = \frac{1}{-21}\begin{bmatrix} -3 & -1 \\ -6 & 5 \end{bmatrix} = \begin{bmatrix} \frac{1}{7} & \frac{1}{21} \\ \frac{2}{7} & -\frac{5}{21} \end{bmatrix}.$$

To check this answer, you can verify directly that $AA^{-1} = I$.

C. Computing Inverses by Elementary Row Operations

We now consider a method that can be used to compute the inverse of an invertible matrix of any size. The method uses *elementary row operations* and *reduced echelon form,* which you may want to review from Section 42B.

It can be proved that if A is an $n \times n$ invertible matrix, then there is a sequence of elementary row operations that will transform A into I_n; moreover, the same sequence of operations will transform I_n into A^{-1}. We can apply this as follows.

> To compute the inverse of an $n \times n$ invertible matrix A, form the $n \times 2n$ matrix
>
> $$[A \mid I_n].$$
>
> Use elementary row operations on this $n \times 2n$ matrix to transform A, the left half, into I_n. Then the result in the right half, which was originally I_n, will be A^{-1}.

EXAMPLE 47.4 The matrix $A = \begin{bmatrix} 1 & 0 & 3 \\ 2 & -5 & 4 \\ 1 & -2 & 2 \end{bmatrix}$ is invertible. Use elementary row operations to compute A^{-1}.

Solution Begin with $[A \mid I_3]$.

$$\left[\begin{array}{ccc|ccc} 1 & 0 & 3 & 1 & 0 & 0 \\ 2 & -5 & 4 & 0 & 1 & 0 \\ 1 & -2 & 2 & 0 & 0 & 1 \end{array}\right]$$

$$\xrightarrow[\text{(Row 3)} - \text{(Row 1)}]{\text{(Row 2)} - 2\text{(Row 1)}} \left[\begin{array}{ccc|ccc} 1 & 0 & 3 & 1 & 0 & 0 \\ 0 & -5 & -2 & -2 & 1 & 0 \\ 0 & -2 & -1 & -1 & 0 & 1 \end{array}\right]$$

$$\xrightarrow{-\frac{1}{5}\text{(Row 2)}} \left[\begin{array}{ccc|ccc} 1 & 0 & 3 & 1 & 0 & 0 \\ 0 & 1 & \frac{2}{5} & \frac{2}{5} & -\frac{1}{5} & 0 \\ 0 & -2 & -1 & -1 & 0 & 1 \end{array}\right]$$

$$\xrightarrow{\text{(Row 3)} + 2\text{(Row 2)}} \left[\begin{array}{ccc|ccc} 1 & 0 & 3 & 1 & 0 & 0 \\ 0 & 1 & \frac{2}{5} & \frac{2}{5} & -\frac{1}{5} & 0 \\ 0 & 0 & -\frac{1}{5} & -\frac{1}{5} & -\frac{2}{5} & 1 \end{array}\right]$$

$$\xrightarrow{-5\text{(Row 3)}} \left[\begin{array}{ccc|ccc} 1 & 0 & 3 & 1 & 0 & 0 \\ 0 & 1 & \frac{2}{5} & \frac{2}{5} & -\frac{1}{5} & 0 \\ 0 & 0 & 1 & 1 & 2 & -5 \end{array}\right]$$

$$\xrightarrow[\text{(Row 2)} - \frac{2}{5}\text{(Row 3)}]{\text{(Row 1)} - 3\text{(Row 3)}} \left[\begin{array}{ccc|ccc} 1 & 0 & 0 & -2 & -6 & 15 \\ 0 & 1 & 0 & 0 & -1 & 2 \\ 0 & 0 & 1 & 1 & 2 & -5 \end{array}\right]$$

The result is $[I_3 \mid A^{-1}]$.

Check

$$AA^{-1} = \begin{bmatrix} 1 & 0 & 3 \\ 2 & -5 & 4 \\ 1 & -2 & 2 \end{bmatrix} \begin{bmatrix} -2 & -6 & 15 \\ 0 & -1 & 2 \\ 1 & 2 & -5 \end{bmatrix} = \begin{bmatrix} 1 & 0 & 0 \\ 0 & 1 & 0 \\ 0 & 0 & 1 \end{bmatrix} = I_3.$$

No sequence of elementary row operations will transform a *noninvertible* matrix into an identity matrix. If the method in Example 47.4 is attempted with a matrix A that is not invertible, then A will be transformed into a matrix with a row of 0's. When that happens we can simply stop.

EXAMPLE 47.5 Use elementary row operations to determine whether $A = \begin{bmatrix} 1 & 1 & 2 \\ 0 & 1 & -4 \\ 3 & 1 & 14 \end{bmatrix}$ is invertible.

Solution

$$\left[\begin{array}{ccc|ccc} 1 & 1 & 2 & 1 & 0 & 0 \\ 0 & 1 & -4 & 0 & 1 & 0 \\ 3 & 1 & 14 & 0 & 0 & 1 \end{array}\right]$$

$$\xrightarrow{\text{(Row 3)} - 3\text{(Row 1)}} \left[\begin{array}{ccc|ccc} 1 & 1 & 2 & 1 & 0 & 0 \\ 0 & 1 & -4 & 0 & 1 & 0 \\ 0 & -2 & 8 & -3 & 0 & 1 \end{array}\right]$$

$$\xrightarrow[\text{(Row 3)} + 2\text{(Row 2)}]{\text{(Row 1)} - \text{(Row 2)}} \left[\begin{array}{ccc|ccc} 1 & 0 & 6 & 1 & -1 & 0 \\ 0 & 1 & -4 & 0 & 1 & 0 \\ 0 & 0 & 0 & -3 & 2 & 1 \end{array}\right]$$

The left half of the third row contains only 0's, so A is not invertible.

D. Application to Linear Systems of Equations

A linear system

$$\begin{aligned} a_1x + b_1y + c_1z &= d_1 \\ a_2x + b_2y + c_2z &= d_2 \\ a_3x + b_3y + c_3z &= d_3 \end{aligned} \tag{47.5}$$

can be represented in matrix form as

$$AX = B, \tag{47.6}$$

where

$$A = \begin{bmatrix} a_1 & b_1 & c_1 \\ a_2 & b_2 & c_2 \\ a_3 & b_3 & c_3 \end{bmatrix}, \quad X = \begin{bmatrix} x \\ y \\ z \end{bmatrix}, \quad \text{and} \quad B = \begin{bmatrix} d_1 \\ d_2 \\ d_3 \end{bmatrix}.$$

The product AX in (47.6) is the matrix product, and the equality in (47.6) is matrix equality. The matrix A is called the **coefficient matrix** of the system. Systems with other variables can be represented similarly.

EXAMPLE 47.6 The matrix form of the system

$$\begin{aligned} x + 3z &= -2 \\ 2x - 5y + 4z &= 4 \\ x - 2y + 2z &= 1 \end{aligned}$$

is

$$AX = B,$$

where

$$A = \begin{bmatrix} 1 & 0 & 3 \\ 2 & -5 & 4 \\ 1 & -2 & 2 \end{bmatrix}, \quad X = \begin{bmatrix} x \\ y \\ z \end{bmatrix}, \quad \text{and} \quad B = \begin{bmatrix} -2 \\ 4 \\ 1 \end{bmatrix}.$$

If the coefficient matrix in Equation (47.6) is invertible, then we can multiply both sides of the equation on the left by A^{-1} to obtain

$$\begin{aligned} A^{-1}(AX) &= A^{-1}B \\ (A^{-1}A)X &= A^{-1}B \\ IX &= A^{-1}B \\ X &= A^{-1}B. \end{aligned} \tag{47.7}$$

Thus, by computing A^{-1} and then $A^{-1}B$ we can solve System (47.5). [*Important:* In general, $A^{-1}B \neq BA^{-1}$. Equation (47.7) requires $A^{-1}B$.]

EXAMPLE 47.7 The inverse of the coefficient matrix for Example 47.6 was computed in Example 47.4:

$$A^{-1} = \begin{bmatrix} -2 & -6 & 15 \\ 0 & -1 & 2 \\ 1 & 2 & -5 \end{bmatrix}.$$

Therefore, the solution of the system in Example 47.6 is given by

$$X = A^{-1}B = \begin{bmatrix} -2 & -6 & 15 \\ 0 & -1 & 2 \\ 1 & 2 & -5 \end{bmatrix} \begin{bmatrix} -2 \\ 4 \\ 1 \end{bmatrix} = \begin{bmatrix} -5 \\ -2 \\ 1 \end{bmatrix}.$$

That is, the solution is $x = -5$, $y = -2$, $z = 1$.

If the coefficient matrix of System (47.5) is not invertible, then the system does not have a unique solution; it may have no solution or infinitely many solutions.

EXERCISES FOR SECTION 47

Verify that $B = A^{-1}$ in each case by computing AB.

1. $A = \begin{bmatrix} 2 & 3 \\ -3 & -5 \end{bmatrix} \quad B = \begin{bmatrix} 5 & 3 \\ -3 & -2 \end{bmatrix}$

2. $A = \begin{bmatrix} 2 & -4 \\ 3 & -7 \end{bmatrix} \quad B = \frac{1}{2}\begin{bmatrix} 7 & -4 \\ 3 & -2 \end{bmatrix}$

3. $A = \begin{bmatrix} 5 & -5 & -5 \\ -4 & 3 & 6 \\ 2 & 1 & -3 \end{bmatrix} \quad B = \frac{1}{5}\begin{bmatrix} 3 & 4 & 3 \\ 0 & 1 & 2 \\ 2 & 3 & 1 \end{bmatrix}$

Determine whether each matrix is invertible by computing its determinant. If the matrix is invertible, compute its inverse by using Equation (47.1). Check each inverse by computing its product with the given matrix.

4. $\begin{bmatrix} 6 & -3 \\ -3 & 2 \end{bmatrix}$ **5.** $\begin{bmatrix} 5 & 7 \\ 2 & 2 \end{bmatrix}$ **6.** $\begin{bmatrix} 15 & 2 \\ -30 & -4 \end{bmatrix}$

7. $\begin{bmatrix} \frac{1}{2} & \frac{1}{4} \\ \frac{1}{3} & \frac{1}{3} \end{bmatrix}$ **8.** $\begin{bmatrix} \frac{3}{2} & -\frac{1}{5} \\ -3 & \frac{2}{5} \end{bmatrix}$ **9.** $\begin{bmatrix} 1.7 & -0.5 \\ -10 & 3 \end{bmatrix}$

10. $\begin{bmatrix} 0.5 & -4 \\ -1.5 & 12 \end{bmatrix}$ **11.** $\begin{bmatrix} 14 & -2 \\ 35 & 5 \end{bmatrix}$ **12.** $\begin{bmatrix} \sqrt{2} & \sqrt{3} \\ 3\sqrt{3} & -2\sqrt{2} \end{bmatrix}$

Use elementary row operations to determine whether each matrix is invertible, and if the matrix is invertible give its inverse.

13. $\begin{bmatrix} 1 & 0 & 1 \\ 5 & -2 & 2 \\ 1 & 3 & 4 \end{bmatrix}$ **14.** $\begin{bmatrix} 1 & 3 & -5 \\ 4 & 10 & 3 \\ 2 & 0 & 2 \end{bmatrix}$ **15.** $\begin{bmatrix} -1 & 2 & 3 \\ 1 & 9 & 3 \\ 5 & 1 & 8 \end{bmatrix}$

16. $\begin{bmatrix} 2 & 2 & -1 \\ 6 & -1 & 7 \\ 3 & 0 & -1 \end{bmatrix}$ **17.** $\begin{bmatrix} 5 & -3 & 6 \\ 1 & 0 & 7 \\ -1 & -2 & 9 \end{bmatrix}$ **18.** $\begin{bmatrix} 1 & 5 & 2 \\ 4 & -2 & -2 \\ 1 & 3 & 8 \end{bmatrix}$

19. $\begin{bmatrix} 1 & 0 & 4 & 0 \\ 0 & -2 & -3 & -1 \\ -2 & 1 & -10 & 1 \\ 1 & 0 & 5 & 2 \end{bmatrix}$ **20.** $\begin{bmatrix} 1 & -2 & 4 & 1 \\ 3 & -5 & 13 & 1 \\ 1 & -2 & 5 & 1 \\ 1 & -2 & 4 & 2 \end{bmatrix}$

21. $\begin{bmatrix} 1 & -1 & 2 & -1 \\ 2 & 1 & -2 & 1 \\ -5 & -2 & 4 & -2 \\ 3 & 1 & -2 & 1 \end{bmatrix}$

Solve each system by first computing the inverse of the coefficient matrix, as in Example 47.7.

22. $2x - 3y = 1$, $x + y = 5$

23. $-x - y = 3$, $4x + 2y = 7$

24. $5x + y = 3$, $7x - 2y = 0$

25. $x + y + z = 1$, $2x - y - z = 3$, $5y + z = 0$

26. $2x - 2y + 3z = 0$, $x - y + 2z = 5$, $2x + 2y - 3z = 1$

27. $x + 2y = 1$, $-3x + z = 1$, $4x - y = 3$

28. To prove that each invertible matrix has a *unique* inverse, suppose that $AB = I$ and $BA = I$ (so that B is an inverse of A) and that $AC = I$ and $CA = I$ (so that C is the inverse of A). Now give a reason for each equality that follows; together, they prove that $B = C$: $B = BI = B(AC) = (BA)C = IC = C$.

29. Verify by direct multiplication that $AA^{-1} = I$ and $A^{-1}A = I$ for A^{-1} as in Equation (47.1) and for A as in the same display.

30. If A is an invertible matrix, then A^{-1} is also invertible. What is the inverse of A^{-1}? Justify your answer.

In Exercises 31–33, first write a linear system of equations and then solve the system by the method of Subsection D.

31. The total world gold production in 1975 was estimated to be 39 million ounces. The production in South Africa was 9 million ounces less than twice the production outside of South Africa. Find the production in South Africa and outside of South Africa. (All figures are given to the nearest million ounces.)

32. The Consumer Price Index increased 27 points from 1960 to 1970 and then increased 113 more points from 1970 to 1980. The index for 1980 was 42 points more than the sum of the indices for 1960 and 1970. Find the index in each of the three years, 1960, 1970, and 1980. (Indices have been rounded to the nearest point for simplicity.) Compute the percentage increases from 1960 to 1970 and from 1970 to 1980.

33. Energy use in the United States in 1980 was 78 quads (quadrillion BTU's). Coal and imported oil provided equal amounts of this energy, and natural gas and domestic oil also provided equal amounts. Natural gas and domestic oil together provided 10 quads more than coal and imported oil together. Nuclear and other sources provided 6 quads. Let C, G, D, and I denote the number of quads provided by coal, natural gas, domestic oil, and imported oil, respectively. Determine C, G, D, and I. (Begin by writing four linear equations in C, G, D, and I. All figures are approximate and adjusted slightly for simplicity.)

SECTION

48 Determinants

A. Second- and Third-Order Determinants

Associated with each square matrix A is a real number called the *determinant* of A, which we denote by either *det* A or $|A|$. Determinants of 2×2 matrices were used in Section 47B to compute matrix inverses. This section and the next will introduce determinants of 3×3 and larger matrices, and will also discuss an application and some of the properties of determinants. We begin by recalling the definition of the determinant of a 2×2 matrix.

DEFINITION

$$\det \begin{bmatrix} a_{11} & a_{12} \\ a_{21} & a_{22} \end{bmatrix} = \begin{vmatrix} a_{11} & a_{12} \\ a_{21} & a_{22} \end{vmatrix} = a_{11}a_{22} - a_{12}a_{21}$$

EXAMPLE 48.1

$$\begin{vmatrix} 12 & 4 \\ 5 & -1 \end{vmatrix} = (12)(-1) - (4)(5) = -12 - 20 = -32$$

To define determinants of 3×3 matrices we use *minors* and *cofactors,* which are defined as follows.

DEFINITION. The **(i, j)-minor** of a 3×3 matrix A, denoted M_{ij}, is the determinant of the 2×2 matrix that remains when the ith row and jth column of A have been deleted.

EXAMPLE 48.2 Assume $A = \begin{bmatrix} 2 & -7 & 0 \\ 3 & -3 & -5 \\ -2 & 6 & 1 \end{bmatrix}$.

(a) To compute M_{11} we first delete the first row and first column of A:

$$\begin{bmatrix} \not{2} & \not{-7} & \not{0} \\ \not{3} & -3 & -5 \\ \not{-2} & 6 & 1 \end{bmatrix}.$$

Thus

$$M_{11} = \begin{vmatrix} -3 & -5 \\ 6 & 1 \end{vmatrix} = (-3)(1) - (-5)(6) = 27.$$

(b) To compute M_{32} we first delete the third row and second column of A:

$$\begin{bmatrix} 2 & -7 & 0 \\ 3 & -3 & -5 \\ -2 & 6 & 1 \end{bmatrix}.$$

Thus

$$M_{32} = \begin{vmatrix} 2 & 0 \\ 3 & -5 \end{vmatrix} = (2)(-5) - (0)(3) = -10.$$

DEFINITION. The **(i, j)-cofactor** of a 3×3 matrix A, denoted C_{ij}, is defined by $C_{ij} = (-1)^{i+j}M_{ij}$.

Since $(-1)^{i+j}$ is $+1$ if $i + j$ is even and -1 if $i + j$ is odd, it follows that either $C_{ij} = M_{ij}$ or $C_{ij} = -M_{ij}$, depending on whether $i + j$ is even or odd, respectively. Here is an easy way to determine the appropriate sign: the minor and cofactor that result from deleting a given row and column will have the same or opposite signs depending on whether there is a $+$ or a $-$ at the intersection for the given row and column in the following matrix.

$$\begin{bmatrix} + & - & + \\ - & + & - \\ + & - & + \end{bmatrix}$$

EXAMPLE 48.3

(a) The (1,1)-cofactor of the matrix in Example 48.2 is

$$C_{11} = +M_{11} = 27.$$

(b) The (3,2)-cofactor of the matrix in Example 48.2 is

$$C_{32} = -M_{32} = -(-10) = 10.$$

DEFINITION

$$\det \begin{bmatrix} a_{11} & a_{12} & a_{13} \\ a_{21} & a_{22} & a_{23} \\ a_{31} & a_{32} & a_{33} \end{bmatrix} = \begin{vmatrix} a_{11} & a_{12} & a_{13} \\ a_{21} & a_{22} & a_{23} \\ a_{31} & a_{32} & a_{33} \end{vmatrix}$$
$$= a_{11}C_{11} + a_{12}C_{12} + a_{13}C_{13} \qquad (48.1)$$

In words: to compute the determinant of a 3 × 3 matrix multiply the entries in the first row of the matrix by their corresponding cofactors and then add the results. Using minors in place of cofactors, the determinant is

$$a_{11}M_{11} - a_{12}M_{12} + a_{13}M_{13}.$$

EXAMPLE 48.4

$$\begin{vmatrix} 2 & -7 & 0 \\ 3 & -3 & -5 \\ -2 & 6 & 1 \end{vmatrix} = 2\begin{vmatrix} -3 & -5 \\ 6 & 1 \end{vmatrix} - (-7)\begin{vmatrix} 3 & -5 \\ -2 & 1 \end{vmatrix} + 0\begin{vmatrix} 3 & -3 \\ -2 & 6 \end{vmatrix}$$
$$= 2(27) - (-7)(-7) + 0(12)$$
$$= 5.$$

It can be proved that the determinant of a 3 × 3 matrix can also be obtained by using the idea in (48.1) with any other row or column in place of the first row. Specifically:

> To compute the determinant of a 3 × 3 matrix, multiply the entries in any row or column by their corresponding cofactors and then add the results.

Using the second column rather than the first row, for example, (48.1) becomes

$$\begin{vmatrix} a_{11} & a_{12} & a_{13} \\ a_{21} & a_{22} & a_{23} \\ a_{31} & a_{32} & a_{33} \end{vmatrix} = a_{12}C_{12} + a_{22}C_{22} + a_{32}C_{32}.$$

This is called *the expansion of the determinant by the second column.* A similar expression is used for the expansion by any other row or column.

If a 3 × 3 matrix has only 0's in some row or column, then we can expand the determinant by using that row or column and the final result will obviously be 0 regardless of the cofactors. (For an illustration, see Example 49.3.)

B. Cramer's Rule

We now look at an application of determinants in solving systems of linear equations. We begin with systems having two equations and two variables.

With the system

$$\begin{aligned} a_1x + b_1y &= c_1 \\ a_2x + b_2y &= c_2 \end{aligned} \qquad (48.2)$$

we associate three determinants:

$$D = \begin{vmatrix} a_1 & b_1 \\ a_2 & b_2 \end{vmatrix}, \quad D_x = \begin{vmatrix} c_1 & b_1 \\ c_2 & b_2 \end{vmatrix}, \quad D_y = \begin{vmatrix} a_1 & c_1 \\ a_2 & c_2 \end{vmatrix}. \tag{48.3}$$

Notice that D is the determinant of the coefficient matrix, that D_x is obtained from D by replacing the coefficients of x (a_1 and a_2) by the constant terms (c_1 and c_2), and that D_y is obtained from D by replacing the coefficients of y by the constant terms. Exercise 37 indicates how to prove the following result.

Cramer's Rule (2 × 2 Case)

If $D \neq 0$ in (48.3), then the system (48.2) has the unique solution

$$x = \frac{D_x}{D}, \quad y = \frac{D_y}{D}.$$

EXAMPLE 48.5 Use Cramer's Rule to solve the system

$$\begin{aligned} -2x + y &= 5 \\ x + 3y &= 7. \end{aligned}$$

Solution

$$D = \begin{vmatrix} -2 & 1 \\ 1 & 3 \end{vmatrix} = -7, \quad D_x = \begin{vmatrix} 5 & 1 \\ 7 & 3 \end{vmatrix} = 8, \quad D_y = \begin{vmatrix} -2 & 5 \\ 1 & 7 \end{vmatrix} = -19.$$

Therefore, the solution is

$$x = \tfrac{8}{-7} = -\tfrac{8}{7}, \quad y = \tfrac{-19}{-7} = \tfrac{19}{7}.$$

Cramer's Rule extends to three equations with three variables as follows. With the system

$$\begin{aligned} a_1x + b_1y + c_1z &= d_1 \\ a_2x + b_2y + c_2z &= d_2 \\ a_3x + b_3y + c_3z &= d_3 \end{aligned} \tag{48.4}$$

we associate four determinants: the determinant of the coefficient matrix,

$$D = \begin{vmatrix} a_1 & b_1 & c_1 \\ a_2 & b_2 & c_2 \\ a_3 & b_3 & c_3 \end{vmatrix}, \tag{48.5}$$

and the three determinants D_x, D_y, and D_z, where D_x is obtained from D by replacing the coefficients of x by the constant terms, and D_y and D_z are obtained similarly.

Cramer's Rule (3 × 3 Case)
If $D \neq 0$ in (48.5), then the system (48.4) has the unique solution

$$x = \frac{D_x}{D}, \quad y = \frac{D_y}{D}, \quad z = \frac{D_z}{D}.$$

EXAMPLE 48.6 Use Cramer's Rule to solve the system

$$\begin{aligned} 2x - 7y &= 1 \\ 3x - 3y - 5z &= -2 \\ -2x + 6y + z &= 0. \end{aligned}$$

Solution The determinant D was evaluated in Example 48.4. The computations for D_x, D_y, and D_z can be carried out in the same way.

$$D = \begin{vmatrix} 2 & -7 & 0 \\ 3 & -3 & -5 \\ -2 & 6 & 1 \end{vmatrix} = 5, \quad D_x = \begin{vmatrix} 1 & -7 & 0 \\ -2 & -3 & -5 \\ 0 & 6 & 1 \end{vmatrix} = 13,$$

$$D_y = \begin{vmatrix} 2 & 1 & 0 \\ 3 & -2 & -5 \\ -2 & 0 & 1 \end{vmatrix} = 3, \quad D_z = \begin{vmatrix} 2 & -7 & 1 \\ 3 & -3 & -2 \\ -2 & 6 & 0 \end{vmatrix} = 8.$$

Therefore, the solution is

$$x = \tfrac{13}{5}, \quad y = \tfrac{3}{5}, \quad z = \tfrac{8}{5}.$$

For a proof of Cramer's Rule in the 3 × 3 case, as well as the statements that follow, see a book on matrices or linear algebra.

In both the 2 × 2 and 3 × 3 cases Cramer's Rule covers all cases where $D \neq 0$. The cases where $D = 0$ are covered by the following statement.

Assume $D = 0$.

(a) If D_x and D_y (and D_z in the 3 × 3 case) all equal zero, then the system is dependent (has infinitely many solutions). (48.6)

(b) If one or more of D_x, D_y, and D_z is nonzero, then the system is inconsistent (has no solution).

A linear system in which the constant terms are all 0 is said to be **homogeneous.** For a homogeneous system each of the determinants D_x and D_y (and D_z in the 3 × 3 case) will contain a column of 0's and will, therefore, have the value 0. It follows, using Cramer's Rule, that if $D \neq 0$ for a homogeneous system, then the only solution of the system is the one in which each variable equals 0. And, using (a) above, if $D = 0$ for a homogeneous system, then the system is dependent.

EXAMPLE 48.7 For the homogeneous system

$$2x - y = 0$$
$$-7x + 2y = 0,$$

$$D = \begin{vmatrix} 2 & -1 \\ -7 & 2 \end{vmatrix} = -3, \quad D_x = \begin{vmatrix} 0 & -1 \\ 0 & 2 \end{vmatrix} = 0, \quad D_y = \begin{vmatrix} 2 & 0 \\ -7 & 0 \end{vmatrix} = 0.$$

Thus (0, 0) is the unique solution.

EXERCISES FOR SECTION 48

Compute the indicated minors and cofactors for the matrix

$$\begin{bmatrix} 2 & 8 & -6 \\ 4 & 1 & 5 \\ 0 & 2 & -2 \end{bmatrix}$$

1. M_{21} **2.** M_{33} **3.** M_{13} **4.** C_{21} **5.** C_{33}
6. C_{13} **7.** C_{22} **8.** C_{31} **9.** C_{12}

Evaluate by expanding by the first row [Equation (48.1)].

10. $\begin{vmatrix} 2 & -1 & 3 \\ 6 & 2 & -5 \\ 0 & 2 & 1 \end{vmatrix}$ **11.** $\begin{vmatrix} 2 & 1 & 0 \\ -5 & 4 & 2 \\ 1 & 0 & 6 \end{vmatrix}$ **12.** $\begin{vmatrix} 3 & 0 & 7 \\ -6 & -3 & -3 \\ 4 & 1 & 2 \end{vmatrix}$

13. $\begin{vmatrix} 1 & 4 & 9 \\ 6 & -1 & 2 \\ 2 & 3 & -1 \end{vmatrix}$ **14.** $\begin{vmatrix} 2 & 1 & 2 \\ 3 & -5 & -1 \\ -2 & 0 & 4 \end{vmatrix}$ **15.** $\begin{vmatrix} -2 & 8 & 3 \\ -7 & 1 & 3 \\ 0 & 3 & -4 \end{vmatrix}$

Evaluate by expanding by the second column.

16. $\begin{vmatrix} 1 & 4 & 1 \\ 3 & -1 & 5 \\ -2 & 6 & 0 \end{vmatrix}$ **17.** $\begin{vmatrix} -3 & 4 & 2 \\ 4 & -1 & 2 \\ 9 & 5 & -6 \end{vmatrix}$ **18.** $\begin{vmatrix} -9 & 2 & 0 \\ 7 & 1 & -5 \\ 7 & 3 & 4 \end{vmatrix}$

Evaluate by expanding by the row or column with the most 0's.

19. $\begin{vmatrix} -1 & 0 & \frac{4}{3} \\ 0 & 7 & 0 \\ 6 & \frac{1}{2} & 2 \end{vmatrix}$ **20.** $\begin{vmatrix} 4 & \frac{1}{5} & 0 \\ \frac{3}{5} & 0 & 2 \\ 3 & 0 & -1 \end{vmatrix}$ **21.** $\begin{vmatrix} 0 & -6 & 2 \\ -5 & \frac{1}{8} & \frac{1}{3} \\ 7 & 0 & 0 \end{vmatrix}$

Solve each system using Cramer's Rule.

22. $x - 2y = 5$
$3x + y = 1$

23. $3x + y = 0$
$-x - 5y = 4$

24. $7x - y = 1$
$-6x + y = 3$

25. $2u - 6v = 1$
$8u - 13v = 5$

26. $8w + z = 5$
$2w - z = 0$

27. $-r + 2s = 3$
$5r - 3s = 2$

28. $\begin{aligned} x + y + z &= 9 \\ 2x - y - z &= 1 \\ 3x \qquad + z &= 2 \end{aligned}$

29. $\begin{aligned} 3y - z &= 1 \\ x \qquad + z &= 5 \\ 2x - y \qquad &= 6 \end{aligned}$

30. $\begin{aligned} x - y + z &= 3 \\ 2x + y + 5z &= 0 \\ x \qquad - z &= 7 \end{aligned}$

Each system is either dependent or inconsistent. Use determinants and (48.6) to determine which in each case.

31. $\begin{aligned} 4x - y &= 5 \\ -8x + 2y &= 1 \end{aligned}$

32. $\begin{aligned} 6x - 4y &= 10 \\ 9x - 6y &= 15 \end{aligned}$

33. $\begin{aligned} 12x - 8y &= 3 \\ -9x + 6y &= 0 \end{aligned}$

34. $\begin{aligned} x - y - z &= 1 \\ 2x + 3y - z &= -2 \\ 2x - 7y - 3z &= 6 \end{aligned}$

35. $\begin{aligned} 5x + y + z &= 2 \\ x - y + z &= 1 \\ 3x + 3y - z &= -1 \end{aligned}$

36. $\begin{aligned} 3x + y + z &= 0 \\ 4x - y - 2z &= 1 \\ -2x - 3y - 4z &= 1 \end{aligned}$

37. Prove the 2×2 case of Cramer's Rule by substitution. [That is, substitute D_x/D for x and D_y/D for y in the System (48.2).]

38. Verify that

$$\begin{vmatrix} 1 & a & a^2 \\ 1 & b & b^2 \\ 1 & c & c^2 \end{vmatrix} = (b - a)(c - a)(c - b).$$

39. Prove that the equation of line through the points (x_1, y_1) and (x_2, y_2) is

$$\begin{vmatrix} x & y & 1 \\ x_1 & y_1 & 1 \\ x_2 & y_2 & 1 \end{vmatrix} = 0.$$

40. Bach, Beethoven, and Mozart lived a total of 157 years. Bach lived 9 years more than Beethoven, and Beethoven lived 20 years more than Mozart. How long did each live? (All ages are to the nearest year.)

41. The United States paid a total of $125 billion for imported oil in the years 1970, 1978, and 1980. The cost in 1978 was 14 times the cost in 1970, and the cost in 1980 was $4 billion less than twice the cost in 1978. Find the cost in each of the three years.

42. In the 50 years from 1927 through 1976, the Montreal Canadiens won the Stanley Cup six more times than the Toronto Maple Leafs, and the Canadiens and the Maple Leafs together won the Cup six more times than all other teams combined. How many times was the Cup won by the Canadiens, by the Maple Leafs, and by all other teams combined?

SECTION

49 More about Determinants

A. General Definition

The definition of *determinant* for 3×3 matrices can be extended in an obvious way to obtain the definition for larger matrices. First, the **(i,j)-minor** of an $n \times n$ matrix A, denoted M_{ij}, is the determinant of the $(n-1) \times (n-1)$ matrix that remains when the ith row and jth column of A have been deleted. The **(i,j)-cofactor** is $C_{ij} = (-1)^{i+j}M_{ij}$. The pattern for the signs relating M_{ij} and C_{ij} is given by simply extending the pattern from the 3×3 case.

$$\begin{bmatrix} + & - & + & - & \cdots \\ - & + & - & + & \cdots \\ + & - & + & - & \cdots \\ - & + & - & + & \cdots \\ \vdots & \vdots & \vdots & \vdots & \ddots \end{bmatrix}$$

EXAMPLE 49.1 If

$$A = \begin{bmatrix} 8 & -2 & 1 & -4 \\ -1 & 0 & 0 & 2 \\ 3 & -3 & 7 & -6 \\ 2 & 1 & -1 & 8 \end{bmatrix}$$

then

$$M_{14} = \begin{vmatrix} -1 & 0 & 0 \\ 3 & -3 & 7 \\ 2 & 1 & -1 \end{vmatrix} = -1 \begin{vmatrix} -3 & 7 \\ 1 & -1 \end{vmatrix} - 0 \begin{vmatrix} 3 & 7 \\ 2 & -1 \end{vmatrix} + 0 \begin{vmatrix} 3 & -3 \\ 2 & 1 \end{vmatrix}$$
$$= -1(-4) = 4.$$

And

$$C_{14} = -M_{14} = -4.$$

DEFINITION. If A is an $n \times n$ matrix ($n \geq 2$), then

$$\det A = |A| = a_{11}C_{11} + a_{12}C_{12} + \cdots + a_{1n}C_{1n}. \tag{49.1}$$

In terms of minors,

$$\det A = |A| = a_{11}M_{11} - a_{12}M_{12} + \cdots + (-1)^{1+n}a_{1n}M_{1n}.$$

As in the 3×3 case, we can expand a determinant by using any other row or column in place of the first row.

> To compute the determinant of an $n \times n$ matrix ($n \geq 2$), multiply the entries in any row or column by their cofactors and then add the results.

EXAMPLE 49.2 To compute $|A|$ for A as in Example 49.1, we can take advantage of the two zeros in the second row; by expanding by the second row in this example we need to compute only two 3×3 cofactors.

$$\begin{vmatrix} 8 & -2 & 1 & -4 \\ -1 & 0 & 0 & 2 \\ 3 & -3 & 7 & -6 \\ 2 & 1 & -1 & 8 \end{vmatrix} = -(-1)\begin{vmatrix} -2 & 1 & -4 \\ -3 & 7 & -6 \\ 1 & -1 & 8 \end{vmatrix} + 0\begin{vmatrix} 8 & 1 & -4 \\ 3 & 7 & -6 \\ 2 & -1 & 8 \end{vmatrix}$$

$$-0\begin{vmatrix} 8 & -2 & -4 \\ 3 & -3 & -6 \\ 2 & 1 & 8 \end{vmatrix} + 2\begin{vmatrix} 8 & -2 & 1 \\ 3 & -3 & 7 \\ 2 & 1 & -1 \end{vmatrix}$$

$$= 1(-66) + 0 - 0 + 2(-57)$$

$$= -180.$$

Exercise 26 asks you to verify that in the 2×2 case the definition in (49.1) agrees with the definition in Sections 47 and 48.

B. Properties of Determinants

In each of the following properties A is assumed to be an $n \times n$ matrix for $n \geq 2$. The examples will indicate some of the proofs for 3×3 matrices. For general proofs consult a book on matrices or linear algebra.

PROPERTY I. If any row or column of A contains only 0's, then $|A| = 0$.

EXAMPLE 49.3 Suppose the second row of a 3×3 matrix A contains only 0's. If we expand $|A|$ by the second row, we get

$$\begin{vmatrix} a_{11} & a_{12} & a_{13} \\ 0 & 0 & 0 \\ a_{31} & a_{32} & a_{33} \end{vmatrix} = 0C_{21} + 0C_{22} + 0C_{23} = 0.$$

The operations in Properties II, III, and V are similar to the elementary row operations on matrices (Section 42B). Here, however, the operations are also applied to columns, and we must pay careful attention to the effect on the determinant.

PROPERTY II. If B is obtained from A by multiplying each element of any row or column of A by the real number k, then $|B| = k|A|$.

EXAMPLE 49.4 Suppose that each element of the third column of a 3×3 matrix A is multiplied by k to produce B. Using expansion by the third column, we have

$$
\begin{aligned}
|B| &= \begin{vmatrix} a_{11} & a_{12} & ka_{13} \\ a_{21} & a_{22} & ka_{23} \\ a_{31} & a_{32} & ka_{33} \end{vmatrix} \\
&= ka_{13}C_{13} + ka_{23}C_{23} + ka_{33}C_{33} \\
&= k(a_{13}C_{13} + a_{23}C_{23} + a_{33}C_{33}) \\
&= k|A|.
\end{aligned}
$$

EXAMPLE 49.5 Property II implies that a common factor of the elements of any row or column can be removed by placing it in front on the determinant. For example,

$$\begin{vmatrix} 0 & 4 & -3 \\ 2 & -6 & 1 \\ -7 & 2 & 5 \end{vmatrix} = \begin{vmatrix} 0 & 2(2) & -3 \\ 2 & 2(-3) & 1 \\ -7 & 2(1) & 5 \end{vmatrix} = 2\begin{vmatrix} 0 & 2 & -3 \\ 2 & -3 & 1 \\ -7 & 1 & 5 \end{vmatrix}.$$

PROPERTY III. If B is obtained from A by interchanging two rows (or two columns) of A, then $|B| = -|A|$.

EXAMPLE 49.6 Exercises 27 and 28 asks you to verify these special cases.

$$\begin{vmatrix} a_1 & b_1 & c_1 \\ a_2 & b_2 & c_2 \\ a_3 & b_3 & c_3 \end{vmatrix} = -\begin{vmatrix} a_3 & b_3 & c_3 \\ a_2 & b_2 & c_2 \\ a_1 & b_1 & c_1 \end{vmatrix} \qquad \text{Interchange rows 1 and 3.}$$

$$\begin{vmatrix} a_1 & b_1 & c_1 \\ a_2 & b_2 & c_2 \\ a_3 & b_3 & c_3 \end{vmatrix} = -\begin{vmatrix} b_1 & a_1 & c_1 \\ b_2 & a_2 & c_2 \\ b_3 & a_3 & c_3 \end{vmatrix} \qquad \text{Interchange columns 1 and 2.}$$

PROPERTY IV. If two rows (or two columns) of A are identical, then $|A| = 0$.

EXAMPLE 49.7 Exercise 29 asks you to verify that

$$\begin{vmatrix} a_1 & b_1 & c_1 \\ a_1 & b_1 & c_1 \\ a_3 & b_3 & c_3 \end{vmatrix} = 0.$$

PROPERTY V. If B is obtained from A by adding k times one row (column) of A to another row (column) of A, then $|A| = |B|$.

EXAMPLE 49.8 Exercise 30 asks you to use Properties II and IV to verify the following special case.

$$\begin{vmatrix} a_1 & b_1 & c_1 \\ a_2 & b_2 & c_2 \\ a_3 & b_3 & c_3 \end{vmatrix} = \begin{vmatrix} a_1 & b_1 + ka_1 & c_1 \\ a_2 & b_2 + ka_2 & c_2 \\ a_3 & b_3 + ka_3 & c_3 \end{vmatrix}$$

Cramer's Rule, stated in Section 48 for systems with two or three variables, can also be extended to systems with more than three variables. For example, if D, the determinant of the coefficient matrix, is nonzero for the system

$$\begin{aligned} a_1w + b_1x + c_1y + d_1z &= k_1 \\ a_2w + b_2x + c_2y + d_2z &= k_2 \\ a_3w + b_3x + c_3y + d_3z &= k_3 \\ a_4w + b_4x + c_4y + d_4z &= k_4, \end{aligned}$$

then the system has a unique solution which is given by

$$w = \frac{D_w}{D}, \quad x = \frac{D_x}{D}, \quad y = \frac{D_y}{D}, \quad z = \frac{D_z}{D},$$

where D_w is obtained from D by replacing the coefficients of w by the constant terms, and D_x, D_y, and D_z are obtained similarly. In general this method of solving a system is not as efficient as the method using augmented matrices and elementary row operations (Section 42); thus we shall not use it.

We computed matrix inverses in Section 47 by using elementary row operations. Matrix inverses can also be computed by using determinants: If A is an invertible $n \times n$ matrix, then

$$A^{-1} = \frac{1}{|A|}\begin{bmatrix} C_{11} & C_{21} & \cdots & C_{n1} \\ C_{12} & C_{22} & \cdots & C_{n2} \\ \vdots & \vdots & & \vdots \\ C_{1n} & C_{2n} & \cdots & C_{nn} \end{bmatrix}. \tag{49.2}$$

As before, each number C_{ij} in (49.2) is the (i, j)-cofactor of A. Notice that the (i, j)-entry on the right in (49.2) is the (j, i)-cofactor of A; for instance, the (1,2)-entry is C_{21}. The method of computing inverses using elementary row operations (Section 47) is more efficient than the method given by (49.2). Equation (49.2) is useful in the general theory of matrices, however. Both Cramer's Rule and Equation (49.2) are discussed more fully in most books on matrices or linear algebra.

EXERCISES FOR SECTION 49

Compute the indicated minors and cofactors for the matrix

$$\begin{bmatrix} 8 & -2 & 3 & 7 \\ 1 & 6 & 1 & 0 \\ 5 & 0 & -1 & 0 \\ 4 & 1 & 9 & 5 \end{bmatrix}.$$

1. M_{41} **2.** M_{33} **3.** M_{12} **4.** C_{41} **5.** C_{33} **6.** C_{12} **7.** C_{22} **8.** C_{34} **9.** C_{13}

Evaluate by expanding by a row or column with the maximum number of 0's.

10. $\begin{vmatrix} 4 & 0 & -1 & 2 \\ 0 & 3 & 1 & 0 \\ 5 & -2 & 6 & 7 \\ 1 & 2 & -1 & 1 \end{vmatrix}$

11. $\begin{vmatrix} 0 & -4 & 3 & 2 \\ 2 & -3 & 0 & 1 \\ 0 & 4 & 0 & 0 \\ 5 & 7 & 6 & -2 \end{vmatrix}$

12. $\begin{vmatrix} 4 & -1 & 7 & -3 \\ 0 & -1 & 6 & 4 \\ 0 & 3 & 7 & 2 \\ 1 & 2 & -2 & 5 \end{vmatrix}$

13. $\begin{vmatrix} 0 & -4 & 2 & 2 & 6 \\ 1 & 3 & -2 & 0 & 1 \\ 0 & 1 & 7 & 0 & 2 \\ 4 & -5 & -3 & 1 & 0 \\ 0 & 1 & 3 & -1 & 0 \end{vmatrix}$

14. $\begin{vmatrix} 1 & 0 & 0 & 2 & 3 \\ 4 & 5 & 0 & 0 & 6 \\ 7 & 8 & 9 & 0 & 0 \\ 0 & 0 & -6 & -5 & -4 \\ -3 & 0 & 0 & -2 & -1 \end{vmatrix}$

15. $\begin{vmatrix} 1 & 0 & -1 & 0 & 1 \\ 0 & -1 & 0 & -2 & 0 \\ -1 & 0 & 2 & 0 & -3 \\ 0 & 1 & 0 & 1 & 0 \\ 1 & 0 & 1 & 0 & 1 \end{vmatrix}$

Each equation in Exercises 16–21 is true because of one of Properties I–V in Subsection B. Provide the reason in each case.

16. $\begin{vmatrix} 2 & 0 & 10 \\ 1 & -3 & 7 \\ 5 & 4 & -6 \end{vmatrix} = -\begin{vmatrix} 2 & 0 & 10 \\ 5 & 4 & -6 \\ 1 & -3 & 7 \end{vmatrix}$

17. $\begin{vmatrix} 12 & 0 & 6 \\ -7 & 4 & 11 \\ 12 & 0 & 6 \end{vmatrix} = 0$

18. $\begin{vmatrix} 6 & -1 & 0 \\ 3 & 2 & 0 \\ 4 & 1 & 1 \end{vmatrix} = \begin{vmatrix} 6 & -1 & 0 \\ 11 & 4 & 2 \\ 4 & 1 & 1 \end{vmatrix}$

19. $\begin{vmatrix} 19 & 7 & 4 \\ -1 & 1 & 1 \\ 0 & 0 & 0 \end{vmatrix} = 0$

20. $\begin{vmatrix} 4 & 7 & 0 \\ -6 & 2 & 5 \\ 10 & 8 & -3 \end{vmatrix} = 2\begin{vmatrix} 2 & 7 & 0 \\ -3 & 2 & 5 \\ 5 & 8 & -3 \end{vmatrix}$

21. $\begin{matrix} \\ -3 \\ \\ \end{matrix}\begin{vmatrix} 1 & 0 & 2 \\ -1 & 4 & 2 \\ 7 & 8 & -3 \end{vmatrix} = \begin{vmatrix} 1 & 0 & 2 \\ 3 & -12 & -6 \\ 7 & 8 & -3 \end{vmatrix}$

Compute the inverse of each matrix using (49.2).

22. $\begin{bmatrix} 2 & -1 & 5 \\ 0 & 2 & 1 \\ 3 & -3 & 1 \end{bmatrix}$

23. $\begin{bmatrix} 4 & 0 & 2 \\ 1 & 6 & -1 \\ 3 & 5 & 4 \end{bmatrix}$

24. $\begin{bmatrix} 10 & 1 & 0 \\ -7 & 2 & 4 \\ -3 & 8 & 1 \end{bmatrix}$

25. Show that

$$\begin{vmatrix} a_{11} & a_{12} & a_{13} & a_{14} \\ a_{21} & a_{22} & a_{23} & a_{24} \\ 0 & 0 & a_{33} & a_{34} \\ 0 & 0 & a_{43} & a_{44} \end{vmatrix} = \begin{vmatrix} a_{11} & a_{12} \\ a_{21} & a_{22} \end{vmatrix} \cdot \begin{vmatrix} a_{33} & a_{34} \\ a_{43} & a_{44} \end{vmatrix}.$$

26. Show that with $n = 2$ the definition in Equation (49.1) agrees with the definition of the determinant of a 2×2 matrix at the beginning of Section 48, if the determinant of a 1×1 matrix is equal to its entry.
27. Verify the first equation in Example 49.6 by expanding both sides.
28. Verify the second equation in Example 49.6 by expanding both sides.
29. Verify the equation in Example 49.7 by expanding the left side.
30. Verify the equation in Example 49.8 by expanding both sides. (Properties II and IV will help.)

REVIEW EXERCISES FOR CHAPTER XI

Perform the indicated operations.

1. $\begin{bmatrix} -3 & 14 \\ 0 & 5 \end{bmatrix} + \begin{bmatrix} 5 & 9 \\ -2 & -2 \end{bmatrix}$

2. $\begin{bmatrix} 0 & 6 \\ -5 & -2 \\ 9 & 11 \end{bmatrix} + \begin{bmatrix} 0 & -3 \\ 2 & 1 \\ 4 & 7 \end{bmatrix}$

3. $\begin{bmatrix} 4 & 1 & 0 \\ -1 & 3 & 5 \end{bmatrix} \begin{bmatrix} -1 & -2 \\ 5 & 6 \\ 8 & 7 \end{bmatrix}$

4. $\begin{bmatrix} 1 & -1 & 0 \\ 4 & 6 & 2 \\ -1 & -2 & 5 \end{bmatrix} \begin{bmatrix} 2 & -1 & 5 \\ 2 & 4 & -3 \\ 1 & -2 & 0 \end{bmatrix}$

Compute the inverse of each matrix by using the determinant [Equation (47.1)].

5. $\begin{bmatrix} 4 & 1 \\ -5 & 2 \end{bmatrix}$

6. $\begin{bmatrix} \frac{1}{2} & \frac{1}{3} \\ \frac{1}{4} & \frac{1}{5} \end{bmatrix}$

Use elementary row operations to deterine whether each matrix is invertible, and if the matrix is invertible give its inverse.

7. $\begin{bmatrix} 1 & 0 & -2 \\ 3 & 2 & 7 \\ -1 & 2 & 6 \end{bmatrix}$

8. $\begin{bmatrix} 4 & 1 & 0 \\ -2 & 3 & 1 \\ 2 & -10 & -3 \end{bmatrix}$

9. $\begin{bmatrix} 3 & 1 & 3 \\ 4 & 0 & -2 \\ 1 & 3 & 13 \end{bmatrix}$

10. $\begin{bmatrix} 1 & 2 & 4 \\ 5 & 0 & -3 \\ -6 & 0 & -1 \end{bmatrix}$

Solve each system by first computing the inverse of the coefficient matrix (that is, use that the solution of $AX = B$ is $X = A^{-1}B$).

11. $\begin{aligned} 0.5x + y &= 2 \\ -0.1x + 3y &= 1.5 \end{aligned}$

12. $\begin{aligned} 2x - y - z &= 0 \\ x + y + z &= 4 \\ x - y \quad &= 1 \end{aligned}$

Evaluate each determinant.

13. $\begin{vmatrix} 2 & -5 & 1 \\ 4 & 0 & 2 \\ -7 & 1 & 5 \end{vmatrix}$

14. $\begin{vmatrix} -\frac{1}{2} & 1 & \frac{1}{2} \\ 2 & \frac{1}{4} & -2 \\ \frac{1}{3} & 1 & -\frac{1}{3} \end{vmatrix}$

15. $\begin{vmatrix} 0 & -1 & 1 & 0 \\ 2 & 5 & 0 & -3 \\ 1 & 0 & -2 & 1 \\ 0 & 2 & -1 & 1 \end{vmatrix}$

16. $\begin{vmatrix} a & 0 & b & 0 \\ 0 & b & 0 & c \\ b & 0 & c & 0 \\ 0 & c & 0 & d \end{vmatrix}$

Use determinants to determine whether each system has a unique solution, is dependent, or is inconsistent. If the system has a unique solution, find it using Cramer's Rule.

17. $\begin{aligned} 4x + y &= 3 \\ -x + y &= 1 \end{aligned}$

18. $\begin{aligned} x - 2y &= 3 \\ -2x + 4y &= 6 \end{aligned}$

19. $\begin{aligned} x - y - z &= 1 \\ 2x - 2y - z &= 3 \\ 4x - 4y - 3z &= 5 \end{aligned}$

20. $\begin{aligned} 2x \quad + z &= 0 \\ x - y - z &= 1 \\ 4x + y + z &= 5 \end{aligned}$

CHAPTER XII
MATHEMATICAL INDUCTION AND SEQUENCES

The first section of this chapter is devoted to an important method of proof for many statements that involve positive integers. The other sections treat ideas about sequences and sums that arise often throughout mathematics and its applications.

SECTION

50 Mathematical Induction

A. Sequences and Series

A collection of real numbers arranged so that there is a first, a second, a third, and so on, is called a **sequence** (or, if we need to be more precise, a **sequence of real numbers**). The successive numbers in the sequence are called its **terms:** the **first term,** the **second term,** the **third term,** and so on. A sequence with a last term is said to be **finite;** if there is no last term the sequence is said to be **infinite.**

EXAMPLE 50.1

(a) $5, 10, 15, \ldots, 100$
(b) $\frac{1}{1}, \frac{1}{3}, \frac{1}{5}, \frac{1}{7}, \ldots$
(c) $1, -1, 1, -1, \ldots$
(d) $4, 4, 4, 4, \ldots$

Sequence (a) is finite. The other sequences are infinite. Sequences (c) and (d) make the point that different terms in the same sequence can be equal.

To represent general sequences we use notation like

$$a_1, a_2, \ldots, a_n \tag{50.1}$$

(for finite sequences) and

$$a_1, a_2, \ldots, a_n, \ldots \tag{50.2}$$

(for infinite sequences). For many sequences a rule or formula can be given for a_n. It's best to give such a rule or formula whenever possible; that eliminates the possibility of uncertainty or ambiguity about the terms that are not listed.

EXAMPLE 50.2 Here are the sequences from Example 50.1 with the nth term specified in each case. If you replace n by 1 in the nth term, you will get the first term, and so on for $n = 2, 3, \ldots$.

(a) $5, 10, 15, \ldots, 5n, \ldots, 100$

(b) $\dfrac{1}{1}, \dfrac{1}{3}, \dfrac{1}{5}, \ldots, \dfrac{1}{2n-1}, \ldots$

(c) $1, -1, 1, \ldots, (-1)^{n+1}, \ldots$
(d) $4, 4, 4, \ldots, 4, \ldots$

Part (b) illustrates that the nth positive odd integer is $2n - 1$. Part (c) illustrates that to introduce alternating signs we can use $(-1)^{n+1}$. [To alternate signs starting with -1, use $(-1)^n$.]

EXAMPLE 50.3 Write the first seven terms of the sequence defined by

$$a_n = \begin{cases} n^2 & \text{if } n \text{ is odd} \\ 1 & \text{if } n \text{ is even.} \end{cases}$$

Solution 1, 1, 9, 1, 25, 1, 49.

A sequence is actually a function. The sequence in (50.1), for example, can be thought of as a function whose domain is the set $\{1, 2, \ldots, n\}$; the output of the function for the input k is a_k. The sequence in (50.2) can be thought of as a function whose domain is the set of all positive integers; and again, the output is a_k for the input k.

Expressions like

$$a_1 + a_2 + a_3 + \cdots + a_n \tag{50.3}$$

and

$$a_1 + a_2 + a_3 + \cdots + a_n + \cdots, \tag{50.4}$$

obtained from adding the terms of a sequence, are called **series.** The series in this section will be **finite,** like (50.3). In Section 52 we'll also consider **infinite series,** like (50.4).

B. Principle of Mathematical Induction

In Example 50.4 we'll prove that if n is a positive integer, then

$$1 + 2 + 3 + \cdots + n = \frac{n(n+1)}{2}. \tag{50.5}$$

We can verify (50.5) for small values of n by direct computation. Here are two examples.

$n = 5$ case:

$$1 + 2 + 3 + 4 + 5 = 15 \quad \text{and} \quad \frac{5 \cdot 6}{2} = 15.$$

$n = 10$ case:

$$1 + 2 + 3 + \cdots + 10 = 55 \quad \text{and} \quad \frac{10 \cdot 11}{2} = 55.$$

To prove (50.5) once and for all, for *every* value of n, we need a different idea. The Principle of Mathematical Induction provides such an idea; it applies not only to (50.5) but to a very broad class of similar statements.

Principle of Mathematical Induction

For each positive integer n, let P_n represent a statement depending on n. If

(a) P_1 is true, and
(b) the truth of P_k implies the truth of P_{k+1} for each positive integer k,

then P_n is true for every positive integer n.

The Principle of Mathematical Induction can be visualized in terms of an infinite sequence of dominoes, originally standing on edge (Figure 50.1): the fall of the nth domino corresponds to the truth of P_n. Part (a) guarantees that the first domino will fall, and part (b) guarantees that if any particular domino [the kth] falls, then the next one [the $(k + 1)$st] will also fall; it follows that every domino will fall.

To apply the Principle of Mathematical Induction we must verify both parts, (a) and (b). Notice that to verify (b) we must prove that *if* P_k is true, *then* P_{k+1} is true; that is, we must establish P_{k+1} based on the assumption of P_k.

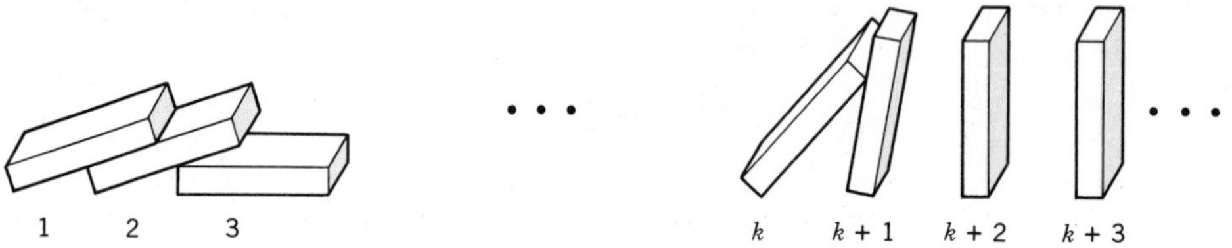

Figure 50.1

C. Examples

EXAMPLE 50.4 Prove that Equation (50.5) is true for every positive integer n.

Proof. Let P_n represent Equation (50.5).

(a) P_1 is true because

$$1 = \frac{1 \cdot 2}{2}.$$

(b) Assume P_k, that is,

$$1 + 2 + 3 + \cdots + k = \frac{k(k + 1)}{2}. \tag{50.6}$$

We must prove P_{k+1}, that is, (50.5) with $k + 1$ in place of n:

$$1 + 2 + 3 + \cdots + (k + 1) = \frac{(k + 1)(k + 2)}{2}. \tag{50.7}$$

If we add $k + 1$ to both sides of (50.6), we obtain

$$1 + 2 + 3 + \cdots + k + (k + 1) = \frac{k(k + 1)}{2} + (k + 1)$$

$$1 + 2 + 3 + \cdots + k + (k + 1) = \frac{k(k + 1) + 2(k + 1)}{2}$$

$$1 + 2 + 3 + \cdots + k + (k + 1) = \frac{(k + 1)(k + 2)}{2}.$$

The last equation is the same as (50.7), which is what we were to prove.

EXAMPLE 50.5 Prove that

$$\frac{1}{1\cdot 3}+\frac{1}{3\cdot 5}+\frac{1}{5\cdot 7}+\cdots+\frac{1}{(2n-1)(2n+1)}=\frac{n}{2n+1} \tag{50.8}$$

for every positive integer n.

Remark To help clarify the meaning of (50.8), here is the special case $n = 5$.

$$\frac{1}{1\cdot 3}+\frac{1}{3\cdot 5}+\frac{1}{5\cdot 7}+\frac{1}{7\cdot 9}+\frac{1}{9\cdot 11}=\frac{5}{11}$$

(Even with a calculator the direct computation of the left-hand side would be tedious, and the answer would only be a decimal approximation of the exact sum $\frac{5}{11}$. More important, a calculator could not provide the general formula $n/(2n + 1)$ for the sum.)

Proof. Let P_n represent Equation (50.8).

(a) P_1 is true because

$$\frac{1}{1\cdot 3}=\frac{1}{2\cdot 1+1}.$$

(b) Assume P_k, that is,

$$\frac{1}{1\cdot 3}+\frac{1}{3\cdot 5}+\frac{1}{5\cdot 7}+\cdots+\frac{1}{(2k-1)(2k+1)}=\frac{k}{2k+1}. \tag{50.9}$$

To find P_{k+1}, which is what we must prove from P_k, substitute $k + 1$ for n in (50.8) and simplify; the result is

$$\frac{1}{1\cdot 3}+\frac{1}{3\cdot 5}+\frac{1}{5\cdot 7}+\cdots+\frac{1}{(2k+1)(2k+3)}=\frac{k+1}{2k+3}. \tag{50.10}$$

To prove (50.10) from (50.9), we add

$$\frac{1}{(2k+1)(2k+3)}$$

to both sides of (50.9) and simplify the right-hand side.

$$\begin{aligned}\frac{1}{1\cdot 3}+\frac{1}{3\cdot 5}+\frac{1}{5\cdot 7}+\cdots+\frac{1}{(2k+1)(2k+3)}\\ &=\frac{k}{2k+1}+\frac{1}{(2k+1)(2k+3)}\\ &=\frac{k(2k+3)+1}{(2k+1)(2k+3)}\end{aligned}$$

$$= \frac{2k^2 + 3k + 1}{(2k + 1)(2k + 3)}$$

$$= \frac{(2k + 1)(k + 1)}{(2k + 1)(2k + 3)}$$

$$= \frac{k + 1}{2k + 3}$$

The next example illustrates a slight variation on the Principle of Mathematical Induction. In place of beginning with P_1 we begin with P_5. That is, we show that

(a) P_5 is true, and
(b) the truth of P_k implies the truth of P_{k+1} for each integer $k \geq 5$;

it will follow that P_n is true for every integer $n \geq 5$. (This corresponds to starting with the fifth domino rather than the first domino in Figure 50.1.) In other problems the same idea can be applied with 5 replaced by another appropriate starting number.

EXAMPLE 50.6 Prove that $2^n > n^2$ for every integer n such that $n \geq 5$.

Proof. Let P_n represent $2^n > n^2$.

(a) P_5 is true because $2^5 > 5^2$, that is, $32 > 25$.
(b) Assume that $k \geq 5$ and that P_k is true, that is,

$$2^k > k^2. \tag{50.11}$$

We must prove P_{k+1}, that is,

$$2^{k+1} > (k + 1)^2. \tag{50.12}$$

If we multiply both sides of (50.11) by 2, we have

$$2^{k+1} > 2k^2.$$

To get (50.12), it will suffice to prove that

$$2k^2 > (k + 1)^2, \tag{50.13}$$

for then we'll have $2^{k+1} > 2k^2 > (k + 1)^2$, from which $2^{k+1} > (k + 1)^2$. Inequality (50.13) is equivalent to

$$2k^2 > k^2 + 2k + 1$$
$$k^2 - 2k > 1$$
$$k(k - 2) > 1. \tag{50.14}$$

The latter inequality is satisfied because $k \geq 5$. This completes the proof.

Remark Notice that (50.14) is true for $k \geq 3$. Thus in part (b) we have actually shown that P_k implies P_{k+1} for $k \geq 3$. However, that does not prove P_n for $n \geq 3$, because P_3 is not true (P_3 would say $2^3 > 3^2$, which is false). What about P_1? P_2? P_4?

The other sections of this chapter will give more proofs by mathematical induction.

EXERCISES FOR SECTION 50

Write the first five terms and the $(n+1)$st term of the sequence whose nth term is a_n.

1. $a_n = 5n$ **2.** $a_n = n^3$ **3.** $a_n = n/(n+1)$

4. $a_n = -2$ **5.** $a_n = (-2)^n$ **6.** $a_n = (-1)^{n+1}[1 + (1/n)]$

Use mathematical induction to prove each statement for every positive integer n or for those n satisfying the stated restriction.

7. $1 + 2 + 2^2 + \cdots + 2^n = 2^{n+1} - 1$

8. $1^2 + 2^2 + 3^2 + \cdots + n^2 = \dfrac{n(n+1)(2n+1)}{6}$

9. $5 + 10 + 15 + \cdots + 5n = \dfrac{5n(n+1)}{2}$

10. $1 + 3 + 5 + \cdots + (2n+1) = n^2$

11. $1^3 + 2^3 + \cdots + n^3 = \left(\dfrac{n(n+1)}{2}\right)^2$

12. $1^3 + 3^3 + \cdots + (2n-1)^3 = n^2(2n^2 - 1)$

13. $(1 \cdot 2) + (2 \cdot 3) + (3 \cdot 4) + \cdots + n(n+1) = \dfrac{n(n+1)(n+2)}{3}$

14. $\dfrac{1}{1 \cdot 2} + \dfrac{1}{2 \cdot 3} + \dfrac{1}{3 \cdot 4} + \cdots + \dfrac{1}{n(n+1)} = \dfrac{n}{n+1}$

15. $\dfrac{1}{1 \cdot 2 \cdot 3} + \dfrac{1}{2 \cdot 3 \cdot 4} + \dfrac{1}{3 \cdot 4 \cdot 5} + \cdots + \dfrac{1}{n(n+1)(n+2)} = \dfrac{n(n+3)}{4(n+1)(n+2)}$

16. $\left(1 + \dfrac{1}{1}\right)\left(1 + \dfrac{1}{2}\right)\left(1 + \dfrac{1}{3}\right) \cdots \left(1 + \dfrac{1}{n}\right) = n + 1$

17. $\left(1 - \dfrac{1}{4}\right)\left(1 - \dfrac{1}{9}\right)\left(1 - \dfrac{1}{16}\right) \cdots \left(1 + \dfrac{1}{n^2}\right) = \dfrac{n+1}{2n}$

18. $\dfrac{1}{2} + \dfrac{2}{2^2} + \dfrac{3}{2^3} + \cdots + \dfrac{n}{2^n} = 2 - \dfrac{n+2}{2^n}$

19. $2n \leq 2^n$

20. $n^3 < 3^n$ for $n \geq 4$.

21. $(1 + \frac{1}{2})^n \geq 1 + (1/2^n)$

22. If $a > 1$, then $a^n > 1$.

23. If $0 < a < 1$, then $a^n < 1$.

24. $\log(a_1 a_2 \cdots a_n) = \log a_1 + \log a_2 + \cdots + \log a_n$

25. The sum of the interior angles of an n-sided convex polygon is $(n - 2) \cdot 180°$ $(n \geq 3)$. (A polygon is *convex* if all of the points on a line segment are inside the polygon whenever the endpoints of the segment are inside the polygon. You may assume that the sum of the interior angles of a triangle is 180°.)
26. The number of lines determined by n points, no three of which are collinear, is $\frac{1}{2}n(n - 1)$ $(n \geq 2)$.
27. The number of diagonals of an n-sided convex polygon is $n(n-3)/2$ $(n \geq 3)$. (See Exercise 25 for the definition of a convex polygon.)

SECTION 51 Arithmetic Sequences and Series

A. Sums of Arithmetic Series

DEFINITION. A finite or infinite sequence $a_1, a_2, a_3, \ldots$ is called an **arithmetic sequence** if there is a constant number d (the **common difference**) such that

$$a_{k+1} - a_k = d \qquad (51.1)$$

for every k. (Arithmetic sequences are sometimes called **arithmetic progressions.**)

By writing (51.1) in the form $a_{k+1} = a_k + d$, we see that each term after the first can be obtained by adding d to the preceding term. Once the first term, the common difference, and the number of terms have been specified, the sequence is completely determined.

EXAMPLE 51.1 A six-term arithmetic sequence has first term 4 and common difference 5. Write the sequence.

Solution 4, 9, 14, 19, 24, 29.

If a_n denotes the nth term of an arithmetic sequence with first term a_1 and common difference d, then

$$a_n = a_1 + (n - 1)d. \qquad (51.2)$$

Proof. We let P_n represent (51.2) and use mathematical induction.

(a) P_1 is true because

$$a_1 + (1 - 1)d = a_1.$$

(b) Assume P_k, that is,

$$a_k = a_1 + (k - 1)d.$$

Then

$$\begin{aligned} a_{k+1} &= a_k + d && \text{by (51.1)} \\ &= a_1 + (k - 1)d + d && \text{by } P_k \\ &= a_1 + kd. \end{aligned}$$

The equation $a_{k+1} = a_1 + kd$ is P_{k+1}, so the proof is complete. □

EXAMPLE 51.2 Write the first three terms and the 20th term of the arithmetic sequence with first term 5 and common difference -3.

Solution Using $a_{k+1} = a_k + d$, we have

$$\begin{aligned} a_1 &= 5 \\ a_2 &= 5 + (-3) = 2 \\ a_3 &= 2 + (-3) = -1. \end{aligned}$$

From (51.2), we have

$$a_{20} = a_1 + 19d = 5 + 19(-3) = -52.$$

The next formulas give the sum of an **arithmetic series,** that is, a series whose terms form an arithmetic sequence.

If S_n denotes the sum of the first n terms of an arithmetic sequence with first term a_1 and common difference d, then

$$S_n = \frac{n}{2}[2a_1 + (n - 1)d] \tag{51.3}$$

and

$$S_n = \frac{n}{2}(a_1 + a_n). \tag{51.4}$$

Proof. We'll prove (51.3) by mathematical induction. Then (51.4) will follow from (51.3) and (51.2) (Exercise 31). Let P_n denote (51.3).

(a) P_1 is true because

$$S_1 = a_1 \quad \text{and} \quad \tfrac{1}{2}[2a_1 + (1 - 1)d] = a_1.$$

(b) Assume P_k, that is,

$$S_k = \frac{k}{2}[2a_1 + (k - 1)d]. \tag{51.5}$$

Because the sum of the first $k + 1$ terms equals the sum of the first k terms added to the $(k + 1)$st term, we can use (51.5) and $a_{k+1} = a_1 + kd$ [from Equation (51.2)] to write

$$S_{k+1} = S_k + a_{k+1}$$

$$S_{k+1} = \frac{k}{2}[2a_1 + (k - 1)d] + a_1 + kd$$

$$S_{k+1} = \frac{k + 1}{2}(2a_1 + kd). \tag{51.6}$$

(Exercise 32 asks you to verify that the right-hand sides of the last two equations are equal.) Equation (51.6) is P_{k+1}, so the proof is complete. □

EXAMPLE 51.3 Find the sum of the first 20 terms of the arithmetic sequence with first term 5 and common difference -3.

First Solution Use (51.3) with $a_1 = 5$, $d = -3$, and $n = 20$:

$$S_{20} = \tfrac{20}{2}[2 \cdot 5 + 19(-3)] = -470.$$

Second Solution Use (51.4) with $a_1 = 5$, $n = 20$, and $a_{20} = -52$ (from Example 51.2):

$$S_{20} = \tfrac{20}{2}(5 - 52) = -470.$$

EXAMPLE 51.4 Find the sum of the positive integers that are multiples of 3 and less than 100.

Solution We must sum the series

$$3 + 6 + 9 + \cdots + 99.$$

Use (51.4) with $a_1 = 3$, $n = 33$, and $a_{33} = 99$:

$$S_{33} = \tfrac{33}{2}(3 + 99) = 1683.$$

B. Summation Notation

The notations

$$\sum_{k=1}^{n} a_k \quad \text{and} \quad \sum_{k=1}^{\infty} a_k \tag{51.7}$$

are used to represent the series

$$a_1 + a_2 + a_3 + \cdots + a_n$$

and

$$a_1 + a_2 + a_3 + \cdots + a_n + \cdots,$$

respectively. (The symbol Σ is a capital Greek *sigma.*) The variable k in (51.7) is called the **index of summation.** The equation $k = 1$ below Σ indicates that the first term is to be a_1. To begin with a_2 rather than a_1, write $k = 2$ below Σ; similarly for other possibilities. An n above Σ indicates that the last term is to be a_n. The symbol ∞ (read *infinity*) is used above Σ to indicate that there is to be no last term. Always begin with the smallest indicated value for the index of summation and repeatedly increase by 1 until reaching the largest indicated value (if there is one). If there is a formula for the kth term, we can write it in place of a_k.

EXAMPLE 51.5

(a) $\sum_{k=1}^{5} a_k = a_1 + a_2 + a_3 + a_4 + a_5$

(b) $\sum_{k=4}^{\infty} a_k = a_4 + a_5 + a_6 + \cdots + a_k + \cdots$

(c) $\sum_{k=0}^{10} (-1)^k(2k + 1) = 1 - 3 + 5 - 7 + \cdots - 19 + 21$

EXAMPLE 51.6 Changing the index of summation does not change the series.

(a) $\sum_{k=1}^{n} k$, $\sum_{j=1}^{n} j$, and $\sum_{t=1}^{n} t$ all represent the series

$$1 + 2 + 3 + \cdots + n.$$

(b) $\sum_{k=1}^{5} 3$, $\sum_{j=1}^{5} 3$, and $\sum_{t=1}^{5} 3$ all represent the series

$$3 + 3 + 3 + 3 + 3.$$

(There are five terms, and each term is 3.)

EXERCISES FOR SECTION 51

Write the arithmetic sequence having the given first term a_1, common difference d, and number of terms n.

1. $a_1 = 5,\ d = 3,\ n = 5$ **2.** $a_1 = -10,\ d = 15,\ n = 6$
3. $a_1 = 40,\ d = -12,\ n = 5$

Write the first three terms and the 10th term of the arithmetic sequence having the given first term a_1 and common difference d.

4. $a_1 = 12,\ d = 0.5$ **5.** $a_1 = \frac{3}{2},\ d = \frac{1}{4}$ **6.** $a_1 = 7,\ d = -1.5$

Find the first three terms and the sum of the arithmetic series having the given first term a_1, common difference d, and number of terms n.

7. $a_1 = 1, d = 7, n = 11$
8. $a_1 = 0, d = -2.1, n = 10$
9. $a_1 = \frac{1}{4}, d = \frac{1}{8}, n = 20$
10. $a_1 = \frac{7}{3}, d = -\frac{1}{3}, n = 15$
11. $a_1 = 120, d = 5, n = 60$
12. $a_1 = 100, d = -\frac{5}{2}, n = 200$

13. (a) Solve Equation (51.2) for a_1.
(b) Use part (a) to find the first term of a 25-term arithmetic sequence having last term 20 and common difference $-\frac{1}{3}$.

14. (a) Solve Equation (51.2) for d.
(b) Use part (a) to find the common difference for a 15-term arithmetic sequence having first term 4 and last term 25.

15. (a) Solve Equation (51.2) for n.
(b) Use part (a) to find the number of terms in an arithmetic sequence having first term 10, last term -60, and common difference $-\frac{1}{5}$.

Find the sum of each series. You may want to use Equation (51.4) and Exercise 15(a).

16. $6 + 10 + 14 + \cdots + 102$
17. $1 - 5 - 11 - 17 - \cdots - 71$
18. $1 + 1.05 + 1.10 + \cdots + 1.95$

Write each sum without using summation notation. (Write all of the terms; don't compute their sum.)

19. $\sum_{k=1}^{5} b_k$
20. $\sum_{k=1}^{4} 2^k$
21. $\sum_{k=1}^{6} (k + 3)$
22. $\sum_{j=3}^{8} (-1)^j 2j$
23. $\sum_{j=5}^{10} \frac{1}{j + 1}$
24. $\sum_{j=2}^{6} j^{-1}$

Represent each series using summation (Σ) notation.

25. $3 + 3^2 + 3^3 + \cdots + 3^{10}$
26. $11 - 12 + 13 - 14 + \cdots - 30 + 31$
27. $1 \cdot 5 + 2 \cdot 5^2 + 3 \cdot 5^3 + \cdots + 20 \cdot 5^{20}$

Each series in Exercises 28–30 is an arithmetic series. Find its sum.

28. $\sum_{k=1}^{n} (2k + 3)$
29. $\sum_{k=1}^{n} 5(3k - 1)$
30. $\sum_{k=1}^{n} (3k + k\sqrt{2})$

31. Use Equations (51.2) and (51.3) to prove (51.4).

32. Verify that Equation (51.6) follows from the equation that precedes it.

33. Use an equation in this section to help justify the following statement: The sum of an arithmetic series with n terms is n times the average of the first and last terms.

34. To raise money for a charity, each of the 50 members of a club agrees to draw a ticket at random from a box containing tickets numbered from 1 through 50, and then contribute an amount of dollars equal to the number on the ticket. (No two members can get the same number.) Compute the total amount the club will contribute.

35. A small auditorium has 20 rows of seats. The first row contains 15 seats and each successive row contains one seat more than the previous row. How many seats are there in all?

36. If you save \$0.50 today, \$1.00 tomorrow, \$1.50 the next day, and so on through 30 days, what will your total savings be?

37. A tank originally contains 1000 gallons of water. Ten gallons are withdrawn one day, 10.5 gallons the next, 11 gallons the next, and so on, so that each day one-half gallon more is withdrawn than the previous day. (a) How much water will remain in the tank after 30 days? (b) On which day will the tank be emptied?

38. You want to put \$50 in an account this month, and then add an additional amount each month, increasing the amount that you add each month by a constant amount in such a way that the account will have \$3250 at the end of 26 months. How much must the payments increase from month to month? (Ignore interest.)

39. You plan to make monthly withdrawals from an account that originally contains \$20,100, and you plan to increase the amount of the withdrawal by \$10 each month. What should the first withdrawal be so that the account will be exhausted at the end of exactly 5 years?

SECTION 52 Geometric Sequences and Series

A. Geometric Sequences

DEFINITION. A finite or infinite sequence $a_1, a_2, a_3, \ldots$ is called a **geometric sequence** if there is a constant number r (the **common ratio**) such that

$$a_{k+1} = a_k r \tag{52.1}$$

for $k \geq 1$. (Geometric sequences are sometimes called **geometric progressions.**)

By applying (52.1) repeatedly, we see that a geometric sequence has the form

$$a_1, a_1 r, a_1 r^2, \ldots, a_1 r^{n-1}, \ldots$$

Exercise 31 asks you to use mathematical induction to prove that

$$a_n = a_1 r^{n-1} \text{ for } n \geq 1. \tag{52.2}$$

EXAMPLE 52.1 Write the first 4 terms, the 15th term, and the nth term of the geometric sequence with first term 5 and common ratio $\frac{2}{3}$.

Solution

$$\begin{aligned}
a_1 &= 5 \\
a_2 &= a_1(\tfrac{2}{3}) = 5 \cdot \tfrac{2}{3} = \tfrac{10}{3} \\
a_3 &= a_2(\tfrac{2}{3}) = \tfrac{10}{3} \cdot \tfrac{2}{3} = \tfrac{20}{9} \\
a_4 &= a_3(\tfrac{2}{3}) = \tfrac{20}{9} \cdot \tfrac{2}{3} = \tfrac{40}{27} \\
a_{15} &= 5(\tfrac{2}{3})^{14} \\
a_n &= 5(\tfrac{2}{3})^{n-1}
\end{aligned}$$

EXAMPLE 52.2 The first term of a geometric sequence is 4 and the sixth term is 8. What is the common ratio?

Solution Use Equation (52.2) with $n = 6$, $a_1 = 4$, and $a_6 = 8$. This gives

$$8 = 4r^5, \quad r^5 = 2, \quad \text{and} \quad r = \sqrt[5]{2}.$$

B. Sums of Finite Geometric Series

A **geometric series** is a series whose terms form a geometric sequence. Thus a geometric series will have the form

$$a_1 + a_1r + a_1r^2 + \cdots + a_1r^{n-1} \qquad \text{(if finite)}$$

or

$$a_1 + a_1r + a_1r^2 + \cdots + a_1r^{n-1} + \cdots \qquad \text{(if infinite).}$$

If S_n denotes the sum of the first n terms of a geometric series with first term a_1 and common ratio $r \neq 1$, then

$$S_n = \frac{a_1 - a_1r^n}{1 - r} \tag{52.3}$$

and

$$S_n = \frac{a_1 - ra_n}{1 - r}. \tag{52.4}$$

Proof. We prove (52.3) by mathematical induction. Then (52.4) will follow from (52.3) and (52.2) (Exercise 32). Let P_n denote Equation (52.3).

(a) P_1 is true because $S_1 = a_1$ and

$$\frac{a_1 - a_1r}{1 - r} = \frac{a_1(1 - r)}{1 - r} = a_1.$$

(b) Assume P_k. Then

$$S_k = \frac{a_1 - a_1 r^k}{1 - r}.$$

Therefore,

$$\begin{aligned} S_{k+1} &= a_1 + a_1 r + a_1 r^2 + \cdots + a_1 r^{k-1} + a_1 r^k \\ &= S_k + a_1 r^k \\ &= \frac{a_1 - a_1 r^k}{1 - r} + a_1 r^k \\ &= \frac{a_1 - a_1 r^k + (1 - r) a_1 r^k}{1 - r} \\ &= \frac{a_1 - a_1 r^{k+1}}{1 - r}, \end{aligned}$$

which is P_{k+1}. □

EXAMPLE 52.3 Find the sum of the first 8 terms of the geometric sequence with first term 5 and common ratio 2.

Solution Use (52.3) with $a_1 = 5$, $r = 2$, and $n = 8$:

$$S_8 = \frac{5 - 5 \cdot 2^8}{1 - 2} = 5(2^8 - 1) = 5 \cdot 255 = 1275.$$

C. Application: Annuities

A common way of saving for some future event such as college or retirement is by the investment of equal amounts of money at equal intervals of time. Any such savings plan is an example of an **annuity,** which is simply a sequence of payments at periodic intervals. By using the formulas from the preceding subsection we now compute the amount accumulated under a **simple annuity,** which is one in which the payment intervals and compounding intervals coincide.

EXAMPLE 52.4 At the end of each year \$1000 is added to an account that pays 10% compounded annually. How much will be in the account at the end of the fifth year?

Solution

	Added excluding interest	Interest earned	Accumulated amount
End of year 1:	\$1000	\$ 0	\$1000
End of year 2:	\$1000	\$100	\$2100
End of year 3:	\$1000	\$210	\$3310
End of year 4:	\$1000	\$331	\$4641
End of year 5:	\$1000	\$464.10	\$6105.10

We see that the accumulated amount is \$6105.10. Notice that the total amount added to the account, excluding interest, is \$5000. Interest adds an additional \$1105.10.

Remark The following observations about Example 52.4 will help us discover a general formula.

The \$1000 added at the end of year 5 earns interest for 0 years.

The \$1000 added at the end of year 4 earns interest for 1 year.

The \$1000 added at the end of year 3 earns interest for 2 years.

The \$1000 added at the end of year 2 earns interest for 3 years.

The \$1000 added at the end of year 1 earns interest for 4 years.

We now consider a general problem of which Example 52.4 is a special case. Suppose that an amount R is added to an account at the end of each of n equal time intervals. Also suppose that the account earns interest at a rate of i percent per time interval (compounding interval).* If we think in terms of the remark at the end of Example 52.4, we find that:

The \$$R$ added at the end of period n earns interest for 0 periods.

The \$$R$ added at the end of period $n - 1$ earns interest for 1 period.

The \$$R$ added at the end of period $n - 2$ earns interest for 2 periods.

• • •

The \$$R$ added at the end of period 2 earns interest for $n - 2$ periods.

The \$$R$ added at the end of period 1 earns interest for $n - 1$ periods.

By results from Section 18, an amount R invested for k compounding periods at a rate of $i\%$ per period accumulates to

$$R(1 + i)^k.$$

Therefore, the accumulated amount for the annuity is

$$R + R(1 + i) + R(1 + i)^2 + \cdots + R(1 + i)^{n-1}. \tag{52.5}$$

To sum (52.5), we apply (52.3) with $a_1 = R$ and $r = 1 + i$. Let S denote the sum. Then

$$S = \frac{R - R(1 + i)^n}{1 - (1 + i)} = R \cdot \frac{(1 + i)^n - 1}{i}.$$

As in previous formulas of this type, the interest rate i must be in decimal form: for example, use 0.06 rather than 6%.

*The letter i here denotes an interest rate, not an imaginary number as in Chapter VIII.

If an amount R is added to an account at the end of each of n equal compounding intervals, and the account earns interest at a rate of $i\%$ per compounding interval, then the total amount accumulated from the n payments will be

$$S = R \cdot \frac{(1 + i)^n - 1}{i} \tag{52.6}$$

EXAMPLE 52.5 Suppose that at the end of each month \$50 is added to an account that pays interest at an annual rate of 6% compounded monthly. How much will be in the account at the end of the tenth year?

Solution The annual rate of 6% converts to a monthly rate of $6/12\% = 1/2\%$, so we use $i = 0.005$ in (52.6). Also, $R = 50$ and $n = 120$ (10 years times 12 months per year).

$$S = 50 \cdot \frac{(1 + 0.005)^{120} - 1}{0.005}$$

$$S = \$8193.97.$$

Ⓒ

The fraction on the right in (52.6) arises so often that it has been given a special symbol, $s_{\overline{n}|i}$. Thus

$$s_{\overline{n}|i} = \frac{(1 + i)^n - 1}{i}.$$

This quantity has been tabulated extensively for different combinations of i and n. Table 52.1 gives the values needed for the exercises and for the following example.

EXAMPLE 52.6 Suppose that at the end of each year \$500 is added to an account that pays 5% compounded annually. How much will be in the account at the end of 10 years?

Solution To the nearest dollar,

$$S = 500\, s_{\overline{10}|0.05} = 500(12.5779) = \$6{,}289.$$

Here we have considered only simple annuities used for saving. Other uses of annuities include repayment of loans by equal installments and equal monthly payments from retirement accounts. These other kinds of annuities are discussed in books on the mathematics of finance.

D. Sums of Infinite Geometric Series

We'll now see that it can make sense to talk about the sum of an *infinite* series of numbers.

EXAMPLE 52.7 Consider the infinite geometric series

$$\frac{1}{2} + \frac{1}{4} + \frac{1}{8} + \cdots + \frac{1}{2^n} + \cdots. \tag{52.7}$$

Since $a_1 = \frac{1}{2}$ and $r = \frac{1}{2}$, we can apply (52.3) to get

TABLE 52.1

n	$s_{\overline{n}\rceil 0.005}$	$s_{\overline{n}\rceil 0.01}$	n	$s_{\overline{n}\rceil 0.05}$	$s_{\overline{n}\rceil 0.10}$
10	10.2280	10.4622	1	1.0000	1.0000
20	20.9781	22.0190	2	2.0500	2.1000
30	32.2800	34.7849	3	3.1525	3.3100
40	44.1588	48.8864	4	4.3101	4.6410
50	56.6452	64.4632	5	5.5256	6.1051
60	69.7700	81.6697	6	6.8019	7.7156
70	83.5661	100.6763	7	8.1420	9.4872
80	98.0677	121.6715	8	9.5491	11.4359
90	113.3109	144.8633	9	11.0266	13.5795
100	129.3337	170.4814	10	12.5779	15.9374
110	146.1759	198.7797	11	14.2068	18.5312
120	163.8793	230.0387	12	15.9171	21.3843

$$S_n = \frac{\frac{1}{2} - \frac{1}{2}(\frac{1}{2})^n}{1 - \frac{1}{2}} = 1 - (\tfrac{1}{2})^n = 1 - \frac{1}{2^n}.$$

Thus

$$S_1 = \frac{1}{2}, S_2 = \frac{3}{4}, S_3 = \frac{7}{8}, \ldots, S_n = 1 - \frac{1}{2^n}, \ldots. \tag{52.8}$$

The sequence in (52.8) is called the *sequence of partial sums* of the series in (52.7). It is clear that the successive terms in (52.8) approach closer and closer to 1 as n increases. We abbreviate this by writing

$$\lim_{n\to\infty} S_n = \lim_{n\to\infty}\left(1 - \frac{1}{2^n}\right) = 1.$$

We say in this case that the series *converges* to 1, and also that 1 is the *sum* of the series.

Now consider an arbitrary geometric series

$$a_1 + a_1r + a_1r^2 + \cdots + a_1r^{n-1} + \cdots. \tag{52.9}$$

By factoring a_1 from the numerator $a_1 - a_1r^n$ of (52.3), we can write the sum of the first n terms of (52.9) as

$$S_n = \frac{a_1}{1 - r}(1 - r^n). \tag{52.10}$$

If $|r| < 1$, then r^n approaches 0 as n increases. (Try some specific examples, like $r = 0.1$, or $r = -\frac{1}{2}$.) Therefore, we can write

$$\lim_{n\to\infty} S_n = \frac{a_1}{1-r} \text{ if } |r| < 1.$$

As in Example 52.7, we say that the series **converges** to $a_1/(1-r)$, and that $a_1/(1-r)$ is the **sum** of the series (52.9). Denoting the sum by S_∞, we can summarize as follows.

> If a geometric series has first term a_1 and common ratio r, and $|r| < 1$, then the series converges with sum
>
> $$S_\infty = \frac{a_1}{1-r}. \qquad (52.11)$$

If $a_1 \neq 0$ and $|r| \geq 1$, then the series (52.9) does not converge; Exercises 34 and 35 give examples. Infinite series are treated more fully in calculus.

EXAMPLE 52.8 Find the sum of the infinite geometric series

$$1 - \tfrac{2}{3} + \tfrac{4}{9} - \tfrac{8}{27} + \cdots + (-\tfrac{2}{3})^{n-1} + \cdots.$$

Solution Use (52.11) with $a_1 = 1$ and $r = -\frac{2}{3}$.

$$S_\infty = \frac{1}{1-(-\frac{2}{3})} = \frac{1}{\frac{5}{3}} = \tfrac{3}{5}.$$

We know from Section 1 that every repeating decimal represents a rational number; that is, every repeating decimal number can be expressed in the form a/b, where a and b are integers and $b \neq 0$. The next example illustrates how to find a fraction a/b representing a given repeating decimal number.

EXAMPLE 52.9 Express $5.\overline{12}$ in the form a/b, where a and b are integers and $b \neq 0$.

Solution

$$\begin{aligned} 5.\overline{12} &= 5.121212\cdots \\ &= 5 + 0.12 + 0.0012 + 0.000012 + \cdots \\ &= 5 + \frac{12}{100} + \frac{12}{10000} + \frac{12}{1000000} + \cdots. \end{aligned}$$

The terms following 5 form an infinite geometric series with $a_1 = \frac{12}{100}$ and $r = \frac{1}{100}$; thus they can be summed by Equation (52.11). Here are the computations

$$\begin{aligned} 5.\overline{12} &= 5 + \frac{\frac{12}{100}}{1-\frac{1}{100}} = 5 + \frac{\frac{12}{100}}{\frac{99}{100}} = 5 + \tfrac{12}{99} \\ &= 5 + \tfrac{4}{33} = \tfrac{169}{33}. \end{aligned}$$

EXERCISES FOR SECTION 52

Write the geometric sequence with the given first term a_1, common ratio r, and number of terms n.

1. $a_1 = \frac{1}{2}$, $r = 4$, $n = 5$
2. $a_1 = 3$, $r = -\frac{1}{2}$, $n = 6$
3. $a_1 = 10^{-1}$, $r = 10^{-3}$, $n = 5$

Write the first three terms and the 10th term of the geometric sequence with the given first term a_1 and common ratio r.

4. $a_1 = -1$, $r = -\frac{1}{2}$
5. $a_1 = 100$, $r = 0.1$
6. $a_1 = 8$, $r = \sqrt{2}$

Find the kth term and the sum of the geometric series with the given first term a_1, common ratio r, and number of terms n.

7. $a_1 = 5$, $r = \frac{2}{5}$, $n = 5$
8. $a_1 = -1$, $r = -1$, $n = 11$
9. $a_1 = 3$, $r = -\sqrt{3}$, $n = 5$
10. $a_1 = 0.3$, $r = 3$, $n = 6$
11. $a_1 = 1 - \sqrt{2}$, $r = 1 + \sqrt{2}$, $n = 3$
12. $a_1 = u$, $r = v^2$, $n = m$

Find the sum of each series.

13. $\frac{1}{4} + \frac{3}{8} + \frac{9}{16} + \cdots + \frac{243}{128}$
14. $2 + 2\sqrt{2} + 4 + \cdots + 32$
15. $\sum_{k=1}^{10} 0.6^k$
16. $\frac{1}{5} + \frac{1}{25} + \frac{1}{125} + \frac{1}{625} + \cdots$
17. $\sum_{k=1}^{\infty} 20(0.1)^{k-1}$
18. $5 - \frac{10}{3} + \frac{20}{3} - \frac{40}{27} + \cdots$

19. (a) Solve Equation (52.2) for a_1.
(b) Use part (a) to find the first term of the five-term geometric sequence with last term 6 and common ratio $\frac{1}{3}$.

20. (a) Solve Equation (52.2) for r.
(b) Use part (a) to find the common ratio for the six-term geometric sequence with first term 2 and last term $-\frac{243}{16}$.

21. (a) Solve Equation (52.3) for a_1.
(b) Use part (a) to find the first term of the five-term geometric sequence with common ratio $-\sqrt{2}$ and sum $1 + 4\sqrt{2}$.

Use Table 52.1 to find the amount accumulated under each annuity.

22. \$100 per month for 30 months at 12% (annual rate) compounded monthly.
23. \$2000 per year for 6 years at 10% (annual rate) compounded annually.
24. \$25 per month for 10 years at 6% (annual rate) compounded monthly.
25. \$500 per year for 12 years at 5% (annual rate) compounded annually.
26. \$50 per month for $7\frac{1}{2}$ years at 6% (annual rate) compounded monthly.
27. \$1200 per year for 10 years for 10% (annual rate) compounded annually.
28. A parent wants \$10,000 available ten years from now for a daughter's college expenses. To carry this out, how much should the parent invest at the end of each month in an account that pays 12% (annual rate) compounded monthly?
29. Suppose you want to save \$3000 for use $2\frac{1}{2}$ years from now. To accomplish this,

how much should you invest at the end of each month in an account that pays 6% (annual rate) compounded monthly?

30. I plan to invest \$1000 at the end of each year for the next 10 years in an account that pays 10% (annual rate) compounded annually. You plan to invest \$75 at the end of each month for the next 10 years in an account that pays 12% (annual rate) compounded monthly. Which of us will accumulate the most, and by how much?

31. Use mathematical induction to prove (52.2).

32. Use (52.2) and (52.3) to prove (52.4).

33. The text proves (52.3) by mathematical induction. The following statements provide the steps in an alternate proof. Justify each step.
(a) $rS_n = a_1r + a_1r^2 + \cdots + a_1r^{n-1} + a_1r^n$.
(b) $S_n - rS_n = a_1 - a_1r^n$.
(c) $S_n = (a_1 - a_1r^n)/(1 - r)$.

34. Determine S_n for a geometric series with first term a_1 and common ratio 1. What happens to S_n as $n \to \infty$?

35. Determine S_n for a geometric series with first term a_1 and common ratio -1. What happens to S_n as $n \to \infty$?

36. Show that (52.2) implies that

$$n = \frac{\log(a_n r/a_1)}{\log r}.$$

(This exercise assume acquaintance with Section 37.)

Express each number in the form a/b where a and b are integers and $b \neq 0$.

37. $3.\overline{4}$ **38.** $0.\overline{18}$ **39.** $21.\overline{12}$ **40.** $1.0\overline{2}$ **41.** $0.\overline{102}$ **42.** $5.1\overline{23}$

43. Prove that if a, b, c is a geometric sequence, then $b = \sqrt{ac}$.

44. Prove that if $a_1, a_2, a_3, \ldots$ is an arithmetic sequence, then $10^{a_1}, 10^{a_2}, 10^{a_3}, \ldots$ is a geometric sequence.

45. Prove that if $a_1, a_2, a_3, \ldots$ is a geometric sequence, then $\log a_1, \log a_2, \log a_3, \ldots$ is an arithmetic sequence.

46. For the geometric sequence 2, 6, 18, 54, . . . , the sequence of the differences of successive terms, 4, 12, 36, . . . , also forms a geometric sequence. Prove that the sequence of the differences of successive terms of every geometric sequence forms a geometric sequence.

47. The frequencies of the semitones of a well-tempered scale in music form a geometric sequence with common ratio $\sqrt[12]{2}$ (Example 39.13). If the frequency of concert A is 440 cycles per second, what is the frequency of middle C, which is 9 semitones lower than concert A? (The higher the semitone, the higher the frequency.)

48. A sequence of squares is constructed by beginning with a square one unit on each side and then, at each step thereafter, connecting the midpoints of the adjacent sides of the square immediately preceding. Prove that the sequence of perimeters forms a geometric sequence. Also prove that the sequence of areas forms a geometric sequence.

SECTION

53 The Binomial Theorem

A. Introduction. Factorial Notation

In this section we show how to extend the following list to include all higher integral powers of $a + b$.

$$(a+b)^0 = 1$$
$$(a+b)^1 = a+b$$
$$(a+b)^2 = a^2 + 2ab + b^2$$
$$(a+b)^3 = a^3 + 3a^2b + 3ab^2 + b^3$$

The theorem that gives $(a+b)^n$ in general is called the *Binomial Theorem*. Before stating and proving this theorem, however, we need some preliminary facts about *factorial notation* and *binomial coefficients.*

The notation $n!$ (read ***n* factorial**) is defined for positive integers as follows:

$$1! = 1$$
$$2! = 2 \cdot 1 = 2$$
$$3! = 3 \cdot 2 \cdot 1 = 6$$
$$4! = 4 \cdot 3 \cdot 2 \cdot 1 = 24$$

and, in general,

$$n! = n(n-1)(n-2) \cdot \cdots \cdot 2 \cdot 1.$$

You can easily verify that $5! = 120$. Fortunately, we won't need larger specific values like

$$20! = 2{,}432{,}902{,}008{,}176{,}640{,}000.$$

Notice that if n is any integer greater than 1, then $n! = n \cdot (n-1)!$. With $n = 4$, for example, we have $4! = 4 \cdot 3! = 4 \cdot 3 \cdot 2 \cdot 1$. If we use $n = 1$ in $n! = n \cdot (n-1)!$, we get $1! = 1 \cdot 0!$. Partly for this reason, and partly to make many formulas simpler, we *define* $0!$ by $0! = 1$.

B. Binomial Coefficients

The coefficients needed to expand $(a+b)^n$ are called **binomial coefficients** (remember from Section 4 that $a + b$ is called a *binomial*). For each pair of integers n and k such that $n \geq k \geq 0$, there is a binomial coefficient denoted by $\binom{n}{k}$ and defined by

$$\binom{n}{k} = \frac{n!}{k!(n-k)!}. \tag{53.1}$$

EXAMPLE 53.1

(a) $\binom{5}{2} = \frac{5!}{2!3!} = \frac{5 \cdot 4 \cdot \cancel{3 \cdot 2 \cdot 1}}{2 \cdot 1 \cdot \cancel{3 \cdot 2 \cdot 1}} = 10$

(b) $\binom{10}{2} = \frac{10!}{2!8!} = \frac{10 \cdot 9 \cdot \cancel{8!}}{2 \cdot 1 \cdot \cancel{8!}} = 45$

(c) $\binom{12}{9} = \frac{12!}{9!3!} = \frac{12 \cdot 11 \cdot 10 \cdot \cancel{9!}}{\cancel{9!} \cdot 3 \cdot 2 \cdot 1} = 220$

(d) $\binom{n}{0} = \frac{n!}{0!n!} = \frac{n!}{1 \cdot n!} = 1 \quad (n \geq 0)$

(e) $\binom{n}{n} = \frac{n!}{n!0!} = 1 \quad (n \geq 0)$

(f) $\binom{n}{1} = \frac{n!}{1!(n-1)!} = \frac{n \cdot (n-1)!}{1 \cdot (n-1)!} = n \quad (n \geq 1)$

Following are two fundamental facts about binomial coefficients. They are valid whenever $n \geq k \geq 0$.

$$\binom{n}{k} = \binom{n}{n-k} \tag{53.2}$$

Proof

$$\binom{n}{k} = \frac{n!}{k!(n-k)!}$$

$$= \frac{n!}{(n-k)!k!}$$

$$= \frac{n!}{(n-k)!(n-(n-k))!} = \binom{n}{n-k}$$ □

EXAMPLE 53.2

(a) $\binom{20}{3} = \binom{20}{17}$ **(b)** $\binom{n}{1} = \binom{n}{n-1}$ **(c)** $\binom{n}{0} = \binom{n}{n}$

$$\binom{n+1}{k} = \binom{n}{k-1} + \binom{n}{k} \tag{53.3}$$

Proof. We add the two terms on the right using a common denominator of $k!(n-k+1)!$. The result will turn out to be the term on the left.

$$\binom{n}{k-1} + \binom{n}{k} = \frac{n!}{(k-1)!(n-(k-1))!} + \frac{n!}{k!(n-k)!}$$

$$= \frac{k \cdot n!}{k \cdot (k-1)!(n-k+1)!} + \frac{n!(n-k+1)}{k!(n-k)!(n-k+1)}$$

$$= \frac{k \cdot n!}{k!(n-k+1)!} + \frac{(n-k+1) \cdot n!}{k!(n-k+1)!}$$

$$= \frac{(k+n-k+1) \cdot n!}{k!(n-k+1)!}$$

$$= \frac{(n+1)!}{k!(n-k+1)!}$$

$$= \binom{n+1}{k} \qquad \square$$

Equation (53.3) provides a convenient scheme for computing binomial coefficients for reasonably small n. This scheme is shown by the triangle in Figure 53.1, named for the French mathematician Blaise Pascal (1623–1662) but known to the Chinese at least 300 years before his time. In Pascal's triangle $\binom{n}{k}$ stands at the intersection of the nth row and the kth diagonal. Equation (53.3) tells us that each entry can be obtained by adding the numbers immediately to the left and right in the row above it. Equation (53.2) is reflected in the left-right symmetry of the table. For

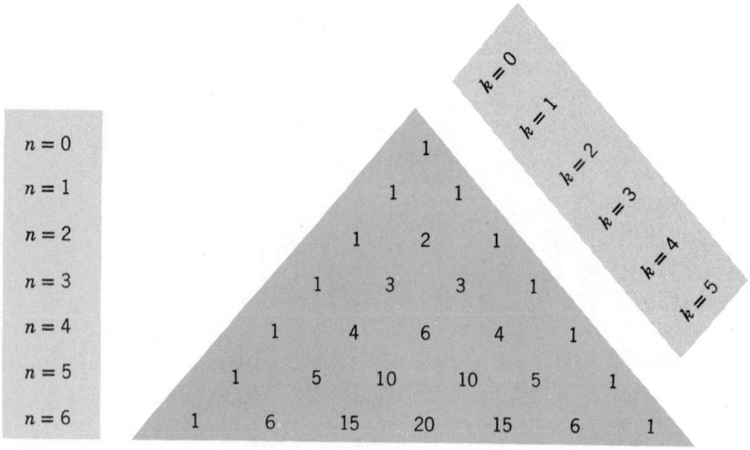

Figure 53.1 Pascal's triangle

example, the pair of 15's in the sixth row represent $\binom{6}{2}$ and $\binom{6}{4}$, which must be equal by (53.2).

Other notation used for the binomial coefficient $\binom{n}{k}$ includes $C(n, k)$, ${}_nC_k$, and C_k^n.

C. The Binomial Theorem

The Binomial Theorem

If n is any nonnegative integer then

$$(a + b)^n = \binom{n}{0}a^n + \binom{n}{1}a^{n-1}b + \binom{n}{2}a^{n-2}b^2 + \cdots + \binom{n}{k}a^{n-k}b^k + \cdots + \binom{n}{n}b^n. \tag{53.4}$$

The proof of the theorem will follow some observations and examples.

Observations

- The top entry in each binomial coefficient is n.
- The bottom entries in the binomial coefficients increase successively from 0 to n as we move to the right.
- In each term, the exponent on b is the same as the bottom entry of the term's binomial coefficient.
- The exponents on a decrease successively from n to 0 as we move to the right, and the exponents on b increase successively from 0 to n.
- In each term, the sum of the exponents on a and b is n.
- There are $n + 1$ terms in the expansion of $(a + b)^n$.

EXAMPLE 53.3

$$(a + b)^4 = \binom{4}{0}a^4 + \binom{4}{1}a^3b + \binom{4}{2}a^2b^2 + \binom{4}{3}ab^3 + \binom{4}{4}b^4$$

The binomial coefficients needed here can be read from the fourth row of Pascal's triangle: 1, 4, 6, 4, 1. Thus

$$(a + b)^4 = a^4 + 4a^3b + 6a^2b^2 + 4ab^3 + b^4.$$

EXAMPLE 53.4 Expand $(x^2 - 2y)^5$.

Solution Use (53.4) with $a = x^2$, $b = -2y$, and $n = 5$. The binomial coefficients can be read from the fifth row of Pascal's triangle. [You may want to write out $(a + b)^5$ first and then substitute x^2 for a and $-2y$ for b.]

$$(x^2 - 2y)^5 = (x^2)^5 + 5(x^2)^4(-2y) + 10(x^2)^3(-2y)^2 + 10(x^2)^2(-2y)^3 + 5(x^2)(-2y)^4 + (-2y)^5$$
$$= x^{10} - 10x^8y + 40x^6y^2 - 80x^4y^3 + 80x^2y^4 - 32y^5$$

EXAMPLE 53.5 Find the coefficient of a^4b^{16} in the expansion of $(a + b)^{20}$.

Solution Use (53.4) with $n = 20$. The coefficient of a^4b^{16} is

$$\binom{20}{16} = \frac{20!}{16!4!} = \frac{20 \cdot 19 \cdot 18 \cdot 17 \cdot \cancel{16!}}{\cancel{16!} \cdot 4 \cdot 3 \cdot 2 \cdot 1} = 4845.$$

EXAMPLE 53.6 Find the coefficient of u^6v^{12} in the expansion of $(2u^2 - v^3)^7$.

Solution Use (53.4) with $n = 7$ and with k chosen so that $(2u^2)^{7-k}(-v^3)^k$ involves v^{12}. This requires $k = 4$, and thus the term involving u^6v^{12} is

$$\binom{7}{4}(2u^2)^3(-v^3)^4 = \frac{7!}{4!3!}8u^6(v^{12})$$
$$= 35 \cdot 8u^6v^{12}$$
$$= 280u^6v^{12}.$$

Therefore, the answer is 280.

Proof of the Binomial Theorem. *Let P_n represent Equation (53.4) and use induction on n.*

(a) P_1 is true because

$$(a + b)^1 = a + b = \binom{1}{0}a^1 + \binom{1}{1}b^1.$$

(b) We now show that P_m implies P_{m+1} for each positive integer m. (We are using m in place of k in the statement of the Principle of Mathematical Induction in Section 50.) Assume P_m, that is,

$$(a + b)^m = \binom{m}{0}a^m + \binom{m}{1}a^{m-1}b + \binom{m}{2}a^{m-2}b^2 + \cdots + \binom{m}{k}a^{m-k}b^k + \cdots + \binom{m}{m}b^m. \tag{53.5}$$

To prove P_{m+1} from (53.5), we multiply both sides of (53.5) by $a + b$. The result on the left will be $(a + b)^{m+1}$, as required. The result on the right will be

$$\left[\binom{m}{0}a^m + \binom{m}{1}a^{m-1}b + \cdots + \binom{m}{k}a^{m-k}b^k + \cdots + \binom{m}{m}b^m\right](a + b)$$
$$= \binom{m}{0}a^{m+1} + \binom{m}{1}a^mb + \cdots + \binom{m}{k}a^{m-k+1}b^k + \cdots$$

$$+ \binom{m}{m} ab^m + \binom{m}{0} a^m b + \binom{m}{1} a^{m-1} b^2 + \cdots$$

$$+ \binom{m}{k} a^{m-k} b^{k+1} + \cdots + \binom{m}{m} b^{m+1}.$$

The term involving b^k in the last expansion is seen to be

$$\binom{m}{k} a^{m-k+1} b^k + \binom{m}{k-1} a^{m-k+1} b^k = \binom{m+1}{k} a^{m-k+1} b^k,$$

where we have used Equation (53.3). But $\binom{m+1}{k} a^{m-k+1} b^k$ is also the term involving b^k in P_{m+1}. Since the terms involving b^k are equal for all k $(0 \le k \le m + 1)$, the proof is complete. □

EXERCISES FOR SECTION 53

Compute

1. $\binom{10}{3}$ **2.** $\binom{12}{1}$ **3.** $\binom{9}{4}$ **4.** $\binom{50}{50}$ **5.** $\binom{20}{3}$ **6.** $\binom{14}{0}$

Simplify as much as possible.

7. $\dfrac{(n + 2)!}{n!}$ **8.** $\dfrac{(n + 1)!}{(n - 1)!}$ **9.** $\dfrac{(n + 1)!(n - 1)!}{(n!)^2}$

10. $\binom{n}{0} + \binom{n}{n}$ **11.** $\binom{n}{1} + \binom{n}{n-1}$ **12.** $\binom{n}{2} + \binom{n}{n-2}$

13. Write rows 7 and 8 of Pascal's triangle.

14. Prove that

$$\binom{n}{k} = \frac{n}{k}\binom{n-1}{k-1}.$$

15. Prove that

$$\binom{n+2}{k} = \binom{n}{k-2} + 2\binom{n}{k-1} + \binom{n}{k}.$$

Expand and simplify.

16. $(x + y)^5$ **17.** $(u + v)^6$ **18.** $(a + 2)^4$

19. $(a - b)^6$ **20.** $\left(x - \dfrac{2}{y}\right)^5$ **21.** $(2u - v^2)^4$

22. Find and simplify the coefficient of u^7v^2 in the expansion of $(u + v)^9$.

23. Find and simplify the coefficient of $x^{23}y^2$ in the expansion of $(x + y)^{25}$.

24. Find and simplify the coefficient of a^3b^9 in the expansion of $(a + b)^{12}$.

25. Find and simplify the first four terms in the expansion of $(a^2 + 2b^3)^{10}$.

26. Find and simplify the first three terms in the expansion of $(2u - 5v)^{20}$.

27. Find and simplify the first four terms in the expansion of $(x + y)^{100}$.
28. Find and simplify the coefficient of a^5b^{10} in the expansion of $(a + b^2)^{10}$.
29. Find and simplify the coefficient of u^6v^{15} in the expansion of $(2u^2 - v^3)^8$.
30. There is a constant term in the expansion of $[x - (1/x^2)]^{30}$. What is it?
31. The sum of the entries in the nth row of Pascal's triangle is 2^n. You can verify this for $1 \le n \le 6$ simply by addition and Figure 53.1. Prove it for every row. [Suggestion: Look at what (53.4) becomes when $a = b = 1$.]
32. The product of the first n positive even integers is $n!$ times a function of n. What is the function of n?
33. Show that if n is a positive integer, then

$$1 \cdot 3 \cdot 5 \cdot \cdots \cdot (2n - 1) = \frac{(2n)!}{2^n(n!)}.$$

REVIEW EXERCISES FOR CHAPTER XII

1. Find and simplify the first four terms and the $(n + 1)$st term of the series

$$\sum_{k=2}^{\infty} 3^{k-1}\binom{k}{2}.$$

2. Represent the infinite series

$$1 - 20 + 300 - 4000 + \cdots$$

using summation (Σ) notation.
3. Use mathematical induction to prove that

$$3 + 7 + 11 + \cdots + (4n - 1) = n(2n + 1)$$

for every positive integer n.
4. Use mathematical induction to prove that

$$1^2 + 3^2 + 5^2 + \cdots + (2n - 1)^2 = \frac{n(2n - 1)(2n + 1)}{3}$$

for every positive integer n.
5. Find the sum of the first ten terms of the sequence $1, 1 - \pi, 1 - 2\pi, 1 - 3\pi, \ldots$.
6. Verify the equation in Review Exercise 3 by using a formula for the sum of an arithmetic sequence.
7. Write the first three terms and the tenth term of the geometric sequence having first term $2\sqrt{3}$ and common ratio $-\sqrt{3}$.
8. A geometric sequence has first term x and fifth term $0.0001xy^8$. What is the common ratio?
9. Find the sum of the first ten terms of the geometric series with first term 5 and common ratio -2.

10. Find the sum of the series

$$a + a^2b^2 + a^3b^4 + \cdots + a^nb^{2n-2}.$$

11. An infinite series has first term 1 and common ratio $5t$. Determine the values of t for which the series converges. Also find the sum of the series when it converges.

12. Find the sum of the series

$$4 + \tfrac{4}{3} + \tfrac{4}{9} + \tfrac{4}{27} + \cdots.$$

13. At the end of each month you place \$25 in an account that pays interest at an annual rate of 6% compounded monthly. How much will be in the account at the end of the 40th month?

14. I plan to deposit a constant amount in an account at the end of each month, and I want \$5000 in the account at the end of 5 years. The account will pay 6% (annual rate) compounded monthly. What is the minimum amount, rounded up to the nearest \$5, that I should deposit each month?

15. Simplify as much as possible:

$$\binom{n+1}{k} \Big/ \binom{n}{k-1}.$$

16. Prove that

$$\binom{2n}{n} + \binom{2n}{n-1} = \frac{1}{2}\binom{2n+2}{n+1}.$$

Expand and simplify.

17. $(a^2 - b)^5$

18. $(10x + 0.1y)^6$

CHAPTER XIII
COMBINATORICS AND PROBABILITY

This chapter introduces the most fundamental ideas of combinatorics and probability: combinatorics is concerned with systematic enumeration and counting; probability is concerned with the mathematics of random behavior or chance. Both subjects form rich branches of mathematics with applications throughout business, technology, and the natural and social sciences.

SECTION

54 Permutations

A. Basic Counting Principle

The list of all possibilities in many enumeration problems can be determined most easily with the aid of a *tree*. Here are some examples.

EXAMPLE 54.1 In how many ways can the three letters A, B, and C be arranged in a row if we require that no letter be repeated?

Solution Although it isn't necessary in an example as simple as this one, all of the different ways can be shown as in Figure 54.1, which is an example of a *tree*. The letters in the first column (reading from the left) indicate the possibilities for the first letter. The letters in the second column indicate the possibilities for the second letter corresponding to each choice for the first letter. The letters in the third column in each case represents the (only) possibility for the third letter. By following the different *paths* through the tree we get the six different ways to arrange the letters, which are indicated to the right.

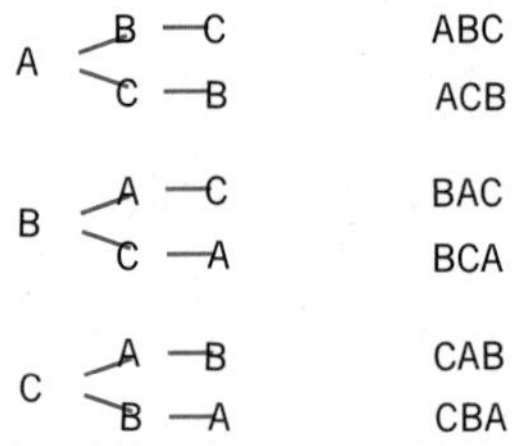

Figure 54.1

EXAMPLE 54.2 For a toss of a coin, let the outcomes heads and tails be denoted by H and T, respectively. How many different possible outcomes are there if a coin is tossed three times?

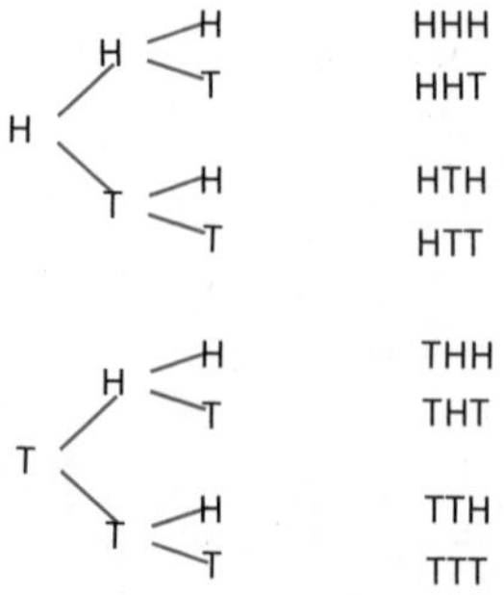

Figure 54.2

Solution The tree is shown in Figure 54.2, with the eight possible outcomes listed to the right.

EXAMPLE 54.3 Assume that students in three departments—English, History, and Philosophy—are to be classified according to sex (F, M), year (1, 2, 3, 4), and department (E, H, P). How many categories are possible?

Solution See Figure 54.3. For each sex there are four possible years, and then there are three possible departments for each combination of sex and year. Thus altogether there are $2 \cdot 4 \cdot 3 = 24$ different categories.

Thinking about the appropriate tree can help even when the size of the problem makes it impractical to write out the tree in full.

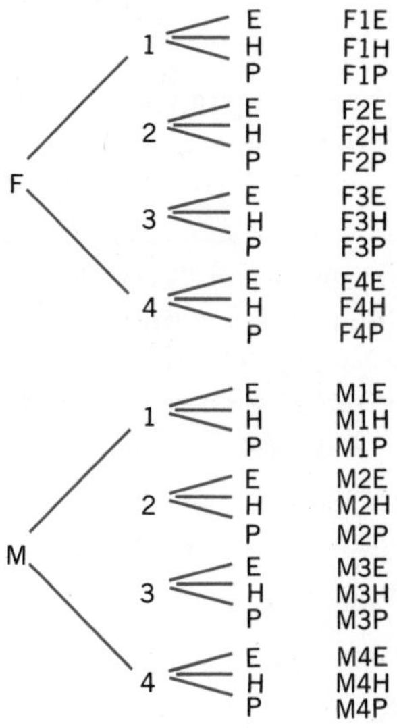

Figure 54.3

EXAMPLE 54.4 In how many ways can ten persons ($P_1, P_2, \ldots, P_{10}$) be seated in a row?

Solution Applying the ideas from the previous examples, we can see that the relevant tree would have each of P_1 through P_{10} in the first column, and for each of these there would be nine possibilities in the second column, and so on, until altogether there would be $10 \cdot 9 \cdot 8 \cdot 7 \cdot 6 \cdot 5 \cdot 4 \cdot 3 \cdot 2 \cdot 1 = 10!$ different paths through the tree. Thus $10!$ ($= 3{,}628{,}800$) is the number of possible seating arrangements.

These examples lead to the following counting principle.

Basic Counting Principle

If one thing can be done in any of n_1 ways, and after it has been done in any of those ways a second thing can be done in any of n_2 ways, and after any of those a third thing can be done in any of n_3 ways, and so on until we get to a kth thing that can be done in any of n_k ways, then the k things can be done together in $n_1 \cdot n_2 \cdot \cdots \cdot n_k$ different ways.

A tree representing the Basic Counting Principle would have k columns, with n_1

places in the first column, n_2 places corresponding to each of those in the second column, and so on.

B. Permutations

EXAMPLE 54.5 In how many ways can three officers—President, Vice President, and Secretary—be selected from a club with ten members?

Solution

First select a President:	10 possibilities
Then select a Vice President:	9 possibilities
Finally, select a Secretary:	8 possibilities

Therefore, by the Basic Counting Principle, there are $10 \cdot 9 \cdot 8$ $(= 720)$ different ways of selecting the officers.

If we think of the officers as being arranged in order (President, Vice President, Secretary), we see that this example is a special case of the idea in the following definition.

DEFINITION. An arrangement of k different elements, selected from a set of n elements, is called a **permutation** of the n elements taken k at a time. We denote the number of such permutations by $P(n, k)$.

Example 54.5 shows that $P(10, 3) = 10 \cdot 9 \cdot 8 = 720$. To obtain a general formula for $P(n, k)$, we first record the number of possibilities for each of the k choices to be made.

1st element: n possibilities

2nd element: $n - 1$ possibilities

3rd element: $n - 2$ possibilities

. . .

kth element: $n - (k - 1) = n - k + 1$ possibilities

Applying the Basic Counting Principle, we see that

$$P(n, k) = n(n-1)(n-2)\cdots(n-k+1). \tag{54.1}$$

The product on the right begins with n and continues till there are k terms. We can write this using factorial notation by multiplying and dividing by $(n - k)!$, as follows.

$$P(n, k) = n(n-1)(n-2)\cdots(n-k+1)$$

$$P(n, k) = \frac{n(n-1)(n-2)\cdots(n-k+1)(n-k)(n-k-1)\cdots 1}{(n-k)(n-k-1)\cdots 1}$$

$$P(n, k) = \frac{n!}{(n-k)!} \qquad (54.2)$$

EXAMPLE 54.6

(a) $P(6, 2) = 6 \cdot 5 = 30$
(b) $P(n, 1) = n \quad (n \geq 1)$
(c) $P(n, n) = n! \quad (n \geq 1)$

EXAMPLE 54.7 How many three-digit numbers can be formed with the digits 1, 2, 3, 4, 5 (a) if no digit can be repeated? (b) if digits can be repeated?

Solution

(a) $P(5, 3) = 5 \cdot 4 \cdot 3 = 60$
(b) Apply the Basic Counting Principle. We must choose three digits, with five possibilities for each choice. Thus the answer is 5^3 or 125.

EXERCISES FOR SECTION 54

Compute.

1. $P(10, 4)$ **2.** $P(6, 4)$ **3.** $P(50, 3)$
4. $P(6, 5)$ **5.** $P(20, 2)$ **6.** $P(10, 5)$

Simplify as much as possible.

7. $P(n, n-1)/P(n, n)$ **8.** $P(n, k)P(n-k, n-k)$
9. $P(n, k)/P(n-1, k-1)$

10. How many possible outcomes are there if a coin is tossed four times? five times? n times?

11. How many possible outcomes are there if a die (one of a pair of dice) is tossed two times? (An outcome will consist of two numbers.) What about three times? n times?

12. There are three routes connecting points A and B. (a) In how many ways can one travel from A to B and back to A?
(b) In how many ways can one travel from A to B and back to A if the route from B to A is required to be different from the route from A to B.

13. A man has three sport coats and five pairs of slacks. He can wear any of the coats with any of the pairs of slacks. How many combinations are possible?

14. In how many ways can ten marching bands be arranged to make up a parade?

15. In how many ways (orders) can the teams in a ten-team soccer league finish if there are no ties?

16. Ten riders are to be assigned randomly to ten horses. In how many ways can this be done?

17. How many even three-digit numbers can be formed with the digits 1, 2, 3, 4, and 5 (a) if no digit can be repeated? (b) if digits can be repeated?

18. In how many ways can three officers—President, Vice President, and Secretary—be selected from a club with 10 male and 15 female members so that the President and Secretary are female and the Vice President is male?
19. In how many ways can 5 men and 4 women be seated in a row so that no men are in adjacent seats?
20. In how many ways can $n + 1$ men and n women be seated in a row under each of the following conditions?
    (a) No restrictions by sex.
    (b) No men are in adjacent seats.
    (c) At least two men are in adjacent seats.
21. In an eight-team basketball league, in how many ways can the top three positions in the final standings be filled?
22. For a given set of pairings in an eight-team basketball tournament, in how many ways can the top three positions in the final standings be filled? (The top two teams must be from different brackets. Compare Exercise 21.)
23. How many positive integers are factors of 4? (Don't forget 1 and 4.) 8? 16? 2^n?
24. How many positive integers are factors of $2^2 \cdot 3^2$? $2^2 \cdot 3^2 \cdot 5^2$? $2^r \cdot 3^s \cdot 5^t$ (r, s, and t positive integers)? (Compare Exercise 23.)

SECTION 55 Combinations

A. Basic Ideas

In any enumeration problem the following question is critical: Does the order in which elements are chosen (or things are arranged, or events happen) matter?

EXAMPLE 55.1 Consider the problem of selecting two letters from A, B, and C.

(a) If order matters, there are six possibilities:

$$AB,\ AC,\ BA,\ BC,\ CA,\ CB.$$

Both AB and BA are listed, for instance, because they give A and B in different orders, and order matters.

(b) If order does not matter, there are only three possibilities:

$$AB,\ AC,\ BC.$$

The number of possibilities in part (a) is $P(3, 2)$, the number of *permutations* of three elements taken two at a time (Section 54). In contrast, the number of possibilities in part (b) is $C(3, 2)$, the number of *combinations* of three elements taken two at a time, as described by the following definition.

DEFINITION. A selection of k different elements from a set of n elements, without regard to order, is called a **combination** of the n elements taken k at a time. We denote the number of such combinations by $C(n, k)$.

Example 55.1(b) shows that $C(3, 2) = 3$. The discussion that follows will lead to a formula for $C(n, k)$ for all n and k.

In Example 54.5 we answered the following question: In how many ways can three officers—President, Vice-President, and Secretary—be selected from a club with ten members? The answer was $P(10, 3)$ $(= 10 \cdot 9 \cdot 8 = 720)$. We now consider a related but different question for which the answer is $C(10, 3)$.

EXAMPLE 55.2 In how many ways can a three-member committee be selected from a club with ten members?

Solution The critical difference between this question and that in Example 54.5 concerns order. Order mattered in Example 54.5: for example, having Smith as President, Jones as Vice-President, and Brown as Secretary is not the same as having Brown as President, Jones as Vice-President, and Smith as Secreary. But order does not matter in the present example: the committee {Smith, Jones, Brown} is the same as the committee {Brown, Jones, Smith}.

The answer in this example is $C(10, 3)$. To obtain a numerical value for this answer we'll redo Example 54.5 and get $C(10, 3)$ as a by-product.

Our first solution for Example 54.5 gave the answer $P(10, 3)$. Here is a second solution. Instead of electing the three officers directly, suppose we elect a slate of three, not specifying who will be President, and so on, but simply agreeing that the persons on the slate will be the officers, in a way to be agreed on among themselves. This method still allows for the same number of possibilities as the first method, but it is conceptually different for counting purposes. To compute the number of possibilities by the second method, use the Basic Counting Principle with two steps: first count the number of ways of selecting three members from ten [this is $C(10, 3)$], then count the number of ways the three chosen members can be given the three different offices, and then multiply. The number of ways three members can be given the three different offices is $3 \cdot 2 \cdot 1$ (any of three for President, then any of the remaining two for Vice-President, then the remaining one for Secretary). Thus the second solution of Example 54.5 gives the answer $C(10, 3) \cdot 3 \cdot 2 \cdot 1$. Since this solution and the original solution must agree, we must have

$$C(10, 3) \cdot 3 \cdot 2 \cdot 1 = P(10, 3).$$

Therefore,

$$C(10, 3) = \frac{10 \cdot 9 \cdot 8}{3 \cdot 2 \cdot 1} = 120.$$

To obtain a general formula for $C(n, k)$ we proceed just as in the example. From Equation (54.2),

$$P(n, k) = \frac{n!}{(n-k)!}. \tag{55.1}$$

But we can also get $P(n, k)$ as follows. First, think of selecting k elements from n elements without regard to order; this can be done in $C(n, k)$ different ways. Second, each of these k-element subsets can be arranged in $k(k-1)(k-2)\cdots 1$ different ways. Therefore, as Example 55.2 showed in a special case,

$$P(n, k) = C(n, k) \cdot k(k-1)(k-2)\cdots 1. \tag{55.2}$$

Thus

$$\boxed{C(n, k) = \frac{P(n, k)}{k!}.}$$

If we use $P(n, k) = n!/(n-k)!$ [Equation (55.1], we find that

$$C(n, k) = \frac{n!}{k!(n-k)!}. \tag{55.3}$$

If you have studied Section 53 you will recognize the expression on the right as a binomial coefficient. Specifically,

$$\boxed{C(n, k) = \binom{n}{k}.} \tag{55.4}$$

B. More Examples

EXAMPLE 55.3 How many different 10-member committees can be formed from the 100 members of the U.S. Senate?

Solution The answer is the number of combinations of 100 elements taken 10 at a time, which is

$$C(100, 10) = \frac{100!}{10!90!}.$$

This is approximately 1.7×10^{13}. [C]

EXAMPLE 55.4 Assume that 52 members of the Senate are Republicans and that 48 are Democrats. How many 10-member committees can be formed subject to the condition that 6 are Republicans and 4 are Democrats?

Solution The Republicans can be chosen in $C(52, 6)$ different ways. The Democrats can be chosen in $C(48, 4)$ different ways. Therefore, by the Basic Counting Principle, the total number of committees is

$$C(52, 6) \cdot C(48, 4) = \frac{52!}{6!46!} \cdot \frac{48!}{4!44!}.$$

This is approximately 4.0×10^{12}. [C]

A *bridge deck* contains 52 cards, 13 in each of four suits: spades, hearts, diamonds, and clubs. Each suit contains an ace, 2, 3, 4, 5, 6, 7, 8, 9, 10, jack, queen, and king.

EXAMPLE 55.5

(a) The number of different 13-card bridge hands is the number of combinations of 52 elements taken 13 at a time, which is $C(52, 13)$. With a calculator you can show that this is approximately 6.4×10^{11}. [C]

(b) The number of different 5-card poker hands that can be dealt from a bridge deck is $C(52, 5)$. This is

$$\frac{52!}{5!47!} = \frac{52 \cdot 51 \cdot 50 \cdot 49 \cdot 48}{5 \cdot 4 \cdot 3 \cdot 2 \cdot 1} = 2{,}598{,}960.$$

EXAMPLE 55.6 In the notation of this section, Equation (53.3) states that

$$C(n + 1, k) = C(n, k - 1) + C(n, k). \tag{55.5}$$

This equation was proved algebraically in Section 53. Following is a combinatorial proof.

Let S be a set with $n + 1$ elements, $x_1, \cdots, x_{n+1}$. Then $C(n + 1, k)$ is the number of k-element subsets of S, and these can be divided into two classes, those that do contain x_1 and those that do not. There are $C(n, k - 1)$ subsets of the first kind (the number of ways we can choose $k - 1$ elements from $x_2, \cdots, x_{n+1}$); there are $C(n, k)$ subsets of the second kind (the number of ways we can choose k elements from $x_2, \cdots, x_n$). Adding these we get Equation (55.5).

EXERCISES FOR SECTION 55

Compute.

1. $C(5, 3)$ **2.** $C(7, 4)$ **3.** $C(10, 9)$

4. $C(n, n - 1)$ **5.** $C(n, 3)$ **6.** $C(n, 2)$

Simplify as much as possible.

7. $P(n, k)/C(n, k)$ **8.** $C(n, k)/C(n, n - k)$ **9.** $C(n + 1, n - 1)/(n + 1)$

10. An organization contains ten men and ten women. How many six-member committees can be chosen under the following conditions?
(a) With three members of each sex.

(b) With four men and two women.
(c) With four of one sex and two of the other.

11. Ten basketball players want to divide themselves into two teams of five players each. In how many ways can this be done?

12. Ten basketball players want to divide themselves into two teams of five players each, in such a way that the two best players are on opposite teams. In how many ways can this be done?

13. How many hands of 13 cards, none of which is a spade, can be dealt from a 52-card bridge deck?

14. How many hands of 13 cards containing all 4 aces can be dealt from a 52-card bridge deck?

15. How many hands of 5 cards containing no aces can be dealt from a 52-card bridge deck?

16. How many hands of 13 cards containing 5 spades, 4 hearts, 3 diamonds, and 1 club can be dealt from a 52-card bridge deck?

17. How many hands of 13 cards containing 4 spades, 3 hearts, 3 diamonds, and 3 clubs can be dealt from a 52-card bridge deck?

18. How many hands of 13 cards containing 4 of any one suit and 3 of each of the other suits can be dealt from a 52-card bridge deck?

19. (a) A baseball league has eight teams, and each team plays six games against each of the other teams. How many games are there altogether?
(b) Repeat part (a) with n in place of eight and k in place of six.

20. In how many ways can a ten-question true-false examination be answered so that there are five true and five false responses? (Any way of answering amounts to a choice of which questions to mark *true*.)

21. A drawer contains seven good light bulbs and three defective light bulbs. In how many ways can four bulbs be chosen under the following conditions?
(a) None is defective.
(b) Two are good and two are defective.
(c) At least two are good.

22. How many lines are determined by n points if no three of the points are collinear?

23. How many planes are determined by n points if no four of the points are coplanar? (Three noncollinear points determine a plane.)

24. How many triangles can be formed if the vertices are chosen from a set of n points all lying on a circle?

25. How many quadrilaterals can be formed if the vertices are chosen from a set of n points all lying on a circle?

26. Ten parallel lines are all perpendicular to ten other parallel lines. How many rectangles are formed?

27. How many diagonals does a regular n-sided polygon have? (A polygon is *regular* if all of its angles are equal and all of its sides are equal. A *diagonal* is a segment joining two nonadjacent vertices.)

28. How many different sums of money can be formed by choosing two coins from a penny, a nickel, a dime, a quarter, and a half dollar?

29. How many different sums of money can be formed by choosing one or more coins from a penny, a nickel, a dime, a quarter, and a half dollar? (Compare Exercise 28.)

30. Explain why the number of subsets of an n-element set is

$$C(n, 0) + C(n, 1) + C(n, 2) + \cdots + C(n, n).$$

[The empty set, which is accounted for by $C(n, 0)$, is included.] Use the Binomial Theorem (Section 53) to show that the indicated sum equals 2^n, so that an n-element set has 2^n subsets. [Suggestion: Use $a = b = 1$ in Equation (53.4). Compare Exercise 31 in Section 53.]

SECTION 56 Probability

A. Introduction

Mathematical probability is concerned with events whose occurrence is subject to chance. The goal is to represent the likelihood of such an event by a real number p, called the *probability* of the event, with $0 \leq p \leq 1$. This is done in such a way that the more likely the event, the larger the number p. If an event is *certain* to occur, then $p = 1$; if an event is certain *not* to occur, then $p = 0$.

Not all events involving uncertainty fall within the province of mathematical probability. A question such as "What is the probability good will win out over evil?," however important, is not one for mathematical probability. Still, mathematical probability is extremely useful, with applications throughout business, science technology, and elsewhere. Unfortunately, all that can be given here is a brief introduction to the subject. To illustrate the basic ideas we begin with three sample questions, which will be answered later in the section.

EXAMPLE 56.1 If a coin is tossed three times, what is the probability a head will occur exactly twice?

EXAMPLE 56.2 If the six persons making up three married couples are seated randomly in a row, what is the probability a particular husband and wife will be in adjacent seats?

EXAMPLE 56.3 In a room of 25 randomly chosen persons, what is the probability they all have different birthdays?

The underlying process in any such example is called an **experiment.** The set of all possible outcomes in an experiment is called the **sample space** of the experiment. Any subset of the sample space is called an **event.**

EXAMPLE 56.1 *(Continued)*. In this example the *experiment* consists of tossing three coins and observing which land heads (H) and which land tails (T). The *sample space* is

$$\{\text{HHH, HHT, HTH, THH, HTT, THT, TTH, TTT}\}. \tag{56.1}$$

Thus there are eight possible outcomes. The subset of those outcomes in which H occurs exactly twice is

$$\{\text{HHT, HTH, THH}\}. \tag{56.2}$$

This subset represents the *event* with which Example 56.1 is concerned; thus there are three possible outcomes corresponding to this event.

In this book we consider only experiments for which the sample space is finite.

DEFINITION. The **probability** of an event E, in an experiment in which all possible outcomes are equally likely, is defined by

$$P(E) = \frac{n(E)}{n(S)}, \tag{56.3}$$

where $n(S)$ denotes the number of elements in the sample space (the total number of possible outcomes) and $n(E)$ denotes the number of possible outcomes corresponding to the event E.

EXAMPLE 56.1 *(Concluded)* For the question in Example 56.1, $n(S) = 8$ [see (56.1)] and $n(E) = 3$ [see (56.2)]. Therefore, $P(E) = \frac{3}{8}$.

The number $P(E)$ in (56.3) corresponds to the number p in the opening paragraph of this section. Our definition of probability applies only to experiments having finitely many possible outcomes, and those outcomes must all be equally likely. The computation of a probability in such cases reduces to the computation of the two numbers $n(E)$ and $n(S)$, and this will very often involve permutations and combinations (Sections 54 and 55). Before we do those computations for Examples 56.2 and 56.3, let's look at the following simpler example. When we say that we *select a number at random* in this example we mean that all of the outcomes are equally likely.

EXAMPLE 56.4 Suppose a number is selected at random from the list 1, 2, 3, 4, 5, 6, 7, 8, 9. Compute the probability of each of the following events.
(a) An even number is selected.
(b) An odd number is selected.
(c) A negative number is selected.
(d) A positive number is selected.

Solution In each part of this example the experiment consists of selecting a number at random from the list 1, 2, 3, 4, 5, 6, 7, 8, 9. The sample space is $S = \{1, 2, 3, 4, 5, 6, 7, 8, 9\}$. Thus $n(S) = 9$.

(a) $E = \{2, 4, 6, 8\}$, so $P(E) = \frac{4}{9}$.

(b) $E = \{1, 3, 5, 7, 9\}$, so $P(E) = \frac{5}{9}$.

(c) E is the empty set, so $P(E) = \frac{0}{9} = 0$.

(d) $E = S$, so $P(E) = \frac{9}{9} = 1$.

Solution for Example 56.2 The experiment consists of arranging the six persons randomly in a row. The sample space S consists of all of the ways this can be done. Thus $n(S)$ is the number of permutations (arrangements) of six elements taken six at a time, so that $n(S) = 6!$ (Section 54).

The event E corresponds to all of the arrangements in which the particular husband and wife are in adjacent seats. Let's compute $n(E)$ in two steps; first we'll compute the number of arrangements in which they are adjacent with the husband to the right of the wife. In this case we have five elements to arrange in a row (the given couple, considered as one, and the other four persons); this can be done in 5! ways. There will be the same number of ways with the husband to the left of the wife. Thus $n(E) = 2(5!)$.

Since $n(S) = 6!$ and $n(E) = 2(5!)$, we arrive at $P(E) = 2(5!)/6! = \frac{2}{6} = \frac{1}{3}$.

Solution for Example 56.3 The experiment consists of selecting 25 persons at random and recording their birthdays. For simplicity we ignore leap years, so that there are 365 possibilities for the birthday of each person. Since these 25 birthdays are mutually independent, the Basic Counting Principle (Section 54) shows that $n(S) = 365^{25}$, where S denotes the sample space.

The event E corresponds to all of the possibilities in which no two birthdays are the same. To compute $n(E)$ we again apply the Basic Counting Principle. If we record the birthdays one at a time, then there are 365 possibilities for the first birthday. Then there are 364 possibilities for the second birthday, which must be different from the first. Then there are 363 possibilities for the third birthday, which must be different from both of the first two. If we continue in this way and apply the Basic Counting Principle we will arrive at $n(E) = 365 \cdot 364 \cdot \cdots \cdot 341$ (25 factors). In the notation of Section 54, $n(E) = P(365, 25)$.

Since $n(S) = 365^{25}$ and $n(E) = P(365, 25)$, we arrive at

$$P(E) = \frac{n(E)}{n(S)} = \frac{365 \cdot 364 \cdot \cdots \cdot 341}{365^{25}} \approx 0.4313.$$ Ⓒ

Notice carefully what this implies: Among 25 randomly chosen persons, the probability is less than $\frac{1}{2}$ that their birthdays will all be different. Put another way, the probability is more than $\frac{1}{2}$ that at least two of the persons will have the same birthday.

Continuation of Example 56.3 The reasoning in the solution for Example 56.3 shows that in a room of n randomly chosen persons the probability they all have different birthdays is

$$P(E) = \frac{n(E)}{n(S)} = \frac{P(365, n)}{365^n} = \frac{365 \cdot 364 \cdot \cdots \cdot (365 - n + 1)}{365^n}. \qquad (56.4)$$

Table 56.1 shows these probabilities (to four decimal places) for selected values of n. The last row shows that more than 94% of the time in a room of 50 randomly

TABLE 56.1

Number of persons	*Probability all birthdays are different*
5	0.9729
10	0.8831
15	0.7471
20	0.6886
22	0.5243
23	0.4927
25	0.4313
50	0.0590

chosen persons at least two of the persons will have the same birthday $(1 - 0.0590 = 0.9410 > 0.94)$. The table also shows that more than 50% of the time in a room of 23 (or more) randomly chosen persons at least two of the persons will have the same birthday.

B. Properties and More Examples

The opening paragraph of this section stated that if p represents the probability of an event, then it should be true that $0 \leq p \leq 1$, with $p = 1$ for an event that is certain to occur and $p = 0$ for an event that is certain not to occur. We can now prove these properties. We use $\varnothing$ (Greek letter *phi*) to denote the empty set, which corresponds to an event that is certain not to occur.

> If S denotes a sample space and E an event, then
>
> $$0 \leq P(E) \leq 1.$$
>
> In particular,
>
> $$P(\varnothing) = 0 \quad \text{and} \quad P(S) = 1.$$

Proof. If E denotes any subset of S, then $0 \leq n(E) \leq n(S)$, so that

$$\frac{0}{n(S)} \leq \frac{n(E)}{n(S)} \leq \frac{n(S)}{n(S)}$$
$$0 \leq P(E) \leq 1.$$

Also, $P(\varnothing) = n(\varnothing)/n(S) = 0/n(S) = 0$ and $P(S) = n(S)/n(S) = 1$. □

Recall that if A and B are sets, then $A \cup B$ denotes the union of A and B, and $A \cap B$ denotes the intersection of A and B (Section 6). If A and B are finite sets, then

$$n(A \cup B) = n(A) + n(B) - n(A \cap B). \tag{56.5}$$

We must subtract $n(A \cap B)$ here because otherwise each element of $A \cap B$ would be counted twice on the right, once as an element of A and again as an element of B.

If A and B represent events, then $A \cup B$ occurs iff either A occurs or B occurs (or both A and B occur); $A \cap B$ occurs iff both A and B occur.

> If A and B denote events from a sample space S, then
>
> $$P(A \cup B) = P(A) + P(B) - P(A \cap B). \tag{56.6}$$

Proof. Apply Equation (56.5):

$$n(A \cup B) = n(A) + n(B) - n(A \cap B)$$

$$\frac{n(A \cup B)}{n(S)} = \frac{n(A)}{n(S)} + \frac{n(B)}{n(S)} - \frac{n(A \cap B)}{n(S)}$$

$$P(A \cup B) = P(A) + P(B) - P(A \cap B). \qquad \square$$

EXAMPLE 56.5 Assume that a die (one of a pair of dice) is rolled twice. Find the probability that either the first number is 1 or the sum of the two numbers is at least 6.

Solution Figure 56.1 shows the sample space. Let A denote the event that the first number is 1, and let B denote the event that the sum of the two numbers is at least 6. The outcomes in A and B are indicated in the figure. We see that

$$n(A) = 6 \text{ so } P(A) = \frac{n(A)}{n(S)} = \frac{6}{36},$$

$$n(B) = 26 \text{ so } P(A) = \frac{n(B)}{n(S)} = \frac{26}{36}, \qquad \text{and}$$

$$n(A \cap B) = 2 \text{ so } P(A \cap B) = \frac{n(A \cap B)}{n(S)} = \frac{2}{36}.$$

Therefore, by (56.6),

$$P(A \cup B) = \tfrac{6}{36} + \tfrac{26}{36} - \tfrac{2}{36} = \tfrac{30}{36} = \tfrac{5}{6}.$$

A	1,1	1,2	1,3	1,4	1,5	1,6
	2,1	2,2	2,3	2,4	2,5	2,6
	3,1	3,2	3,3	3,4	3,5	3,6
	4,1	4,2	4,3	4,4	4,5	4,6
B	5,1	5,2	5,3	5,4	5,5	5,6
	6,1	6,2	6,3	6,4	6,5	6,6

Figure 56.1

Events A and B are said to be **mutually exclusive** if $A \cap B = \varnothing$. Since $P(\varnothing) = 0$, we have the following special case of (56.6).

> If the events A and B are mutually exclusive, then
>
> $$P(A \cup B) = P(A) + P(B). \qquad (56.7)$$

If E denotes an event from a sample space S, then the **complement** of E, denoted $\bar{E}$, represents all of the outcomes that are in S but not in E. For example, in rolling a die, the complement of the event "the outcome is at least 3" is the event "the outcome is less than 3." We have the following convenient result between the probability of an event and the probability of its complement.

> If E is any event with complement $\bar{E}$, then
>
> $$P(\bar{E}) = 1 - P(E). \qquad (56.8)$$

Proof. From the definition of $\bar{E}$ it follows that $E \cup \bar{E} = S$ and $E \cap \bar{E} = \varnothing$. Therefore, by (56.7) with $A = E$ and $B = \bar{E}$, we have $P(S) = P(E) + P(\bar{E})$, so that $1 = P(E) + P(\bar{E})$, which gives (56.8). □

One convenient use of (56.8) comes by realizing that the complement of "at least one" is "none." Here is an example.

EXAMPLE 56.6 A hand of 13 cards is chosen randomly from a bridge deck of 52 cards. What is the probability the hand contains at least 1 ace? (The makeup of a bridge deck is explained preceding Example 55.5.)

Solution The sample space S consists of all possible 13-card hands from a 52-card bridge deck. Therefore, from Example 55.5, $n(S) = C(52, 13)$.

Let E represent the event "the hand contains at least one ace." Then $\bar{E}$ represents the event "the hand contains no ace." The number $n(\bar{E})$ is the number of 13-card hands from the 48-card deck that remains when the 4 aces have been removed. Thus $n(\bar{E}) = C(48, 13)$. Therefore,

$$\begin{aligned} P(E) &= 1 - P(\bar{E}) \\ &= 1 - \frac{C(48, 13)}{C(52, 13)} \\ &= 1 - \left[\left(\frac{48!}{13!35!}\right)\Big/\left(\frac{52!}{13!39!}\right)\right] \\ &= 1 - \frac{39 \cdot 38 \cdot 37 \cdot 36}{52 \cdot 51 \cdot 50 \cdot 49} \\ &\approx 0.70. \end{aligned}$$

Ⓒ

EXERCISES FOR 56

1. If a coin is tossed three times, what is the probability a head will occur exactly once? Begin by listing all of the elements of the sample space.
2. If a coin is tossed four times, what is the probability a tail will occur at least twice? Begin by listing all of the elements of the sample space.
3. If a coin is tossed three times, what is the probability of an odd number of heads? Begin by listing all of the elements of the sample space.
4. A number is formed by arranging the digits 1, 2, and 3 in random order (with no digit repeated). Find the probability of each of the following events.
   (a) The number is greater than 100.
   (b) The number is greater than 200.
   (c) The number is greater than 220.
   (d) The number is greater than 330.
5. A number is chosen randomly from the set $\{17, 18, 19, 20, 21, 22\}$. Find the probability of each of the following events.
   (a) The number is even.
   (b) The sum of the digits is even.
   (c) Either the number is even or the sum of its digits is even.
   (d) The number is a perfect square (that is, the square of an integer).
6. A number is selected randomly from the set of eight primes less than 20. Find the probability of each of the following events.
   (a) The number is even.
   (b) The number is less than 10.
   (c) The number is two more than another prime.
   (d) The number is a sum of three different primes.
7. Assume that a pair of dice is rolled. Find the probability of each of the following events. (An array such as in Figure 56.1 should help.)
   (a) Double 4 is obtained.
   (b) Neither die falls 3.
   (c) Neither 3 nor 4 appears.
   (d) Each die shows more than 3.
   (e) At least one die shows more than 3.
   (f) Exactly one die shows more than 3.
8. A card is chosen randomly from a bridge deck of 52 cards. Find the probability of each of the following events.
   (a) The card is a spade.
   (b) The card is not a spade.
   (c) The card is an ace.
   (d) The card is not an ace.
   (e) The card is either a spade or an ace.
   (f) The card is a jack, queen, or king.
9. Assume that in families with three children, all of the eight arrangements GGG, GGB, GBG, . . . , BBB are equally likely, where, for example, GGB denotes that the oldest is a girl, the next is also a girl, and the youngest is a boy. Compute the probability of each of the following events.
   (a) There is exactly one boy.

(b) There is at least one boy.
(c) There is at most one boy.
(d) All are girls.
(e) The oldest is a boy.
(f) There is no boy older than a girl.

10. Six dice are rolled. What is the probability that no number occurs twice?

11. Four friends enter an ice cream parlor, and each one chooses a flavor randomly from among the 20 that are offered. What is the probability that no two choose the same flavor?

12. Four cards are drawn at random from a bridge deck of 52 cards. What is the probability no two of the cards are from the same suit?

13. Two shoes are chosen randomly from the twelve shoes making up six distinguishable pairs. What is the probability the two shoes belong to the same pair?

14. A committee contains six men and six women. What is the probability that a randomly chosen three-member subcommittee will contain only women?

15. Ten basketball players, three of whom are brothers, are divided randomly into two five-member teams. What is the probability the three brothers are on the same team?

16. A 25-member university committee contains 7 female faculty members, 8 male faculty members, 6 female student members, and 4 male student members. A 5-member subcommittee is chosen at random. Find the probability of each of the following events. How does Equation (56.6) apply?
(a) The committee contains only males.
(b) The committee contains only faculty members.
(c) The committee is made up entirely of male faculty members.
(d) The committee contains only males or only faculty members.

17. A die is rolled twice. Find the probability of each of the following events. (An array such as that in Figure 56.1 will help.)
(a) The first number is odd.
(b) The sum of the numbers is odd.
(c) The first number is odd and the sum of the numbers is odd.
(d) The first number is odd or the sum of the numbers is odd.

18. A subset is chosen randomly from the 15 nonempty subsets of $\{A, B, C, D\}$. Find the probability of each of the following events. How does Equation (56.6) apply?
(a) The subset contains D.
(b) The subset contains exactly three elements.
(c) The subset contains exactly three elements, one of which is D.
(d) The subset contains exactly three elements or the subset contains D.

19. A hand of 13 cards is chosen randomly from a bridge deck of 52 cards. Compute the probability of each of the following events.
(a) The hand contains no spades.
(b) The hand contains only spades.
(c) The hand contains no aces.
(d) The hand contains all four aces.

20. Ninety lottery tickets are numbered from 1 to 90. Assume that two tickets are chosen at random from the 90. Compute the probability of each of the following events.

(a) Each of the tickets has a number less than 10.
(b) Each of the tickets drawn has an even number.
(c) The first ticket has an even number and the second ticket has an odd number.
(d) One of the tickets has an even number and one of the tickets has an odd number.

21. Imagine a game in which a player either wins, in which case he receives a certain prize, or loses, in which case he receives nothing. The player's **mathematical expectation** for such a game is defined to be his probability of winning times the value of the prize. This can be thought of as the reasonable amount to pay for the privilege of playing. For example, in 100 rounds of coin tossing with prize \$1 for each tail, a player can reasonably expect to win \$50, or, on the average, \$0.50 (the mathematical expectation) per round.
(a) If a die is to be thrown once and a prize of \$3 is to be awarded if it turns up 6, what is the mathematical expectation?
(b) How much should a gambling house charge for each turn at a certain game if the player has probability $\frac{1}{5}$ of winning, the prize is \$10, and the house wants a profit of \$0.50 per turn on the average?
(c) Would you pay \$1 for a chance to win \$1,000,000 if your chances of winning were only 1 in 5 million? Compute the relevant mathematical expectation. (Be honest—lotteries thrive because commercial value can exceed mathematical expectation.)
(d) Would you pay \$10,000 for a chance to win \$100,000 if your chance of winning were 1 in 5? Compute the relevant mathematical expectation. (The point here is the opposite of that in (c)—mathematical expectation can exceed commercial value. Of course, your answer here may depend on how much \$10,000 means to you.)

REVIEW EXERCISES FOR CHAPTER XIII

Compute.

1. $P(12, 3)$ **2.** $P(8, 4)$ **3.** $C(7, 5)$ **4.** $C(11, 5)$

5. Five flutists audition for the first, second, and third chairs in the flute section of a school orchestra. In how many ways can the chairs be filled (ignoring ability)?

6. The four faces on Mount Rushmore are in the order Washington, Jefferson, Roosevelt, Lincoln. How many other orders would have been possible?

7. A menu lists three appetizers, four main courses, and five desserts. How many different ways are there to choose one of each? (Assume all combinations are acceptable.)

8. A school play will have five roles: a king, a queen, a prince, a princess, and a (male) court jester. At the auditions, 10 boys try for the 3 male roles, and 12 girls try for the 2 female roles. In how many different ways can the 5 roles be filled?

9. A sergeant must choose four "volunteers" for a clean-up crew. He has ten recruits from which to choose. How many different crews are possible?

10. To fulfill her graduation requirements, a student must choose four courses: two electives from a group of six courses in the humanities, and two other electives from a group of five courses in the sciences. How many different combinations are possible?

11. The three numbered volumes of a three-volume set of books are placed on a shelf in random order. What is the probability they are in the correct order?
12. You are one of 20 students in a class in which the teacher asks questions of three randomly chosen students each day. What is the probability that you will be chosen on a particular day?
13. A box contains 20 blue cards numbered 1–20, and 15 white cards numbered 1–15. One card is chosen randomly from the box.
    (a) What is the probability the card is white?
    (b) What is the probability the number on the card is odd?
    (c) What is the probability the card is either white or has an odd number?
14. A hand of 5 cards is chosen randomly from a bridge deck of 52 cards. What is the probability the cards are all diamonds? What is the probability they all belong to the same suit?
15. An investor randomly chooses two stocks from a set of five that look equally attractive. Over the next year, two of the original five increase in value while the other three decrease. What is the probability that the two stocks that increase are those the investor chose? If three of the five increase and the other two decrease, what is the probability that both of the investor's choices increase?
16. A three-member subcommittee is chosen randomly from a ten-member committee of five males and five females. What is the probability the committee contains at least one male?

APPENDIX

A Real Numbers

Order Properties

In Chapter I, *positive* real numbers were defined as those to the right of 0 on a real line directed to the right. To define *positive* without appealing to a real line, we amend the axioms for a field (Section 1C) by adding the following *order axiom.*

Order Axiom. A field F is said to be an **ordered field** if there is a subset F^p of F satisfying the following three conditions.

If $a \in F^p$ and $b \in F^p$, then $a + b \in F^p$.	**Closure under addition**
If $a \in F^p$ and $b \in F^p$, then $ab \in F^p$.	**Closure under multiplication**
If $a \in F$, then exactly one of the following is true: $a = 0$, $a \in F^p$, or $-a \in F^p$.	**Law of trichotomy**

The elements of F^p are called the **positive elements** of F. Elements that are neither zero nor positive are said to be **negative.**

The system of real numbers is an ordered field. So is the system of rational numbers. It can be proved, however, that the system of complex numbers is a field that is *not* ordered (regardless of what one tries for the set of positive elements). For a more extensive discussion of fields and related algebraic systems, see any introductory book on modern algebra, such as the author's *Modern Algebra: An Introduction* (John Wiley & Sons, New York, 1979).

Decimal Numbers

Section 1 states that a decimal number represents a rational number iff it either terminates or repeats. To understand why, first consider a rational number a/b, where a and b are integers and $b \neq 0$. In computing the decimal representation of a/b, the remainders in the division process will be either 0, 1, 2, . . . , or $b - 1$. If one of these remainders is 0 after we start bringing down zeros in the division process, the decimal terminates. If not, one of the numbers 1, 2, . . . , or $b - 1$ must repeat as a remainder (in fewer than b steps after we start bringing down zeros in the division process); this means that the decimal has started to repeat. Thus, if a decimal number represents a rational number, then it either terminates or repeats.

Now consider the converse, namely, if a decimal number either terminates or repeats, then it represents a rational number. For a terminating decimal this can be verified by multiplying and dividing by an appropriate power of 10. For example, $7.321 = 7.321(10^3/10^3) = 7321/1000$, which is a quotient of two integers. For repeating decimals see Example 52.9, which shows how to convert each repeating decimal to a quotient of two integers.

Irrationality of $\sqrt{2}$

Section 1 states that $\sqrt{2}$ *is irrational.* We can prove this by showing that the alternative—$\sqrt{2}$ is rational—leads to a contradiction. Thus assume that $\sqrt{2}$ is rational, so that $\sqrt{2} = a/b$ for some integers a and b with $b \neq 0$. We can also assume that a/b is reduced to lowest terms, so that a and b have no common factor except ± 1. Since $\sqrt{2} = a/b$, we have $2 = a^2/b^2$ and $2b^2 = a^2$. The left side of this equation, $2b^2$, is even; therefore the right side, a^2, must also be even. But if a^2 is even, then a must be even, and hence $a = 2k$ for some integer k. Substituting $a = 2k$ in $2b^2 = a^2$, we obtain $2b^2 = (2k)^2$, $2b^2 = 4k^2$, and $b^2 = 2k^2$. The last equation shows that since $2k^2$ is even, b^2 must also be even, and thus b must be even. Thus we have deduced that a and b are even, and they therefore have 2 as a common factor. This contradicts the assumption that a/b was reduced to lowest terms, and completes the proof.

APPENDIX B Sets

The text uses a number of ideas about sets. This appendix gives more examples of these ideas.

Set-Builder Notation

Section 4B introduces the *set-builder notation*

$$\{x: \ \cdots\},$$

which means

the set of all x such that $\cdots$.

The notation $\{a, b, \cdots\}$ denotes the set whose members are $a, b, \cdots$.

EXAMPLE B.1

(a) $\{x: \ x \text{ is a positive integer}\}$ denotes the set of all positive integers.
(b) $\{x: \ x \text{ is a positive even integer}\}$ denotes the set $\{2, 4, 6, \cdots\}$.
(c) $\{a/b: \ a \text{ and } b \text{ are integers with } b \neq 0\}$ denotes the set of all rational numbers (Section 1).

When the context implies that a variable represents a number, and no other restriction is specified (such as "x is an integer"), then (in this book) x is assumed to represent a real number.

EXAMPLE B.2

(a) $\{x: \ x < 0\}$ denotes the set of all negative real numbers.
(b) $\{x: \ x^2 > 2\}$ denotes the set of all real numbers whose squares are greater than 2.

Other Set Notation

The following notation was introduced in Section 6C.

$x \in A$ denotes that x *is an element of* (or is a member of, or belongs to) the set A.

$\varnothing$ denotes the *empty-set*—that is, the set that contains no elements.

$A \cup B$ denotes $\{x: \ x \in A \text{ or } x \in B\}$, which is called the *union* of A and B.

$A \cap B$ denotes $\{x: \ x \in A \text{ and } x \in B\}$, which is called the *intersection* of A and B.

EXAMPLE B.3 Let $S = \{a, b, c\}$, $T = \{c, d, e\}$, and $U = \{d, e\}$. Then

$$S \cup T = \{a, b, c, d, e\} \qquad S \cap T = \{c\}$$
$$S \cup U = \{a, b, c, d, e\} \qquad S \cap U = \varnothing$$
$$T \cup U = T \qquad T \cap U = U.$$

EXAMPLE B.4 If A is any set, then $A \cup \varnothing = A$ and $A \cap \varnothing = \varnothing$.

The following ideas are not used elsewhere in the book, but they are so basic that they are included here for completeness.

To indicate that x is *not* an element of a set A, we write $x \notin A$. Thus, for example,

$x \notin \varnothing$ is always true, and

$x \in \varnothing$ is always false.

If A and B are sets and each element of A is an element of B, then A is called a **subset** of B; this is denoted by

$$A \subseteq B \quad \text{or} \quad B \supseteq A.$$

Notice that $A \subseteq B$ does not preclude the possibility that $A = B$. In fact,

$$A = B \text{ iff } A \subseteq B \text{ and } A \supseteq B.$$

The following three statements are equivalent:

(i) $A \subseteq B$.
(ii) If $x \in A$, then $x \in B$.
(iii) If $x \notin B$, then $x \notin A$.

If A is not a subset of B, we write $A \not\subseteq B$. This is true, of course, iff A contains at least one element that is not in B. Set inclusion ($\subseteq$) has the following properties:

$\varnothing \subseteq A$ for every set A.

$A \subseteq A$ for every set A.

If $A \subseteq B$ and $B \subseteq C$, then $A \subseteq C$.

EXAMPLE B.5 If S, T, and U are as in Example B.3, then $S \not\subseteq T$ and $T \supseteq U$.

Cartesian Products, Functions, and Relations

The **Cartesian product** of sets S and T is denoted by $S \times T$, and is defined by

$$S \times T = \{(x, y): \quad x \in S \text{ and } y \in T\}$$

Here each (x, y) is an *ordered pair* (see the footnote on page 152).

EXAMPLE B.6 If $S = \{1, 2\}$ and $T = \{u, v, w\}$, then

$$S \times T = \{(1, u), (1, v), (1, w), (2, u), (2, v), (2, w)\}.$$

Notice that in this case $S \times T \neq T \times S$. In general, $S \times T = T \times S$ iff $S = T$.

Coordinate geometry (Section 19) begins with a *Cartesian plane,* which is a one-to-one correspondence between the points of a plane and the set of ordered pairs of real numbers—that is, the elements of the Cartesian product $R \times R$, where R denotes the set of all real numbers. This example explains the choice of the name "Cartesian product."

The notion of Cartesian product can be used to give a definition of *function* that is preferred by some to that given in Section 22: A *function* from a set S to a set T is a subset of $S \times T$ such that each $x \in S$ is a first member of precisely one pair in the subset. The connection between this definition and that given in Section 22 is that if f is a function from S to T (in the sense of Section 22), then each $x \in S$ contributes the pair $(x, f(x))$ to the subset of $S \times T$ in the definition of function as a subset of $S \times T$.

EXAMPLE B.7 If f denotes the function such that $f(x) = x^2$ for each real number x, then in the sense introduced above f is

$$\{(x, x^2): \quad x \text{ is a real number}\}.$$

A more general idea than that of a function is that of a *relation:* A **relation** is any subset of a Cartesian product. Although relations occur often in mathematics, we have no need to dwell on them here.

APPENDIX C Tables

TABLE I Squares and Square Roots

n	n^2	$\sqrt{n}$	n	n^2	$\sqrt{n}$
1	1	1.000	51	2,601	7.141
2	4	1.414	52	2,704	7.211
3	9	1.732	53	2,809	7.280
4	16	2.000	54	2,916	7.348
5	25	2.236	55	3,025	7.416
6	36	2.449	56	3,136	7.483
7	49	2.646	57	3,249	7.550
8	64	2.828	58	3,364	7.616
9	81	3.000	59	3,481	7.681
10	100	3.162	60	3,600	7.746
11	121	3.317	61	3,721	7.810
12	144	3.464	62	3,844	7.874
13	169	3.606	63	3,969	7.937
14	196	3.742	64	4,096	8.000
15	225	3.873	65	4,225	8.062
16	256	4.000	66	4,356	8.124
17	289	4.123	67	4,489	8.185
18	324	4.243	68	4,624	8.246
19	361	4.359	69	4,761	8.307
20	400	4.472	70	4,900	8.367
21	441	4.583	71	5,041	8.426
22	484	4.690	72	5,184	8.485
23	529	4.796	73	5,329	8.544
24	576	4.899	74	5,476	8.602
25	625	5.000	75	5,625	8.660
26	676	5.099	76	5,776	8.718
27	729	5.196	77	5,929	8.775
28	784	5.292	78	6,084	8.832
29	841	5.385	79	6,241	8.888
30	900	5.477	80	6,400	8.944
31	961	5.568	81	6,561	9.000
32	1,024	5.657	82	6,724	9.055
33	1,089	5.745	83	6,889	9.100
34	1,156	5.831	84	7,056	9.165
35	1,225	5.916	85	7,225	9.220
36	1,296	6.000	86	7,396	9.274
37	1,369	6.083	87	7,569	9.327
38	1,444	6.164	88	7,744	9.381
39	1,521	6.245	89	7,921	9.434
40	1,600	6.325	90	8,100	9.487
41	1,681	6.403	91	8,281	9.539
42	1,764	6.481	92	8,464	9.592
43	1,849	6.557	93	8.649	9.644
44	1,936	6.633	94	8,836	9.695
45	2,025	6.708	95	9,025	9.747
46	2,116	6.782	96	9,216	9.798
47	2,209	6.856	97	9,409	9.849
48	2,304	6.928	98	9,604	9.899
49	2,401	7.000	99	9,801	9.950
50	2,500	7.071	100	10,000	10.000

TABLE II Common Logarithms

n	0	1	2	3	4	5	6	7	8	9
1.0	0.0000	0043	0086	0128	0170	0212	0253	0294	0334	0374
1.1	.0414	0453	0492	0531	0569	0607	0645	0682	0719	0755
1.2	.0792	0828	0864	0899	0934	0969	1004	1038	1072	1106
1.3	.1139	1173	1206	1239	1271	1303	1335	1367	1399	1430
1.4	.1461	1492	1523	1553	1584	1614	1644	1673	1703	1732
1.5	.1761	1790	1818	1847	1875	1903	1931	1959	1987	2014
1.6	.2041	2068	2095	2122	2148	2175	2201	2227	2253	2279
1.7	.2304	2330	2355	2380	2405	2430	2455	2480	2504	2529
1.8	.2553	2577	2601	2625	2648	2672	2695	2718	2742	2765
1.9	.2788	2810	2833	2856	2878	2900	2923	2945	2967	2989
2.0	.3010	3032	3054	3075	3096	3118	3139	3160	3181	3201
2.1	.3222	3243	3263	3284	3304	3324	3345	3365	3385	3404
2.2	.3424	3444	3464	3483	3502	3522	3541	3560	3579	3598
2.3	.3617	3636	3655	3674	3692	3711	3729	3747	3766	3784
2.4	.3802	3820	3838	3856	3874	3892	3909	3927	3945	3962
2.5	.3979	3997	4014	4031	4048	4065	4082	4099	4116	4133
2.6	.4150	4166	4183	4200	4216	4232	4249	4265	4281	4298
2.7	.4314	4330	4346	4362	4378	4393	4409	4425	4440	4456
2.8	.4472	4487	4502	4518	4533	4548	4564	4579	4594	4609
2.9	.4624	4639	4654	4669	4683	4698	4713	4728	4742	4757
3.0	.4771	4786	4800	4814	4829	4843	4857	4871	4886	4900
3.1	.4914	4928	4942	4955	4969	4983	4997	5011	5024	5038
3.2	.5051	5065	5079	5092	5105	5119	5132	5145	5159	5172
3.3	.5185	5198	5211	5224	5237	5250	5263	5276	5289	5302
3.4	.5315	5328	5340	5353	5366	5378	5391	5403	5416	5428
3.5	.5441	5453	5465	5478	5490	5502	5514	5527	5539	5551
3.6	.5563	5575	5587	5599	5611	5623	5635	5647	5658	5670
3.7	.5682	5694	5705	5717	5729	5740	5752	5763	5775	5786
3.8	.5798	5809	5821	5832	5843	5855	5866	5877	5888	5899
3.9	.5911	5922	5933	5944	5955	5966	5977	5988	5999	6010
4.0	.6021	6031	6042	6053	6064	6075	6085	6096	6107	6117
4.1	.6128	6138	6149	6160	6170	6180	6191	6201	6212	6222
4.2	.6232	6243	6253	6263	6274	6284	6294	6304	6314	6325
4.3	.6335	6345	6355	6365	6375	6385	6395	6405	6415	6425
4.4	.6435	6444	6454	6464	6474	6484	6493	6503	6513	6522
4.5	.6532	6542	6551	6561	6571	6580	6590	6599	6609	6618
4.6	.6628	6637	6646	6656	6665	6675	6684	6693	6702	6712
4.7	.6721	6730	6739	6749	6758	6767	6776	6785	6794	6803
4.8	.6812	6821	6830	6839	6848	6857	6866	6875	6884	6893
4.9	.6902	6911	6920	6928	6937	6946	6955	6964	6972	6981

TABLE II (Continued)

n	0	1	2	3	4	5	6	7	8	9
5.0	.6990	6998	7007	7016	7024	7033	7042	7050	7059	7067
5.1	.7076	7084	7093	7101	7110	7118	7126	7135	7143	7152
5.2	.7160	7168	7177	7185	7193	7202	7210	7218	7226	7235
5.3	.7243	7251	7259	7267	7275	7284	7292	7300	7308	7316
5.4	.7324	7332	7340	7348	7356	7364	7372	7380	7388	7396
5.5	.7404	7412	7419	7427	7435	7443	7451	7459	7466	7474
5.6	.7482	7490	7497	7505	7513	7520	7528	7536	7543	7551
5.7	.7559	7566	7574	7582	7589	7597	7604	7612	7619	7627
5.8	.7634	7642	7649	7657	7664	7672	7679	7686	7694	7701
5.9	.7709	7716	7723	7731	7738	7745	7752	7760	7767	7774
6.0	.7782	7789	7796	7803	7810	7818	7825	7832	7839	7846
6.1	.7853	7860	7868	7875	7882	7889	7896	7903	7910	7917
6.2	.7924	7931	7938	7945	7952	7959	7966	7973	7980	7987
6.3	.7993	8000	8007	8014	8021	8028	8035	8041	8048	8055
6.4	.8062	8069	8075	8082	8089	8096	8102	8109	8116	8122
6.5	.8129	8136	8142	8149	8156	8162	8169	8176	8182	8189
6.6	.8195	8202	8209	8215	8222	8228	8235	8241	8248	8254
6.7	.8261	8267	8274	8280	8287	8293	8299	8306	8312	8319
6.8	.8325	8331	8338	8344	8351	8357	8363	8370	8376	8382
6.9	.8388	8395	8401	8407	8414	8420	8426	8432	8439	8445
7.0	.8451	8457	8463	8470	8476	8482	8488	8494	8500	8506
7.1	.8513	8519	8525	8531	8537	8543	8549	8555	8561	8567
7.2	.8573	8579	8585	8591	8597	8603	8609	8615	8621	8627
7.3	.8633	8639	8645	8651	8657	8663	8669	8675	8681	8686
7.4	.8692	8698	8704	8710	8716	8722	8727	8733	8739	8745
7.5	.8751	8756	8762	8768	8774	8779	8785	8791	8797	8802
7.6	.8808	8814	8820	8825	8831	8837	8842	8848	8854	8859
7.7	.8865	8871	8876	8882	8887	8893	8899	8904	8910	8915
7.8	.8921	8927	8932	8938	8943	8949	8954	8960	8965	8971
7.9	.8976	8982	8987	8993	8998	9004	9009	9015	9020	9025
8.0	.9031	9036	9042	9047	9053	9058	9063	9069	9074	9079
8.1	.9085	9090	9096	9101	9106	9112	9117	9122	9128	9133
8.2	.9138	9143	9149	9154	9159	9165	9170	9175	9180	9186
8.3	.9191	9196	9201	9206	9212	9217	9222	9227	9232	9238
8.4	.9243	9248	9253	9258	9263	9269	9274	9279	9284	9289
8.5	.9294	9299	9304	9309	9315	9320	9325	9330	9335	9340
8.6	.9345	9350	9355	9360	9365	9370	9375	9380	9385	9390
8.7	.9395	9400	9405	9410	9415	9420	9425	9430	9435	9440
8.8	.9445	9450	9455	9460	9465	9469	9474	9479	9484	9489
8.9	.9494	9499	9504	9509	9513	9518	9523	9528	9533	9538
9.0	.9542	9547	9552	9557	9562	9566	9571	9576	9581	9586
9.1	.9590	9595	9600	9605	9609	9614	9619	9624	9628	9633
9.2	.9638	9643	9647	9652	9657	9661	9666	9671	9675	9680
9.3	.9685	9689	9694	9699	9703	9708	9713	9717	9722	9727
9.4	.9731	9736	9741	9745	9750	9754	9759	9763	9768	9773
9.5	.9777	9782	9786	9791	9795	9800	9805	9809	9814	9818
9.6	.9823	9827	9832	9836	9841	9845	9850	9854	9859	9863
9.7	.9868	9872	9877	9881	9886	9890	9894	9899	9903	9908
9.8	.9912	9917	9921	9926	9930	9934	9939	9943	9948	9952
9.9	.9956	9961	9965	9969	9974	9978	9983	9987	9991	9996

TABLE III Powers of e

x	e^x	e^{-x}
0.00	1.0000	1.00000
0.01	1.0101	0.99005
0.02	1.0202	.98020
0.03	1.0305	.97045
0.04	1.0408	.96079
0.05	1.0513	.95123
0.06	1.0618	.94176
0.07	1.0725	.93239
0.08	1.0833	.92312
0.09	1.0942	.91393
0.10	1.1052	.90484
0.11	1.1163	.89583
0.12	1.1275	.88692
0.13	1.1388	.87809
0.14	1.1503	.86936
0.15	1.1618	.86071
0.16	1.1735	.85214
0.17	1.1853	.84366
0.18	1.1972	.83527
0.19	1.2092	.82696
0.20	1.2214	.81873
0.21	1.2337	.81058
0.22	1.2461	.80252
0.23	1.2586	.79453
0.24	1.2712	.78663
0.25	1.2840	.77880
0.26	1.2969	.77105
0.27	1.3100	.76338
0.28	1.3231	.75578
0.29	1.3364	.74826
0.30	1.3499	.74082
0.31	1.3634	.73345
0.32	1.3771	.72615
0.33	1.3910	.71892
0.34	1.4049	.71177
0.35	1.4191	.70469
0.36	1.4333	.69768
0.37	1.4477	.69073
0.38	1.4623	.68386
0.39	1.4770	.67706
0.40	1.4918	.67032
0.41	1.5068	.66365
0.42	1.5220	.65705
0.43	1.5373	.65051
0.44	1.5527	.64404
0.45	1.5683	.63763
0.46	1.5841	.63128
0.47	1.6000	.62500
0.48	1.6161	.61878
0.49	1.6323	.61263
0.50	1.6487	.60653

TABLE III (Continued)

x	e^x	e^{-x}
0.51	1.6653	.60050
0.52	1.6820	.59452
0.53	1.6989	.58860
0.54	1.7160	.58275
0.55	1.7333	.57695
0.56	1.7507	.57121
0.57	1.7683	.56553
0.58	1.7860	.55990
0.59	1.8040	.55433
0.60	1.8221	.54881
0.61	1.8404	.54335
0.62	1.8589	.53794
0.63	1.8776	.53259
0.64	1.8965	.52729
0.65	1.9155	.52205
0.66	1.9348	.51685
0.67	1.9542	.51171
0.68	1.9739	.50662
0.69	1.9937	.50158
0.70	2.0138	.49659
0.71	2.0340	.49164
0.72	2.0544	.48675
0.73	2.0751	.48191
0.74	2.0959	.47711
0.75	2.1170	.47237
0.76	2.1383	.46767
0.77	2.1598	.46301
0.78	2.1815	.45841
0.79	2.2034	.45384
0.80	2.2255	.44933
0.81	2.2479	.44486
0.82	2.2705	.44043
0.83	2.2933	.43605
0.84	2.3164	.43171
0.85	2.3396	.42741
0.86	2.3632	.42316
0.87	2.3869	.41895
0.88	2.4109	.41478
0.89	2.4351	.41066
0.90	2.4596	.40657
0.91	2.4843	.40252
0.92	2.5093	.39852
0.93	2.5345	.39455
0.94	2.5600	.39063
0.95	2.5857	.38674
0.96	2.6117	.38289
0.97	2.6379	.37908
0.98	2.6645	.37531
0.99	2.6912	.37158
1.00	2.7183	.36788

TABLE IV Natural Logarithms

x	*ln x*	*x*	*ln x*
0.1	−2.3026	5.1	1.6292
0.2	−1.6094	5.2	1.6487
0.3	−1.2040	5.3	1.6677
0.4	−0.9163	5.4	1.6864
0.5	−0.6931	5.5	1.7047
0.6	−0.5108	5.6	1.7228
0.7	−0.3567	5.7	1.7405
0.8	−0.2231	5.8	1.7579
0.9	−0.1054	5.9	1.7750
1.0	0.0000	6.0	1.7918
1.1	0.0953	6.1	1.8083
1.2	0.1823	6.2	1.8245
1.3	0.2624	6.3	1.8405
1.4	0.3365	6.4	1.8563
1.5	0.4055	6.5	1.8718
1.6	0.4700	6.6	1.8871
1.7	0.5306	6.7	1.9021
1.8	0.5878	6.8	1.9169
1.9	0.6419	6.9	1.9315
2.0	0.6931	7.0	1.9459
2.1	0.7419	7.1	1.9601
2.2	0.7885	7.2	1.9741
2.3	0.8329	7.3	1.9879
2.4	0.8755	7.4	2.0015
2.5	0.9163	7.5	2.0149
2.6	0.9555	7.6	2.0281
2.7	0.9933	7.7	2.0412
2.8	1.0296	7.8	2.0541
2.9	1.0647	7.9	2.0669
3.0	1.0986	8.0	2.0794
3.1	1.1314	8.1	2.0919
3.2	1.1632	8.2	2.1041
3.3	1.1939	8.3	2.1163
3.4	1.2238	8.4	2.1282
3.5	1.2528	8.5	2.1401
3.6	1.2809	8.6	2.1518
3.7	1.3083	8.7	2.1633
3.8	1.3350	8.8	2.1748
3.9	1.3610	8.9	2.1861
4.0	1.3863	9.0	2.1972
4.1	1.4110	9.1	2.2083
4.2	1.4351	9.2	2.2192
4.3	1.4586	9.3	2.2300
4.4	1.4816	9.4	2.2407
4.5	1.5041	9.5	2.2513
4.6	1.5261	9.6	2.2618
4.7	1.5476	9.7	2.2721
4.8	1.5686	9.8	2.2824
4.9	1.5892	9.9	2.2925
5.0	1.6094	10	2.3026

Answers to Odd-numbered Exercises

Section 1

1. *Natural numbers:* 17, 6. *Integers:* 17, $-6/\sqrt{4}$ $(=-3)$, 6. *Rational numbers:* 17, -3.14, $-16\frac{2}{3}$, $4.\overline{07}$, $-6/\sqrt{4}$, 6. *Irrational numbers:* π, $\sqrt{10}$.

3. *Natural numbers:* $\sqrt{64}$ $(=8)$. *Integers:* $0\sqrt{2}$ $(=0)$, $\sqrt{64}$, $-15/\sqrt{9}$ $(=-5)$. *Rational numbers:* $0\sqrt{2}$, $4\frac{1}{7}$, $-13.\overline{5}$, $\sqrt{64}$, -2.81, $-15/\sqrt{9}$. *Irrational numbers:* $\sqrt{3}$, 2π.

5. -1.9375 **7.** $-0.\overline{384615}$ **9.** -0.035

11.

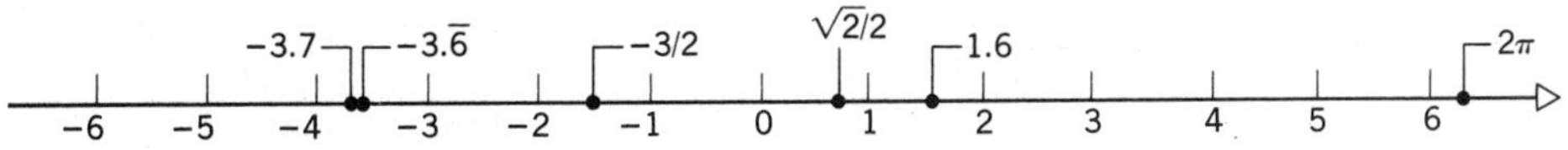

13. $<$	**15.** $<$	**17.** $<$	**19.** $>$	**21.** $<$	**23.** 35
25. -9	**27.** -5	**29.** -15	**31.** -4	**33.** -3	**35.** 0
37. 0	**39.** -0.3	**41.** 0.45	**43.** -24	**45.** -2	**47.** 23
49. -49.4	**51.** -3.706	**53.** $2a-2b$	**55.** $-12b$	**57.** $5a+b-2c$	**59.** $-8b$

61. (b) is true. Any choice of a, b, and c with $c \neq 0$ will show (a) false.

63. (a) is true. The choice $a = b = \sqrt{2}$ will show (b) false.

65. (b) is true. Any choice with $r < 0$ will show (a) false.

Section 2

1. $2 \cdot 3^2$	**3.** $2 \cdot 3 \cdot 5$	**5.** 17	**7.** 31	**9.** 83
11. 30	**13.** 210	**15.** 1400	**17.** $-\frac{35}{6}$	**19.** 18
21. $\frac{1}{6}$	**23.** $\frac{1}{15}$	**25.** $\frac{5}{18}$	**27.** $-\frac{59}{24}$	**29.** $\frac{15}{2}$
31. $\frac{14}{9}$	**33.** $-\frac{4}{3}$	**35.** $-\frac{3}{8}$	**37.** $\frac{172}{75}$	**39.** $\frac{21}{10}$
41. $\frac{3}{4}$	**43.** $(b-a)/ab$	**45.** $(3a+2b)/6$	**47.** $(2x-3y)/x^2y^2$	**49.** rs
51. $(x^2+y^2)/2$	**53.** $-3/a$	**55.** $r-1$	**57.** $1/(1-a)$	**59.** $a/(2a-1)$

61. $a = b = 1$ will do. **63.** $a = b = 1$ and $c = 2$ will do.

65. $a = 2$ and $b = c = 1$ will do.

Section 3

1. $-\frac{8}{81}$	**3.** $-\frac{49}{125}$	**5.** 100	**7.** $\frac{1}{64}$
9. 729	**11.** -25	**13.** a^{11}	**15.** x^{45}
17. $-27v^6$	**19.** $9/b^6$	**21.** $64x^6/y^3$	**23.** $-x^3y^{10}$
25. $-3b$	**27.** $-50t$	**29.** $9x^5y^3$	**31.** $10a^7$
33. $4b^8/3a^{10}$	**35.** u^n/v^{7n}	**37.** $3^ny^{4n}/x^{2n}$	**39.** a^nb^{4n}
41. 1.454×10^3	**43.** 2.9028×10^4	**45.** 2.52×10^{13}	**47.** 6.7×10^{-4}
49. 0.356	**51.** 20,000	**53.** 20	**55.** 9

57. 11 **59.** 4.3589 **61.** 0.6164 **63.** 360.6
65. 0.06 **67.** $2\sqrt{3}$ **69.** $6\sqrt{2}$ **71.** $5a^2\sqrt{3b}$
73. $7b^2\sqrt{2abc}$ **75.** $6mn^4\sqrt{10}$

Section 4

1. $4x^2 - 2x - 2$ **3.** $t^2 + 3$ **5.** $2x^3 + 2x^2 - 2x - 1$
7. $-u^3v^4$ **9.** $-4x^2yz^3$ **11.** $u^4 - \frac{3}{2}u^3$
13. $3x^3 + 2x^2 - 7x + 2$ **15.** $4y^3 - 5y^2 - 11y + 3$ **17.** $0.1x - 0.4$
19. $2xy - 5y$ **21.** $mn + m - 2n$ **23.** $-dm$
25. $2x^2 + 3xy + y^2$ **27.** $3u^2 + 10uv + 3v^2$ **29.** $1.5a^2 + ab - 0.5b^2$
31. $\frac{1}{4}u^2 - \frac{1}{9}v^2$ **33.** $m^2 - \frac{2}{3}mn + \frac{1}{9}n^2$ **35.** $x^2 - 1$
37. $t^2 - 2tc + c^2$ **39.** $a^2 + 4ab + 4b^2$ **41.** $(u + 6)^2$
43. $(t - 3)^2$ **45.** $(m + 7)^2$ **47.** $(t + \sqrt{3})(t - \sqrt{3})$
49. $(2x + 1)^2$ **51.** $(\sqrt{2}y + 1)^2$ **53.** $(3y - 1)^2$
55. $(3z + 0.1)(3z - 0.1)$ **57.** $(w + 0.7)^2$ **59.** $(z + 1)(z + 2)$
61. $(2y - 1)(y + 1)$ **63.** $(3w + 1)(w + 1)$ **65.** $2(y + 1)(2y - 3)$
67. $(3v - 2)(2v + 3)$ **69.** $(4n - 1)(n - 3)$ **71.** $\{x: \ x \geq 4\}$
73. $\{w: \ w \geq -5\}$ **75.** $\{z: \ z \geq 0 \text{ and } z \neq 2\}$

Section 5

1. Yes **3.** Yes **5.** No **7.** No **9.** Yes **11.** 7
13. $\frac{25}{14}$ **15.** $\frac{1}{12}$ **17.** -6 **19.** $\frac{6}{5}$ **21.** $-\frac{1}{2}$ **23.** -2
25. No solution. **27.** $\frac{1}{5}$
29. (a) $y = (x - 1)/2$ (b) $x = 2y + 1$
31. (a) $y = (6 - 3x)/4$ (b) $x = (6 - 4y)/3$
33. (a) $y = (8x - 5)/3$ (b) $x = (3y + 5)/8$
35. $K = R^3/T^2$ **37.** $t = (v - v_0)/g$ **39.** $h = (T - 2\pi r^2)/2\pi r$
41. $h = 2V/(a + b)$ **43.** $r = r_1r_2/(r_1 + r_2)$ **45.** $K = Nr_m/(r_m - r)$

Section 6

1. $>$ **3.** $>$ **5.** $<$ **7.** $>$ **9.** $>$ **11.** $>$
13. $\{x: \ x > -2\}$ **15.** $\{z: \ z < \frac{1}{6}\}$ **17.** $\{u: \ u > -7\}$
19. $\{v: \ v \geq -0.75\}$ **21.** $\{y: \ y \geq 3.5\}$

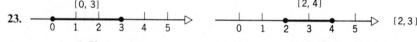

23.
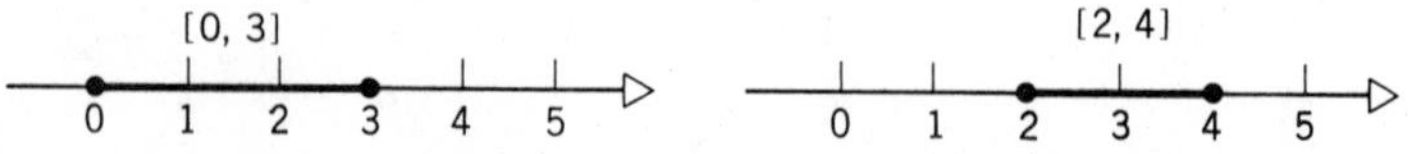

[2, 3]

25.
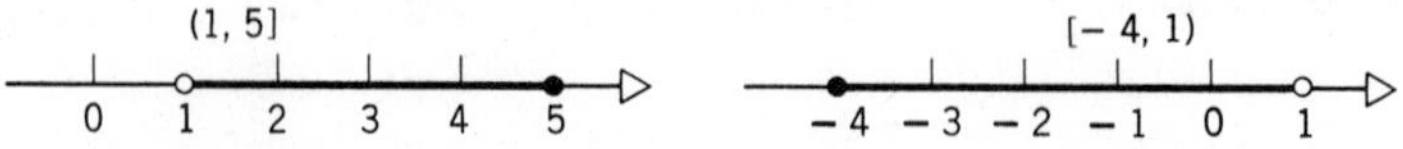

ϕ

27.
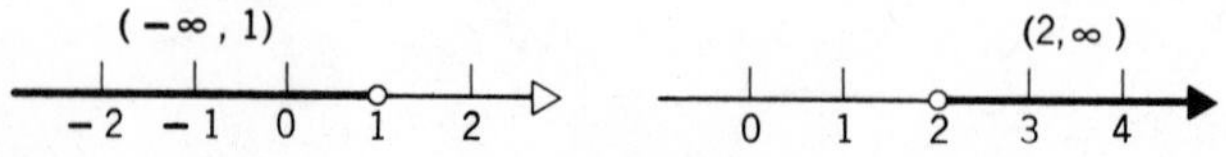

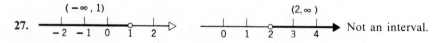

Not an interval.

29. 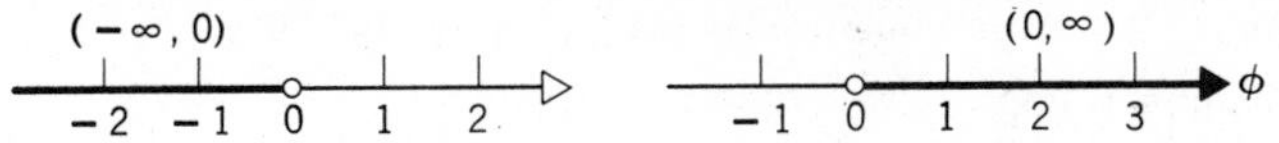

31. 5 **33.** -2 **35.** 3 **37.** 10 **39.** -10 **41.** 5

43. 4 **45.** -1 **47.** 1.9 **49.** $\pi/2$ **51.** $5/(2\pi)$

53. $(-\infty, -7) \cup (7, \infty)$ **55.** $(-\infty, -\frac{1}{2}) \cup (\frac{1}{2}, \infty)$ **57.** $(-\infty, -2) \cup (2, \infty)$

59. $(-\infty, -2] \cup [\frac{4}{3}, \infty)$ **61.** $(-\infty, -1] \cup [5, \infty)$ **63.** $(-\infty, 1) \cup (2, \infty)$

65. $1 \le 5$ **67.** $1.1 \le 3.5$ **69.** $\frac{7}{12} \le \frac{7}{12}$

71. $a < b$, $0 < b - a$, $0 < (b - c) - (a - c)$, $a - c < b - c$.

73. With $b = 0$, (6.6) becomes: If $a < 0$ and $c < 0$, then $ac > 0$.

75. Use the suggestion.

77. If $a > b > 0$, then $ab > 0$, so $a/ab > b/ab$, $1/b > 1/a$.

CHAPTER I REVIEW

1. *Natural numbers:* $\sqrt{36}$. *Integers:* $0/\sqrt{2}$, $-12/\sqrt{9}$, $\sqrt{36}$, $-\sqrt{2}\sqrt{2}$. *Rational numbers:* $20.\bar{1}$, $0/\sqrt{2}$, $-12/\sqrt{9}$, $\sqrt{36}$, -3.14, $-\sqrt{2}\sqrt{2}$. *Irrational numbers:* 3π, $\sqrt{5}$.

2. See Section 1. **3.** 0.2 **4.** -0.88 **5.** -0.4

6. 0.4 **7.** 0.66 **8.** -0.43 **9.** -46

10. -21 **11.** $2a - 14b$ **12.** $3a - 4b$ **13.** $\frac{8}{15}$

14. $-\frac{2}{21}$ **15.** 4 **16.** $\frac{1}{4}$ **17.** $(x + 1)/(x - 1)$

18. $(3y + 2)/(y + 2)$ **19.** $-1/a$ **20.** $2bc$ **21.** 49

22. $-\frac{1}{1000}$ **23.** $-11.\bar{1}$ **24.** -125 **25.** $x^3/(27y^3)$

26. $32a^{10}/b^5$ **27.** $1/(36a^2)$ **28.** $y^8/256$ **29.** $1000a^6b^3$

30. $125x^9y^{15}$ **31.** a^{6n}/b^{9n} **32.** $2^n u^{2n} v^{4n}/9^n$ **33.** 7.72×10^8

34. 4.8×10^{-6} **35.** 547.72 **36.** 0.017321 **37.** $2x^3 + x^2 + x - 1$

38. $y^3 - y^2 - 8y + 6$ **39.** $a^2 - \frac{2}{3}a + \frac{1}{9}$ **40.** $b^2 + 0.4b + 0.04$

41. $(t + 3)(t - 3)$ **42.** $(u - 11)^2$ **43.** $(v + 6)^2$

44. $(2w + 1)(2w - 1)$ **45.** $(5x + 2)(2x - 1)$ **46.** $(3y + 2)(3y - 1)$

47. $\{x\colon\ x \ne -\frac{1}{2}\}$ **48.** $\{x\colon\ x \ge -6\}$ **49.** $\frac{13}{11}$

50. $-\frac{9}{2}$ **51.** $(3y + 4)/4$ **52.** $(y - 9)/4$ **53.** -4

54. No solution. **55.** $\{x\colon\ x > -9\}$ **56.** $\{y\colon\ y \ge \frac{1}{15}\}$ **57.** 4

58. $-\frac{4}{3}$ **59.** $[-1, 2)$ **60.** $(-\pi, \sqrt{3}]$ **61.** $(-2, 7]$

62. $(-\infty, 6]$ **63.** 1.99 **64.** 10 **65.** $(-5, 7)$

66. $(-\infty, -2] \cup [\frac{1}{2}, \infty)$ **67.** $(-\infty, -2] \cup [1.2, \infty)$ **68.** $(-\frac{4}{3}, 2)$

Section 7

Other choices for the variables are possible in 1–15.

1. $S = 4\pi r^2$ **3.** $V = \frac{1}{3}Bh$ **5.** $C = F + V$

7. $M = (x + y)/2$ **9.** $\sqrt{xy} \le (x + y)/2$ **11.** $f = 1/p$

13. $||a| - |b|| \le |a - b|$ **15.** $d = \sqrt{l^2 + w^2 + h^2}$ **17.** 24

19. $l = 14.5$ ft, $w = 10.5$ ft **21.** $l = 110$ m, $w = 75$ m **23.** 15, 16, 17

25. Approximately 75,000. **27.** Approximately 55 million tons.
29. 6.34 m/sec **31.** 360 mph **33.** 8.5 m **35.** $W_3 = \frac{15}{16} W_1$
37. All sets with the first integer greater than 10. **39.** $x < 50\pi$ m

Section 8

1. (a) 81 (b) 240 (c) 2.4×10^5 (d) 0.16
3. (a) 273.5 (b) 4.724×10^{-7} (c) 0.01358 (d) 628,800
5. 500 dm^2 = 5 m^2, 11.5 yd^2 $\approx$ 9.6 m^2, 105 ft^2 $\approx$ 9.8 m^2, 10 m^2, 500,000 cm^2 = 50 m^2
7. (a) 16,000 lira (b) \$1.25 **9.** (a) 100 francs (b) \$200
11. 4.5 tblsp **13.** 3 m/min **15.** 0.1524 m/sec
17. (a) Approximately 150 million. (b) Approximately 19.6 persons/km^2.
19. 1.76 cm^3 **21.** 271 g **23.** 35 **25.** $133.\overline{3}$ **27.** 6000 **29.** 839 million
31. 12.5% **33.** 5% **35.** \$5000 **37.** \$32 **39.** \$30 **41.** \$1000
43. $1\frac{1}{3}$ ℓ **45.** \$750 at 6% and \$1750 at 8%
47. 46% **49.** \$159.65 **51.** 11,700,000

Section 9

1. 1.25 **3.** 13.5 **5.** 1 : 2,000,000 **7.** 4.8 in.
9. 115 yd **11.** \$39.00 per share **13.** $44.\overline{4}$% **15.** $66.\overline{6}$%
17. $\frac{50}{7}$ ft and $\frac{60}{7}$ ft **19.** 23 ft 4 in. **21.** 1.8 m **23.** 50
25. (a) 6.9 land miles (b) 8.7 nautical miles
27. 23 mi/hr **29.** 1.4 kg **31.** $55.\overline{5}$ in^3 **33.** Increase 9 lb/in.^{2}
35. 0.00056 watts/m^2 **37.** $F' = \frac{9}{4}F$ **39.** $F' = 12F$
41. $a : b = c : d$ iff $a/b = c/d$ iff $ad = bc$ iff $d/b = c/a$ iff $d : b = c : a$
43. 52.7% **45.** 50.4%

CHAPTER II REVIEW

1. $\frac{4}{5}$ **2.** 20, 21, 22 **3.** It's 3.5% more **4.** \$285 billion
5. 37.5 mi/hr **6.** 12.42 mi/hr **7.** 50 ft/min **8.** 315 ℓ/m
9. 77.2 g **10.** 3.69 cm^3 **11.** 1.35 **12.** 147
13. 208.3% **14.** 6400 **15.** \$210 **16.** \$150
17. $1\frac{1}{3}$ℓ **18.** \$350 at 8% and \$850 at 10%
19. 300 km **20.** $4.\overline{4}$ cm **21.** 12.5 cm and 15 cm **22.** 12 ft
23. $2\frac{1}{3}$ ounces **24.** $\frac{1}{16}$ **25.** 1.4 cm **26.** 22 lb/in.^{2}

Section 10

1. ± 6 **3.** ± 10 **5.** ± 1.5 **7.** No real solution.
9. No real solution. **11.** $\pm 3\sqrt{3}/A$ **13.** 0 **15.** $\pm\frac{2}{7}$
17. 0, 3 **19.** 1 **21.** 3 **23.** -1, 4 **25.** 0, 3
27. 0.5 **29.** 5, 6 **31.** $-1, \frac{1}{5}$ **33.** $\frac{3}{5}$, 1 **35.** $-3 \pm \sqrt{10}$

37. $-1, \frac{1}{3}$ **39.** $(1 \pm \sqrt{6})/5$ **41.** $(-1 \pm \sqrt{37})/6$ **43.** $\sqrt{30}$ m **45.** $\sqrt{5/\pi}$ cm
47. $\frac{1}{2}$ **49.** 3, 4, 5 **51.** 5 in. **53.** ± 0.266

Section 11

1. Two. **3.** None. **5.** One. **7.** None.
9. One. **11.** Two. **13.** $(1 \pm \sqrt{17})/4$ **15.** $-\frac{4}{5}, 1$
17. $1, \frac{10}{3}$ **19.** 0.1, 0.2 **21.** $-0.1, 0.4$
23. $(-q \pm \sqrt{q^2 + 4q})/2 \quad (q^2 + 4q \geq 0)$
25. 0, 7 **27.** $-\frac{1}{4}, 0$ **29.** $\pm\sqrt{3}$ **31.** $1 \pm \sqrt{3}$
33. $(-1 \pm \sqrt{5})/2$ **35.** ± 0.5 **37.** $-2, 1$ **39.** $-3, -2$
41. $\pm\sqrt{R^2 - (V/k)}$ **43.** $-14, -12$ and 12, 14 **45.** $(-1 + \sqrt{13})/2$
47. Replace x in (11.1) by the given expression and simplify.
49. $[(-b + \sqrt{b^2 - 4ac})/2a] + [(-b - \sqrt{b^2 - 4ac})/2a] = -2b/2a = -b/a$
51. $[(-b + \sqrt{b^2 - 4ac})/2a) - (-b - \sqrt{b^2 - 4ac})/2a]^2 =$
$(2\sqrt{b^2 - 4ac}/2a)^2 = |b^2 - 4ac|/a$

Section 12

1. 7 **3.** $-\frac{2}{3}$ **5.** 1, 4 **7.** 5
9. 1, 9 **11.** $\frac{1}{4}$ **13.** 2 **15.** $-\frac{3}{4}$
17. 2 **19.** -3 **21.** No solution. **23.** No solution.
25. $-\frac{1}{2}, -\frac{1}{5}$ **27.** $-\frac{6}{5}, \frac{24}{5}$ **29.** $\pm\sqrt{3}$ **31.** $A = \pi r^2$
33. $g = 4\pi^2 l/T^2$ **35.** $h = \sqrt{S^2 - \pi^2 r^4}/\pi r$ **37.** $\frac{1}{4}$ **39.** $(1 + \sqrt{5})/2$
41. 7.5 hours

Section 13

1. $11\sqrt{10}$ yd **3.** $s = (2\sqrt{3}/3)h$ **5.** $d = s\sqrt{2}$
7. $\sqrt{83375}/2 \approx 144$ m **9.** $\sqrt{47821} \approx 219$ m **11.** $\sqrt{65700} \approx 256$ mi
13. No. **15.** 80th **17.** $t = \sqrt{s}/4$
19. 400 ft, 3.75 sec **21.** $3\sqrt{70}/4 \approx 6.3$ sec later
23. 36 ft/sec **25.** 24 sec **27.** $\frac{60}{7}$ sec

CHAPTER III REVIEW

1. $-5, 2$ **2.** $-\frac{3}{2}, -1$ **3.** $(-3 \pm \sqrt{17})/4$
4. $(1 \pm \sqrt{13})/6$ **5.** $\pm\frac{11}{3}$ **6.** $(-p \pm \sqrt{p^2 - 4})/2$
7. $(3 \pm \sqrt{2})/7$ **8.** $\pm\frac{7}{5}$ **9.** Two.
10. Two. **11.** None. **12.** One.
13. $-\frac{3}{2}, -\frac{1}{2}$ **14.** -3 **15.** $\frac{1}{2}$
16. 2, 6 **17.** No real solution. **18.** No solution.
19. $T = 4df^2L^2$ **20.** $K = -1 \pm \sqrt{r + 1}$ **21.** $3 + \sqrt{11}$
22. 10 **23.** $s = \sqrt{h^2 + (x/2)^2}$

24. First, show that the length of OD is $\sqrt{r^2 - (s/2)^2}$.
25. $\sqrt{10}/4 \approx 0.79$ sec **26.** 3 sec
27. 20 min **28.** $13\frac{7}{11}$ sec

Section 14

1. $(x + 6)^2$ **3.** $(z + 1)(z - 1)$ **5.** $z^2(z + 1)^2$
7. $(2z - 1)^2$ **9.** $(3y + 1)^2$ **11.** $(v - 1)(v^2 + v + 1)$
13. $v^2(v - 1)(v - 3)$ **15.** $u^3(u + 1)(u + 6)$
17. $(u - 10v)(u^2 + 10uv + 100v^2)$ **19.** $r(r - 5)(r^2 + 5r + 25)$
21. $t(4 - t)(16 + 4t + t^2)$ **23.** $(t + 1)^2(t - 1)^2$
25. $(t - 0.1)(t^2 + 0.1t + 0.01)$ **27.** $(s + 0.2)(s^2 - 0.2s + 0.04)$
29. $(b^n + 1)(b^n - 1)$ **31.** $(b + 1)^3$
33. $(a - 2)^3$ **35.** $(a + 3b)^3$ **37.** $xy(x - 5y)^2$
39. $rs(r - 3s)^2$ **41.** $(\frac{1}{3} - n)^3$ **43.** $(a - b)(x + y)$
45. $(a + b)(x + y)$ **47.** $(c + d)(u + v)(u - v)$
49. $t^3 + 3t^2 + 3t + 1$ **51.** $v^3 + 15v^2 + 75v + 125$
53. $x^3 + \frac{3}{2}x^2 + \frac{3}{4}x + \frac{1}{8}$ **55.** $\frac{1}{8}x^6 + \frac{3}{4}x^4 + \frac{3}{2}x^2 + 1$
57. $y^3 - 2y^2 + \frac{4}{3}y - \frac{8}{27}$

Section 15

1. $\frac{1}{5}$ **3.** 4 **5.** $2/(y - 3)$
7. $(z^2 + z + 1)/(z + 1)$ **9.** $(z + 1)^2/(z^2 - z + 1)$ **11.** $2/(y^2 + y)$
13. $a^2 - a$ **15.** $c - 1$ **17.** $s^4 + s^2 + 1$
19. $v/(4u^3 - 2u^2v + uv^2)$ **21.** 5 **23.** $3/z$
25. $a^2 - a + 1$ **27.** $(100 + 10c + c^2)/(10 + c)$
29. $(-2hx - h^2)/[x^2(x + h)^2]$ **31.** $x^2(x + 1)(x - 1)$ **33.** $(x + 4)(x + 5)^2$
35. $a^2b^2(a - b)(a + b)$ **37.** $a/(a^2 - b^2)$ **39.** 1
41. $1/[2(c - d)]$ **43.** $x/(x - 2)$ **45.** $-z - 2$
47. $(z^2 + 2z + 4)/(z^2 + 2z)$ **49.** $(z^2 + z + 1)/(z + 1)$ **51.** $2y/(y + 1)$
53. $-2/(cd)$ **55.** $-1/[x(x + h)]$
57. $-(3x^2 + 3hx + h^2)/[(x + h)^3x^3]$

Section 16

1. 2 **3.** 4 **5.** -2 **7.** -5 **9.** -10
11. -0.2 **13.** $-\frac{1}{2}$ **15.** $-\frac{2}{3}$ **17.** 20 **19.** -13
21. $-\frac{1}{12}$ **23.** -1.7 **25.** 0.027 **27.** 0.729 **29.** $\sqrt{3}$
31. 0 **33.** $-7\sqrt{3}$ **35.** 30 **37.** a^2 **39.** c^2
41. $y^3\sqrt[3]{y}$ **43.** $\sqrt{u}$ **45.** $8\sqrt{w}$ **47.** $3s\sqrt{2s}$ **49.** $2\sqrt{7}/7$
51. $\sqrt{3}/3$ **53.** $-1 - \sqrt{3}$ **55.** $(\sqrt{x} + \sqrt{y})/(x - y)$

57. $(\sqrt{a}-1)/(a-1)$ **59.** $1/(7+4\sqrt{3})$ **61.** $-1/(\sqrt{x-h}+\sqrt{x})$
63. $1/(\sqrt{1-x+h}+\sqrt{1-x})$

Section 17

1. 32 **3.** 27 **5.** $\frac{1}{128}$ **7.** 0.0001 **9.** 0.09
11. $-\frac{1}{8}$ **13.** 900 **15.** $\frac{1}{27}$ **17.** 343 **19.** $\frac{1}{512}$
21. 0.001 **23.** $y^{8/3}$ **25.** $\sqrt[12]{a}$ **27.** $1/\sqrt[20]{c}$ **29.** 81
31. $25x^2$ **33.** $8z^6$ **35.** s^5 **37.** $1/(a^{16}b^4)$ **39.** m
41. b^{n+1} **43.** $x-y$ **45.** $c+d$ **47.** $\dfrac{x^2+4x}{2(x+1)^{5/2}}$ **49.** a^n
55. 1.5848932 **57.** 0.30285343 **59.** 0.14777213

Section 18

1. $V=(S/6)^{3/2}$ **3.** $P=4\sqrt{A}$ **5.** $S=\sqrt[3]{36\pi V^2}$
7. $w=(S/k)^{3/2}$ **9.** $(10^6\text{ miles})^3/(\text{Earth years})^2$
11. 0.99 m^2 **13.** 36 million miles **15.** 484 million miles
17. 83.89 Earth years **19.** \$126.68 **21.** \$670.05
23. \$470.95 **25.** 8.24% **27.** 12.55%
29. 10.57% **31.** \$1161.40 **33.** \$598.21
35. \$1194.05 **37.** 8.3% **39.** \$2117.13
41. 150.9

CHAPTER IV REVIEW

1. $x(3x-1)^2$ **2.** $y^2(y+1)(y^2-y+1)$
3. $(5a-2b)(3a+2b)$ **4.** $(c^2+d^2)(c+d)(c-d)$
5. $(m-n)^3$ **6.** $(u+v+2)(u+v-2)$
7. $(x+y)(x^n+y^n)(x^n-y^n)$ **8.** $(x^n-y^n)(x^{2n}+x^ny^n+y^{2n})$
9. $a^3+6a^2b+12ab^2+8b^3$ **10.** $c^6-3c^4d^2+3c^2d^4-d^6$
11. $2m+2$ **12.** $v/(u^2+2uv)$ **13.** $(x+1)/4$ **14.** $4x/(x-1)$
15. $1/(x^2-1)$ **16.** $2/[x(x+1)(x-2)]$ **17.** $-ab$ **18.** $(d+c)/d$
19. $2a^2$ **20.** $2a^2b^3$ **21.** $3x^2\sqrt{y}$ **22.** $x-4y$
23. 64 **24.** $\frac{9}{4}$ **25.** $125b^6/(8a^9)$ **26.** c^{10}
27. $5\sqrt{3}/3$ **28.** $(a^{1/2}+b^{1/2})/(a-b)$ **29.** $x=\sqrt[3]{y^2/2}$ **30.** $a=b\sqrt{b}$
31. $\$2000(1.02)^{12}=\$2536.48$ **32.** $\$5000(1.005)^{30}=\$5807.80$
33. $\$10{,}000(1.02)^{-40}=\$4528.90$ **34.** $\$2000(1.025)^{-16}=\$1347.25$
35. $1.015^4-1\approx 6.14\%$ **36.** $12(\sqrt[12]{1.10}-1)\approx 9.57\%$

Section 19

1.

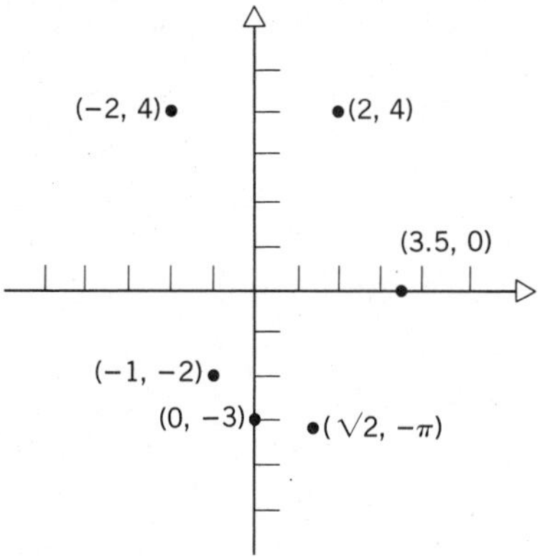

3.

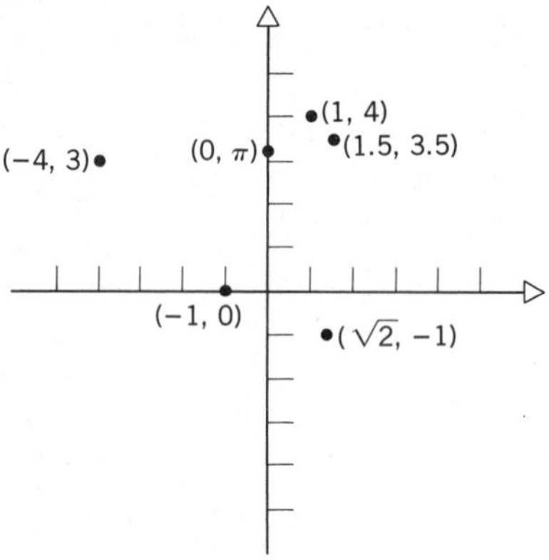

5. (2.5, −4) **7.** (0, −4) **9.** (0, 3.2)

11. (0, 2) and (4, 2) are. **13.** (1, 1) and (−2, 4) are. **15.** None is.

17. Straight line parallel to and 1 unit below the x-axis.

19. Straight line parallel to and 1.5 units below the x-axis.

21. Straight line parallel to and 2 units to the left of the y-axis.

23. Circle centered at the origin with radius 4.

25. Circle centered at (1, 0) with radius $\sqrt{10}$.

27. Circle centered at (−2, −1) with radius $\sqrt{12}$.

29. $y = -3$ **31.** $x = 5$ **33.** $y = 5$

35. $(x + 2)^2 + (y + 3)^2 = 2$ **37.** $x^2 + y^2 = 25$

39. $(x + 3)^2 + (y - 1)^2 = 9$ **41.** $\sqrt{34}$

43. $\sqrt{389}/20$ **45.** $\sqrt{29}/2$

47. Circle centered at (−6, 1) with radius $\sqrt{74}$.

49. The empty set. **51.** The point (10, 0). **53.** The empty set.

55. The points to the left of the y-axis.

57. The points in the second and fourth quadrants.

59. The points outside the circle centered at the origin with radius $\sqrt{5}$.

61. $(\frac{1}{2}, \frac{9}{2})$

63. Show that the midpoints of P_1P_3 and P_2P_4 coincide.

Section 20

1. -3 **3.** $\frac{1}{9}$ **5.** 0

7. $x - 2y + 5 = 0$ **9.** $x - 4y = 0$ **11.** $10x - 4y - 15 = 0$

13. $2x - 5y + 8 = 0$ **15.** $x - y + 3 = 0$ **17.** $6x - y - 5 = 0$

19. $y = -x + 2$ **21.** $y = 2$ **23.** $y = 0$

25. $m = -2, b = 3$ **27.** $m = \frac{1}{2}, b = -\frac{3}{2}$ **29.** $m = 1, b = 0$

31.

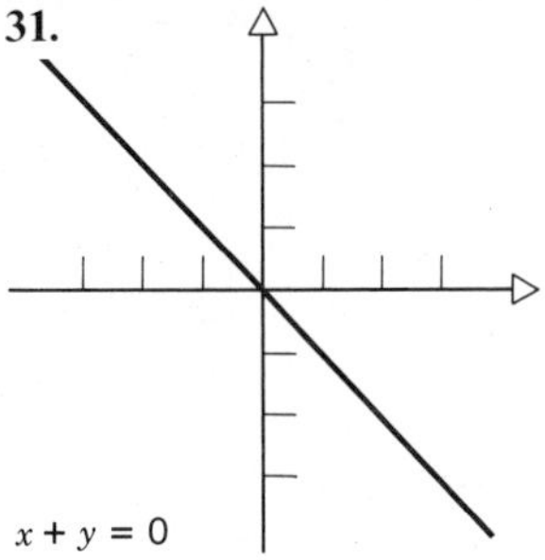

33.

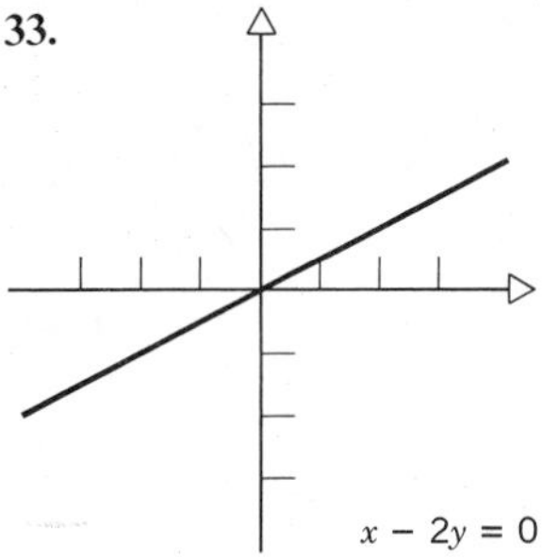

35.

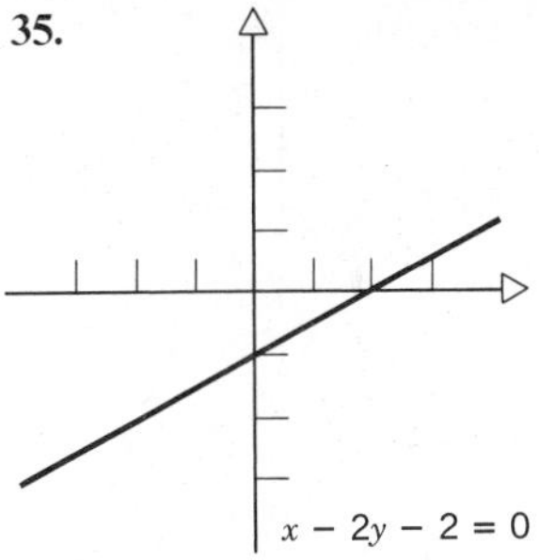

37.

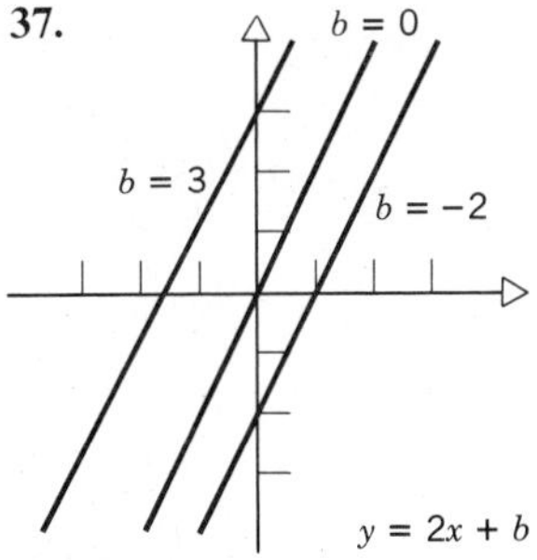

39.

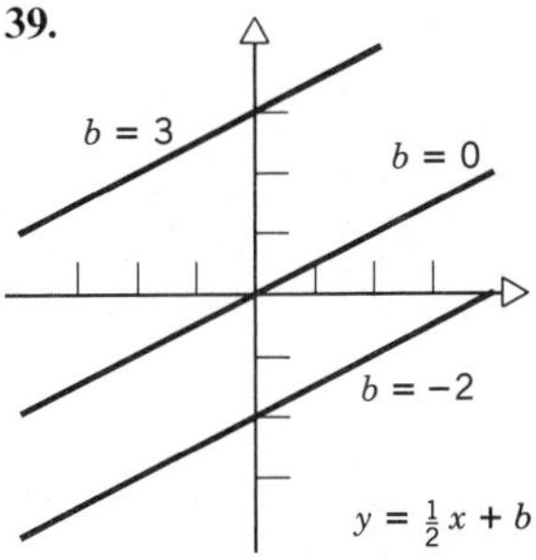

41.

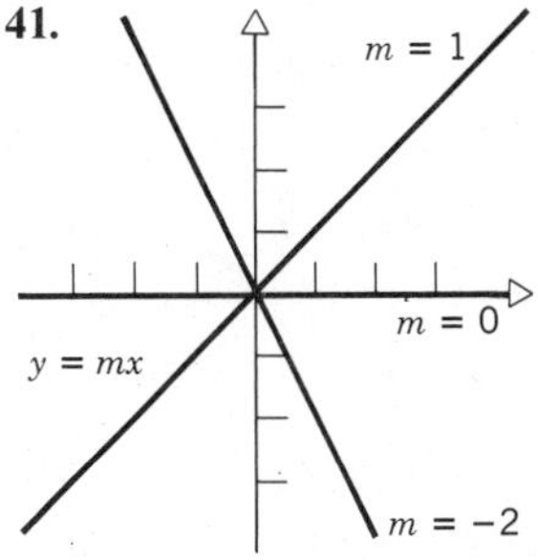

43. $y = \frac{2}{3}x$ **45.** $y = x + 2$

47. Use $(x_1, y_1) = (a, 0)$ and $(x_2, y_2) = (0, b)$ in Equation (20.2) to get m. Then use Equation (20.4)

49. Substitute $x = 1$, $y = 1$ in Equation (20.5), and then compare m and b.

51. $2x - y - 7 = 0$

Section 21

1.

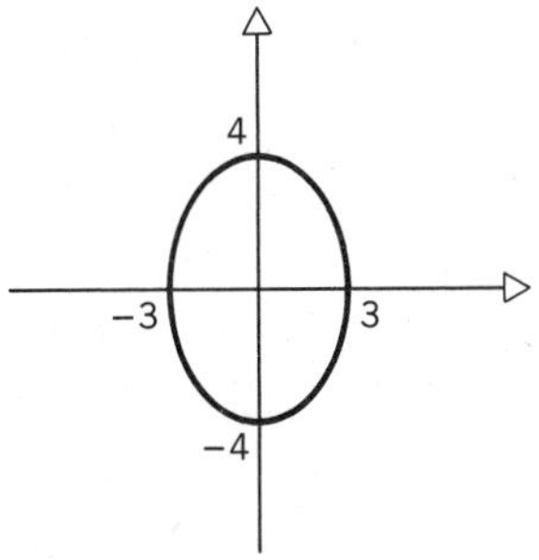

3.

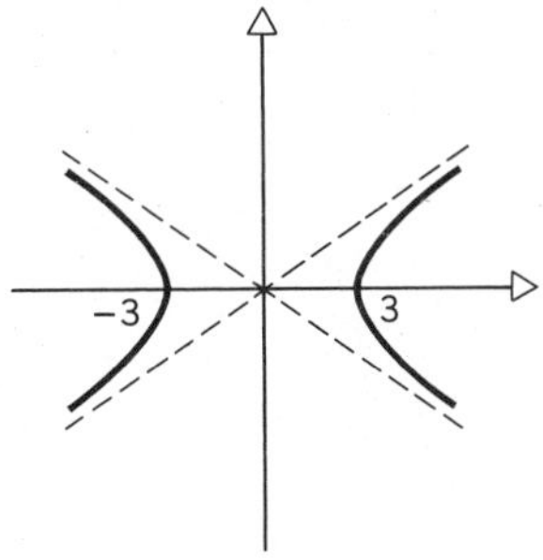

5.

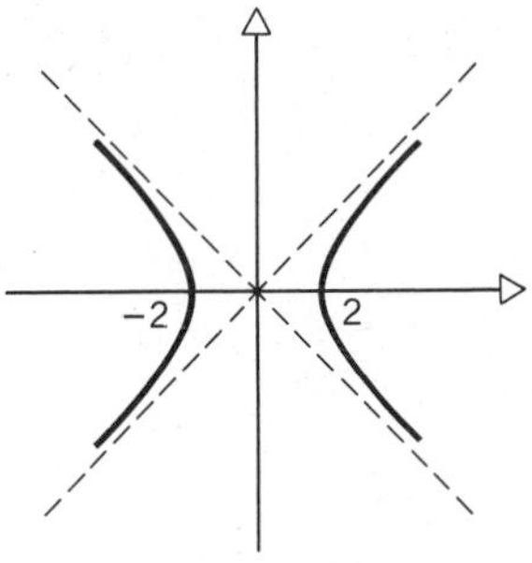

7.

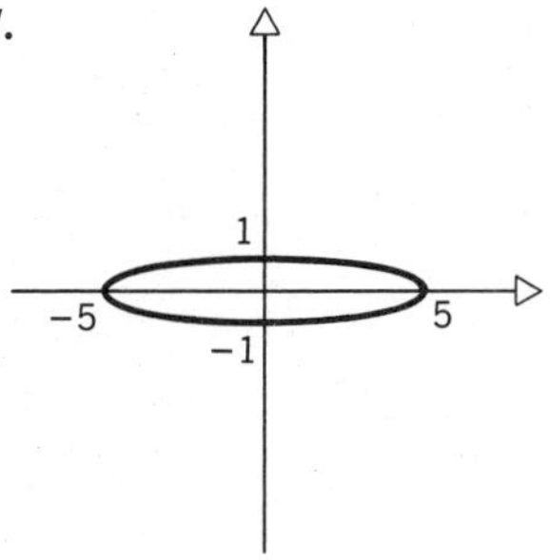

9.

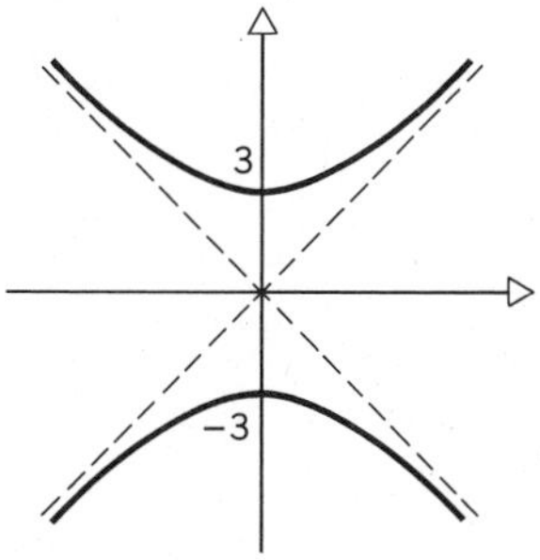

11.

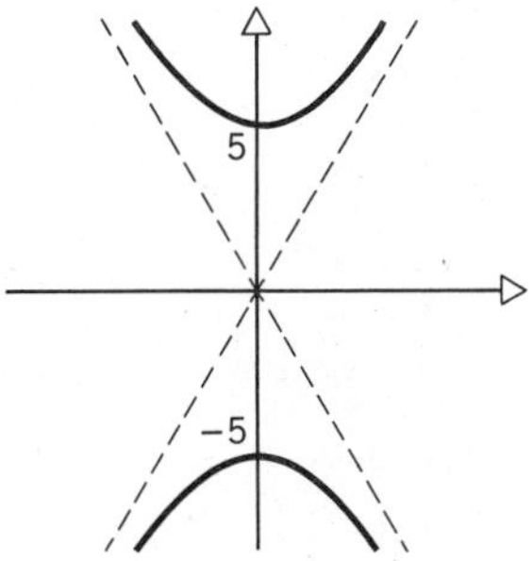

13.

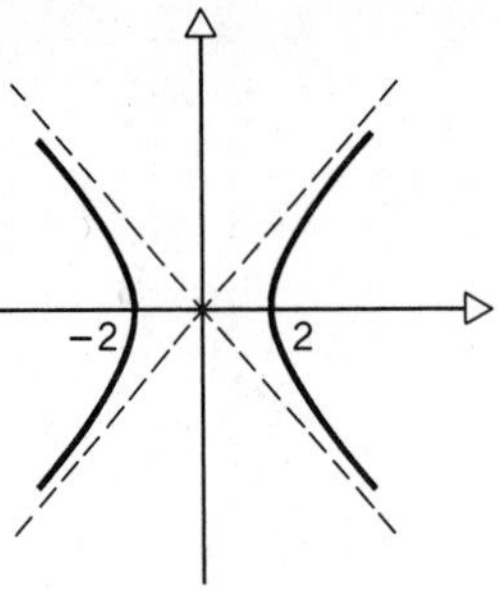

15.

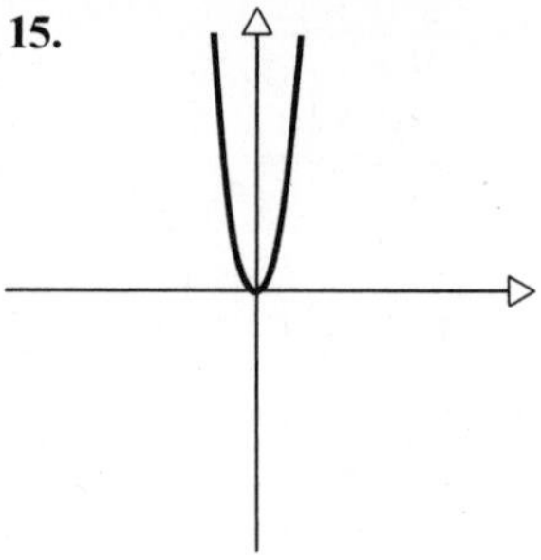

17.

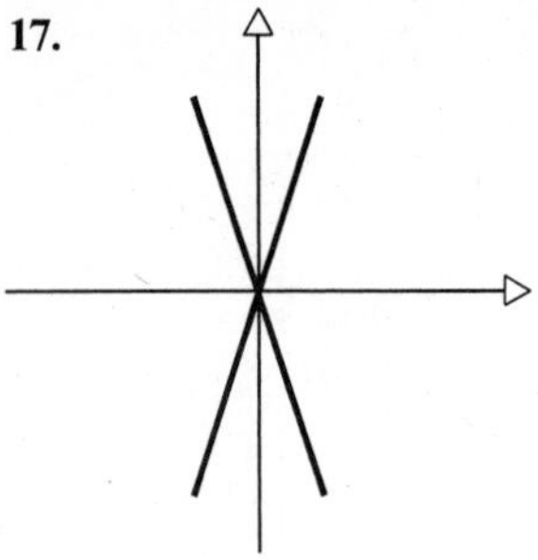

19.

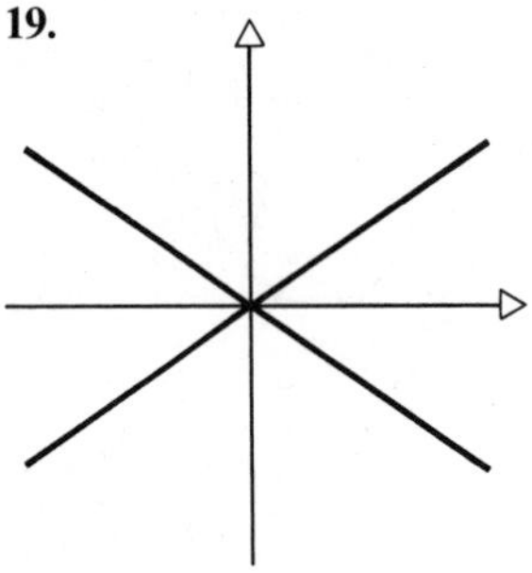

21.

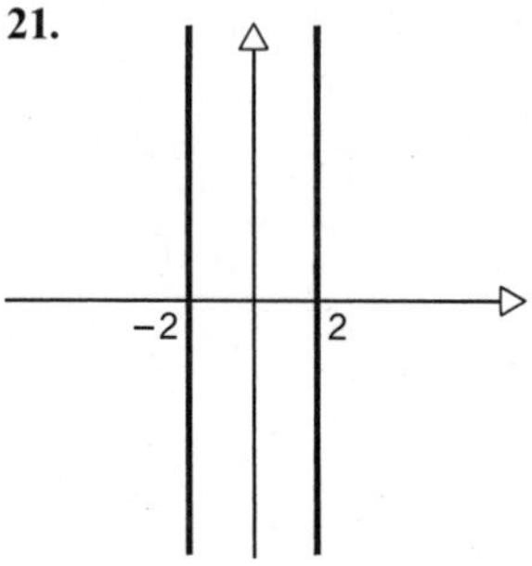

23.

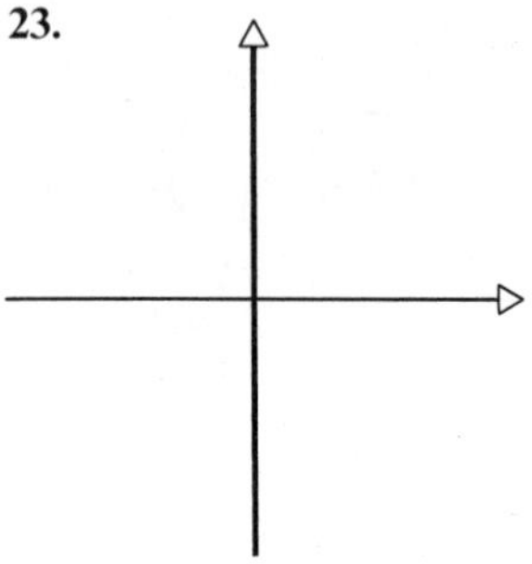

25.

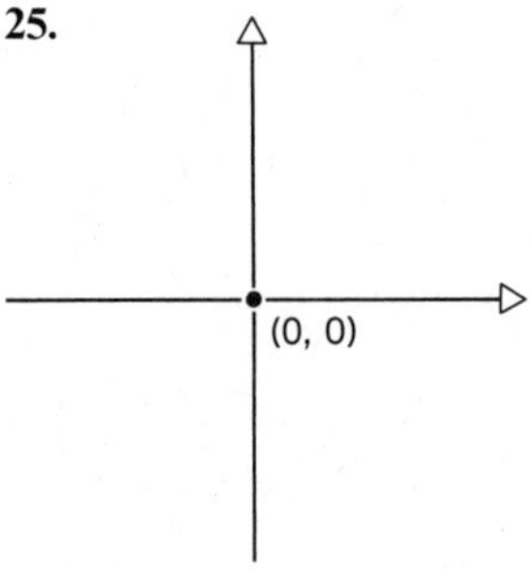

27.

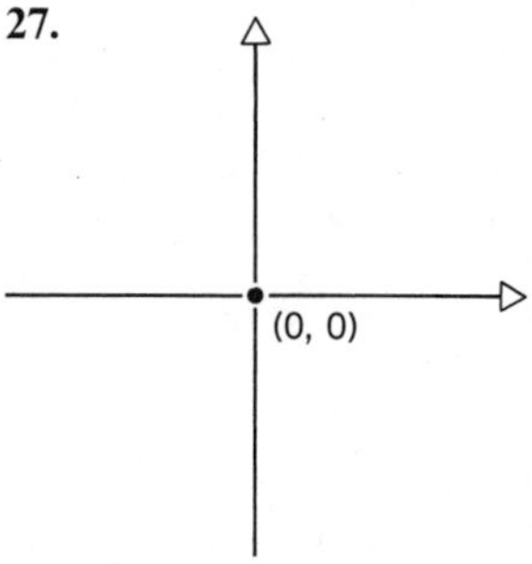

29.

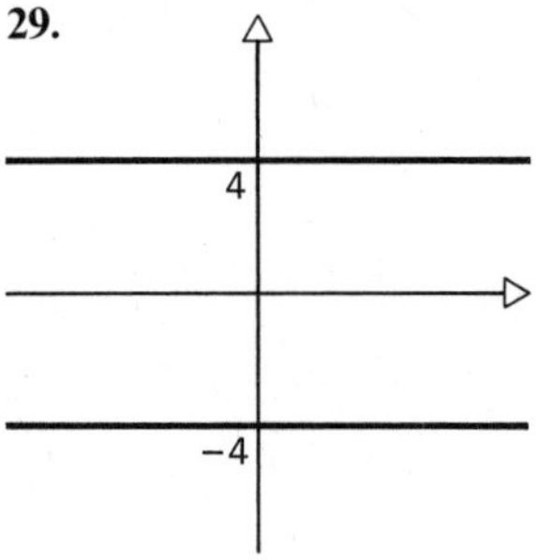

31.

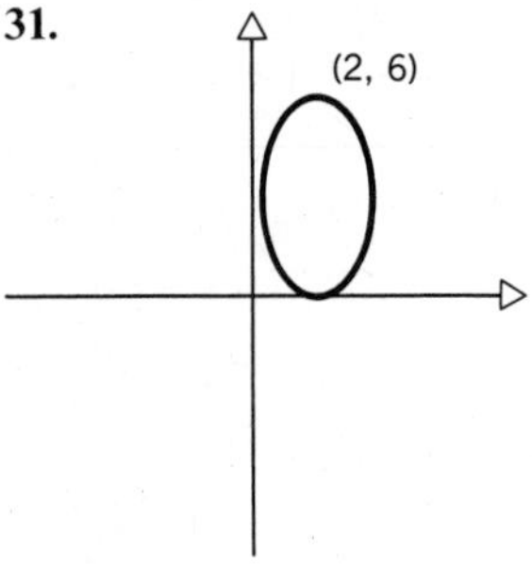

33.

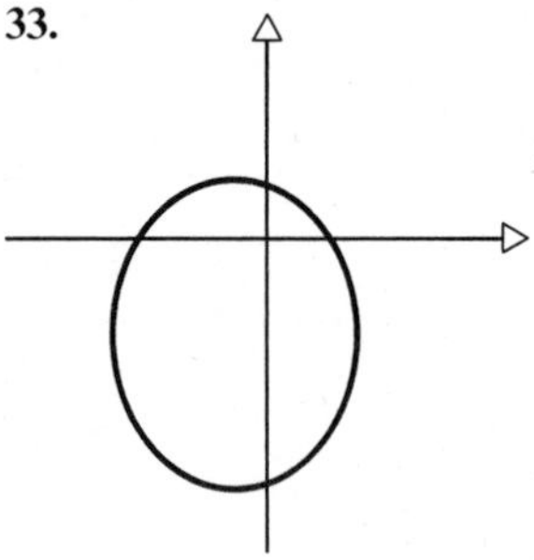

35.

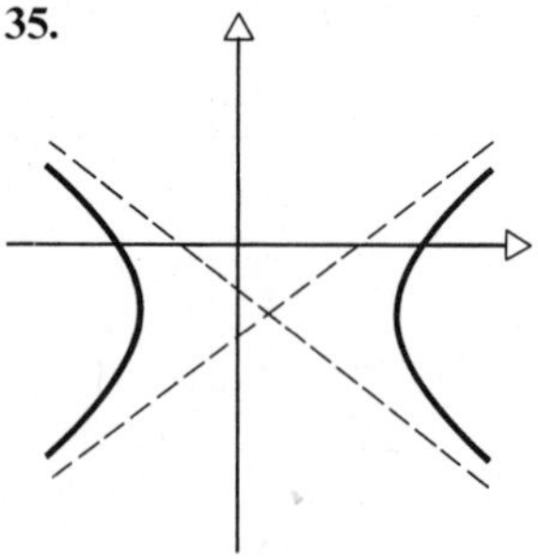

CHAPTER V REVIEW

1. $2\sqrt{5}$ **2.** $\sqrt{29}$ **3.** $7x + 5y - 3 = 0$

4. $2x - 3y + 9 - 0$ **5.** $y = -4x + 2$ **6.** $x - y - 3 = 0$

7.

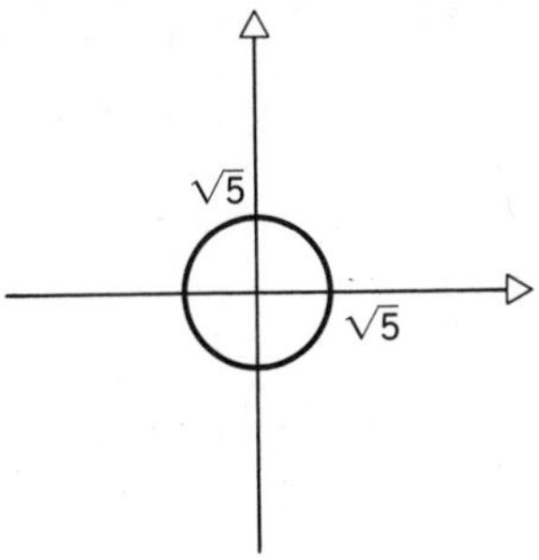

8.

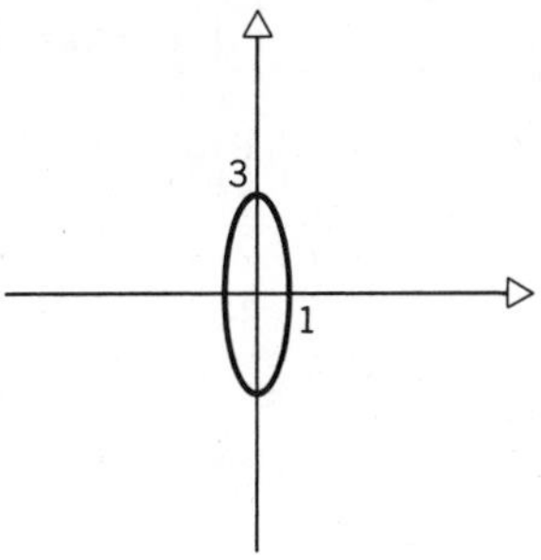

9.

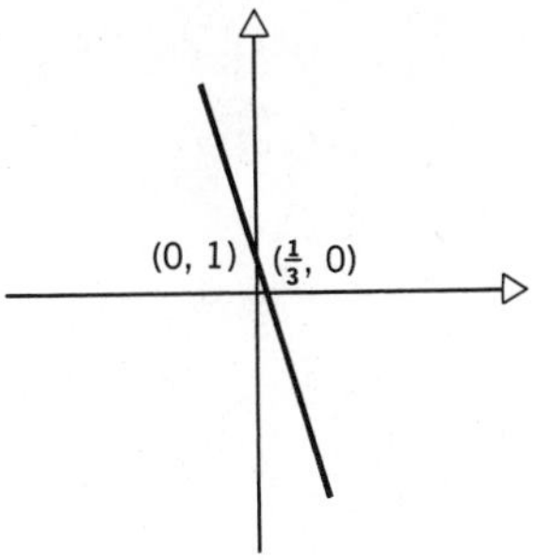

10.

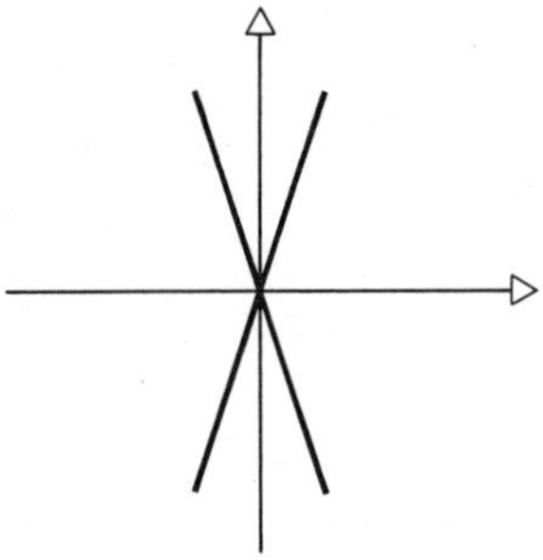

11.

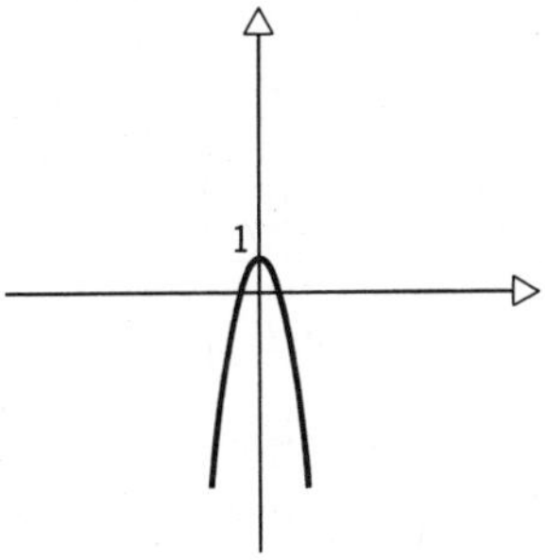

12.

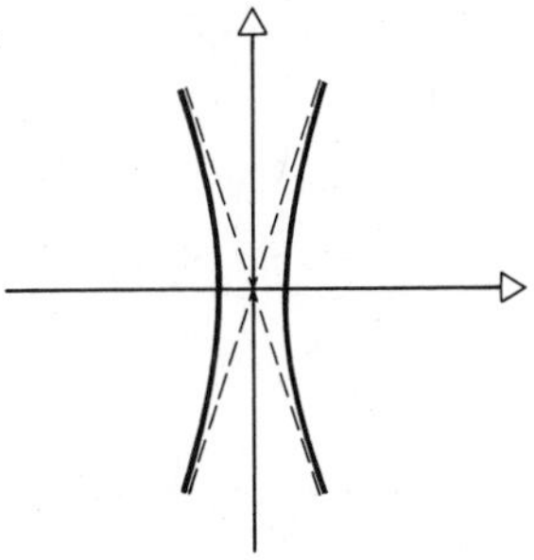

13.

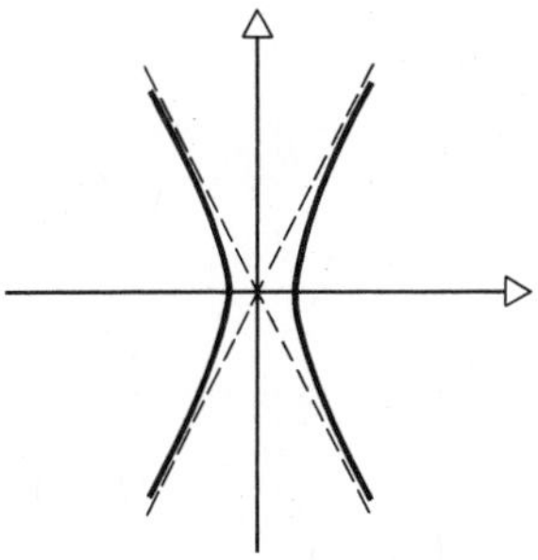

14.

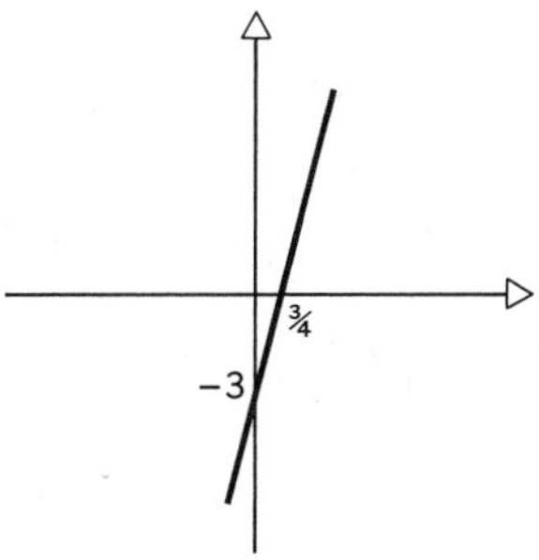

15.

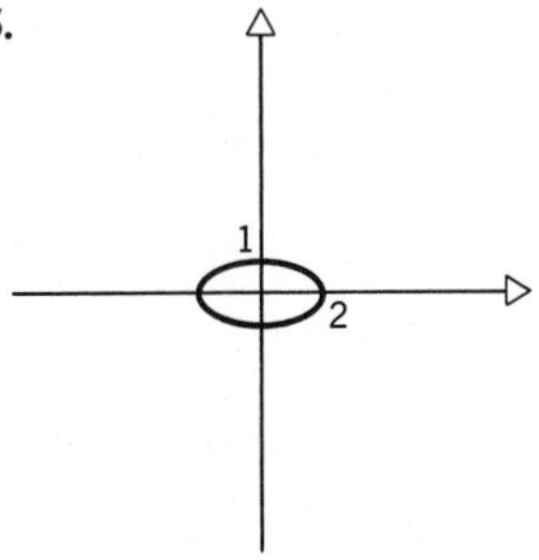

16.

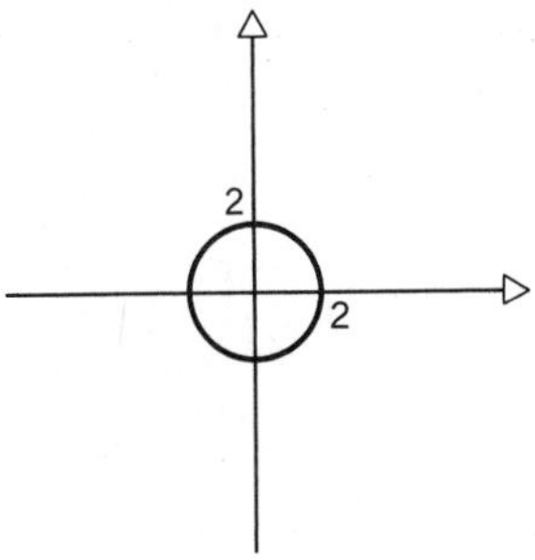

17.

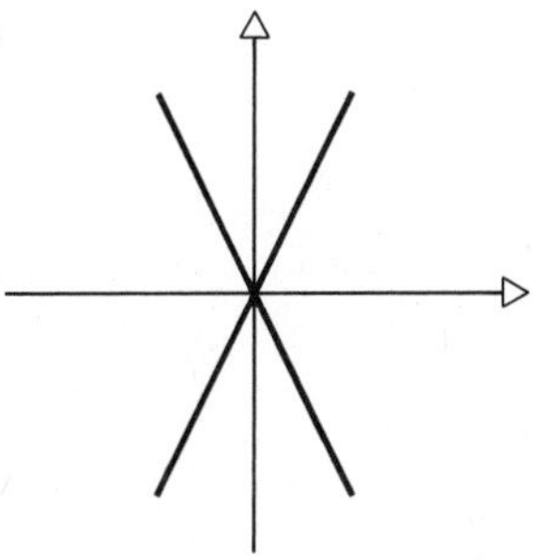

18.

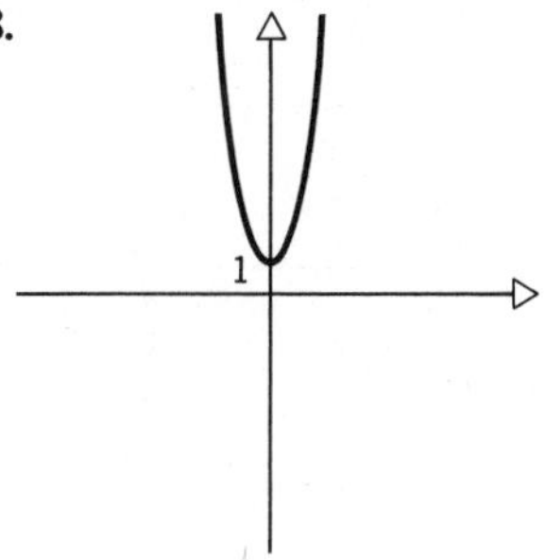

19.

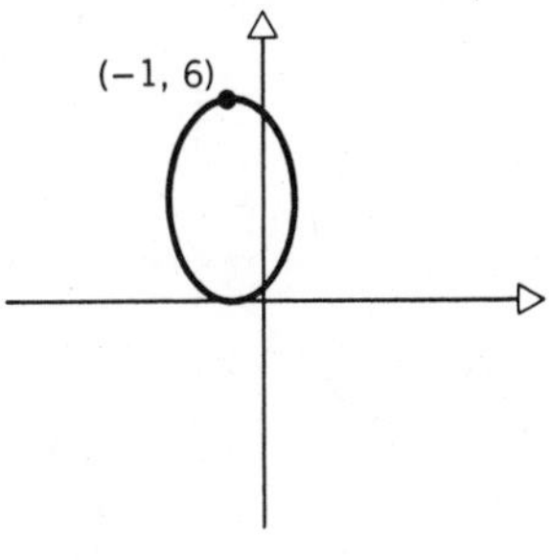

20.

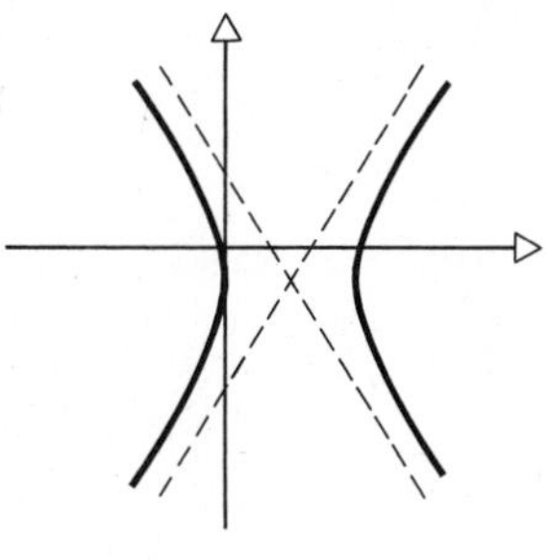

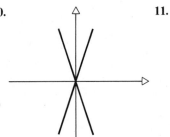

Section 22

1. 1, -3, $-2\sqrt{5}-3$, $2t-3$, $a-3$, $2b-1$, $2x^2-3$

3. 3, 5, $\sqrt{5}+5$, $-t+5$, $(10-a)/2$, $-b+4$, $-x^2+5$

5. 5, 1, 6, t^2+1, $(a^2+4)/4$, b^2+2b+2, x^4+1

7. $\{x:\ x \neq 1\}$ **9.** $\{x:\ x \neq -1\}$ **11.** $\{x:\ x \leq 1\}$

13. $\{x:\ x \geq -1 \text{ and } x \neq 0\}$ **15.** $\{x:\ x \leq 4 \text{ and } x \neq -1\}$

17. (a) x^2+x (b) $-x^2+x+4$ (c) x^3+2x^2-2x-4
(d) $(x+2)/(x^2-2)$ (e) 1 (f) 5 (g) $a+2$ (h) $-(t+2)/2$

19. (a) $2x+a+b$ (b) $b-a$ (c) $x^2+ax+bx+ab$ (d) $(x+b)/(x+a)$
(e) $a+b+1$ (f) $3-a+b$ (g) $(a+b)(\sqrt{3}+a)$ (h) $(t+b)/a$

21. (a) $ax+x+2b$ (b) $ax-x$ (c) $ax^2+abx+bx+b^2$ (d) $(ax+b)/(x+b)$
(e) $2b+1$ (f) $2a+1$ (g) $(a^2+b)(\sqrt{3}+b)$ (h) $(at+b)/b$

23. (a) $-4x+2$ (b) $-4x-8$ (c) -6 (d) 4

25. (a) $-27x^3$ (b) $-3x^3$ (c) -216 (d) 81

27. (a) $-2(x+1)^3$ (b) $-2x^3+1$ (c) -54 (d) 55

29. (a) -2 (b) 4 (c) -2 (d) 4

31. (a) $|x|$ (b) $-|x|$ (c) 2 (d) -3

33. (a) $x/(2-x)$ (b) $2x-2$ (c) Undefined. (d) -8

35. $g(x)=x+3$, $h(x)=\sqrt{x}$

37. $g(x)=x^2$, $h(x)=x+1$

39. $g(x)=\sqrt{x}$, $h(x)=2x$

41. $\{y:\ y \geq 1\}$

Section 23

1.

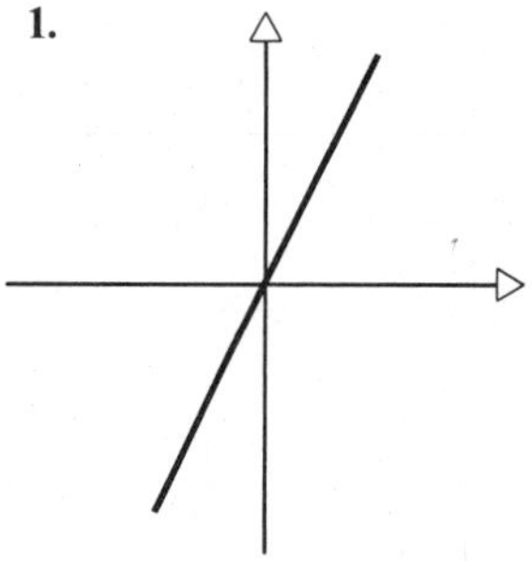

3.

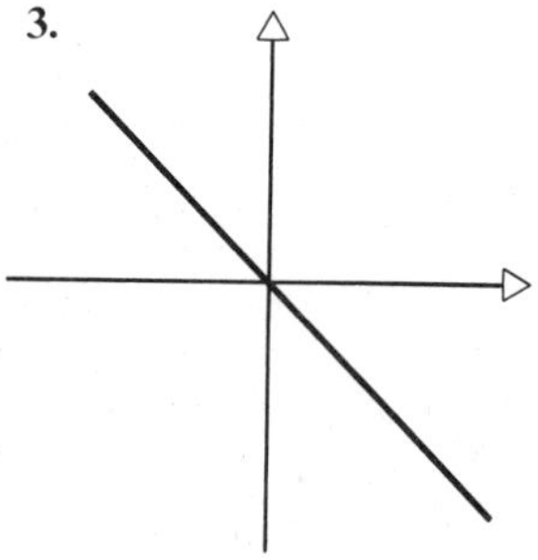

5.

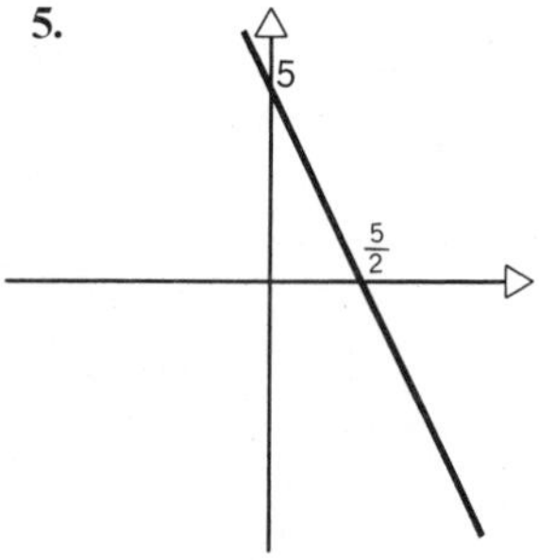

7.

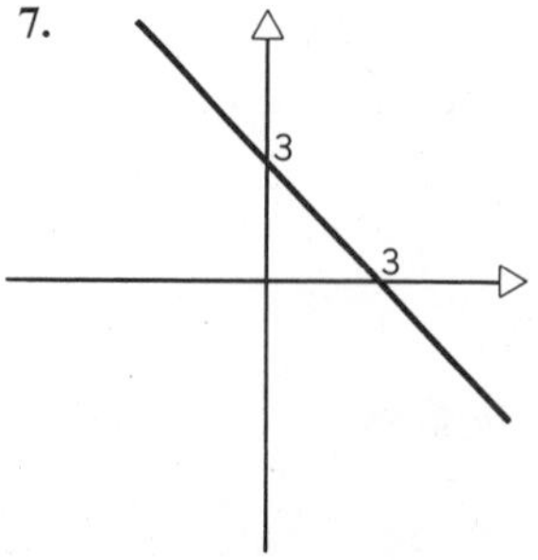

9.

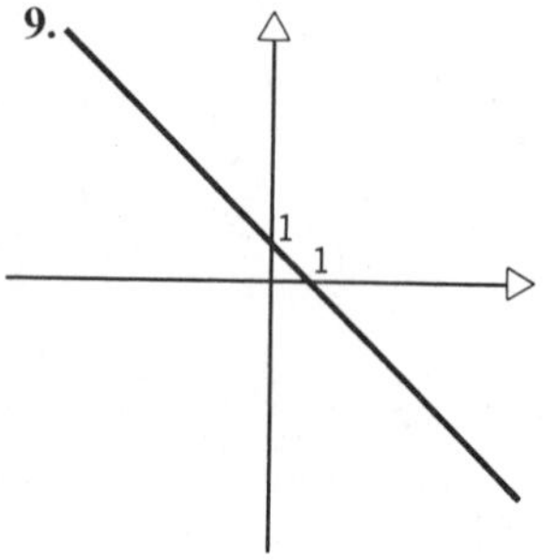

11.

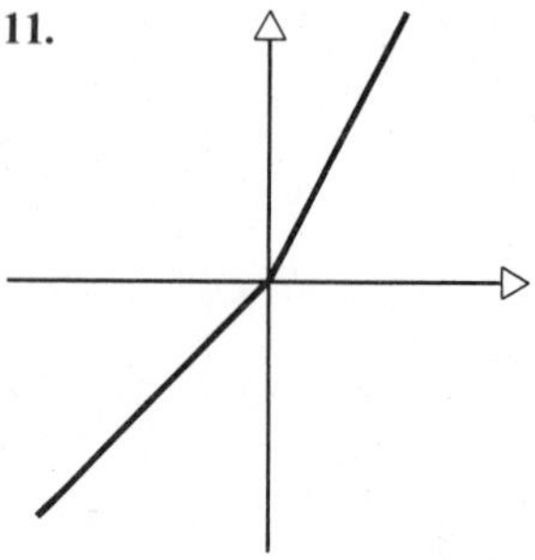

13. 1 **15.** 1 **17.** 2

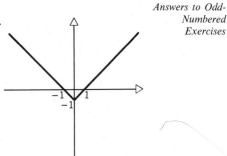

19. 4 **21.** 3

23.

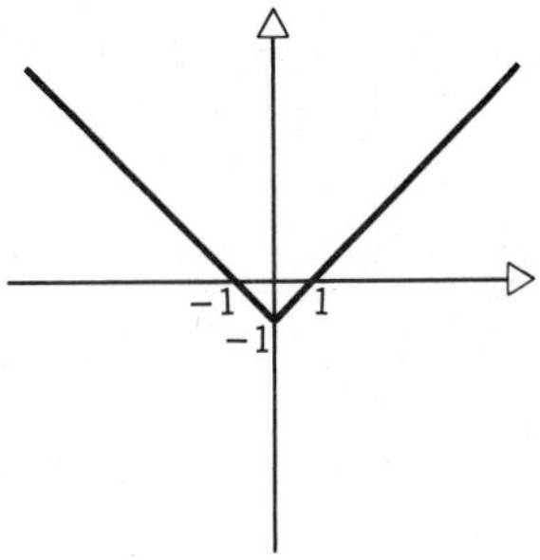

25.

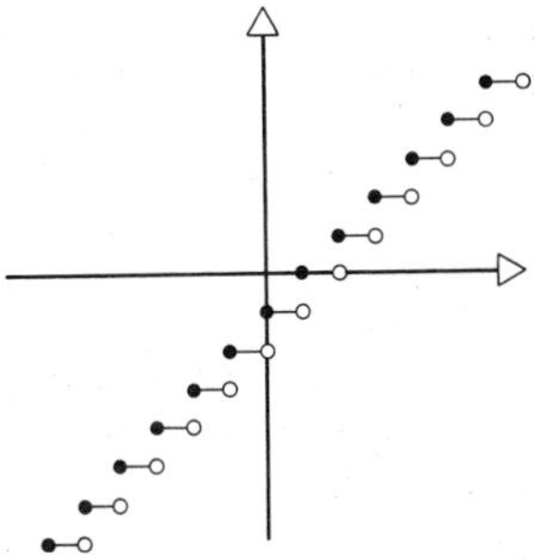

27.

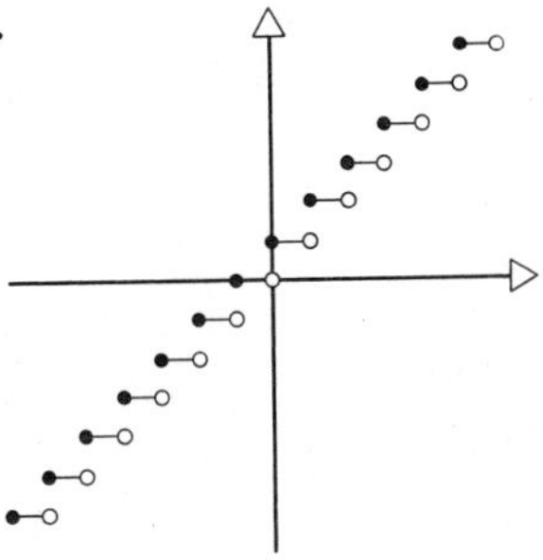

29. (a) $f(V, b) = 3V/b^2$ (b) $g(V, h) = \sqrt{3V/h}$

31. $f(P, r, n) = P[1 + (r/n)]^n$, $f(100, 0.08, 4) = \$108.24$

33. $h(r_N, n) = [1 + (r_N/n)]^n - 1$, $h(0.09, 4) = 0.0931$

35. $2\pi r(\Delta r) + \pi(\Delta r)^2$

37. (a) $4\,\Delta x$ (b) 12 (c) -6

39. (a) $-\Delta x$ (b) -3 (c) 1.5

41. (a) $3x^2(\Delta x) + 3x(\Delta x)^2 + (\Delta x)^3$ (b) 117 (c) -1.125

43. 4 **45.** -1 **47.** $4x + 2(\Delta x)$

Section 24

1.

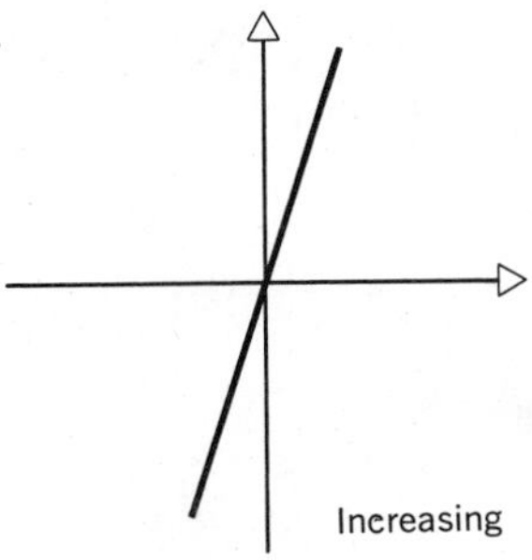

3.

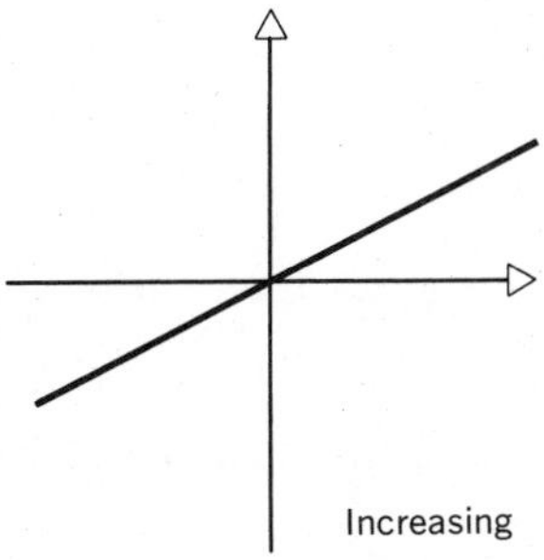

5.

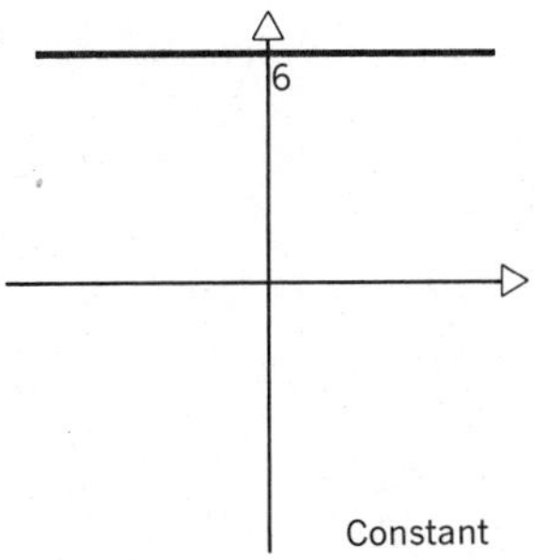

7.

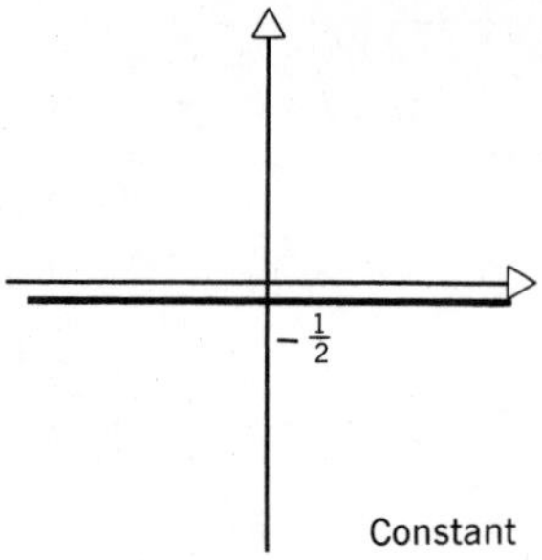

9.

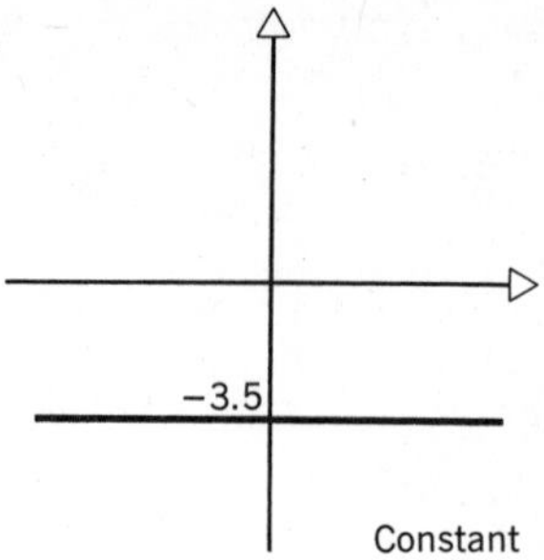

11. $f(x) = 2x - 1$ **13.** $f(x) = -\frac{2}{3}x - 3$ **15.** $f(x) = -\frac{3}{2}x + 5$ **17.** 2.286
19. 7.73 **21.** 5.45 **23.** $f(x) = 5.6x$ **25.** $f(x) = 3x$
27. $g(D) = D$ **29.** $f(t) = 2000 + 160t$ dollars
31. $f(x) = 25 + 0.2x$. The slope is the cost (in dollars) per mile.
33. m

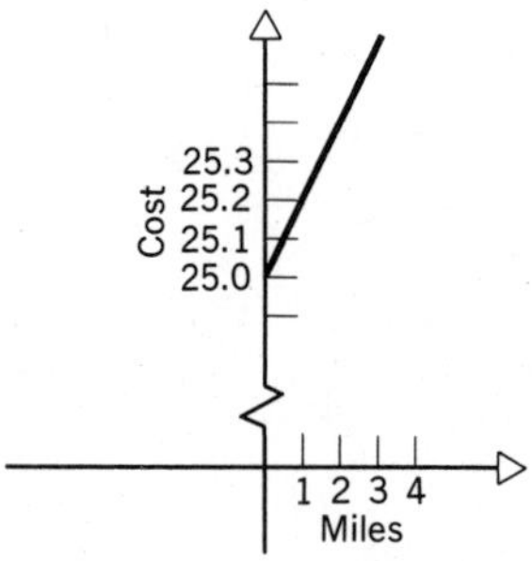

Section 25

1.

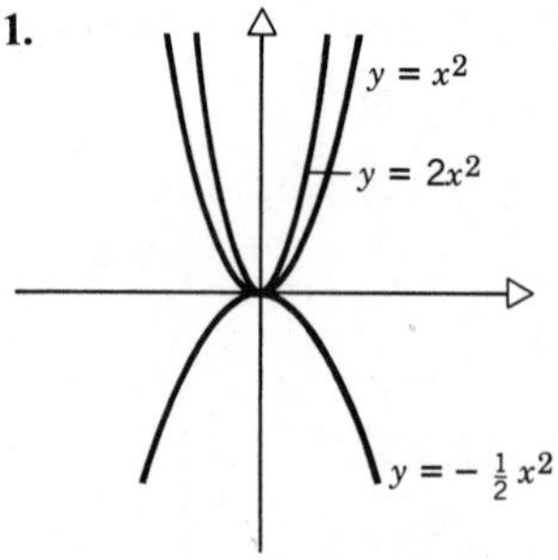

3.

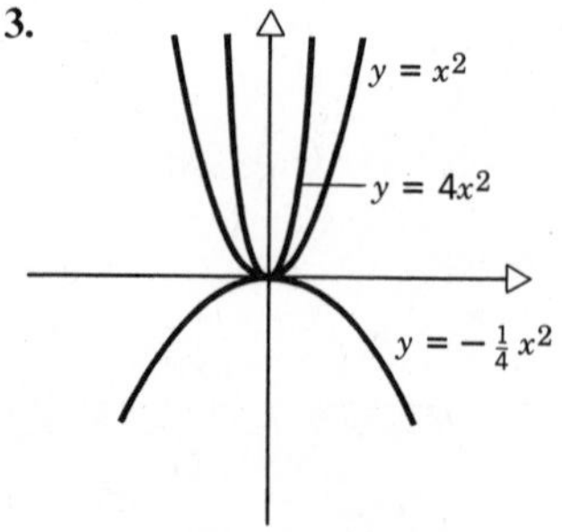

5.

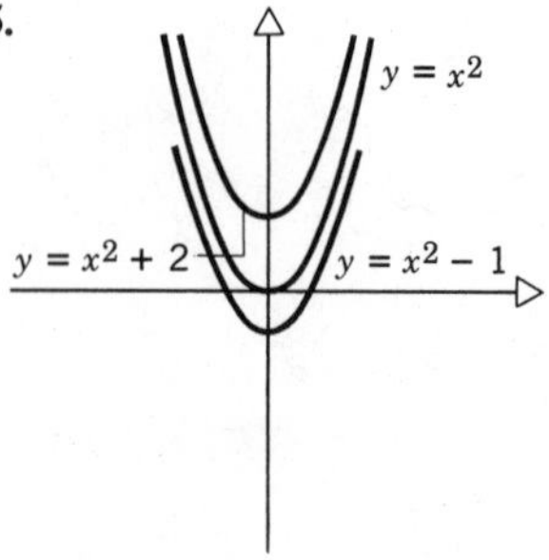

7. $f(-\frac{1}{4}) = \frac{7}{8}$, minimum. **9.** $h(3) = -9$, minimum.
11. $g(\frac{1}{6}) = \frac{13}{12}$, maximum.

13.

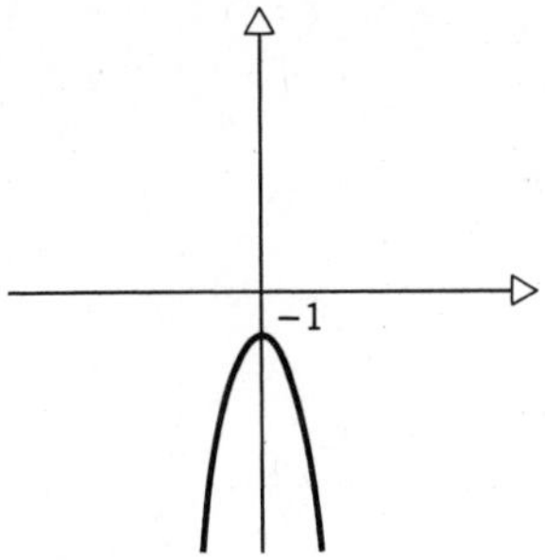

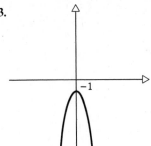

Vertex $(0, -1)$; axis $x = 0$; no x-intercepts.

15.

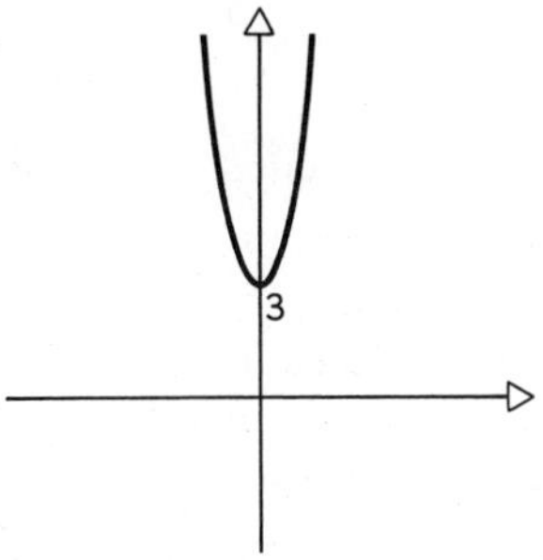

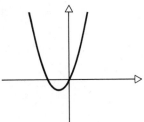

Vertex $(0, 3)$; axis $x = 0$; no x-intercepts.

17.

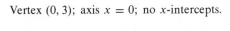

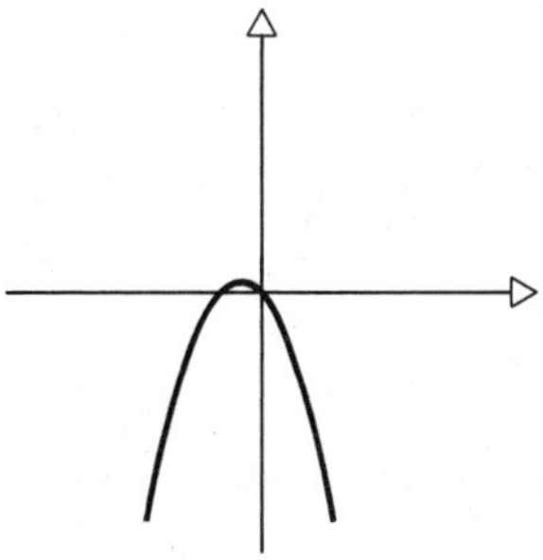

Vertex $(-\frac{1}{2}, \frac{1}{4})$; axis $x = -\frac{1}{2}$; x-intercepts -1 and 0.

19.

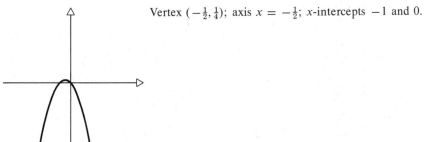

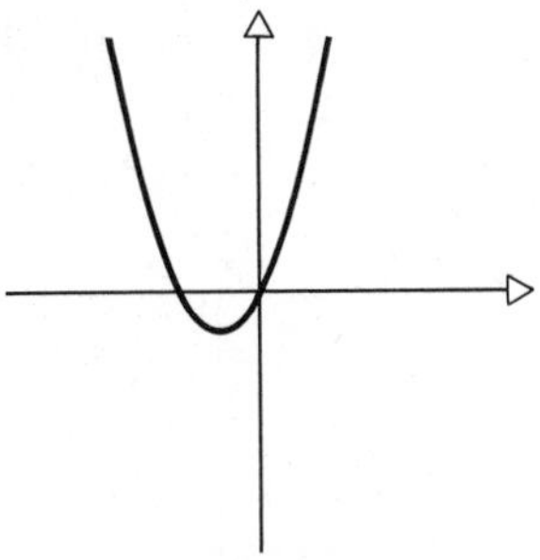

Vertex $(-1, -1)$; axis $x = -1$; x-intercepts -2 and 0.

21.

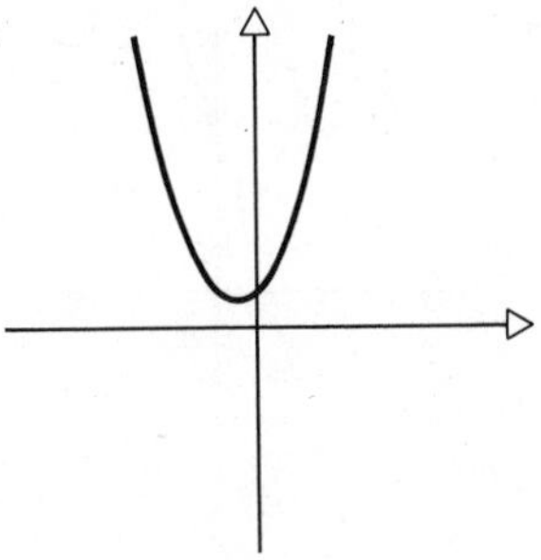

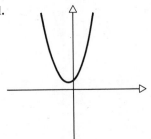

Vertex $(-\frac{1}{2}, \frac{3}{4})$; axis $x = -\frac{1}{2}$; no x-intercepts.

23.

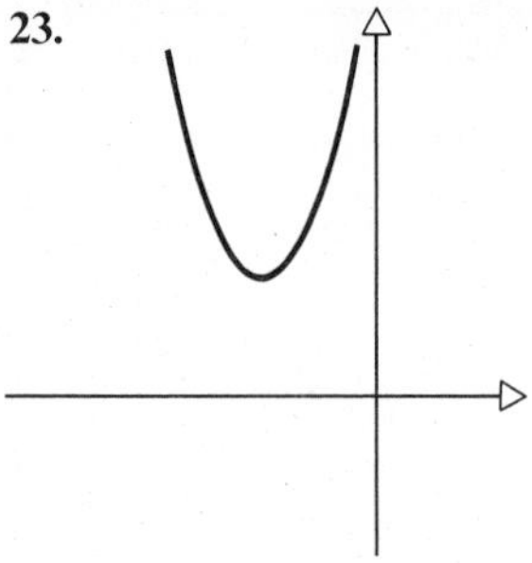

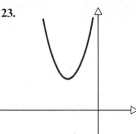

Vertex $(-3, 3)$; axis $x = -3$; no x-intercepts.

25. Maximum altitude: 100 feet above the ground. It will reach the ground 4 seconds after it is fired.

27. Maximum altitude: 256 feet above the ground. It will reach the ground 6 seconds after it is fired.

29. 31,250 square yards **31.** 3.5, 3.5

33. $y = s - x$, the product is $x(s - x)$, and the latter is maximum when $x = s/2$, which implies $y = s/2$ also.

35. $2ax + b + a(\Delta x)$

Section 26

1. One-to-one. **3.** One-to-one. **5.** One-to-one.

7. One-to-one. **9.** One-to-one.

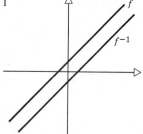

11. $f^{-1}(x) = \frac{1}{4}x$

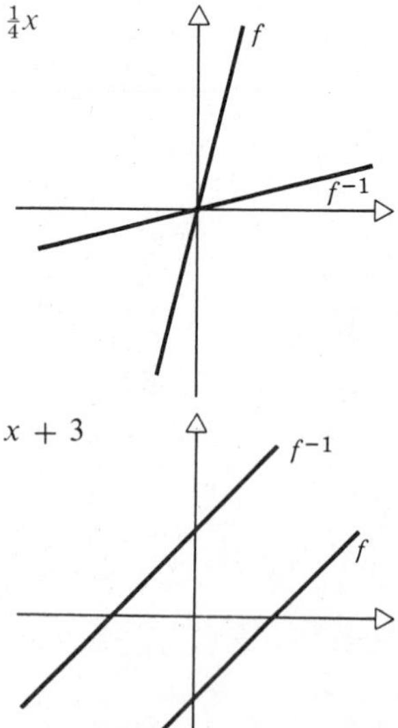

13. $f^{-1}(x) = x - 1$

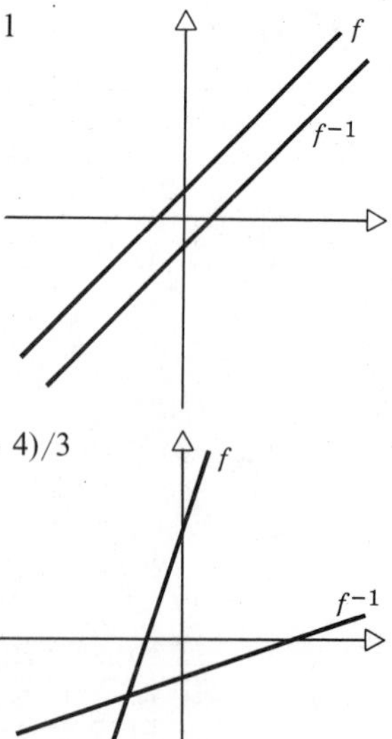

15. $f^{-1}(x) = x + 3$

17. $f^{-1}(x) = (x - 4)/3$

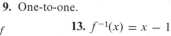

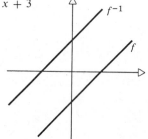

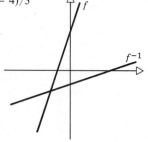

19.

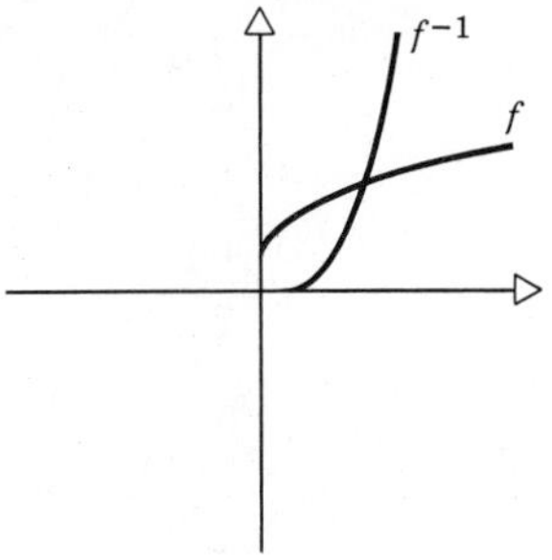

$f^{-1}(x) = (x - 1)^2,\ x \geq 1$

21.

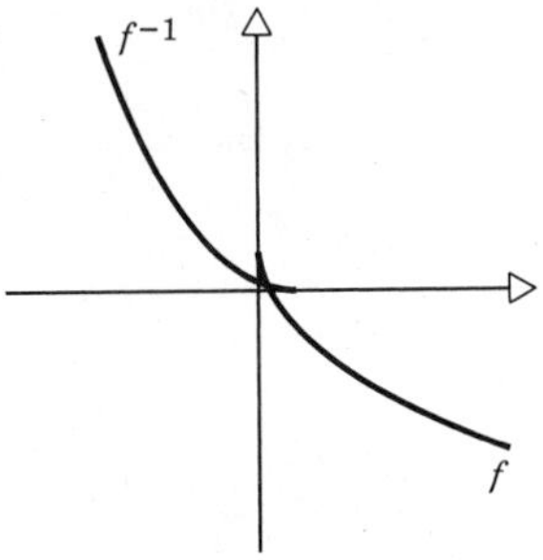

$f^{-1}(x) = (1 - x)^2/4,\ x \leq 1$

23.

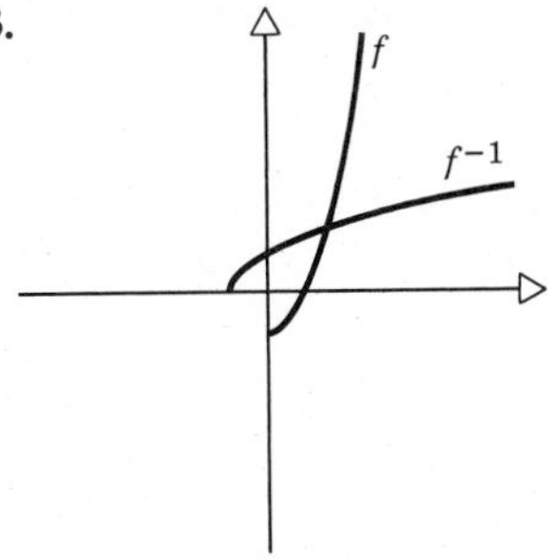

$f^{-1}(x) = \sqrt{x + 1},\ x \geq -1$

25. $f^{-1}(x) = (x - b)/a$

CHAPTER VI REVIEW

1. $\{x\colon\ x \geq -1 \text{ and } x \neq 2\}$ **2.** $\{x\colon\ x \geqslant \frac{1}{2} \text{ and } x \neq -1\}$

3. (a) $x^2 + 3x + 4$ (b) $3x^3 + 5x^2 - 3x - 5$ (c) 2 (d) 0 (e) $3x^2 + 2$ (f) 195

4. (a) $x^2 - 2x + 4$ (b) $-2x^3 + 3x^2 - 2x + 3$ (c) 2 (d) $\frac{1}{2}$ (e) $4x^2 - 12x + 10$ (f) -17

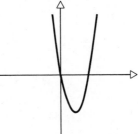

5. -3 **6.** $2x + \Delta x$

7.

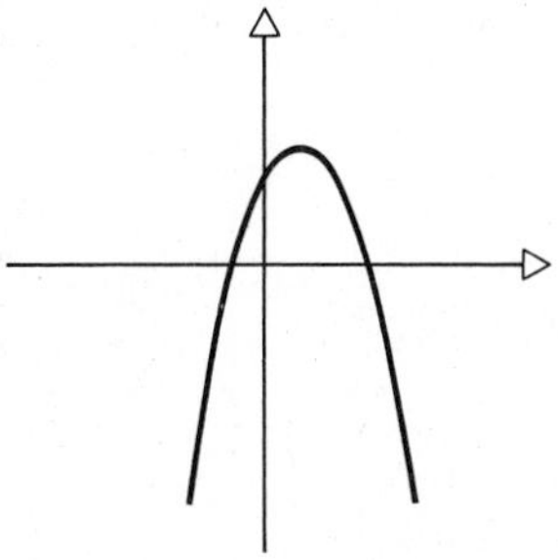

8.

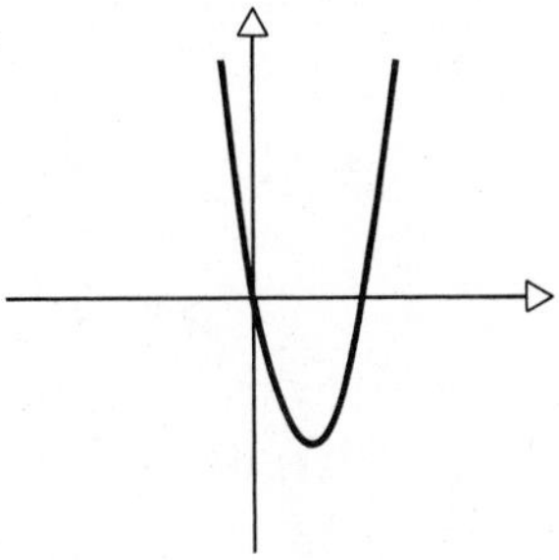

9.

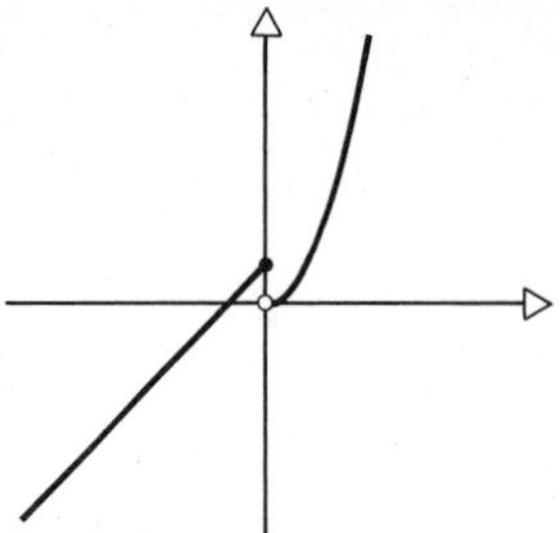

10.

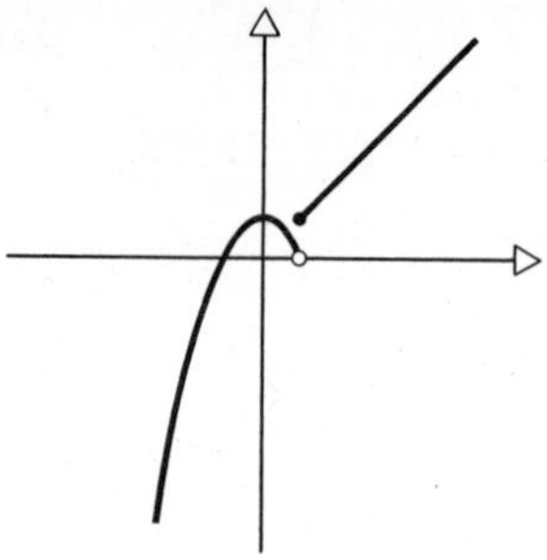

11. $\frac{1}{4}$ **12.** $-\frac{1}{4}$ **13.** 5 and -5 **14.** 5 and 2.5 **15.** 9.21 **16.** 17.5

17. $f^{-1}(x) = (5 - x)/3$ **18.** $f^{-1}(x) = \sqrt{x - 2},\ x \geq 2$

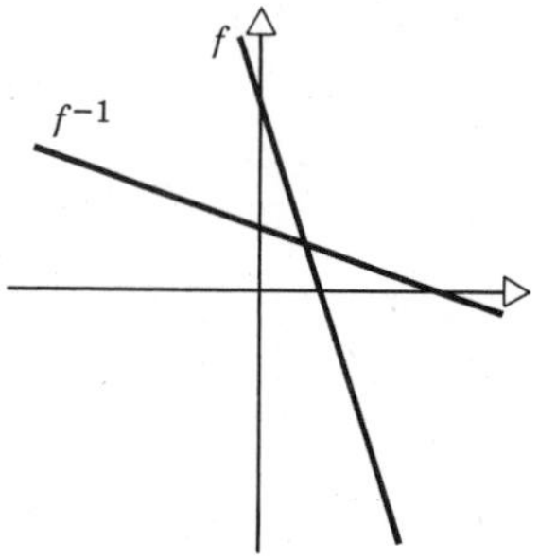

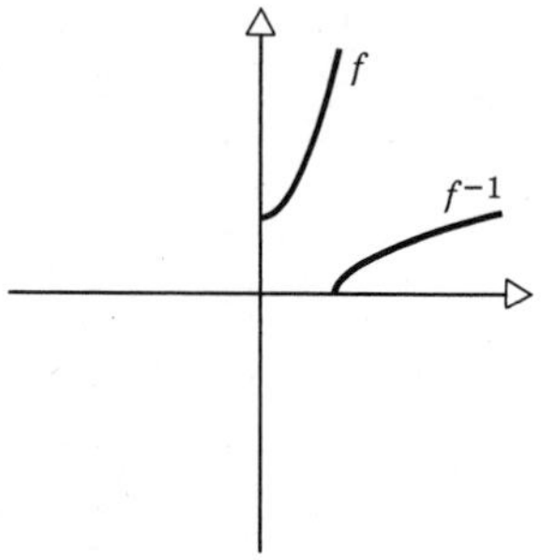

19. $f(x) = \frac{5}{4}x^2$ **20.** $g(v) = 30/v$

21. $f(n) = 1000 + 50n$. The variable cost is the slope.

22. $f(I) = 0.03I$ for $0 \leq I \leq 2000$, $f(I) = 0.04I - 20$ for $2000 < I \leq 4000$, $f(I) = 0.05I - 60$ for $4000 < I \leq 10{,}000$.

23. (a) $f(m_1, m_2, r, F) = Fr^2/(m_1 m_2)$ (b) $h(m_1, m_2, G, F) = \sqrt{Gm_1 m_2/F}$

24. (a) $g(d_0, d_i) = d_0 d_i/(d_0 + d_i)$ (b) $h(f, d_i) = d_i f/(d_i - f)$

Section 27

For Exercises 1–23, the first polynomial is the quotient and the second is the remainder.

1. $x^2 - x + 2,\ 5$ **3.** $x^2 - 2x - 3,\ 4$

5. $3x - 6,\ x + 2$ **7.** $2,\ 3x^2 + x - 2$

9. $\frac{1}{5}x^2 + \frac{6}{25}x + \frac{6}{125},\ \frac{131}{125}x + 1$ **11.** $0,\ x^3 + x^2 - 4$

13. $2x^2 + x + 6,\ 7$ **15.** $x^2 + 4x + 8,\ 15$

17. $x^2 - 5x + 24,\ -120$ **19.** $x^3 + \frac{1}{2}x^2 - \frac{1}{4}x + \frac{1}{8},\ -\frac{1}{16}$

21. $x^2 - 0.5x + 1.25,\ -0.625$ **23.** $x^4 + \frac{3}{2}x^3 + \frac{7}{4}x^2 + \frac{15}{8}x + \frac{31}{16},\ \frac{63}{32}$

25. $7,\ -119$ **27.** $3,\ -45$ **29.** $-\frac{94}{81},\ -1.0195$

Section 28

1. Factor. $q(x) = 6x^2 + x - 2$ **3.** Factor. $q(x) = x^3 - 2x^2 + 1$

5. Factor. $q(x) = x^3 - x + 4$ **7.** Not a factor.

9. Not a factor. **11.** 0 (one), -1 (one), 5 (two)

13. 0 (three), 10 (two), -0.1 (five) **15.** 0 (one), -2 (one), -0.4 (two)

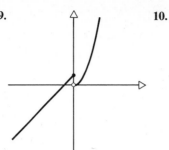

17. $x(x - \frac{1}{2})^2(x + 5)^3$ **19.** $x^4(x + 0.2)^2(x - 2)$

21. $(x + \frac{1}{4})^3(x - 1.2)^2(x - 5)$

23. 1 (one), -2 (one), 2 (one); $(x - 1)(x + 2)(x - 2)$

25. $\frac{1}{2}$ (one), 1 (two); $(2x - 1)(x - 1)^2$

27. -0.4 (one), -1 (two); $(5x + 2)(x + 1)^2$

29. 5 (one), $\sqrt{5}$ (one), $-\sqrt{5}$ (one); $(x - 5)(x - \sqrt{5})(x + \sqrt{5})$

31. -2 (one); $(x + 2)(x^2 + 1)$ **33.** $-\frac{1}{3}$ (one); $(3x + 1)(x^2 - 3x + 3)$

35. $2^6 + 64 \neq 0$ **37.** None. **39.** n odd.

Section 29

1. $\pm 1, \pm 2, \pm 3, \pm 6, \pm \frac{1}{5}, \pm \frac{2}{5}, \pm \frac{3}{5}, \pm \frac{6}{5}$ **3.** $\pm 1, \pm 2, \pm \frac{1}{2}, \pm \frac{1}{7}, \pm \frac{2}{7}, \pm \frac{1}{14}$

5. Upper bound 4. Lower bound -3.

7. Between -3 and -2, between 0 and 1, and between 2 and 3.

9. Between -2 and -1, between -1 and 0, and between 2 and 3.

11. $-1, -\sqrt{3}, \sqrt{3}$ **13.** $-1, -\frac{1}{2}, \frac{3}{2}$ **15.** $-\frac{1}{2}, 2, 2$

17. $\frac{1}{2}$ **19.** 0 **21.** $0, 5 + \sqrt{3}, 5 - \sqrt{3}$

23. $-2, \frac{1}{2}, \frac{1}{2}, 2$ **25.** 0, 5 **27.** $-\frac{4}{3}, 0, -\sqrt{3}, \sqrt{3}$

29. 3.5 **31.** -3 **33.** -6 **35.** 1.5 m

Section 30

1.

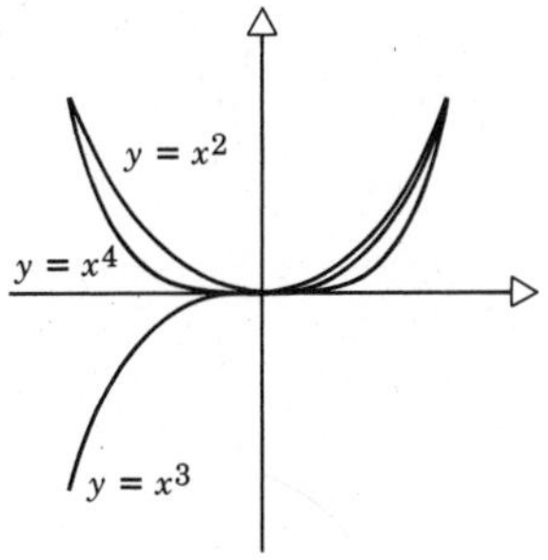

3.

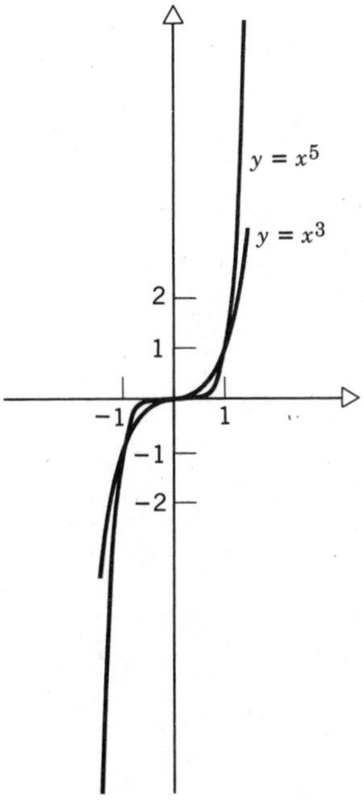

5.

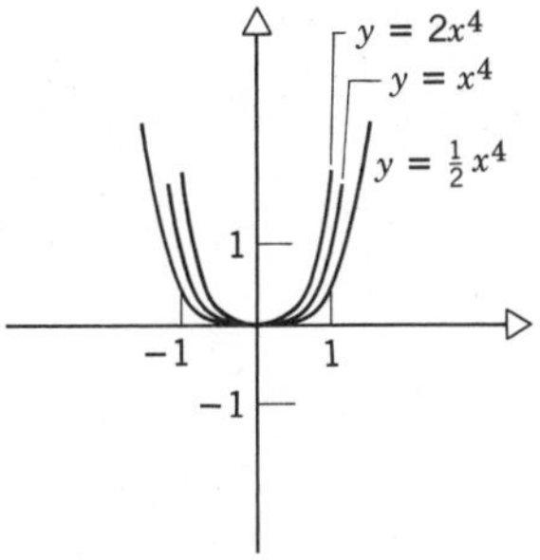

7. Odd. Symmetric about the origin.

9. Even. Symmetric about the y-axis.

11. Odd. Symmetric about the origin.

13. Even. Symmetric about the y-axis.

15. Neither.

17. Neither.

19.

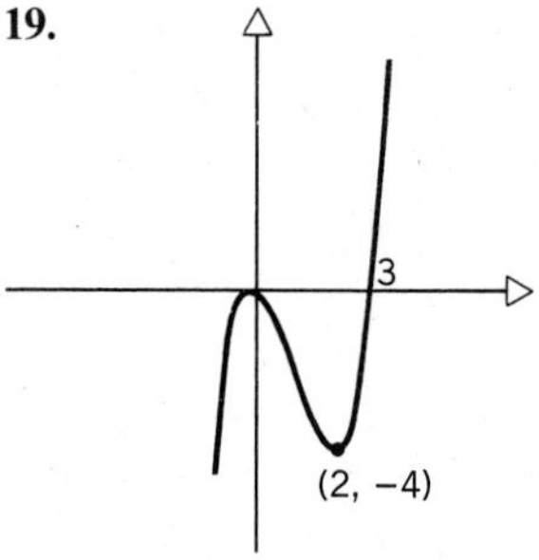

21.

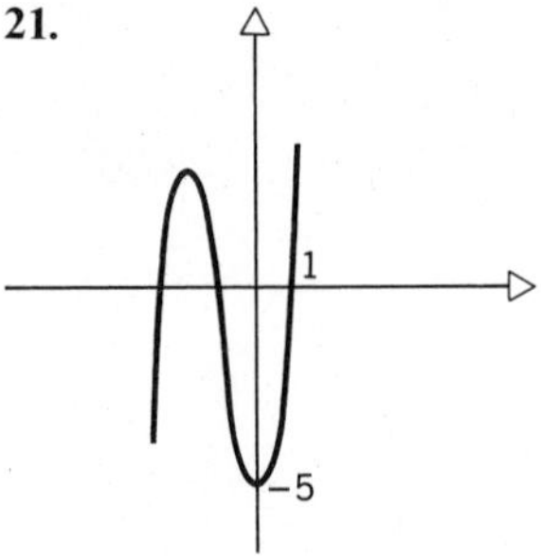

23.

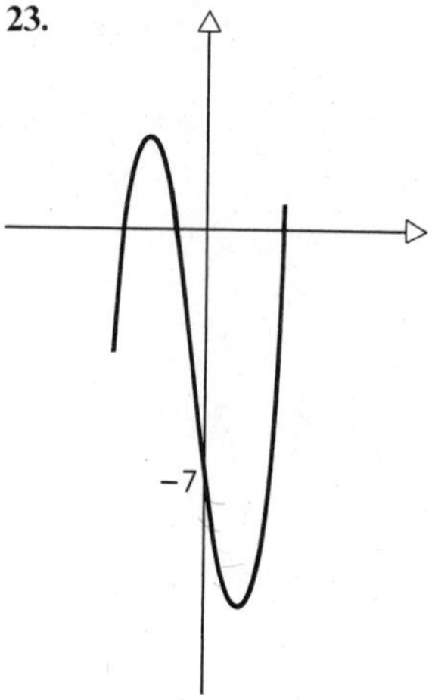

25.

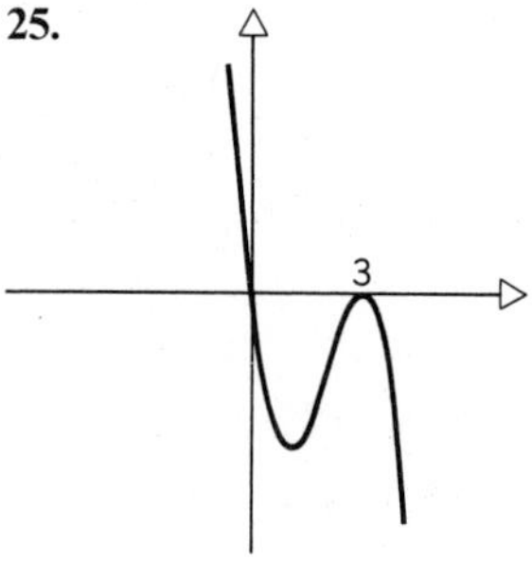

27.

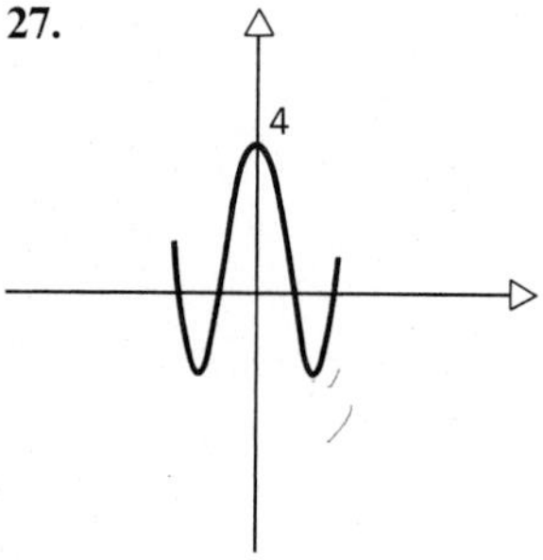

29.

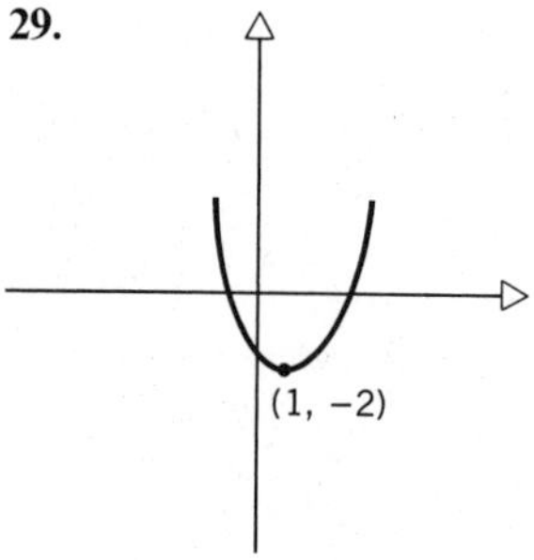

31. $x^3 - 5x^2 - 2x + 24$

33. $3x^3 - 3x^2 - 6x$

Section 31

1.

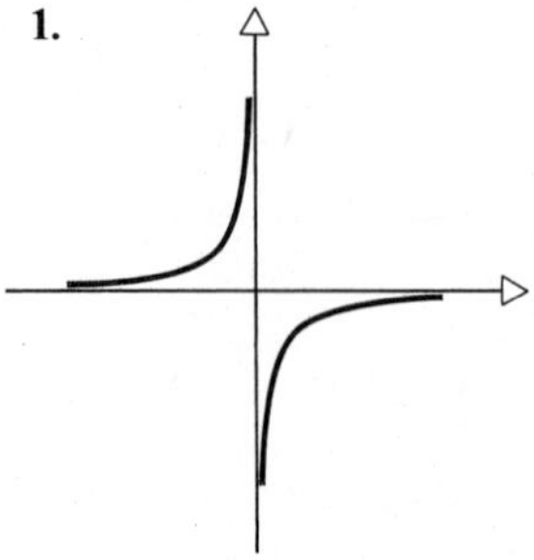

3.

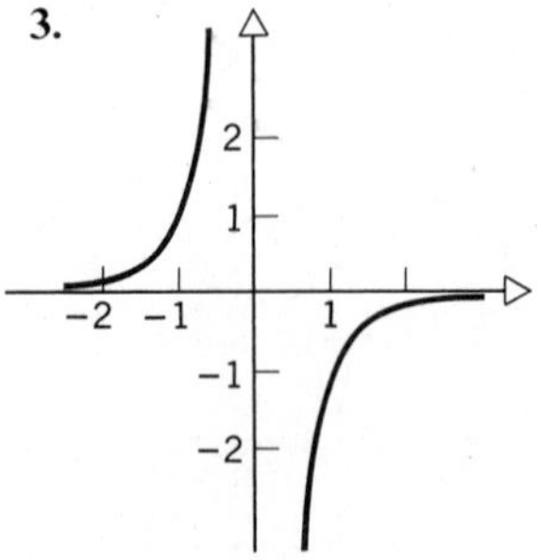

5.

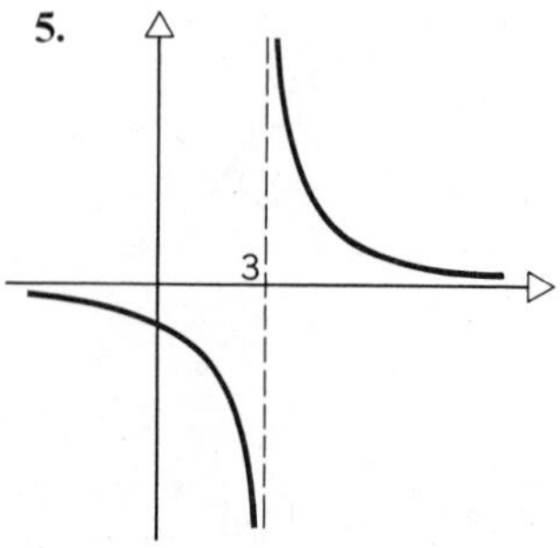

7.
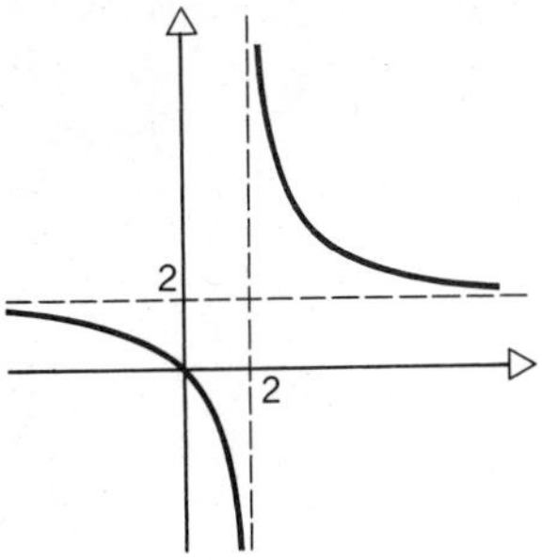

9.
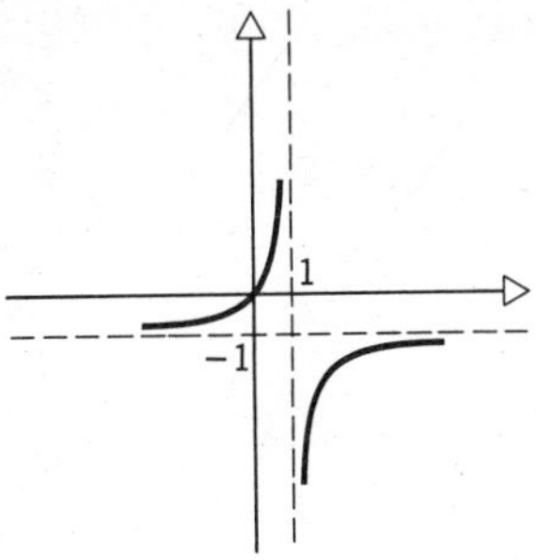

11.
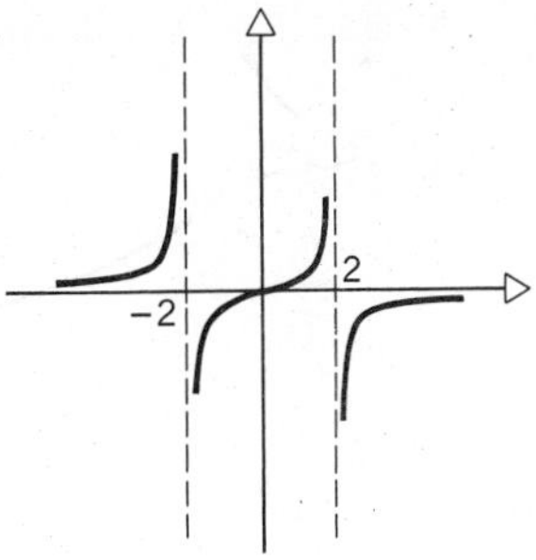

13.
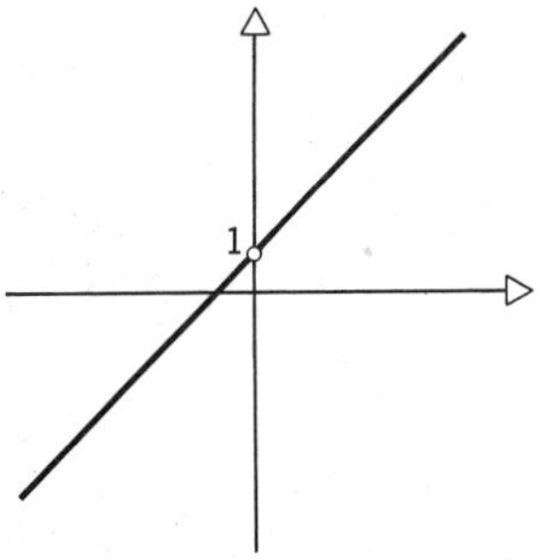

15.
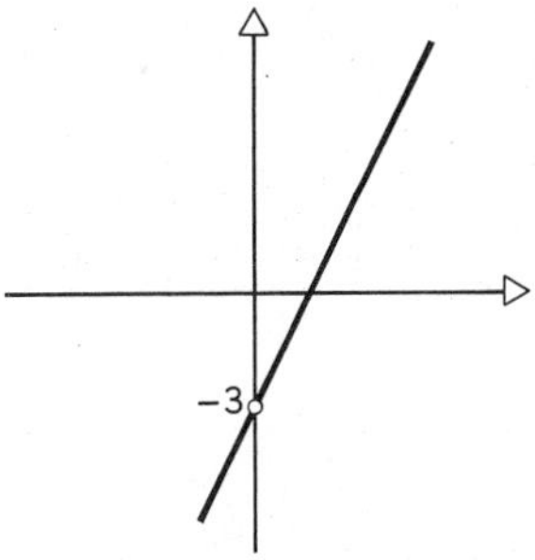

17.
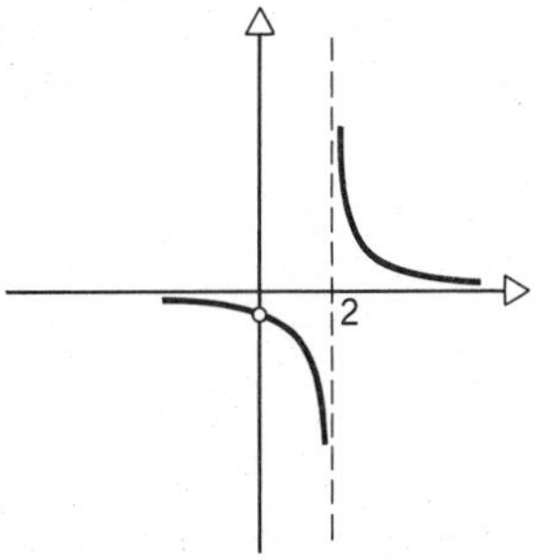

19.
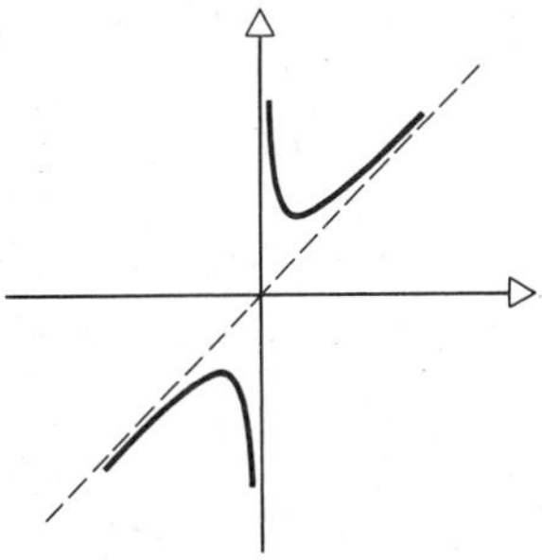

21.
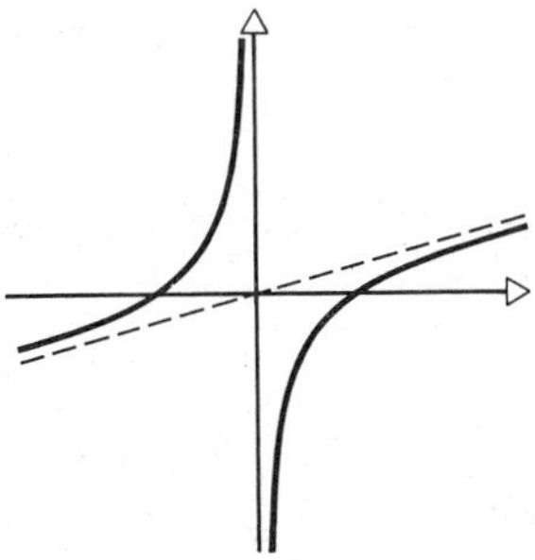

23.
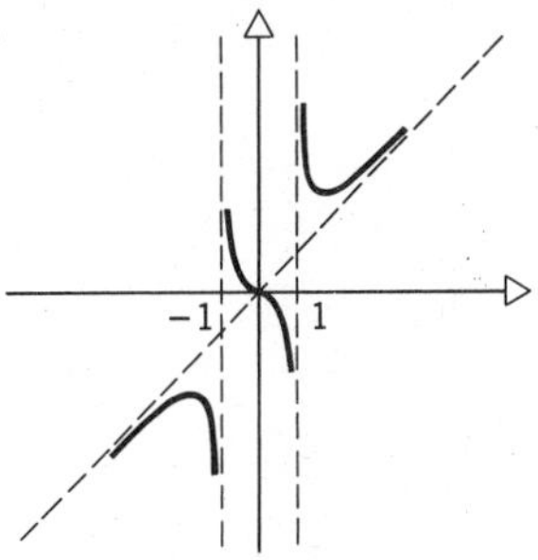

25.
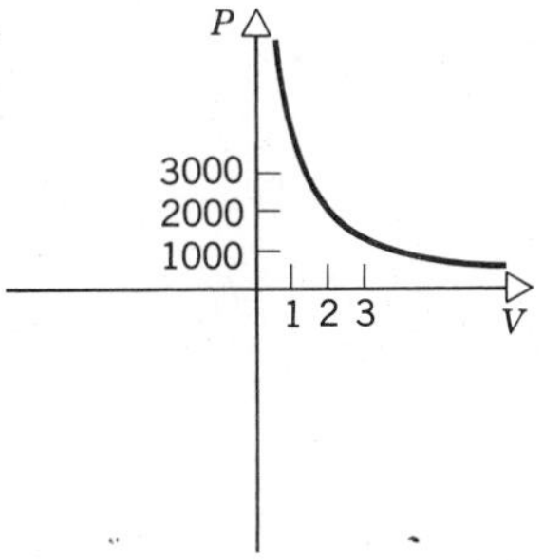

27.
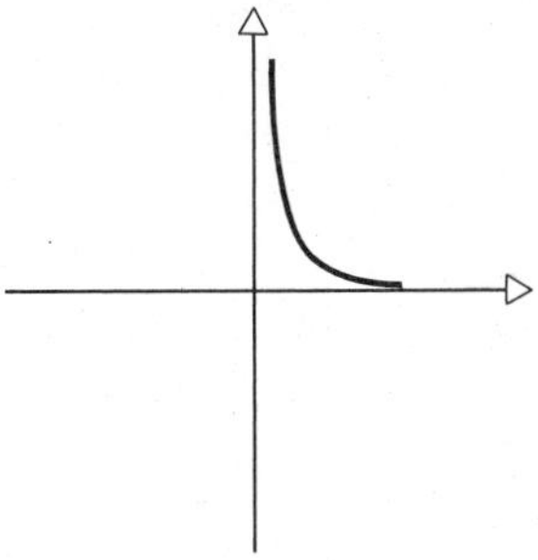

Section 32

1. $\dfrac{3}{x+1} - \dfrac{2}{x-2}$

3. $\dfrac{5}{x+2} + \dfrac{1}{x-4}$

5. $\dfrac{1}{3} \cdot \dfrac{1}{x+1} - \dfrac{2}{3} \cdot \dfrac{1}{2x-1}$

7. $\dfrac{3}{x-1} - \dfrac{1}{(x-1)^2}$

9. $\dfrac{-2}{2x-3} + \dfrac{5}{(2x-3)^2}$

11. $\dfrac{3}{x} - \dfrac{1}{x-1} + \dfrac{3}{(x-1)^2}$

13. $\frac{4}{x} - \frac{3}{x^2 + x + 1}$

15. $\frac{-3}{x} + \frac{4}{x^2 + 2}$

17. $\frac{-2}{x^2} + \frac{3}{x^2 + 1}$

19. $\frac{x}{x^2 + 1} - \frac{2x - 1}{(x^2 + 1)^2}$

21. $\frac{2x - 1}{x^2 + x + 1} + \frac{3x}{(x^2 + x + 1)^2}$

23. $\frac{4}{x - 1} - \frac{3}{(x - 1)^2} + \frac{1}{(x - 1)^3}$

25. $3x - 1 + \frac{1}{x} - \frac{2}{x + 1}$

27. $2x - 1 + \frac{x - 1}{x^2 + 1} - \frac{2x}{(x^2 + 1)^2}$

CHAPTER VII REVIEW

1. $q(x) = x - 3,\ r(x) = x + 2$
2. $q(x) = x^2 - x + 1,\ r(x) = x$
3. $q(x) = x^3 + 2x^2 + x + 3,\ r(x) = 10$
4. $q(x) = 3x^2 - 8x + 8,\ r(x) = -6$
5. $f(3) = 58,\ f(-1) = -6$
6. $f(2) = 26,\ f(-2) = 30$
7. $q(x) = x^2 + x + 2$
8. $q(x) = 4x^3 - 2$
9. $-1, 2, 3;\ (x + 1)(x - 2)(x - 3)$
10. $-\sqrt{3}, \sqrt{3}, 2;\ (x + \sqrt{3})(x - \sqrt{3})(x - 2)$
11. $-\sqrt{2}, \sqrt{2}, \frac{1}{2}, 1;\ (x + \sqrt{2})(x - \sqrt{2})(2x - 1)(x - 1)$
12. $-\sqrt{3}, \sqrt{3}, -1, \frac{2}{3};\ (x + \sqrt{3})(x - \sqrt{3})(x + 1)(3x - 2)$
13. Between -4 and -3, between -3 and -2, and between 2 and 3. (The zeros are $-\frac{7}{2}$, $-\sqrt{5}$, and $\sqrt{5}$.)
14. Between -4 and -3, between -2 and -1, and between 3 and 4. (The zeros are $-\frac{5}{3}$, $-\sqrt{10}$, and $\sqrt{10}$.)
15. Neither. Not symmetric about either the y-axis or the origin.
16. Even. Symmetric about the y-axis.
17. Even. Symmetric about the y-axis.
18. Odd. Symmetric about the origin.

19.

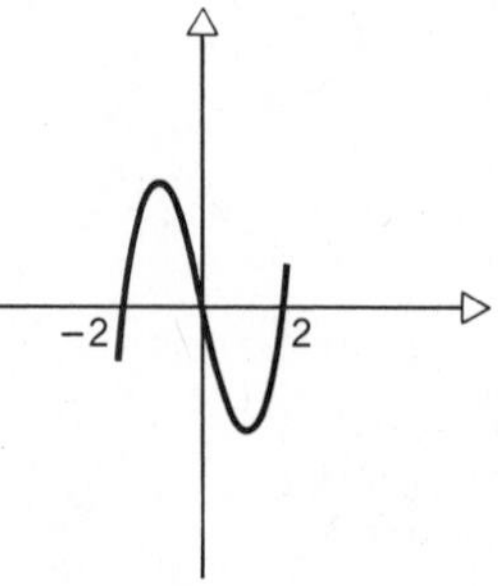

20.

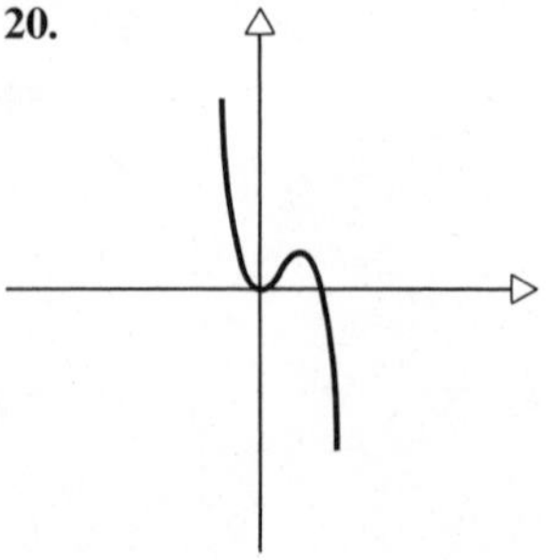

21.

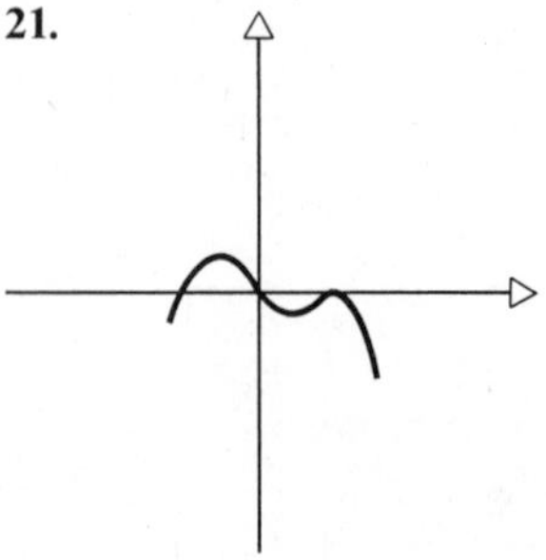

22.

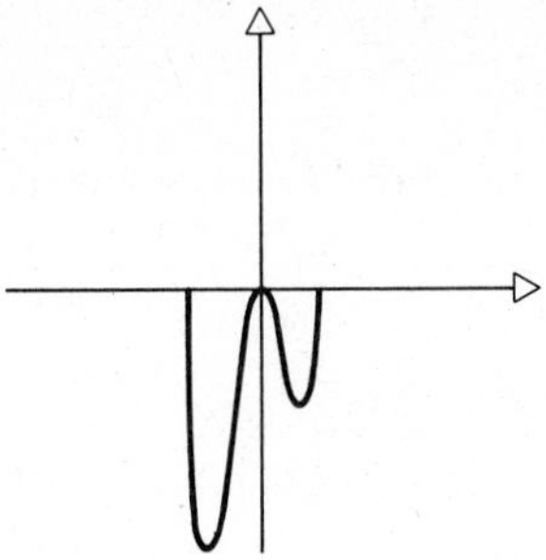

23.

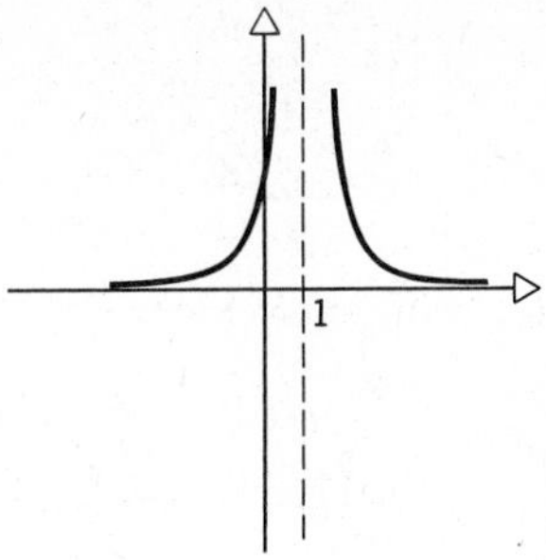

24.

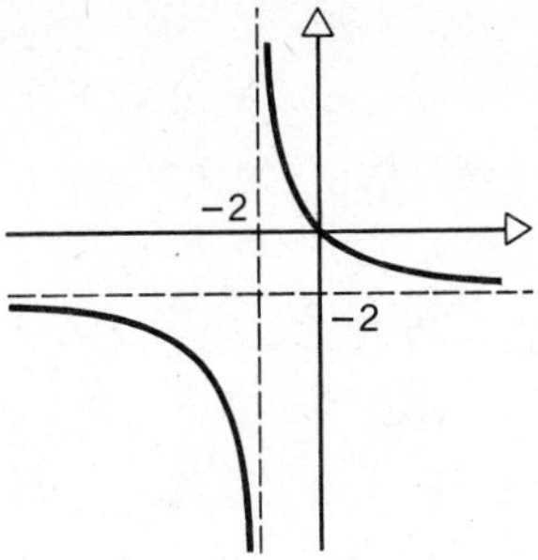

25. $\dfrac{3}{x+2} - \dfrac{4}{(x+2)^2}$

26. $\dfrac{4}{x} - \dfrac{3}{x^2+1}$

27. $x + \dfrac{3}{x-1} + \dfrac{1}{x+2}$

28. $x^2 + 1 - \dfrac{1}{x-1} + \dfrac{5}{x^2+2}$

29. $\frac{2}{3}$

30. 7 cm, 9 cm, 21 cm

Section 33

1. *Real numbers:* $\sqrt{3}$, 0. *Imaginary numbers:* $4 - i$, πi, $2 + \sqrt{-4}$. *Pure imaginary numbers:* πi.

3. *Real numbers:* π, -1. *Imaginary numbers:* $3 + \sqrt{-15}$, $\sqrt{2} + i$, $4i$. *Pure imaginary numbers:* $4i$.

5. 1
7. $-4 + 6i$
9. $5 - \sqrt{2}i$
11. $9 + 3i$
13. $-3 + 2i$
15. $3 - 24i$
17. $-9 - 2i$
19. 2
21. -10
23. -5
25. $-i$
27. 1
29. $\frac{3}{10} + \frac{1}{10}i$
31. $-\frac{1}{2} + \frac{1}{2}i$
33. $\frac{4}{5} - \frac{3}{5}i$
35. $-1 - 5i$
37. $1 - i$
39. $1 + i$
41. $-11 - 2i$
43. $(1 \pm \sqrt{7}i)/2$
45. $(-1 \pm \sqrt{11}i)/2$
47. $(5 \pm \sqrt{11}i)/6$
49. $\pm 3i$
51. $(3 \pm \sqrt{7}i)/4$
53. $x = -2$, $y = -1$
55. No.
57. Yes.
59. Substitute and simplify.
61. (a) $(a + bi) + (a - bi) = 2a$
(b) $(a + bi) - (a - bi) = 2bi$.
63. $\pm i$
65. $\overline{\overline{(a + bi)}} = \overline{(a - bi)} = a + bi$
67. $\overline{(a + bi) + (c + di)} = \overline{(a + c) + (b + d)i} = (a + c) - (b + d)i = (a - bi) + (c - di)$.
69. Use $\overline{(z/w)} = \overline{(z\overline{w})}/(w\overline{w})$, $\overline{z}/\overline{w} = (\overline{z}w)/(\overline{w}w)$. $z = a + bi$, $w = c + di$, and simplify.
71. Replace z by $a + bi$ and simplify both sides.
73.

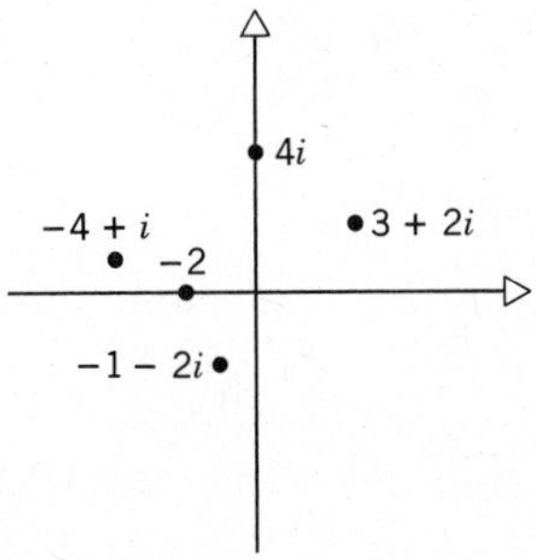

75.

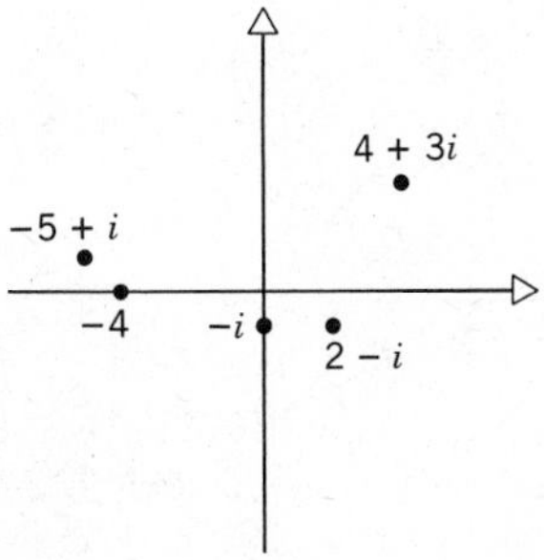

77. Use $\sqrt{a^2 + b^2} = \sqrt{a^2 + (-b)^2}$.

79. Use $z\bar{z} = a^2 + b^2$.

81. One method: Use $z = a + bi$, $w = c + di$, $|z/w| = |z\bar{w}/w\bar{w}|$, compute, and simplify.

Section 34

1. $q(z) = z^2 - iz + 2$, $r(z) = 3i$

3. $q(z) = 2z^3 + (-1 + 4i)z^2 + (-8 - i)z + (1 - 16i)$, $r(z) = 33 + 3i$

5. $f(i) = -1 - 2i$, $f(1 + i) = -4 + 6i$

7. $q(z) = z^3 + 2iz^2 - 3z - 6i$

9. $q(z) = z^3 - \sqrt{5}iz^2 - z + \sqrt{5}i$

11. $z^4 + (-1 + 3i)z^3 - 3iz^2 + 4iz - 4i$

13. $x^4 - 4x^3 + 8x^2 - 16x + 16$

15. $x^6 - 2x^5 + 20x^4 - 36x^3 + 117x^2 - 162x + 162$

17. (a) 5, $(-1 + \sqrt{3}i)/2$, and $(-1 - \sqrt{3}i)/2$, each of multiplicity one.
(b) $(z - 5)\left(z + \frac{1}{2} - \frac{\sqrt{3}}{2}i\right)\left(z + \frac{1}{2} + \frac{\sqrt{3}}{2}i\right)$
(c) $(z - 5)(z^2 + z + 1)$

19. (a) i, $-i$, $2i$, and $-2i$, each of multiplicity one.
(b) $(z - i)(z + i)(z - 2i)(z + 2i)$
(c) $(z^2 + 1)(z^2 + 4)$

21. (a) $1 + i$, $1 - i$, and -3, each of multiplicity one.
(b) $(z - 1 - i)(z - 1 + i)(z + 3)$
(c) $(z^2 - 2z + 2)(z + 3)$

23. $1, -1, -i$

CHAPTER VIII REVIEW

1. $7 - i$	**2.** $-1 - 5i$	**3.** 3	**4.** $24 + 6i$
5. $-8 + i$	**6.** $-10 + 10i$	**7.** $1 + 2i$	**8.** $-\frac{15}{26} - \frac{3}{26}i$
9. $4 + 2i$	**10.** $11 + 2i$	**11.** $(1 \pm i)/2$	**12.** $(-3 \pm \sqrt{3}i)/2$

13. $z^4 - iz^3 + 2z^2$ **14.** $z^4 + (-3 + i)z^3 + (1 - 3i)z^2 + (-3 + i)z - 3i$

15. $z^4 - 4z^3 + 8z^2$ **16.** $z^6 + 11z^4 + 19z^2 + 9$

17. (a) $2i$, $-2i$, and $\frac{3}{2}$, each of multiplicity one.
(b) $(z - 2i)(z + 2i)(2z - 3)$
(c) $(z^2 + 4)(2z - 3)$

18. (a) $i/2$, $-i/2$, $\sqrt{2}i$, and $-\sqrt{2}i$, each of multiplicity one.
(b) $(2z - i)(2z + i)(z - \sqrt{2}i)(z + \sqrt{2}i)$
(c) $(4z^2 + 1)(z^2 + 2)$

Section 35

The entries for the tables in Exercises 1, 3, and 5 were computed directly with a calculator and then rounded; they will not agree completely with entries computed using the approximations given in the exercises.

1.

x	-3	$-\frac{5}{2}$	-2	$-\frac{3}{2}$	-1	$-\frac{1}{2}$	0	$\frac{1}{2}$	1	$\frac{3}{2}$	2	$\frac{5}{2}$	3
5^x	0.008	0.018	0.040	0.090	0.200	0.447	1	2.24	5	11.2	25	55.9	125

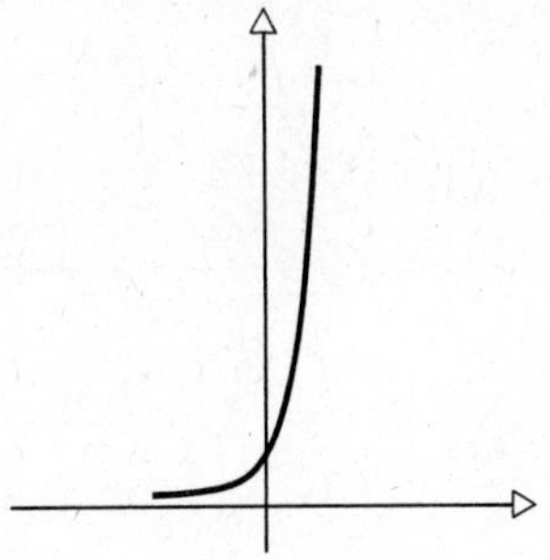

3.

x	-3	$-\frac{5}{2}$	-2	$-\frac{3}{2}$	-1	$-\frac{1}{2}$	0	$\frac{1}{2}$	1	$\frac{3}{2}$	2	$\frac{5}{2}$	3
3.5^x	0.023	0.044	0.082	0.153	0.286	0.535	1	1.87	3.50	6.55	12.25	22.9	42.9

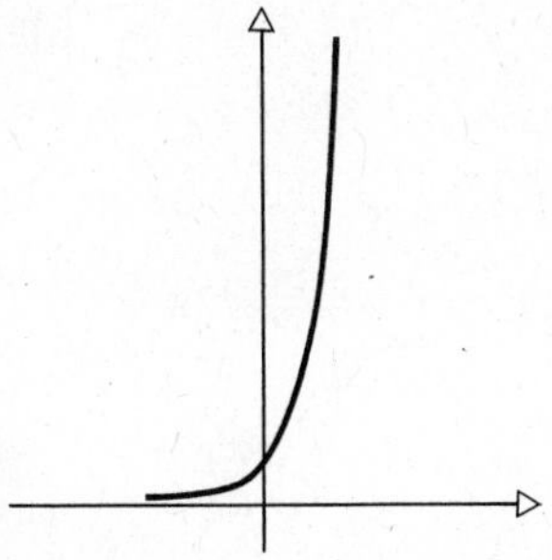

5.

x	-3	$-\frac{5}{2}$	-2	$-\frac{3}{2}$	-1	$-\frac{1}{2}$	0	$\frac{1}{2}$	1	$\frac{3}{2}$	2	$\frac{5}{2}$	3
0.4^x	15.62	9.88	6.25	3.95	2.5	1.58	1	0.63	0.4	0.25	0.16	0.10	0.06

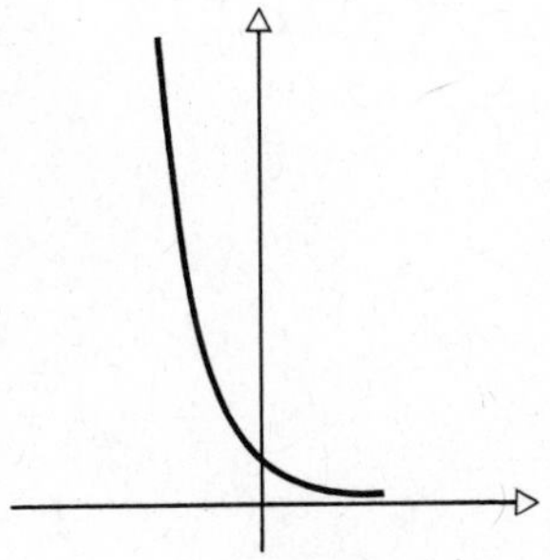

7. 1 **9.** 1 **11.** 1 **13.** -2 **15.** $\frac{1}{3}$

17. 0 **19.** 0 **21.** -4 **23.** $\frac{3}{2}$ **25.** -3

27. $\frac{3}{4}$ **29.** $x = \log_u v$ **31.** $x = \frac{1}{3}\log_4 2 = \frac{1}{6}$ **33.** $x = \frac{1}{4}\log_y z$ **35.** $x = \log_5 2$

37. $x = 3^{\pi}$ **39.** $x = 10^{\sqrt{2}}$ **41.** $x = \frac{1}{2} \cdot 3^{2y}$ **43.** $x = a^4$ **45.** $x = 2^{y/3}$

47.

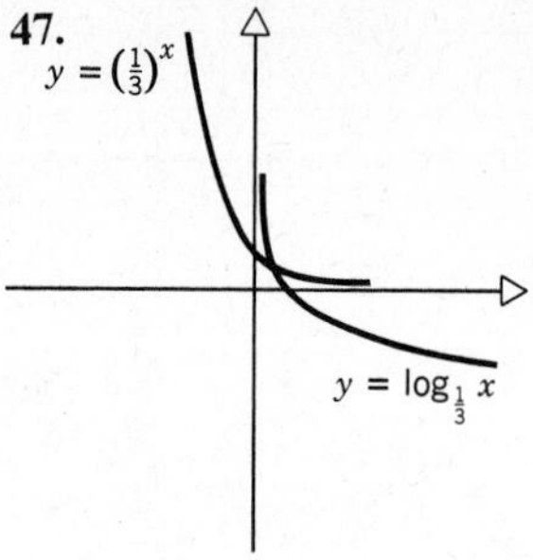

49.

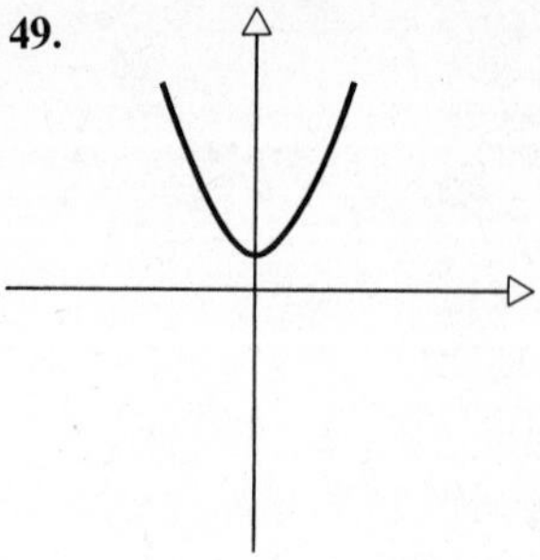

51.

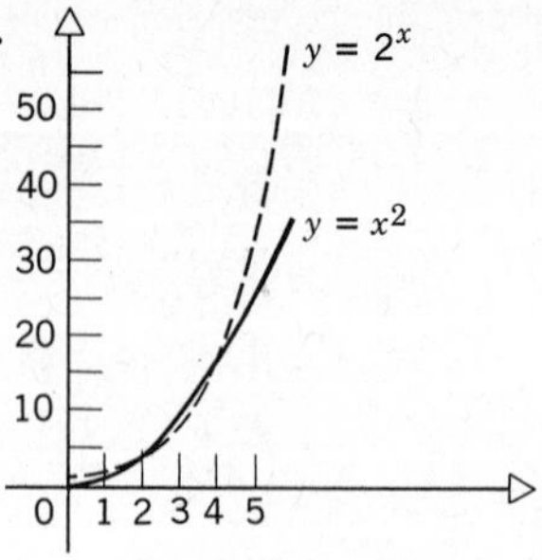

$2^x \leq x^2$ for $2 \leq x \leq 4$

Section 36

1. $\log_b x + \log_b y - \log_b z$ **3.** $-\log_b x - \log_b y$

5. $\log_b x + \frac{1}{2}\log_b z$ **7.** $\frac{3}{2}(\log_b x - \log_b y - \log_b z)$

9. $\frac{1}{3}\log_b x + \frac{1}{4}\log_b y + \frac{1}{5}\log_b z$ **11.** $\frac{1}{2}\log_b x - \frac{3}{2}\log_b y - \frac{3}{2}\log_b z$

13. $\log_b(uv)^2$ **15.** $\log_b(1/v)$ **17.** $\log_b v^3u$ **19.** $\log_b 4$ **21.** 0

23. 2 **25.** 0.9208 **27.** 1.1292 **29.** 0.1781 **31.** -1.0686

33. -0.0937 **35.** 3 **37.** π **39.** 1.73 **41.** $-x$

43. 0 **45.** 0 **47.** -1 **51.** Let $c = a$.

53. (a) $0 < x < 1$ (b) $x > 1$

Section 37

1. (a) 0.9299 (b) 2.9299 (c) $-2 + 0.9299 = -1.0701$

3. (a) 0.5315 (b) 3.5315 (c) $-1 + 0.5315 = -0.4685$

5. (a) 4.2095 (b) 0.2095 (c) $-2 + 0.2095 = -1.7905$

7. $(\log 2)/(\log 3) \approx 0.6309$ **9.** $(\log 10)/(\log 2) \approx 3.322$

11. $(\log 18)/(\log 4.5) \approx 1.922$ **13.** $(\log 2.5)/(\log 9) \approx 0.4170$

15. $(\log 1.5)/(5 \log 4) \approx 0.0585$ **17.** 8 years

19. 11 years **21.** 23 years

23. (a) 9.006 years (b) 9 years (c) 0.07%

25. (a) 69.66 years (b) 72 years (c) 3.4%

27. (a) 4.19 years (b) 4 years (c) 4.5%

29. 756 million **31.** 549 million **33.** 327 million **35.** 1986 **37.** 2015

39. 1991 **41.** 35 years **43.** 117 years **45.** 70 years **47.** 1987

Section 38

1. \$24.41, \$26.13, \$27.15 **3.** \$2613.04, \$2714.57, \$2718.28

5. \$8243.50 **7.** \$210.26

9. \$6749.30 **11.** (a) and (b) 4.5218

13. (a) -3.2188 (b) -3.2189 **15.** (a) and (b) -7.9576

17. (a) -7.2645 (b) -7.2645 **19.** $Q = 0.1e^{-[(\ln 2)/11]t} = 0.1e^{-0.063t}$ gm

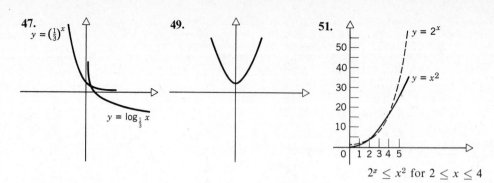

21. $Q = 0.001e^{-[(\ln 2)/(2.4\times10^4)]t} = 0.001e^{-0.0000289t}$ gm

23. (a) 0.0625 mg (b) 0.751 mg (c) At the end of 14.9 years.

25. 5.2 billion **27.** 7.5 billion

29. 784,000 **31.** Use $e^{\ln 2} = 2$.

Section 39

1. 10,000 **3.** 1 **5.** 1.16 **7.** 33.3 **9.** 94,900

11. 0.550 **13.** 0.200 **15.** 0.502 **17.** 0.4510 **19.** 1.4704

21. −1.3490 **23.** 1.8803 **25.** 4.553 **27.** 8.542 **29.** 90.58

31. 0.006446 **33.** 0.6967 **35.** 59.47 **37.** 5.42×10^2 **39.** 1.76×10^{-1}

41. 7.787 **43.** 3.68 **45.** 44.5 **47.** 0.447 **49.** 51.5 cm^2

51. 3.94 in.

CHAPTER IX REVIEW

1. 3 **2.** −2 **3.** 1.5 **4.** −1

5. 49 **6.** 5 **7.** 75 **8.** $(b^{d/a})/c$

9. $\log_b x + \log_b z - \log_b y$ **10.** $\frac{1}{5}\log_b y + \frac{1}{5}\log_b z$

11. $\log_b(v^3/u)$ **12.** $\log_b \sqrt{15}$

13. $(\log 4)/(\log 5) \approx 0.861$ **14.** $(\log 0.5)/(\log 9) \approx -0.315$

15. $(\log 9)/(\log \frac{5}{3}) \approx 4.30$ **16.** $-(\log 1.25)/(\log 7) \approx -0.115$

17. 4.77×10^3 **18.** 8.58×10^{-4} **19.** 9.28 **20.** 5.30×10^2

21. 1.26×10^{30} **22.** 1.82 **23.** (a) 3.4341 (b) 3.4340 **24.** (a) 5.9000 (b) 5.8999

25. $3.61e^{(0.019)(25)} \approx 5.80$ billion

26. (a) 7.27 years (b) 7.2 years (c) 0.96%

27. $1271

28. The suggestion will lead to the equation $(e^y)^2 - 2x(e^y) - 1 = 0$.

Section 40

1. (2, 1) **3.** (0, 3) **5.** (−2, −3)

7. (1, −1) **9.** (0, −2) **11.** (4, 0)

13. $\{(x, 3x - 1)$: x is a real number$\}$

15. $\{(x, -\frac{1}{2}(x + 3))$: x is a real number$\}$

17. If you add −2 times the second equation to 3 times the first equation, the result will be $-23 = 0$.

19. (1, 2) **21.** $\{(x, 1 - 2x)$: x is a real number$\}$

23. $(\frac{1}{3}, \frac{1}{3})$ **25.** No solution. **27.** (−12, 6)

29. 65 for children and 270 for adults.

31. Ground speed 290 mph, wind speed 10 mph.

33. Runner 10 mph, walker 5 mph.

35. 12 gm of 14-karat gold, 8 gm of 24-karat gold.

37. 47 **39.** American League 44, National League 31.

41. Use the fact that the graphs coincide iff they have equal slopes and equal y-intercepts.

Section 41

1. $(5, -1, -2)$

3. $(3, \frac{1}{2}, 3)$

5. $\{(5t + 1, -2t, t)$: t is a real number$\}$

7. Inconsistent.

9. $\{(t + 4, -2t - 10, t)$: t is a real number$\}$

11. Inconsistent.

13. $\{(\frac{7}{11}t + \frac{6}{11}, \frac{1}{11}t + \frac{4}{11}, t)$: t is a real number$\}$

15. Inconsistent.

17. $\{(\frac{3}{4}t + \frac{1}{8}, \frac{1}{4}t + \frac{11}{8}, t)$: t is a real number$\}$

19. $x = -2, y = 4, z = 3, w = 7$

21. Inconsistent.

23. 500 pounds each of B and D.

25. National League teams 20, Yankees 19, other American League teams 11.

27. Cleveland 35 inches, Mobile 67 inches, Phoenix 7 inches.

29. $y = 2x^2 + x$

31. $x^3 - 2x^2 + x + 2$

33. $v_0 = 24$ ft/sec, $s_0 = 216$ ft

Section 42

1. Inconsistent.

3. $\{(4, -3t - 1, -2t + 7, t)$: t is a real number$\}$

5. Inconsistent.

7. $(-3, 8, 6, 6)$

9. Inconsistent.

11. $(-4, 6)$

13. $(0, 2, 4)$

15. $\{(5t - \frac{4}{3}, -3t + \frac{2}{3}, t)$: t is a real number$\}$

17. Inconsistent.

19. $\{(\frac{4}{7}t + \frac{2}{7}, -\frac{1}{7}t + \frac{10}{7}, t)$: t is a real number$\}$

21. Inconsistent.

23. $(-4, 0, 2, 1)$

25. $(\frac{1}{2}, \frac{3}{2})$

27. $(4, 0)$

29. $\{(t - 3, -2t + 1, t, 2)$: t is a real number$\}$

31. 7 cm, 17 cm, 21 cm

33. $\frac{1}{3}, 1, \frac{5}{3}$

35. 9 four-game series, 11 five-game series, 8 six-game series, 22 seven-game series.

37. $-x^3 + x + 4$

39. Use the suggestion.

Section 43

1. $(-3, 3)$

3. $(\frac{5}{3}, \frac{4}{3})$

5. $(\frac{3}{2}, -1)$ and $(-\frac{11}{6}, \frac{7}{3})$

7. $(2\sqrt{2}, \frac{2}{3})$, $(-2\sqrt{2}, \frac{2}{3})$, and $(0, -2)$

9. No real solution.

11. $(0, 0)$ and $(2^{-1/5}, 2^{-3/5})$

13. $(3, \frac{7}{3})$ and $(-2, -1)$

15. $(-1 + \sqrt{2}, 1 + \sqrt{2})$ and $(-1 - \sqrt{2}, 1 - \sqrt{2})$

17. $(\frac{2}{3}, 2)$

19. (0.54, 1.46)

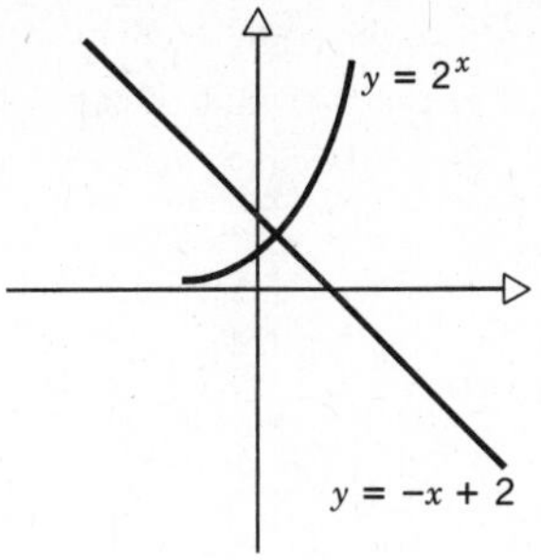

21. (0.4, −0.4)

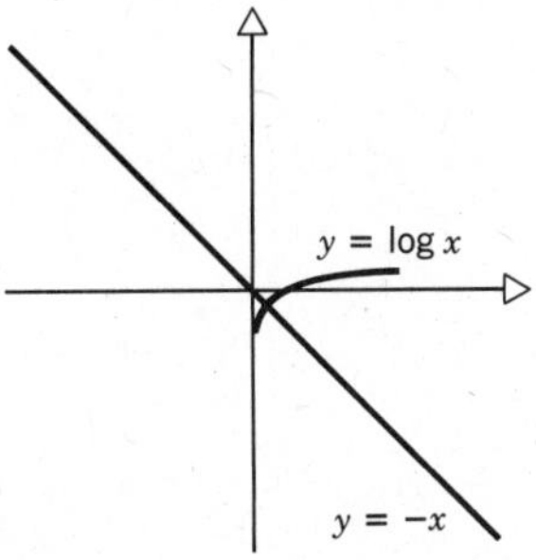

23. (0.64, 0.64)

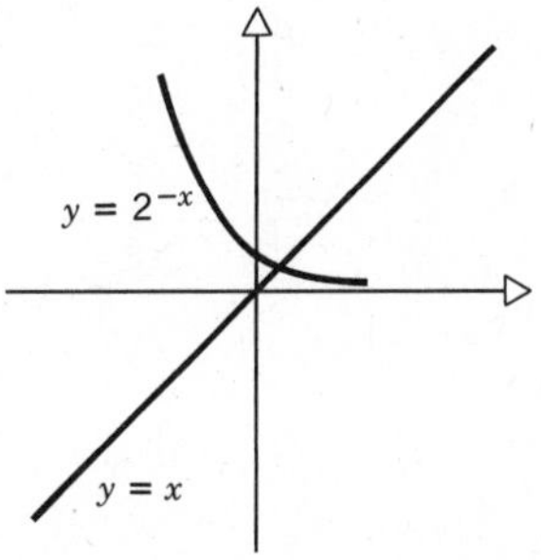

25. 2.5 and 1.5 **27.** 15 cm and 9 cm

29. 360 mph, 1 hr 10 min

31. Two solutions for $k^2 < 18$, one solution for $k^2 = 18$, and no solution for $k^2 > 18$.

33. a and b are given by $c(1 \pm \sqrt{2k - 1})$.

Section 44

1.

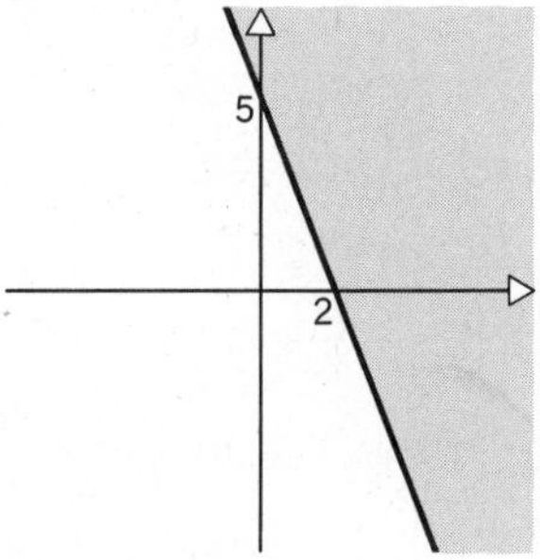

3.

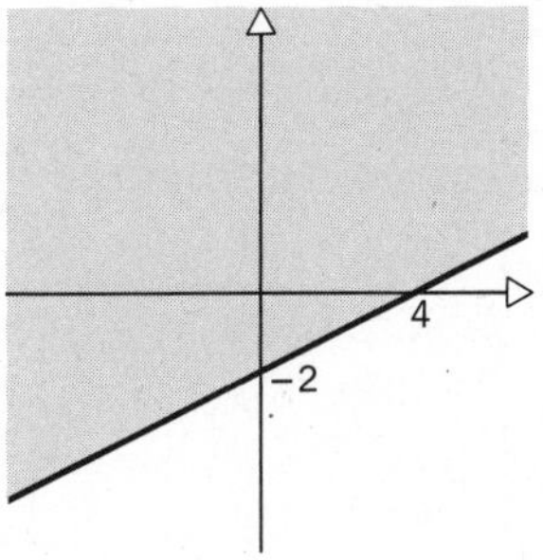

5.

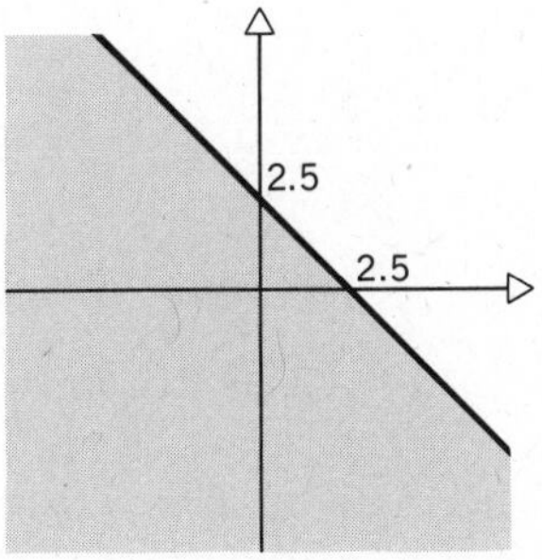

7.

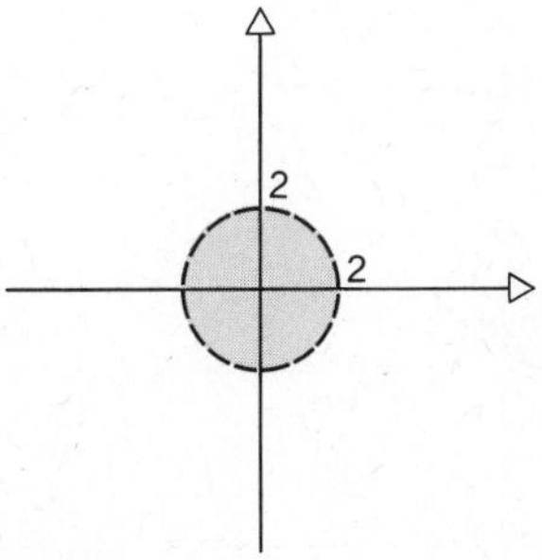

9.

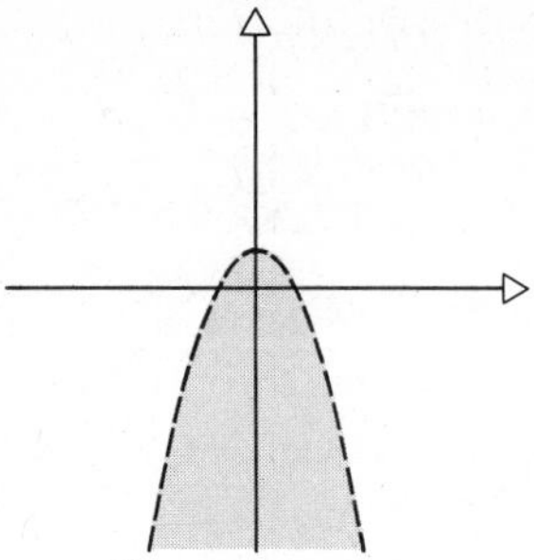

11.

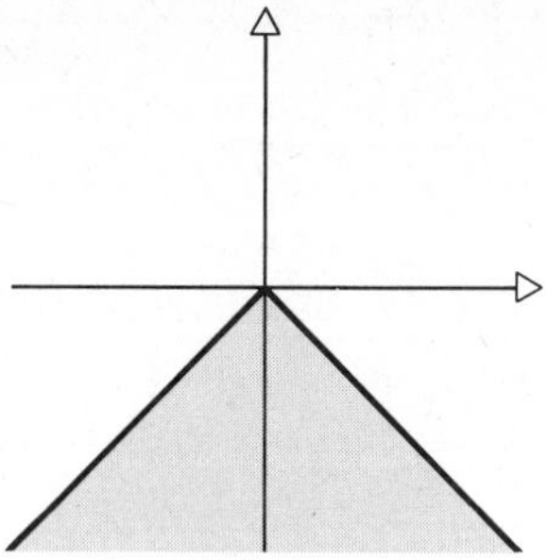

13.

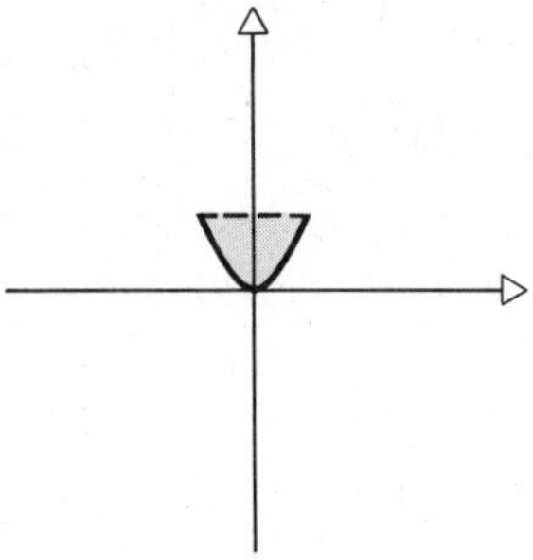

15.

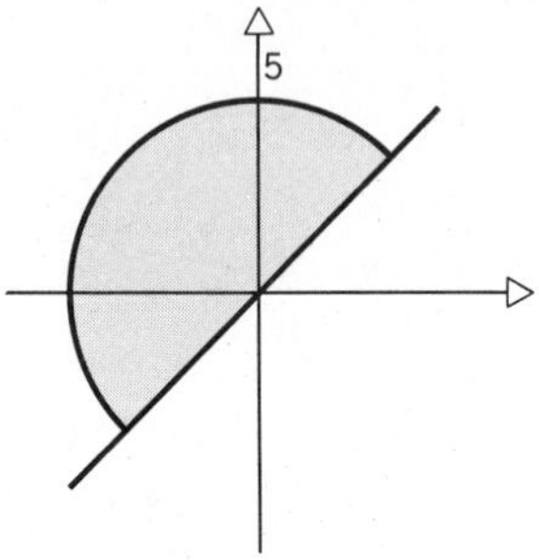

17.

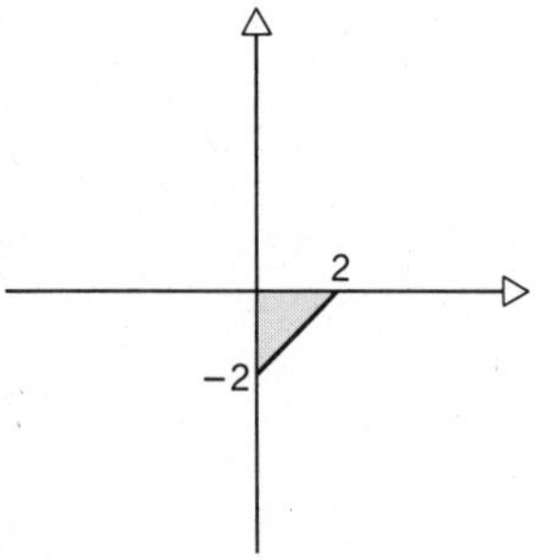

19.

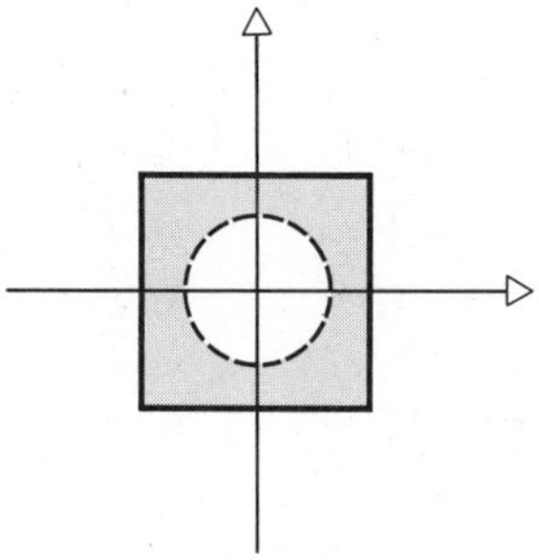

21.

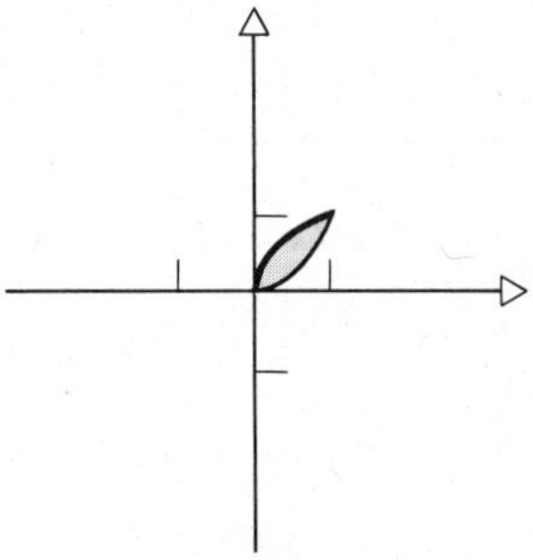

23.

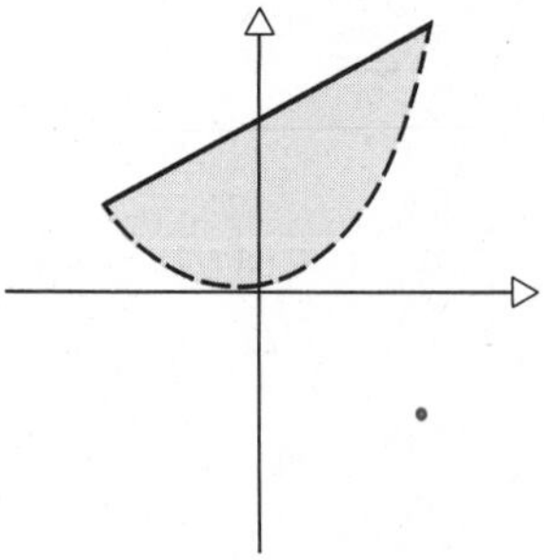

25.

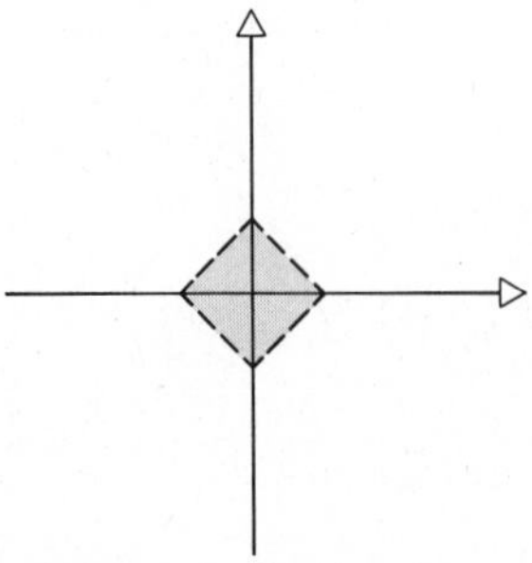

27.

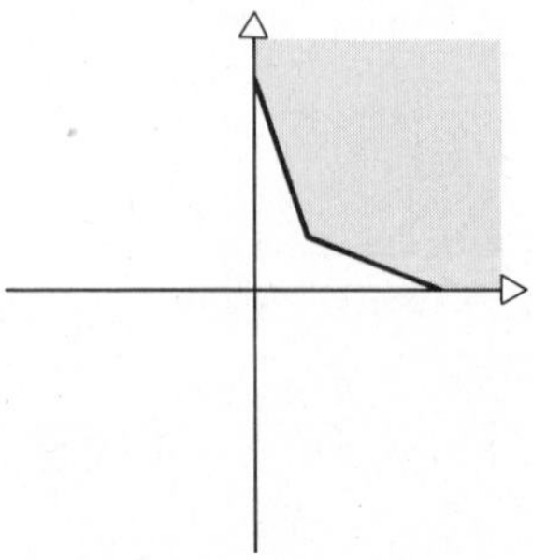

29. $x < 0, y > 0$ **31.** $x < 0, y < 0, x^2 + y^2 > 100$ **33.** $|y| < |x|$

Section 45

In Exercises 1–6, the vertices of the feasible set are at (0, 0), (0, 4), and (2, 0).

1. Maximum $f(0, 4) = 11$, minimum $f(0, 0) = -1$.

3. Maximum $f(0, 0) = -2$, minimum $f(2, 0) = -8$.

5. Minimum $f(0, 0) = 0$, maximum, $f(x, y) = 8$ for $\{(x, 4 - 2x)\colon\ 0 \le x \le 2\}$.

In Exercises 7–12, the vertices of the feasible set are at (0, 0), (6, 0) , (5, 4), and (0, 3).

7. Maximum, $f(6, 0) = 13$, minimum $f(0, 3) = -5$.

9. Maximum $f(6, 0) = 28$, minimum $f(0, 0) = -2$.

11. Maximum $f(x, y) = 48$ for $\{(x, 24 - 4x)\colon\ 5 \le x \le 6\}$, minimum $f(0, 0) = 0$.

13. $\frac{5}{6}$ units of chicken and no tuna.

15. Use 3 bags of bıand A and 2 bags of brand B. The total cost will be \$100.

17. No orange juice and 6 cups of tomato juice. 13.2 mg.

19. Use the suggestions.

21. Any combination with $a = 3b$ and $b > 0$.

CHAPTER X REVIEW

1. $\{(x, \frac{1}{3}(x - 2))\colon\ x$ is a real number$\}$

2. (4, 7) **3.** $(\frac{1}{3}, -\frac{1}{2})$ **4.** Inconsistent.

5. Inconsistent. **6.** $(x, \frac{1}{2},(4x - 3))\colon\ x$ is a real number$\}$

7. $(0, -2, 4)$ **8.** $(\frac{1}{2}, \frac{1}{4}, \frac{3}{4})$

9. $\{(\frac{4}{3}, t - \frac{5}{3}, t)\colon\ t$ is a real number$\}$

10. $(\frac{23}{3}, \frac{26}{3}, \frac{5}{3})$ **11.** (8, 2, 0)

12. Inconsistent. **13.** $(4, -2, 5)$

14. $(\frac{3}{13}, -\frac{5}{13}, \frac{53}{13})$ **15.** $(\frac{5}{3}, -1, -\frac{2}{3})$

16. $\{(\frac{1}{2}t + \frac{5}{4}, \frac{1}{2}t + \frac{3}{4}, t)\colon\ t$ is a real number$\}$

17. $(5, 6, -2)$ **18.** (17, 6, 10)

19. $(-1, 4)$ and (2, 13) **20.** $(-1, 2)$ and (2, 8)

21. $\left(\frac{3\sqrt{42}}{4}, \frac{\sqrt{22}}{4}\right), \left(\frac{3\sqrt{42}}{4}, -\frac{\sqrt{22}}{4}\right), \left(-\frac{3\sqrt{42}}{4}, \frac{\sqrt{22}}{4}\right), \left(-\frac{3\sqrt{42}}{4}, -\frac{\sqrt{22}}{4}\right)$

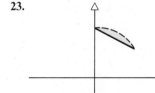

22. $(3, 2), (3, -2), (-3, 2), (-3, -2)$

23.

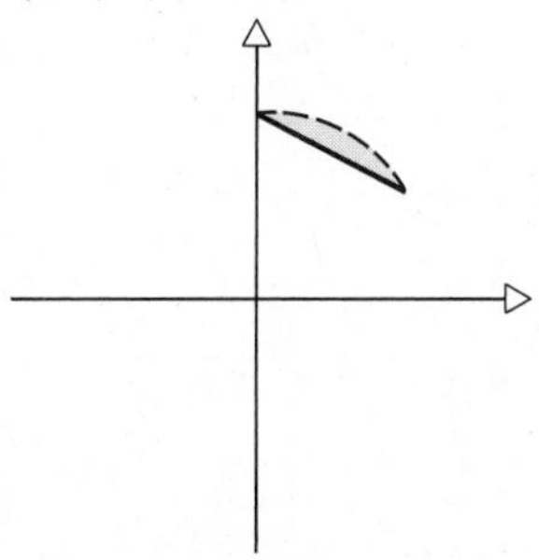

24.

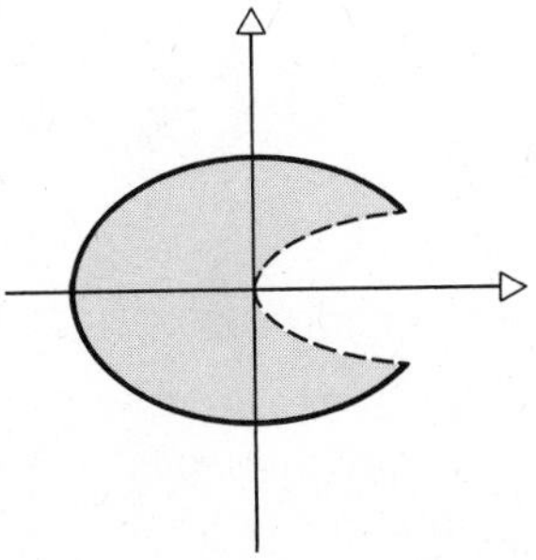

25. Maximum $f(3, 0) = 9$, minimum $f(0, 1) = -9$.

26. Maximum $f(x, y) = 2$ at $\{(x, \frac{1}{2}(5 - x))$: x is a real number$\}$, minimum $f(0, 0) = -3$.

Section 46

1. $x = -5, y = -4$

3. $c = \frac{3}{4}, d = 7, x = 1, y = -\frac{1}{2}, z = \frac{7}{8}$

5. $\begin{bmatrix} -1 & 28 \\ 14 & -8 \end{bmatrix}$

7. $\begin{bmatrix} 2 & 6 & -12 \\ -4 & -4 & 16 \end{bmatrix}$

9. $\begin{bmatrix} 13 \\ 3 \\ -5 \end{bmatrix}$

11. $[10 \quad 1 \quad 15]$

13. 3×1

15. $m = 6, p = 7$

17. $\begin{bmatrix} -23 & -52 \\ 9 & -18 \end{bmatrix}$

19. $\begin{bmatrix} -16 & -69 & 34 \\ 2 & 27 & -41 \end{bmatrix}$

21. $\begin{bmatrix} -51 & 37 \\ 0 & 0 \\ 46 & 31 \end{bmatrix}$

23. $\begin{bmatrix} -41 & 15 & 34 \\ -3 & 5 & 12 \\ -1 & -1 & 10 \end{bmatrix}$

25. $\begin{bmatrix} 52 & 18 \\ 26 & 9 \\ 4 & 22 \end{bmatrix}$

27. $\begin{bmatrix} -20 & 24 \\ -59 & -23 \\ 21 & 15 \end{bmatrix}$

29. $\begin{bmatrix} -14 & -10 \\ 0 & 0 \\ 10 & -12 \end{bmatrix}$

31. $\begin{bmatrix} -25 & 59 \\ 0 & 0 \\ 36 & -23 \end{bmatrix}$

33. $\begin{bmatrix} -49 & 26 & 3 \\ -42 & 9 & -9 \end{bmatrix}$

Section 47

1. $AB = I_2$

3. $AB = I_3$

5. $d = -4$, invertible, $\begin{bmatrix} -\frac{1}{2} & \frac{7}{4} \\ \frac{1}{2} & -\frac{5}{4} \end{bmatrix}$

7. $d = \frac{1}{12}$, invertible, $\begin{bmatrix} 4 & -3 \\ -4 & 6 \end{bmatrix}$

9. $d = 0.1$, invertible, $\begin{bmatrix} 30 & 5 \\ 100 & 17 \end{bmatrix}$

11. $d = 140$, invertible, $\begin{bmatrix} \frac{1}{28} & \frac{1}{70} \\ -\frac{1}{4} & \frac{1}{10} \end{bmatrix}$

13. $\begin{bmatrix} -\frac{14}{3} & 1 & \frac{2}{3} \\ -6 & 1 & 1 \\ \frac{17}{3} & -1 & -\frac{2}{3} \end{bmatrix}$

15. $\begin{bmatrix} -\frac{69}{187} & \frac{13}{187} & \frac{21}{187} \\ -\frac{7}{187} & \frac{23}{187} & -\frac{6}{187} \\ \frac{4}{17} & -\frac{1}{17} & \frac{1}{17} \end{bmatrix}$

17. $\begin{bmatrix} \frac{7}{53} & \frac{15}{106} & -\frac{21}{106} \\ -\frac{8}{53} & \frac{51}{106} & -\frac{29}{106} \\ -\frac{1}{53} & \frac{13}{106} & \frac{3}{106} \end{bmatrix}$

19. $\begin{bmatrix} \frac{17}{5} & \frac{8}{15} & \frac{16}{15} & -\frac{4}{15} \\ 1 & -\frac{1}{3} & \frac{1}{3} & -\frac{1}{3} \\ -\frac{3}{5} & -\frac{2}{15} & -\frac{4}{15} & \frac{1}{15} \\ -\frac{1}{5} & \frac{1}{15} & \frac{2}{15} & \frac{7}{15} \end{bmatrix}$

21. Not invertible.

23. $x = \frac{13}{2}, y = -\frac{19}{2}$

25. $x = \frac{4}{3}, y = \frac{1}{12}, z = -\frac{5}{12}$

27. $x = \frac{7}{9}, y = \frac{1}{9}, z = \frac{21}{9}$

31. 23 million ounces in South Africa, 16 million ounces outside.

33. $C = I = 15.5$ quads. $G = D = 20.5$ quads.

Section 48

1. -4 **3.** 8 **5.** -30 **7.** -4

9. 8 **11.** 80 **13.** 215 **15.** -261

17. 238 **19.** -70 **21.** $-\frac{63}{4}$ **23.** $x = \frac{2}{7}, y = -\frac{6}{7}$

25. $u = \frac{17}{22}, v = \frac{1}{11}$ **27.** $r = \frac{13}{7}, s = \frac{17}{7}$ **29.** $x = \frac{24}{7}, y = \frac{6}{7}, z = \frac{11}{7}$

31. Inconsistent. **33.** Inconsistent. **35.** Inconsistent.

41. \$3 billion in 1970, \$42 billion in 1978, \$80 billion in 1980.

Section 49

1. -42 **3.** -30 **5.** -150 **7.** 228

9. -150 **11.** -204 **13.** -584 **15.** 4

17. Property IV **19.** Property I **21.** Property II

23. $\begin{bmatrix} \frac{29}{90} & \frac{1}{9} & -\frac{2}{15} \\ -\frac{7}{90} & \frac{1}{9} & \frac{1}{15} \\ -\frac{13}{90} & -\frac{2}{9} & \frac{4}{15} \end{bmatrix}$

CHAPTER XI REVIEW

1. $\begin{bmatrix} 2 & 23 \\ -2 & 3 \end{bmatrix}$ **2.** $\begin{bmatrix} 0 & 3 \\ -3 & -1 \\ 13 & 18 \end{bmatrix}$ **3.** $\begin{bmatrix} 1 & -2 \\ 56 & 55 \end{bmatrix}$

4. $\begin{bmatrix} 0 & -5 & 8 \\ 22 & 16 & 2 \\ -1 & -17 & 1 \end{bmatrix}$ **5.** $\begin{bmatrix} \frac{2}{13} & -\frac{1}{13} \\ \frac{5}{13} & \frac{4}{13} \end{bmatrix}$ **6.** $\begin{bmatrix} 12 & -20 \\ -15 & 30 \end{bmatrix}$

7. $\begin{bmatrix} \frac{1}{9} & \frac{2}{9} & -\frac{2}{9} \\ \frac{25}{18} & -\frac{2}{9} & \frac{13}{18} \\ -\frac{4}{9} & \frac{1}{9} & -\frac{1}{9} \end{bmatrix}$ **8.** Not invertible. **9.** Not invertible.

10. $\begin{bmatrix} 0 & \frac{1}{23} & -\frac{3}{23} \\ \frac{1}{2} & \frac{1}{2} & \frac{1}{2} \\ 0 & -\frac{6}{23} & -\frac{5}{23} \end{bmatrix}$ **11.** $x = \frac{45}{16}, y = \frac{19}{32}$ **12.** $x = \frac{4}{3}, y = \frac{1}{3}, z = \frac{7}{3}$

13. 170 **14.** 0 **15.** -14

16. $(ac - b^2)(bd - c^2)$ **17.** $x = \frac{2}{5}, y = \frac{7}{5}$ **18.** Inconsistent.

19. Dependent. **20.** $x = \frac{6}{5}, y = \frac{13}{5}, z = -\frac{12}{5}$

Section 50

1. 5, 10, 15, 20, 25, $5n + 5$ **3.** $\frac{1}{2}, \frac{2}{3}, \frac{3}{4}, \frac{4}{5}, \frac{5}{6}, \frac{n+1}{n+2}$

5. $-2, 4, -8, 16, -32, (-2)^{n+1}$

For Exercises 7–23, only the induction step (P_k implies P_{k+1}) is given.

7. $1 + 2 + 2^2 + \cdots + 2^{k+1} = 2^{k+1} - 1 + 2^{k+1} = (2 \cdot 2^{k+1}) - 1 = 2^{k+2} - 1$

9. $5 + 10 + 15 + \cdots + 5(k+1) = \dfrac{5k(k+1)}{2} + 5(k+1)$

$$= \frac{5k(k+1) + 10(k+1)}{2} = \frac{5(k+1)(k+2)}{2}$$

11. $1^3 + 2^3 + \cdots + (k+1)^3 = \left[\dfrac{k(k+1)}{2}\right]^2 + (k+1)^3$

$$= \frac{k^2(k+1)^2 + 4(k+1)^3}{4}$$

$$= \left[\frac{(k+1)(k+2)}{2}\right]^2$$

13. $(1 \cdot 2) + (2 \cdot 3) + (3 \cdot 4) + \cdots + (k+1)(k+2)$

$$= \frac{k(k+1)(k+2)}{3} + (k+1)(k+2) = \frac{(k+1)(k+2)(k+3)}{3}$$

15. $\dfrac{1}{1 \cdot 2 \cdot 3} + \dfrac{1}{2 \cdot 3 \cdot 4} + \dfrac{1}{3 \cdot 4 \cdot 5} + \cdots + \dfrac{1}{(k+1)(k+2)(k+3)} = \dfrac{k(k+3)}{4(k+1)(k+2)}$

$$+ \frac{1}{(k+1)(k+2)(k+3)} = \frac{k(k+3)^2 + 4}{4(k+1)(k+2)(k+3)} = \frac{(k+1)(k+4)}{4(k+2)(k+3)}$$

17. $\left(1 - \frac{1}{4}\right)\left(1 - \frac{1}{9}\right)\left(1 - \frac{1}{16}\right) \cdots \left(1 - \dfrac{1}{(k+1)^2}\right) = \left(\dfrac{k+1}{2k}\right)\left(1 - \dfrac{1}{(k+1)^2}\right)$

$$= \frac{k+1}{2k} - \frac{1}{2k(k+1)} = \frac{(k+1)^2 - 1}{2k(k+1)} = \frac{k+2}{2(k+1)}$$

19. $2k \le 2^k$ implies $2(2k) \le 2^{k+1}$, which implies $2(k+1) \le 2^{k+1}$ because $k + 1 \le 2k$.

21. $\left(1 + \frac{1}{2}\right)^k \ge 1 + \frac{1}{2^k}$ implies $\left(1 + \frac{1}{2}\right)^{k+1} \ge \left(1 + \frac{1}{2^k}\right)\left(1 + \frac{1}{2}\right) \ge 1 + \frac{1}{2^{k+1}}$, which implies $\left(1 + \frac{1}{2}\right)^{k+1} \ge 1 + \frac{1}{2^{k+1}}$.

23. $a^k < 1$ and $0 < a < 1$ imply $a^{k+1} < a < 1$, which implies $a^{k+1} < 1$.

25. By connecting appropriately chosen vertices, a $(k+1)$-sided polygon can be divided into a k-sided polygon and a triangle. From this, the sum of the interior angles will be $(k-2)180° + 180° = (k-1)180°$.

27. Induction step: The number of diagonals of a $(k+1)$-sided polygon is $\dfrac{k(k-3)}{2} + (k-1) = \dfrac{(k+1)(k-2)}{2}$.

Section 51

1. 5, 8, 11, 14, 17 **3.** 40, 28, 16, 4, −8 **5.** $\frac{3}{2}, \frac{7}{4}, 2, \ldots, \frac{15}{4}$

7. 1, 8, 15, 396 **9.** $\frac{1}{4}, \frac{3}{8}, \frac{1}{2}, \frac{115}{4}$ **11.** 120, 125, 130, 16,050

13. (a) $a_1 = a_n - (n-1)d$ (b) 28

15. (a) $n = \dfrac{a_n - a_1}{d} + 1$ (b) 351 **17.** −455

19. $b_1 + b_2 + b_3 + b_4 + b_5$ **21.** $4 + 5 + 6 + 7 + 8 + 9$

23. $\frac{1}{6} + \frac{1}{7} + \frac{1}{8} + \frac{1}{9} + \frac{1}{10} + \frac{1}{11}$ **25.** $\sum_{k=1}^{10} 3^k$

27. $\sum_{k=1}^{20} k \cdot 5^k$ **29.** $\frac{5n}{2}(3n + 1)$ **33.** $S_n = \frac{n}{2}(a_1 + a_n)$ **35.** 490

37. (a) 482.5 gal (b) 47th day **39.** \$40

Section 52

1. $\frac{1}{2}$, 2, 8, 32, 128 **3.** $10^{-1}, 10^{-4}, 10^{-7}, 10^{-10}, 10^{-13}$

5. 100, 10, 1, 10^{-7} **7.** $\frac{2^{k-1}}{5^{k-2}}, \frac{1031}{125}$

9. $3(-\sqrt{3})^{k-1}$, $39 - 12\sqrt{3}$ **11.** $(1 - \sqrt{2})(1 + \sqrt{2})^{k-1}$, $-1 - 2\sqrt{2}$

13. $\frac{665}{128}$ **15.** $1.5(1 - 0.6^{10}) \approx 1.49$ **17.** $22.\bar{2}$

19. (a) $a_1 = \frac{a_n}{r^{n-1}}$ (b) 486

21. (a) $a_1 = \frac{S_n(1 - r)}{(1 - r^n)}$ (b) $1 + \sqrt{2}$

23. \$15,431.20 **25.** \$7958.55 **27.** \$19,124.88

29. \$3000/32.38 $\approx$ \$92.94 **31.** $a_{k+1} = (a_1 r^{k-1})r = a_1 r^k$

35. $\frac{a_1 - a_1(-1)^n}{2}$. S_n alternates between 0 and a_1.

37. $\frac{31}{9}$ **39.** $\frac{697}{33}$ **41.** $\frac{34}{333}$

43. Suggestion: If $b = ar$ and $c = br$, then $b/a = c/b$.

45. Suggestion: If $a_{k+1} = a_k r$, what is $\log a_{k+1} - \log a_k$?

47. 262 cycles per second

Section 53

1. 120 **4.** 126 **5.** 1140

7. $(n + 2)(n + 1)$ **9.** $\frac{n + 1}{n}$ **11.** $2n$

13.

	1		7		21		35		35		21		7		1	
1		8		28		56		70		56		28		8		1

15. A common denominator for the right side is $k!(n - k + 2)!$. Or use (53.3) three times.

17. $u^6 + 6u^5v + 15u^4v^2 + 20u^3v^3 + 15u^2v^4 + 6uv^5 + v^6$

19. $a^6 - 6a^5b + 15a^4b^2 - 20a^3b^3 + 15a^2b^4 - 6ab^5 + b^6$

21. $16u^4 - 32u^3v^2 + 24u^2v^4 - 8uv^6 + v^8$

23. 300 **25.** $a^{20} + 20a^{18}b^3 + 180a^{16}b^6 + 960a^{14}b^9$

27. $x^{100} + 100x^{99}y + 4950x^{98}y^2 + 161{,}700x^{97}y^3$ **29.** -448

31. With $a = b = 1$, the left side of (53.4) equals 2^n, and the right side is the desired sum.

33. Suggestion: Multiply and divide the left side by the product of the first n positive even integers.

CHAPTER XII REVIEW

1. 3, 27, 162, 810, $3^n(n+1)n/2$ **2.** $\sum_{k=1}^{\infty} k(-10)^{k-1}$

3. $3 + 7 + 11 + \cdots + (4k+3) = k(2k+1) + (4k+3) = 2k^2 + 5k + 3 = (k+1)(2k+3)$

4. $1^2 + 3^2 + 5^2 + \cdots + (2k+1)^2 = \dfrac{k(2k-1)(2k+1)}{3} + (2k+1)^2$

$= \dfrac{4k^3 + 12k^2 + 11k + 3}{3} = \dfrac{(k+1)(2k+1)(2k+3)}{3}$

5. $10 - 45\pi$ **6.** Use $a_1 = 3$ and $a_n = 4n - 1$ in Equation (51.4).

7. $2\sqrt{3}, -6, 6\sqrt{3}, -486$ **8.** $0.1y^2$ **9.** -1705

10. $\dfrac{a - a^{n+1}b^{2n}}{1 - ab^2}$ **11.** $-\frac{1}{5} < t < \frac{1}{5}, \dfrac{1}{1-5t}$ **12.** 6

13. \$1103.97 **14.** \$75 **15.** $\dfrac{n+1}{k}$

17. $a^{10} - 5a^8b + 10a^6b^2 - 10a^4b^3 + 5a^2b^4 - b^5$

18. $1{,}000{,}000x^6 + 60{,}000x^5y + 1500x^4y^2 + 20x^3y^3 + 0.15x^2y^4 + 0.0006xy^5 + 0.000001y^6$

Section 54

1. 5040 **3.** 117,600 **5.** 380 **7.** 1

9. n **11.** 36, 216, 6^n **13.** 15 **15.** $10! = 3{,}628{,}800$

17. (a) 24, (b) 50 **19.** 2880 **21.** 336 **23.** 3, 4, 5, $n+1$

Section 55

1. 10 **3.** 10 **5.** $\dfrac{n(n-1)(n-2)}{6}$

7. $k!$ **9.** $n/2$ **11.** 126

13. $C(39, 13)$ **15.** $C(48, 5)$ **17.** $C(13, 4)C(13, 3)^3$

19. (a) 168 (b) $\dfrac{k(n-1)n}{2}$

21. (a) 35 (b) 63 (c) 203 **23.** $C(n, 3)$

25. $C(n, 4)$ **27.** $\dfrac{n(n-3)}{2}$ **29.** 31

Section 56

1. $\frac{3}{8}$ **3.** $\frac{1}{2}$

5. (a) $\frac{1}{2}$ (b) $\frac{2}{3}$ (c) $\frac{5}{6}$ (d) 0

7. (a) $\frac{1}{36}$ (b) $\frac{25}{36}$ (c) $\frac{4}{9}$ (d) $\frac{1}{4}$ (e) $\frac{3}{4}$ (f) $\frac{1}{2}$

9. (a) $\frac{3}{8}$ (b) $\frac{7}{8}$ (c) $\frac{1}{2}$ (d) $\frac{1}{8}$ (e) $\frac{1}{2}$ (f) $\frac{1}{2}$

11. 0.72675 **13.** $\frac{1}{11}$ **15.** $\frac{1}{6}$

17. (a) $\frac{1}{2}$ (b) $\frac{1}{2}$ (c) $\frac{1}{4}$ (d) $\frac{3}{4}$

19. (a) $C(39, 13)/C(52, 13)$ (b) $1/C(52, 13)$ (c) $C(48, 13)/C(52, 13)$ (d) $C(48, 9)/C(52, 13)$

21. (a) \$0.50 (b) \$2.50 (c) \$0.20 (d) \$20,000

CHAPTER XIII REVIEW

1. 1320 **2.** 1680 **3.** 21 **4.** 462 **5.** 60 **6.** 23

7. 60 **8.** 95,040 **9.** 210 **10.** 150 **11.** $\frac{1}{6}$ **12.** $\frac{3}{20}$

13. (a) $\frac{3}{7}$ (b) $\frac{18}{35}$ (c) $\frac{5}{7}$ **14.** $C(13, 5)/C(52, 5)$, $4 \cdot C(13, 5)/C(52, 5)$

15. $\frac{1}{10}$, $\frac{3}{10}$ **16.** $\frac{11}{12}$

INDEX

INDEX